MANUAL OF
METHODS FOR SOIL
AND
LAND EVALUATION

MANUAL OF METHODS FOR SOIL AND LAND EVALUATION

Editor

Edoardo A.C. Costantini

CRA-Centro di ricerca per l'agrobiologia e la pedologia
Florence
Italy

CRC Press
Taylor & Francis Group
Boca Raton London New York

CRC Press is an imprint of the
Taylor & Francis Group, an **informa** business

A SCIENCE PUBLISHERS BOOK

CRC Press
Taylor & Francis Group
6000 Broken Sound Parkway NW, Suite 300
Boca Raton, FL 33487-2742

First issued in paperback 2017

© 2009 by Taylor & Francis Group, LLC
CRC Press is an imprint of Taylor & Francis Group, an Informa business

No claim to original U.S. Government works

General enquiries : *info@scipub.net*
Editorial enquiries: *editor@scipub.net*
Sales enquiries   : *sales@scipub.net*

ISBN 13: 978-1-138-11398-5 (pbk)
ISBN 13: 978-1-57808-571-2 (hbk)

Library of Congress Cataloging-in-Publication Data

Manual of methods for soil and land evaluation / editor, Edoardo
    A.C. Costantini.
        p. cm.
    Includes bibliographical references and index.
    ISBN 978-1-57808-571-2 (hardcover)
1. Land capability for agriculture. 2. Land capability for
    wildlife. 3. Soils--Quality--Measurement--Technique. 4.
    Soils--Quality--Italy. 5. Plants, Cultivated--Italy. I.
    Costantini, Edoardo A. C.
    S591.M325 2009
    631.4'7--dc22

                                                    2009032721

Preface

This manual is revised from a text published in Italian (*Manuale dei metodi di valutazione dei suoli e delle terre*), which belongs to a series of publications about methods of agricultural analysis supervised by Prof. Paolo Sequi (Centro di ricerca per lo studio delle relazioni tra pianta e suolo, Rome) and formulated during the support activities of the National Observatory for Pedology and Agriculture and Forest Soil Quality of the Italian ministry of agriculture (Ministero delle politiche agricole, alimentari e forestali, Rome). The work was endorsed by the VI Commission for Soil Use and Conservation of the Italian Society of Soil Science.

The goal of the manual is to supply an operational tool for pedologists, agronomists, environmentalists, and all of the other specialists who carry out land evaluation for agriculture and forestry or, more generally, stakeholders and policy makers who make decisions at the local level based on the knowledge of the nature of soil. Discussion of the topics is not only technical and operational, but also in-depth and didactic; therefore, the text may also be used as a valid complement for students majoring in subjects that involve soil use, management and conservation. Such knowledge has gradually increased in recent years, thanks to research, experimentation and many programmes of pedological survey done on a detailed scale. In particular, the studies to determine land capability and suitability for cultivation have assumed an integrated and interdisciplinary character, involving aspects pertaining to environmental factors (soil and climate foremost), ecology, agro-techniques and genetics, and the transformation and analytic and organoleptic evaluation of products. Thus was born the concept of crop "vocation" for a land, superseding the concept of suitability. Crop "vocation" connotes qualitative excellence of the crop/land relationship, which is founded on the peculiarities of the land. The identification of crop vocation led to the development of "crop zoning", i.e., the subdivision of a land by its eco-pedological characteristics and the evaluation of crop response. Crop zoning is proving to be one of the most effective tools for making the most of product quality and typicality, which is universally accepted to be the best way to safeguard the rural landscape, especially in the Mediterranean countries.

It was not possible to discuss all of the soil and land evaluation methods in this manual; therefore, reference was made chiefly to experiments carried out in Italy, maintaining a methodology that reflects the indications of the international literature. As stated in the text, the literature offers a wide choice of possible soil and land evaluation methods, while knowledge of the relationships existing between the physical characteristics of lands, particularly those of soils, and the requirements of specific uses is limited. In this manual we have attempted a presentation of the state of the art on these relationships.

The introduction supplies an outline of methodological references, useful for orientation in the choice of the evaluation procedure to use, but also as a source of ideas for further research. Then there is a large collection of selected evaluations and commentaries on examples mostly drawn from Italy. These involve some of the most important applications of soil knowledge for general planning, for crop choice and crop zoning, for the evaluation of land suitability to slurry spreading, and for the reclamation and requalification of contaminated soils.

The formulation of the matching tables is the fruit of the collaboration of specialists from many different sectors, particularly agronomy and pedology. This collaboration is credited with having supplied an updated survey outline of the subject themes and an articulated and organic vision of the Italian and international frameworks.

Even though the manual does not have a rigid setup, each thematic chapter is organized in different sections that follow a common outline. The first paragraphs supply a wide agronomic picture that summarizes the characteristics of the species, cultivation techniques, and edaphic and agrotechnical requirements. The second series of paragraphs illustrate the functional factors of the soils and the environment with respect to the crop under examination, i.e., factors whose significant influence on agronomic results has been proved by research. Then follows a close examination of the indications that may be gleaned from the knowledge of the relationships that exist between soil nature and crop species.

Besides acknowledging all the contributors who made possible the publication of this manual, special thanks are due to Dr. Marcello Pagliai, former President of the Italian Soil Science Society and Director of the Research Centre for Agrobiology and Pedology, of the National Council of Research in Agriculture (CRA-Centro di ricerca per l'agrobiologia e la pedologia, Florence), who encouraged its publication, and to Dr. Elena Capolino, of the same research centre, for her invaluable contribution to the editing of the book.

The editor
Edoardo A.C. Costantini
CRA—Centro di ricerca per l'agrobiologia e la pedologia
Piazza M. D'Azeglio 30
50121 Firenze, Italy

Contents

PART III
LAND SUITABILITY AND LAND ZONING
Land suitability for row crops

Land suitability for small-scale niche cultivations

Land suitability for tree cultivations

Land suitability for forestry and grazing

List of Contributors

Serafina Andiloro
Università Mediterranea di Reggio Calabria—Dipartimento di Scienze e Tecnologie Agroforestali e Ambientali, Reggio Calabria, Italy.

Giovanni Aramini
ARSSA Calabria—Servizio di Agropedologia, Catanzaro Lido, Italy.

Paolo Baldaccini
Formerly, Università di Sassari, Italy.

Stefano Barbieri
Università di Udine—Dipartimento di Scienze Agrarie e Ambientali, Udine, Italy.

Devis Bartolini
CNR-IRPI—Istituto di Ricerca per la Protezione Idrogeologica, Unità Operativa di Supporto, Pedologia Applicata, Firenze, Italy.

Elvio Bellini
Università di Firenze-Facoltà di Agraria, Dipartimento di Ortoflorofrutticoltur a, Firenze, Italy.

Enrico Biancardi
CRA-CIN—Centro di Ricerca per le Colture Industriali, Sede distaccata di Rovigo, Rovigo, Italy.

Claudio Bini
Università di Venezia—Dipartimento di Scienze Ambientali, Venezia, Italy.

Sandro Bolognesi
Azienda Agraria Sperimentale M. Marani, Ravenna, Italy.

Lorenzo Borselli
CNR-IRPI—Istituto di Ricerca per la Protezione Idrogeologica, Unità Operativa di Supporto, Pedologia Applicata, Firenze, Italy.

Gilberto Bragato
CRA—Centro di Ricerca per lo Studio delle Relazioni tra Pianta e Suolo, Gorizia, Italy.

Pierluigi Bucelli
CRA-ABP—Centro di Ricerca per l'Agrobiologia e la Pedologia, Firenze, Italy.

Francesco Calabrese
Università di Palermo—Facoltà di Agraria, Palermo, Italy.

Costanza Calzolari
CNR-IRPI—Istituto Ricerca Protezione Idrogeologica, Unità Operativa di Supporto, Pedologia Applicata, Firenze, Italy.

Stefano Campagnolo
CRA-ABP—Centro di Ricerca per l'Agrobiologia e la Pedologia, Firenze, Italy.

Adele Maria Caruso
CRA-ABP—Istituto Sperimentale per lo Studio e la Difesa del Suolo, Catanzaro, Italy.

Paola Cassi
CNR-IRPI—Istituto di Ricerca per la Protezione Idrogeologica, Unità Operativa di Supporto, Pedologia Applicata, Firenze, Italy.

Fabio Castelli
CRA-CAT—Unità di Ricerca per le Colture Alternative al Tabacco, Scafati (Salerno), Italy.

Giancarlo Chisci
Università di Firenze, Firenze, Italy.

Antonio Cimato
CNR- IVaLSA—Istituto per la Valorizzazione del Legno e delle Specie Arboree, Sesto Fiorentino (Firenze), Italy.

Caterina Colloca
ARSSA Calabria—Servizio di Agropedologia, Catanzaro Lido, Italy.

Roberto Colombo
CISA M. Neri, Imola (BO), Italy.

Anna Maria Corea
ARSSA Calabria—Servizio di Agropedologia, Catanzaro Lido, Italy.

Edoardo A.C. Costantini
CRA-ABP—Centro di Ricerca per l'Agrobiologia e la Pedologia, Firenze, Italy.

Carmelo Dazzi
Università di Palermo—Facoltà di Agraria, Palermo, Italy.

Giovanni Fatta Del Bosco
Università di Palermo—Facoltà di Agraria, Palermo, Italy.

Elena Franchini
CNR- IVaLSA—Istituto per la Valorizzazione del Legno e delle Specie Arboree, Sesto Fiorentino (Firenze), Italy.

Lorenzo Gardin
Freelance Forester, Firenze, Italy.

Andrea Giapponesi
Regione Emilia Romagna—SSSA, Bologna, Italy.

Edgardo Giordani
Università di Firenze-Facoltà di Agraria, Dipartimento di Ortoflorofrutticoltura, Firenze, Italy.

Andrea Giordano
Università degli Studi di Torino—Facoltà di Agraria, Dipartimento di Economia e Ingegneria Agraria Forestale e Ambientale, Torino, Italy.

Marina Guermandi
Regione Emilia Romagna—Servizio Geologico, Sismico e dei Suoli, Bologna, Italy.

Paolo Inglese
Università degli Studi di Palermo-Facoltà di Agraria, Dipartimento di Colture Arboree, Palermo, Italy.

Luciano Lulli
Formerly, Istituto Sperimentale per lo Studio e la Difesa del Suolo, Firenze, Italy.

J. Mitchell McGrath
USDA-ARS Sugarbeet and Bean Research, Michigan State University East Lansing, MI, USA.

Roberto Mercurio
Università degli Studi di Reggio Calabria—Dipartimento di Gestione dei Sistemi Agrari e Forestali, Località Feo di Vito (RC), Italy.

Fabiano Miceli
Università di Udine—Dipartimento di Scienze Agrarie e Ambientali, Udine, Italy.

Gianfranco Minotta
Università degli Studi di Torino-Dipartimento AGROSELVITER , Torino, Italy.

Daniele Morelli
Università di Firenze—Facoltà di Agraria, Dipartimento di Ortoflorofrutticoltura, Firenze, Italy.

Rosario Napoli
CRA-ABP—Centro di Ricerca per l'Agrobiologia e la Pedologia, Firenze, Italy.

Luca Ongaro
IAO—Istituto Agronomico per l'Oltremare, Firenze, Italy.

Roberto Oppedisano
ARSSA Calabria—Servizio di Agropedologia, Catanzaro Lido, Italy.

Ugo Palara
CISA M. Neri, Imola (BO), Italy.

Enrico Palchetti
Università di Firenze—Facoltà di Agraria, Dipartimento di Scienze Agronomiche e Gestione del Territorio Agroforestale, Firenze, Italy.

Raffaele Paone
ARSSA Calabria—Servizio di Agropedologia, Catanzaro Lido, Italy.

Sergio Pellegrini
CRA-ABP—Centro di Ricerca per l'Agrobiologia e la Pedologia, Firenze, Italy.

Mauro Piazzi
IPCA—Istituto per le Piante da Legno e l'Ambiente, Settore Pedologia, Torino, Italy.

Paola Pirazzini
CISA M. Neri, Imola (BO), Italy.

Marcello Raglione
CRA—Unità per i Sistemi Agropastorali dell'Appennino Centrale, Rieti, Italy.

Vanna Maria Sale
ERSAF—Dipartimento dei servizi all'Agricoltura, Milano, Italy.

Pilar Salvador Sanchis
CNR-IRPI—Istituto di Ricerca per la Protezione Idrogeologica, Unità Operativa di Supporto, Pedologia Applicata, Firenze, Italy.

Paolo Sarfatti
IAO—Istituto Agronomico per l'Oltremare, Firenze, Italy.

Giampaolo Sarno
Regione Emilia Romagna—SSSA, Bologna, Italy.

Riccardo Scalenghe
Università degli Studi di Palermo—Facoltà di Agraria, Dipartimento di Colture Arboree, Palermo, Italy.

Carla Scotti
I.TER—Cooperative, Bologna, Italy.

Piergiorgio Stevanato
Università di Padova-Dipartimento di Biotecnologie Agrarie, Legnaro (Pd), Italy.

Vincenzo Tamburino
Università Mediterranea di Reggio Calabria—Dipartimento di Scienze e Tecnologie Agroforestali e Ambientali, Reggio Calabria, Italy.

Fabrizio Ungaro
CNR-IRPI—Istituto di Ricerca per la Protezione Idrogeologica, Unità Operativa di Supporto, Pedologia Applicata, Firenze, Italy.

Andrea Vacca
Università di Cagliari-Dipartimento di Scienze della Terra, Cagliari, Italy.

Giuseppe Vecchio
CRA—ABP Istituto Sperimentale per lo Studio e la Difesa del Suolo, Catanzaro, Italy.

Letizia Venuti
CNR-IRPI—Istituto di Ricerca per la Protezione Idrogeologica, Unità Operativa di Supporto, Pedologia Applicata, Firenze, Italy.

Santo Marcello Zimbone
Università Mediterranea di Reggio Calabria—Dipartimento di Scienze e Tecnologie Agroforestali e Ambientali, Reggio Calabria, Italy.

PART-I

INTRODUCTION

1. Soil and Land Evaluation: History, Definitions and Concepts

Costanza Calzolari,[1] Edoardo A.C. Costantini,[2] Fabrizio Ungaro[1] and Letizia Venuti[1]

1.1. Introduction

Land and soil evaluation is a classification system that assesses the best use for a given portion of territory, while evidencing the existing limitations for more or less specific types of use. Although land evaluation dates back to the introduction of agricultural use of land, it has only been since the beginning of the 20th century that, with the spread of pedology, and soil mapping, as a science in its own right, soil and land evaluation has taken on a purely territorial perspective. The term "land" is defined by the United Nations Food and Agriculture

[1] CNR-IRPI—Istituto di Ricerca per la Protezione Idrogeologica, Unità Operativa di Supporto, Pedologia Applicata, Firenze, Italy.
[2] CRA-ABP—Centro di Ricerca per l'Agrobiologia e la Pedologia, Firenze, Italy.

Organisation (FAO, 1985) as "an area of the earth's surface, the characteristics of which embrace all reasonably stable, or predictably cyclic, attributes of the biosphere ... including those of the atmosphere, the soil and underlying geology, the hydrology, the plant and animal populations, and the results of past and present human activity.…"

However, the objective of the evaluation may also be the soil itself, or rather the "soil type", meaning a group of soils with similar features and properties; so it is that we talk about soil evaluation.

According to the Soil Taxonomy definition (1999), soil is a natural body, present on the land surface, formed by solid materials (minerals and organic matter), liquids and gases. Soil occupies space and is characterized by one or both of the following elements: (1) horizons or layers, which can be distinguished from the initial material as the consequence of additions, losses, transfers and transformations of energy and matter and (2) the capacity to support plants with root systems in a natural environment. The upper limit of soil borders the air, shallow water, live plants or undecayed plant matter.[1]

The land surface is not considered to have soil if permanently covered with water too deep for plants to develop a root system (usually more than 2.5 m). The soil's lower boundary is more difficult to define. Soil consists of horizons that, unlike the underlying parent material, have been altered over time by the interactions of climate, relief and living organisms. Typically, at its lower boundary, soil grades to coherent or incoherent bedrock, virtually devoid of biological activity.

Coming back to land evaluation, this is defined by FAO as "the process of assessment of land performance when used for specific purposes, involving the execution and interpretation of surveys and studies of land forms, soils, vegetation, climate and other aspects of land in order to identify and make a comparison of promising kinds of land use in terms applicable to the objectives of the evaluation" (FAO, 1985). Another definition (Van Diepen et al., 1991) directly involves a variety of possible uses, describing land evaluation as a collection of all kinds of methods explaining or predicting potential land use. Indeed, according to Dent and Young (1981), the purpose of land evaluation is fundamentally to anticipate the consequences of change.

The main reasons for the success of land evaluation must be underlined. The first is linked to the transparency, simplicity and linearity of the evaluation process. The second is that the thematic cartography derived from soil maps has turned out to be a very useful instrument for presenting pedological information in an easily comprehensible way for lay people.

The limitations persisting in the land evaluation process will also be highlighted as the text unfolds. An outstanding limitation is the lack of knowledge about the functional characteristics of the soil and environment. Indeed, while soil and land evaluation techniques and methods have increased considerably over the years, especially through growth in possibilities of acquiring and processing digital data, knowledge of the conditions and processes that determine responses from plants and the soil itself to different environmental conditions and to human activities have not progressed at the same rate, especially because of the lack of specific experimental trials.

---

[1]The following definition of soil is given in the Communication "Thematic Strategy for Soil Protection" (COM(2006)231 final): "Soil is generally defined as the top layer of the earth's crust, formed by mineral particles, organic matter, water, air and living organisms. It is the interface between earth, air and water and hosts most of the biosphere."

Besides, it is undoubtedly difficult to update the panorama of our knowledge linking the environment to plant life, to agrotechniques and to farm management, in a situation that is ever more dynamic, continually producing important innovations in the fields of genetics, agrotechnique, mechanization, etc., and, above all, in a socio-economic context such as that in Europe, increasingly conditioned by European community regulations and the effects of the global market, rather than by local agricultural suitability.

1.2. Land Evaluation in Different Cultural and Operational Contexts

Despite the fact that land evaluation is culturally based on the understanding of soil, and consequently of its characteristics, quality and geographical distribution, the methodologies of land evaluation may have an almost infinite series of variations, depending on the cultural and practical context in which they are carried out. The performance evaluation of a certain area of land is present in all planning processes of the activities engaged on it. Indeed, the many functions of the soil and the different disciplines involved in its evaluation must be considered. With regard to agriculture and forestry sciences, the most important soil function is the production of biomass, food and fibre; in environmental sciences, the filtering and buffering function is most important; in geology, the soil is considered a regulator of hydrogeological balances and a substrate from which raw materials and water are extracted; in ecology, soil serves as biological habitat and gene pool; in disciplines dealing with land planning, soil is evaluated as the physical base for the development of settlements and infrastructures; further, the interaction between scientific and humanistic disciplines leads to the acknowledgement that soil has the function of preserving cultural heritage linked to agricultural landscapes, and to paleo-ecological, paleontological and archaeological heritages.

When considering soil evaluation for agriculture[1] and forestry, it operates at every level and geographical scale: from farm scale to more general ones. Evaluation is carried out at the farm level because of the needs of farmers, who then decide how to exploit their lands after weighing economic, social and cultural considerations. On a municipal, regional or national scale, it is the collectivity, otherwise defined and constituted, that adopts the most suitable forms of land planning for the purposes established for it.

The purpose of evaluation determines what aspect of assessment is focused on. So it is that numerous other concepts, often used as synonyms, are found in the field of land evaluation. The term "land capability", for example, pertains not to specific crops or agricultural practices, but rather to broad agriculture and forestry systems, in relation to their sustainable use. The term specifically refers to the American Land Capability Classification, which is dealt with in detail in the second part of this handbook. On the other hand, the term "land suitability" is intended to mean the suitability of land for a specific land use. Land vocation is in some way

---

[1] According to the OECD definition, "agriculture includes households engaged in farming, herding, livestock production,fishing and aquaculture. Also included are other producers and individuals employed incultivating and harvesting food resources from salt and fresh water and cultivating trees and shrubs and harvesting non-timber forest products—as well as processors, small-scale traders, managers, extension specialists, researchers, policy makers and others engaged in the food, feed and fibre system and its relationships with natural resources. This system also includes processes and institutions, including markets, that are relevant to the agriculture sector" (OECD 2006—"Promoting Pro-Poor Growth: Policy Guidance for Donors").

a synonym for suitability of use, even though the term "vocation" also has a "positive" connotation that is not necessarily contained in the term "suitability", often based on the concept of "use limitation". "Vocation" is a term coming from a purely agronomic context and encompasses the "peculiarity" of the environmental characteristics of a soil, thus determining both the qualitative and productive success of a crop. Recognizing that a soil has a vocational trait for a particular crop goes beyond technical bounds and also involves the social sphere. This happens when the vocational trait becomes universally recognized and connotes the image of an entire district. In such a way, other quality crops produced on that territory also acquire an added value, expendable on the market. A well-known example is the land in which great wines are produced and land evaluation activities are outlined as "land zoning" (see, for example, the chapter on vine growing).

When land and soil evaluation are tackled in the context of themes of a more environmental importance, for example the risk of pollution linked to some agricultural ractices, terms frequently used include "sustainable use", "sensitivity", "risk", "vulnerability" and "resilience" of the soil.

Sustainable land use is a form of use that can be practised for an indefinite period without causing the degradation or impoverishment of resources.[1] In examining possible alternative land uses, whether agricultural, forestry or otherwise, it is clearly necessary to avoid "pillaging" types of land use, which quickly provide a high income but may lead to accelerated erosion, water table pollution, changes in hydrological regimes or other forms of degradation.

The term "soil sensitivity" evokes the tendency the soil may have to undergo rapid change, depending on natural or anthropic pressure applied to it. For example, a soil is sensitive to compaction if easily compacted following cultivation with heavy farm machinery.

The term "vulnerability" indicates the tendency of soil to undergo irreversible modification. Shallow soils, for example, can be completely removed by accelerated water erosion. In this case, the "risk" of soil erosion is determined by interaction between its vulnerability and the probability that conditions will occur that set off an intense erosive process.

Finally, soil "resilience" is the capacity of soil to regenerate when faced with external disturbances, for example, the capacity it has to self-structure and reconstruct natural aggregates after compaction, or the pedogenesis rate, i.e., the formation of new soil in place of that which has been eroded.

1.3. Types and Logic of Land Evaluation

There are many land evaluation systems with various conceptual sources that use different techniques. An exhaustive discussion can be found in Van Diepen et al. (1991) and in Rossiter (1996). The first article presents a critical review of different methods; Rossiter's article (1996) for the first time puts forward an attempt at a theoretical, unifying and systematic framework of types of land evaluation. To put it simply, the classification proposed by Rossiter (1996) subdivides different types of land evaluation on the basis of the spatial approach

---

[1] In the Thematic Strategy on the sustainable use of natural resources (COM(2005) 670 final) adopted by the European Commission in 2005, sustainable use is defined as use capable of "ensuring that the consumption of resources and their associated impacts do not exceed the carrying capacity of the environment and breaking the linkages between economic growth and resource use".

(spatial vs. not spatial models) and temporal approach (dynamic vs. static models) and of the land quality concept (also discussed in Van Diepen et al., 1991).

From the operational point of view, the fundamental structure of the land evaluation process may be schematized as follows: (i) identification of users (administrators, technicians, farmers, consortia, etc.) and definition of aims; (ii) identification of forms of use of the area under consideration and the technological level of reference; (iii) definition of a geographical area or of the spatial entity being evaluated; (iv) definition of cultural, management and environmental sustainability requirements of the considered land use types, whether existing or potential; (v) definition of the considered area's physical properties, which as a rule can be measured and which are important for the forms of use detected; (vi) definition of models for the evaluation; (vii) carrying out of the evaluation, calibration and presentation of results.

Land evaluation, defined as above, may be accomplished directly by evaluating the results obtained over a particular area, or indirectly by analysing soil characteristics, interpreting them in either a positive or a negative way on consideration of the specific or general use proposed. In the first case, evaluation will be based on experiments in the field, farm economic analyses or agricultural statistics, depending on the scale of evaluation. Since these data are not always available or are discontinuous in time and space and difficult to extrapolate out of the context in which they were surveyed, indirect systems are often used, based on soil and land characteristics, presupposing a correlation between these and the crop yields, for the same level of energetic and technological inputs of the management. This division between direct and indirect methods is not, actually, a strict one, as indirect evaluation often takes into consideration the economic nature of crops, such as costs for agronomic intervention.

Some systems refer to agricultural or forestry land uses, while others relate to engineering uses or to uses aimed at land protection (e.g., from pollution or erosion). Within land evaluation systems for agriculture, those with a general intent are distinguished from those for specific ends: in the former, the environmental characteristics are interpreted only so as to indicate the potential and limits of agricultural land use and forestry; in the latter, the evaluation takes into consideration a particular land use type or a particular agro-technique, such as the production of winter wheat or irrigated maize, or the application of animal slurries. In the agricultural context, another distinction may be made between a type of land evaluation that considers the current use or else a potential use, i.e., the possibility of introducing new crops, perhaps after reclamation (e.g., irrigation, drainage).

Land and soil evaluation may be expressed in qualitative, semi-quantitative or quantitative terms. A qualitative approach may simply consist in a description of the land suitability for different land use types, or in a subdivision into suitability classes or levels (hierarchical structure). Qualitative approaches may have a quantitative component, usually expressed in terms of percentage of yield reduction from that obtained in the first class conditions.

Some methods use qualitative data that are weighed separately and combined to obtain a result expressed numerically. The semi-quantitative evaluation is the more frequent, for which a reference crop is considered and evaluation classes are established that are expressed in percentage fractions of the target production. An example of this evaluation system is the "single-factor system". Such systems quantitatively represent the influence a single land feature has, for example, on a crop's yield, by means of the yield curve response to single-factor variations. This approach is well adapted to the case in which there is a single parameter that has a clearly positive or negative effect on a determinate use (e.g., soil depth). Indeed, the interaction of many factors is not considered, even though single response curves may combine together to form just one suitability index (De La Rosa, 2002).

Even the so-called arithmetic or parametric methods represent a semi-quantitative approach to land evaluation, inasmuch as they are half-way between qualitative methods, based only on expert opinion, and quantitative ones, which consist in the application of mathematical models. These quantitatively define the effects of the various land features on one land use type, expressing the final result numerically as the solution to a mathematical formula. Parametric methods may be additive (index = A + B + C + D + ...) or multiplicative (index = A * B * C * D * ...). In the multiplicative method, the various ratings attributed to the different properties are multiplied by one another, reaching a final rating. This method has the advantage that every factor is given weight in the final rating. The disadvantage is that the overall land value may also end up considerably lower than the rating of each factor considered (De La Rosa, 2002).

The first example of the multiplicative system is the Storie Index Rating (SIR), developed in 1933 by R.E. Storie (Storie, 1933, 1978), which assesses the soil on the grounds of its suitability for a generic type of intensive cropping. This actually was the first application of land evaluation.

In the additive method, as in the multiplicative one, each feature is assigned a rating based on its impact on the use considered. However, instead of the various factors being multiplied by one another, they are summed up or else subtracted from a maximum amount equal to 100, thus obtaining the final rating. There are related advantages and disadvantages in this case as well. Among the advantages is the fact that there is no limit to the number of parameters that may be taken into consideration, and that no factor can strongly influence the final rating. Among the disadvantages, on the other hand, the system becomes rather complex as the factors increase and, from a strictly arithmetic point of view, there is the possibility of getting negative values.

A combined method does exist, using both approaches (multiplicative and additive). In such an instance, the additive method is used to obtain the factor rating and the multiplicative method to put them together for the final score. The advantage of this approach is that the effect each factor has on the final value is neither played down nor amplified. On the other hand, the calculations are more complex (see, for example, De La Rosa, 2002).

In their simplicity, the so-called parametric methods are easy to comprehend, apply and adapt to local needs. All the same, it must be emphasized that their objectivity and precision may turn out to be an illusion inasmuch as the rating attribution is, in any case, a subjective choice, requiring careful validation on site. Furthermore, although they can be useful in estimating land qualities to be used in the evaluation process, they give few indications as to the consequences of the choice of a determinate land use.

In a quantitative approach to land evaluation, reference is made to quantities, for example of achievable biomass production, or at any rate to measurable data. In this case, it is necessary to have field-surveyed data, collected in several representative environmental conditions. As this approach is much more demanding, models can be adopted that can simulate the environmental processes. In this sense, modelling can be considered a quantitative approach when supported by a sufficient amount of reliable data. Indeed, the level of detailed data at hand permits the construction of a model of soil response to the different land uses, and therefore an evaluation complying with the traditional definition of land evaluation can be carried out (FAO, 1985).

The application of models to the process of land evaluation often translates into the estimation of the potential production of a crop system. At times, however, models are merely used for the indirect estimation of the land qualities, which are then implemented in the evaluation.

The main difficulty in completely applying quantitative methods is the need to have a considerable amount of continuous data in space. Nowadays, this is facilitated by employing survey systems that permit more or less continuous data collection (such as remote sensing) and by the availability of personal computers capable of processing a high quantity of data.

Progress in the field of informatics has brought about evolution in land evaluation methodologies, boosting the use of ever more quantitative approaches, essentially modelling. Hoosbeek and Bryant (1992) hold that models may be classified according to three characteristics: degree of generalization, complexity and hierarchical level of scale. Analogous classification can be used for land evaluation models (Fig. 1.1) (Bouma, 2000). The ordinate gives the reference scale in which the processes may be described, from global to molecular, passing through the origin, represented by the pedon/plot. The two perpendicular abscissas represent the level of knowledge: one relative to the degree of generalization, going from the quantitative to the qualitative approach, the other representing the degree of complexity of the process description, which moves from the empirical-statistical approach to the physical-mechanistic.

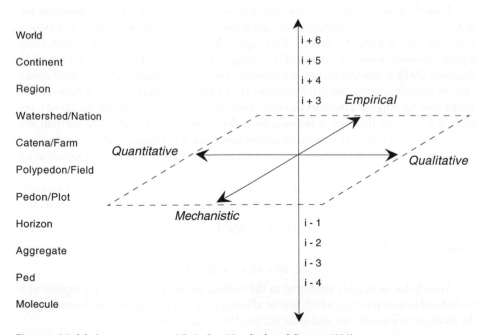

Fig. 1.1. Model characteristics (modified after Hoosbeek and Bryant, 1992).

There may be many kinds of calculation logic on which the mathematical models are based for the land evaluation. The main ones are Boolean logic, fuzzy logic and artificial neural networks.

Boolean Logic

In Boolean logic, the limits of a set are clearly defined, so that an element does or does not belong to a determinate set. It is the logic of true or false, traditionally used in the applicative sciences and, in the case of land evaluation, the logic that permeates the FAO method (1976). According to that approach, a soil may be very suitable, moderately suitable, marginally suitable or not suitable. There is no possibility of describing the slight distinctions between the classes, as intermediate classes are not contemplated.

Fuzzy Logic

Although traditional land evaluation is based on an approach of the Boolean type, that approach may be inadequate to represent a situation such as that of natural resources, which are subjected to a type of continuous rather than discrete variability. Fuzzy logic permits logical operations to be carried out without using the rigidity typical of Boolean logic.

In Boolean logic, an object does or does not belong to a determinate set, the limits of which are strictly defined: it takes value 1 in the first case, in the second 0. In fuzzy logic, the set limits are blurred and an object is defined by a certain degree of belonging, the degree being defined by a number between zero and one. A fuzzy set is, therefore, a set in which there is no precise, well-defined border between the objects belonging to it and those that do not (Faggiani, 1999), but the borders are blurred and fuzzy.

Fuzzy logic is particularly appropriate in those cases where one wishes to minimize the influence of a property, its characteristics and quality on land evaluation as a whole, having a value just outside a specific interval (Burrough, 1989; Burrough et al., 1992). With fuzzy logic in land evaluation, the concept of belonging to a class, represented by the "membership function" (MF), is introduced. Individuals with a value under a defined class are attributed a value of belonging to the class that is equivalent to 1 (MF=1). Individuals with a value outside such a class are assigned a membership value lower than 1 and greater than 0 (0<MF<1); the lower it is the more the value draws away from that of the class. Therefore, rather than as a class, the value of land characteristic or quality is expressed as a degree of class membership. Furthermore, single land characteristics or qualities are attributed with a weight representing their relative importance in the final assessment as to overall suitability. Hence, a land unit's overall suitability is expressed as the degree of joint membership function (JMF), which is the sum weight of the functions of the various characteristics or qualities considered. Therefore:

$$JMF = aA*MFA + aB*MFB + ...+ aZ*MFZ \qquad (1)$$

where:

$$aA + aB +...+ aZ = 1$$

The choice of weights attributed to the various parameters (aA, aB, ...) represents a critical and important stage, which may be affected by an expert's judgement, knowledge of the place, experimental data, and other factors (De La Rosa, 2002).

Fuzzy logic was applied to the evaluation of soil erodibility (Wischmeier's K factor) in the Fusle programme (Borselli, 1995). It was also used at Cochabamba in Bolivia for an urban development land evaluation procedure (IAO, 1999), as described in the chapter concerning land evaluation in developing countries.

Neural Networks

Neural networks are algorithms made up of a grid of nodes linked to one another, the structure of which is like that of biological nervous systems. Neural networks do not need a knowledge of the relationships existing between input and output variables, such as in the case of regression models. They are non-linear and can consequently manipulate much more complex data than those manageable by simple mathematical models. Furthermore, it is possible to process all types of data, from continuous or almost continuous ones, up to those expressed in categories. Moreover, neural networks have the capacity for generalizing and abstracting, simulating a property of the human brain, which allows the treatment of data and models that cannot be processed by traditional methods.

Systems based on neural networks represent a type of continuous simulation, founded on statistical study of the phenomena considered, by means of which the value of some either qualitative or quantitative dependent variables is calculated from other qualitative or quantitative independent variables (e.g., how much a pollutant varies at a certain point, as some wind and traffic vary at adjacent points). Neural network methodology provides precise results starting from a training stage carried out with historical measurements of the dependent and independent variables, expressed as either quantities or classes. It is used especially in operational planning, which, even more than verifying choices, needs a high speed of response even in unpredictable situations that cannot be codified or exactly measured. Indeed, this methodology allows the "training" of the system itself with known and resolved cases, leaving the system with the task of giving coherent answers to such learning in situations that have not yet been verified and resolved.

The logic of neural networks was used in producing software programs for land evaluation, such as ImpelERO (De La Rosa et al., 1999) for assessing soil erosion risk, and, integrated into an geographical information system, for evaluating the suitability of soils to different crops in Greece (Yialouris et al., 1997) and in the Philippines (Bandibas, 1998).

1.4. Main International References

Many land evaluation systems are used at an international level. Among them, some constitute an acknowledged reference.

1.4.1. Land Capability Classification

The Land Capability Classification was worked out by the Soil Conservation Service of the U.S. Department of Agriculture (Klingebiel and Montgomery, 1961) and, although developed to sustain specific soil conservation projects in the central part of the United States, it has been widely used throughout the world with numerous adaptations. In the simplicity of application and approximation of the system lie both its strength and its drawbacks. This is a very well-known example of a system of the categorical type, based on criteria of qualitative estimate, the main purpose being to render collected data on natural resources (pedological data specifically) easily readable and comprehensible. The classification system envisages eight classes defined from the combination of land use potential and intensity of limitations present: typically, the number and degree of limitations increases from the first to the eighth class and the potential for alternative uses narrows down. For the first four classes, agricultural use is possible, with increasing limitations and consequently less choice of possible crops. The subsequent three classes have no possible use other than forestry and grazing and the eighth class is intended exclusively for conservation of the soil and environment. Subclasses and units are catered for, depending on the type of limitation present.

1.4.2. The US Bureau of Reclamation (USBR) Land Classification System for Irrigation

The classification of suitability for irrigation was developed in the 1950s in the United States (USBR, 1953), with the aim of classifying lands according to their capability for sustaining agricultural irrigation. This is the first example of classification of land potential for a specific use (Beek, 1978). In the USBR frame, the classes are defined according to their relative capacity to yield a return, or their "capacity for return" in financial terms. That is, they are classified according to net income that the entrepreneur gets after paying operation and maintenance costs. In this sense, it is an example of land evaluation heavily based on economic criteria. The system envisages six classes, defined for the specific characteristics of the area studied. The first three are irrigable, with progressively lower ability to repay costs. The sixth class is not irrigable. The fifth is a "provisional" class, comprising lands that are currently not irrigable but deserve more thorough study. The fourth class is defined as "special", inasmuch as the economic gain and, therefore, the advantage of irrigation, depend on its specific use.

1.4.3. Soil Fertility Capability Classification (FCC)

Put forward by Buol et al. (1975) and modified by Sanchez et al. (1982), the Soil Fertility Capability Classification (FCC) has been reviewed and updated (Sanchez et al., 2003). It is a system aimed at reclassifying soils into homogeneous groups with regard to fertility management, particularly chemical fertility. Three levels are envisaged in this system: the first level is defined by the texture of the surface horizon; the second is defined by the texture of the subsurface horizon (within 50 cm); and the third is identified by "specifiers" representing the chemical and physical conditions that may negatively affect soil fertility and that range from the moisture regime to the presence of toxic elements, salinity, and other factors. In the revision proposed by Sanchez et al. (2003), the FCC is suggested as a semi-quantitative system for estimating soil quality.

1.4.4. The FAO Land Evaluation Scheme

As already mentioned, since the 1970s the FAO (1976, 1991) has developed a land evaluation methodology with the objective of establishing a theoretical framework for the evaluation procedure, rather more than just a formal class description (Davidson, 1980). The FAO framework is considered to be standard reference in land evaluation (Dent and Young, 1981; Van Diepen el al., 1991). The classification structure consists of four hierarchical levels: order, class, subclass, and unit.

There are two orders, only indicating whether a portion of land is suitable (S) or not (N) for sustainable use. The use of orders on their own is generally restricted to studies on a small scale or summary tables. The suitable S order defines an area of land in which the use taken into consideration provides crops or results justifying the employment of human, economic and technological resources, without any risk of degradating the environment. The unsuitable N order, on the other hand, defines a piece of land with qualities unsuitable for an economic and environmental sustainable use, given a particular land utilization type.

The classes define the degree of suitability. For the S order, three classes are recommended: very suitable (S1), moderately suitable (S2), scarcely or marginally suitable (S3); for the N order, two are recommended: at present not suitable (N1) and permanently unsuitable (N2). If there are valid reasons, the number of classes may be increased or decreased. The subclasses define the type of limitation present, while the unit defines the degree. The use of subclasses and units is similar to that of the land use capability classification.

The classification structure allows, in some limited case, the use of a "Sc" phase, defined as suitable under condition (FAO, 1984). If, for example, a land unit is classified as N2o/Sc1, this means that it is not at the moment suitable due to problems of oxygen availability in the rooting zone, but it could become very suitable if the drainage problems were to be resolved.

In 2007, FAO published the text "Land evaluation, towards a revised framework" (FAO, 2007), in which the concepts expressed in the 1976 framework are substantially confirmed. One of the principal innovations in the land evaluation process pertains to the need to develop a more active, significant role for stakeholders. The latter should be approached in both the initial stage, when the objectives, problems and relative causes are defined, and the final one when results are assessed. Also discussed is the contribution that local populations can make with soil maps, and ways in which this can be integrated into international classification. Examples of pedological mapping are then given, where local knowledge on soil behaviour is combined with the genetic characteristics of soil and applicative features, investigated by standard reference methods.

Further methodological progress described in the revised text consists in the highlighting of the need to accompany land evaluation with agronomic research activity, principally so as to establish cultural and biophysical requirements, in environmental modelling, in particular.

1.4.5. *Agro-ecological Zonation: FAO Agro-ecological Zones*

For small-scale applications (national or continental), the reference evaluation system of natural resources for land evaluation is the FAO's agro-ecological zonation (Agro-ecological Zones, AEZ) (1978–1981). Originated as a planning, management and monitoring system of resources on continental and national scales, it envisages the representation of land in distinct layers of spatial information and their consequent integration through geographical information systems. A key concept is the length of the growing period, which is based on rainfall and temperature regimes. The growing period forms the basis for a quantitative climatic classification for each chosen crop, assuming rain-fed agriculture. An agro-climatic adaptability classification matches each crop with climate and soil resources. Agro-ecological zones created with these intersections are thus added to topographical-administrative and demographic information, land use data, etc. Databases linked to the agro-ecological zones provide inputs for the models used for the different applications such as models for calculating the crops' growing period, irrigation scheduling, and soil suitability for different crops and for land planning.

This approach for agro-ecological zones was widely used by FAO in studies of a general nature in developing countries. Also available are software packages containing the system (AEZWIN, ftp://ftp.fao.org/agl/agll/misc/aezwin.exe). A continental assessment was carried out for Africa, Southeast and Southwest Asia and Central and South America (FAO 1978–1981).

1.5. Basic Concepts of Land Evaluation

It has been mentioned that land evaluation is not always geared towards agriculture, but also geared toward protecting the various ecological functions of the soil from possible degrading agents (such as erosion, desertification, loss of organic matter, pollution). We have also seen that, apart from the purposes of evaluation with the model chosen and the method adopted, many systems have some basic principles in common. The basic concepts of the land evaluation process are described in greater detail below, taking the FAO system (1976, 2007) as reference.

1.5.1. Definition of the Spatial Entity under Evaluation

The process of land evaluation traditionally produces maps. It is understood that the subject of evaluation is the mapping unit, whether a pedological unit or landscape unit, at different levels of detail. Furthermore, it is implicit that the output of the evaluation process is a map that generally derives directly from the soil map.

Some land evaluation applications also exist that are not necessarily linked to soil mapping in the traditional sense, which envisages the subdivision of an area into cartographical units characterized by internal variability. For example, just consider precision agriculture, or procedures using continuous spatial models (De Gruijter, 1996). The object of evaluation may, therefore, be any cartographical unit or other spatial entity topologically defined (profile, transect, raster map pixel). The evaluation may also pertain to a soil type not topologically defined. In this case, the cartographical representation will be a function of the probability of encountering the typology itself in the geographical polygon. The spatial accuracy and the quantity of information available for the entity to be evaluated determine the scale of reference and the detail of the evaluation process.

1.5.2. Land Utilization Types

In the FAO framework and in all the models referring to it, the definition of the Land Utilization Type (LUT), i.e., the detailed description of land use in accordance with the scale of reference adopted, is an extremely delicate and important step. It specifies for what type of agricultural or forestry management system or for what crop system the process of evaluation is implemented. The LUT concept requires that the context be explicitly specified from both the socio-economic and technical points of view. This means that, in the definition of LUT, it is necessary to give a complete technical description appropriate for the reference scale of evaluation (e.g., what crop in what agricultural system, and with what energetic and technological input). This is extremely important because modifying the reference scenario means modifying the logical presuppositions of interpretation. If, for example, an area is classified according to its capacity of producing a completely mechanized crop, the land characteristics must be considered in a completely different way compared to the case in which partly manual cultivation is practised.

The LUT specification level depends on the working scale and the level of detail required. The description of a given LUT is often limited to an indication of the crop, or to a succession of crops, with relative cultural practices and levels of technological and energetic input. In the evaluation carried out over limited areas of the territory, explicit reference is usually made to local conditions, often functioning as a variable guide in the evaluation process (see, for example, the chapter on evaluation for the flowering ash). In any case, in evaluation processes designed for agricultural planning, and more generally for territorial planning, it is evident that the evaluation context must be clearly defined. Consequently, the evaluator is required to prepare a list of specifics identifying a determinate LUT.

In the case of evaluations for non-agricultural purposes, the land is assessed for the capability to fulfil functions other than agriculture or forestry production. Soil protective capacity, or attenuation capacity, as regards water and suitability for animal slurry application are possible examples. In such cases, the type of utilization must be adequately specified, so that the outlines of the problem are clear. For example, when the soil protective capacity is to be evaluated, it must be clear whether the evaluation refers to waters, to the soil or to the crops being cultivated on it. This step is fundamental in order to identify the soil and land features and qualities to be considered.

1.5.3. Land Characteristics and Quality

A land characteristic is "an attribute of land that can be measured or estimated, and which can be employed for distinguishing between land units of differing suitabilities for use and as a means of describing land qualities" (FAO, 1985). On the other hand, land quality is defined as "a complex attribute of land which acts in a manner distinct from the actions of other land qualities in its influence on the suitability of land for a specific kind of use" (FAO, 1983). As a rule, the characteristics are described during surveys and the basic studies for evaluation, and they include the following: soil features such as particle size distribution, stoniness, pH values, calcium carbonate content; climatic characteristics such as rainfall and temperature; geomorphological features such as susceptibility to landslide; topographical characteristics such as elevation, aspect and gradient; hydrological characteristics such as depth of permanent and temporary water tables; and anything else that could be of use in defining land units and their evaluation. Consequently, land characteristics are attributes that may be measured or estimated in ordinary surveys, and that influence soil quality, but, as a rule, do not directly influence the degree of capability for a certain soil use. Soil particle size distribution, for example, has an effect on many soil qualities, such as capacity for air, capacity for water, cultivability and practicability, risk of compacting and risk of erosion. The effect may also differ: coarse soils ensure good capacity for air, but they are wanting from a nutritional point of view, or, in sub-arid regimes, in holding water.

In practice, soil characteristics are often directly used to evaluate the degree of capacity for a certain type of use, but it would be more correct to use land characteristics to estimate land quality. Indeed, land qualities cannot generally be measured directly, but they can be estimated through measurement of land characteristics.

1.5.4. Land Use Requirements

Any type of land use, as defined in the previous paragraphs, has certain needs or requirements (Land Use Requirements), which can be defined: (1) in physiological terms, regarding crop growth, (2) technologically and in terms of management, and (3) in terms of soil conservation. In the first case, one has to establish the optimum edaphic conditions for crop growth (crop requirements); in the second, the requirements for realizing crop management techniques (management requirements); in the third, the conditions for controlling erosion processes and soil degradation (conservation requirements). Not included in the requirements are specifics regarding socio-economic aspects such as labour or capital, even though almost all physical features of a territory have a bearing on them. Requirements of the kinds of use may be expressed in terms of both characteristics and quality. In 2007, the FAO introduced the term "sustainability", broadening the concept of conservation requirements and including productivity, social equity and environmental aspects (FAO, 2007).

In particular, knowledge of crop requirements is fundamental for success in evaluation. The availability of experiences, field trials and experimentation pertinent to the object of evaluation and the territory under investigation allows evaluators to select and rank soil and land characteristics or qualities in the most appropriate way. Otherwise, information retrievable from manuals or other texts has to be relied on to describe (at times in very general terms) crop requirements or suitable conditions for some agricultural practices.

1.5.5. Evaluation Procedure: Matching Stage

Once the aim of the evaluation is established, the data to process have been retrieved, and the requirements for a specific use have been defined, in order to apply the evaluation procedure in the territory under investigation, it is necessary to treat these elements so as to attribute suitability classes to the objects in the evaluation, for example soil typological units or soil mapping units. This objective is traditionally reached by means of a table (matching or lookup table) in which the requirements of a determinate type of utilization are compared with the land unit qualities. This is truly a delicate process, because giving a different weight to this or that quality means varying, sometimes notably, the outcome of the evaluation. There are many methods of attributing suitability class by using the comparison table, but the most frequently used is the maximum limitation method. Following this method, the land units are classified according to the most severe limitation for a determinate use. The method is analogous to the "law of the minimum" postulated in 1840 by Liebig, who stated that *"yield is proportional to the amount of the most limiting nutrient, whichever nutrient it may be"*. This concept can be (and in fact was) extended to populations and ecosystems, where growth is controlled by the most limiting factor. The result of the comparison allows the quality with the lowest value for each soil to be detected and attributed with the relative class. For example, for a given land use type, five qualities are taken into consideration, which for a certain soil take on values s1, s2, s1, s3, s1. As the s3 value is the lowest, that land unit will assume the overall suitability class s3.

In order to calibrate the correctness of evaluation, it is indispensable to establish the benchmark soil where the soil characteristics and qualities are known in detail, along with the estimated cultural or agro-technical response. However, the tables are not the only way to proceed for comparing requirements and qualities or for assessing the qualities from diagnostic characteristics. The "decision tree" type of approach (see below) allows the comparison structure to be hierarchically arranged and possible interactions to be taken into account at the same time. In recent times, the comparison stage has been achieved, to an increasing extent, by alternative procedures rather than the traditional ones, such as statistical analysis, neural networks, or simulation models.

1.5.6. Classification Structure

As has been mentioned, there are many evaluation methods, each connected to different purposes, but all follow the same logical outline, based on the principles below:

1) The classes are attributed with a decreasing value, i.e. the first classes express the best conditions for the use considered, the last the worst.
2) The judgement expressed by the evaluation must be justifiable; indeed, all classifications envisage that the limiting conditions that have affected the final assessment be expressed.

1.5.7. Spatialization of Evaluation Results

The spatialization of land evaluation results is traditionally obtained on the basis of the soil map. The evaluation object may be the soil mapping unit or the soil typological unit. In the first case, a correspondence between the soil and the cartographical unit is presupposed, in the second, the soil benchmarks or the modal typology soils are evaluated and the result of the evaluation is compared with the single or associated presence of the typologies in the mapping units. There will, therefore, be cartographical units of S1 type, or S1 and S2, or S1 and S3 etc., with a certain percentage for each class. Alternatively, instead of evaluating the benchmark profile alone or the typology modal profile, it is possible to assess all single observations pertaining to the typology present in the database, and to establish the typology modal class or even the more representative classes, to then be able to proceed as described.

Recently, spatialization methods other than the soil map in the classic sense have been put forward (De Gruijter, 1996). The use of spatial analysis techniques, like geostatistics, and the availability of continuous informational layers, like satellite images or digital elevation model, permit the continuous characterization of a determinate area of land for some attributes and therefore the carrying out of the evaluation by pixel in a raster-type representation.

1.5.8. Land Evaluation and GIS

The extensive use of GIS has changed the methodology of land evaluation. From an evaluation that essentially used the soil typological unit or the soil mapping unit as the only container of information and unit of evaluation, over recent years an integrated use of geographical data bases has been preferred, among which are the soil data bases. The information required from the soil data bases only concerns a few soil characteristics and qualities, essentially drawn from the profile description and analysis, which are placed in relation to soil functional aspects related to land use and to morphological, geological and hydrological features. In this way, the evaluation process has the pixel as its object, associated with information received from the various original informational layers (e.g.: slope, height and drainage from the DEM; spectral response in the various wavelengths from the satellite; available water, depth and stoniness from the soil data bank; substratum from the lithological data bank; precipitation, temperature and their processing by climatic spatialization). The advantages of this methodological approach are:

- a more detailed spatialization of the evaluation;
- automation of evaluation procedures;
- the possibility of modifying evaluation parameters and immediately verifying the result; and
- integration of many informational layers.

However, the risks inherent in this method must also be considered, particularly relative to the coherence between the nominal scale of the themes/thematic maps used, the corresponding informational content and the evaluation reference scale. If, for example, a certain evaluation is to be based upon the realization of a map of a given detail, the functional land qualities must also necessarily be established at that detail. If a high resolution DEM is used with pedological or climatical information at a reconnaissance scale, an evaluation will be obtained that is only seemingly detailed.

Then there is the problem of error propagation. Each thematic map/theme displayed on map brings with it a series of typological and topological errors (e.g., of measurement, of processing, of spatialization) that are then multiplied by the errors in each single (thematic) map. Several informational layers are used and the greater their degree of approximation, the greater the error in the evaluation. Since propagation of the error is generally exponential, it can be easily appreciated how important it is to make a cautious choice of which themes to use.

Land evaluation by means of GIS could, however, also have a polygon instead of the pixel as the object. Indeed, each single delineation present in the GIS may have a different distribution from the pedological typologies within it, thus a more or less articulated combination of evaluation classes. In this case, it is possible to provide a probabilistic evaluation of the presence of various aptitude classes within each polygon. An example of such an evaluation spatialization procedure is reported in this manual in the chapter on olive cultivation.

1.5.9. Reliability of Land Evaluation

It can clearly be seen from previous sections how the same final evaluation can be reached by more or less complex means, as well as methodological and technical tools, which are also very different. A more exhaustive panorama of the various approaches to land evaluation is laid out in subsequent sections. However, like any assessment, land evaluation is influenced by a certain degree of error, deriving particularly from any approximations and generalizations there may be. It does not always apply that the greater the approximation, the more defective and less reliable the outcome; all the same, the degree of uncertainty in the evaluation is rarely defined and clarified in quantitative terms, or, at least, expressed clearly in descriptive terms. That can be a serious defect of the evaluation, especially when the users must take decisions on the basis of the most precise information possible and the degree of accuracy must be the highest possible.

Awareness of the doubtful reliability of an evaluation process has resulted in the concepts of "purity" and "confidence" becoming apparent in the evaluation (King and Le Bas, 1996;

Napoli et al., 1999; Costantini and Sulli, 2000). Such criteria, which produce cartography associated with the evaluation, provide instruments capable of defining the informational reliability of the cartography produced. The "purity" of the evaluation represents the degree of geographic validity of the thematic information and is a function of the presence of "impurities" in the single polygon, i.e., of soils with a different evaluation from that described in the legend. The "confidence" of the evaluation expresses the concept of reliability of the assessment. This may refer to the reliability of the evaluation model adopted for the soils under assessment or to the presence of specific experimental trials with regard to benchmark soils. Numerous examples of cartography realized from "confidence" and "purity" evaluations are listed in the European soil database (JRC, 2007).

In the land evaluation process, sources of uncertainty may substantially be ascribed to: (i) data uncertainty, linked to sampling and measuring problems, (ii) enlargement of data from sampling points to map unit; and (iii) rule uncertainty. Measuring errors are generally considered independent, normally distributed and more or less easily identifiable; for these reasons they have reduced weight in affecting the reliability of the final evaluation. Sampling errors, on the other hand, are more difficult to eliminate: the sampled object never coincides with the object of the evaluation, and the gap between the two "realities" critically depends on the sampling strategy adopted. The scope of such errors can only be determined by repeatedly carrying out sub-sampling or by tightening the sampling mesh. In both cases, the uncertainty of the data may be expressed by associating the sampling variation to the expected value (i.e., the average).

Extending punctual data to a cell or polygon necessarily entails a certain degree of approximation, variable according to the strategy adopted: (i) attribution to the unit of a single "typical" or "representative" value; (ii) attribution to the unit of a range of values (or classes, defined by the minimum and maximum values); (iii) description of the unit through parameters of a statistical distribution (e.g., normal, uniform, triangular, exponential); and (iv) description of the unit through a non-parametric distribution (i.e., through a frequency distribution that completely specifies the probability of a certain datum value occurring). Once the data that the evaluation is based on are known, together with their degree of uncertainty and methods of spatialization, it remains to verify the formal correctness of the process that leads to the combination being attributed a determinate class. Indeed, in this passage, too, there are sources of more or less obvious uncertainty: if the continuous transfer functions are adopted (e.g., multiple regressions, polynomials), their "true" form is never known but only inferred by experimental data, or furthermore, in the case of an "expert's assessment", the outcome of the evaluation may change from individual to individual, at the same time retaining (even for the same individual) a certain degree of ambiguity.

Should the data be entirely described by the distribution of probabilities and these probabilities in turn combined in continuous functions for the definition of suitability classes, it is possible to apply the error propagation theory so as to statistically determine the reliability of the final result. It is, therefore, possible in this case to define precisely the statistical confidence limits for the results of the land evaluation procedure. An example is land evaluation based on the yield forecast of a relevant crop estimated by multiple regression, the function of a certain number of land attributes, each characterized by their own probability distributions. In such cases, reference can be made to the following general operational indications, to be adhered to so as to reduce the estimation error as much as possible: (i) avoid indexes obtained when multiplying the factors by one another, since the errors of each factor are thus also multiplied; (ii) reduce the number of factors as much as possible; and (iii) detect those factors constituting the main source of error with their greater weight and attempt to control them as far as possible (Rossiter, 1994).

In order to define evaluation results in statistical terms of confidence limits, a commonly used method is the Monte Carlo simulation technique applied to input parameter values of a given evaluation model (e.g., a regression equation). Given the distribution characteristics of a parameter population (e.g., average and standard deviation of the clay content or soil depth), the Monte Carlo simulation allows a high number of values to be generated for use as the chosen evaluation model input. Therefore, the output obtained from the model itself could be described statistically, quantifying the uncertainty, for example through the confidence interval of the average value obtained.

An aspect of the land evaluation process that plays a critical role in determining reliability is that the characters of the soils vary in space according to a pattern that, in most cases, is not fortuitous but shows a spatial dependence allowing the extrapolation of available information in correspondence to surrounding points that are not sampled. The existence of spatial dependence between the values of a given variable implicates the existence of a spatial structure that, once the experimental semivariogram is computed and modelled, allows the assessment of the target variable in the whole study area, at the same time quantifying the spatial uncertainty by resorting to geostatistical approaches (multiple indicator kriging) or to sequential simulation techniques (Goovaerts, 2001).

A further possibility for expressing evaluation reliability comes from *post hoc* validation. Once the evaluation has been carried out (or at least using a procedure completely independent from it), a series of soil observations must be made in the territory under investigation, preferably over a regular grid, or sampling the crops at farm level, in order to verify whether the evaluation carried out in those situations corresponds to what has actually been found. The data thus obtained may then be processed statistically, obtaining various overall index estimations of evaluation reliability, or else geostatistically, for example through a multiple indicator kriging, so as to ascertain the distribution of those areas in which evaluation is more uncertain and consequently needs further survey and verification. An application in Tuscany of such a methodology is described in Costantini et al. (2002) and in Castrignanò et al. (2009).

1.6. Recent Evolution in Land Evaluation

With the increased availability of ever more powerful instruments of calculation, land evaluation procedures have in recent times considerably evolved from techniques used in the early 1970s. In the present section, a few techniques that were introduced into the evaluation process a short while ago are briefly illustrated.

1.6.1. Expert Systems

Expert Systems are computer programs that simulate the capacity of one or more experts to resolve a problem. Thus, they are instruments that back up decisions taken (decision support tools) and they simplify the finding of solutions to the more frequent and complex problems.

It is necessary to choose the most suitable way to reflect specific knowledge, as well as to represent the type of result expected. As for land evaluation, this system is fundamentally based on using a "decision tree", representing the stage that matches land properties with requirements for the type of use. A tree structure is constructed by experts and therefore in itself reflects their level of knowledge. The decision tree structure is a pyramid or hierarchical structure, with the "leaves" representing land features or qualities (expressed as value intervals or classes) and inner nodes representing the value attributed for a certain use. A tree structure may be used effectively both in the quality assessment process and in the attribution to a determinate suitability class. In itself, an expert's system does not add information, but rather structures it logically. When enough experimental data can be accessed, it is possible to construct a statistically based decision tree, characterized by a good predictive capacity. At times, both types of expert system are applied (theoretical and statistical decision trees) with the aim of obtaining better results.

The *Automated Land Evaluation System* or ALES (Rossiter, 1990; http://www.css.cornell.edu/landeval/ales/ales.htm) is a computer program that supports users in constructing their own decision tree, usable in the following capability or suitability assessment stage (physical and economical) of determinate uses of the soil in relation to the features of the land. It represents an external container within which an "expert system" can be constructed and consequently does not in itself provide any kind of knowledge. Inside this empty container, possible soil uses must be described, in both physical and economic terms, and keyed into the data base containing the land qualities to be assessed. Therefore, ALES actually represents the software implementation of the FAO approach (1976), with the advantage that it is implemented along with a well-defined method of economic analysis, as well as facilitating and accelerating its application. Furthermore, employment of the decision tree allows the logical process of anyone constructing the model to be made explicit, and the results of evaluation used in the successive decisional planning stage, as well.

1.6.2. Physico-mathematical Modelling

The main utility of simulation models in the process of land evaluation for agricultural and forestry purposes lies in yield forecast: land value is directly assessed through the production obtained from it, and in this sense, evaluation by means of modelling is quantitative by definition. In evaluations for non-agricultural purposes, such as in assessments for environmental purposes, modelling could help in simulating the conduct of a determinate environmental system with external input; erosion modelling is a well-known example. Physico-mathematical models are also used for assessing single qualities, particularly important in the evaluation process: for example, the field water balance of the soil-plant system (which determines the process of crop growth as well as the buffer function of soils against pollution) may be simulated with a certain degree of accuracy by means of mathematical models. According to Rossiter (1996), there are fundamentally two kinds of quantitative modelling approaches to land evaluation: the first is empirical (or statistical) and the second is deterministic and dynamic.

Statistical or Empirical Models

For land evaluation purposes, the aim of statistical models is to quantify the relationships observed between the land characteristics (or qualities) and the potential production for a given area, and to use them in making previsions with regard to other areas. An indispensable requirement for the application of such a model is a sufficient supply of data from which to statistically deduce the desired variables. The employment of statistical models in evaluating land goes back quite far in time, as far as the origins of this discipline. As previously mentioned, the result obtained by the application of such models consists in evaluating the potential production of a given cultural system, index of its suitability for that specific use. Therefore, production factors must be defined *a priori*, as they coincide with measurable land properties and they constitute independent variables (x) from which to obtain the dependent variable (y), i.e., the potential production. The aim is to achieve a causal relationship from the production data observed (y) and productive factors present (e.g., quantities of fertilizer used, quantities of nutrients in the soil).

Two different kinds of data relative to production factors can be used: control data, retrieved from ad hoc experimentation, or data derived from direct observation. In the first case, more reliable cause-and-effect relationships are obtained, but at a greater cost than in the second case. In the domain of land evaluation, experimental plots are often used to achieve the relationship developing between diagnostic characteristics and land qualities, while direct observation is resorted to in order to assess potential production (land suitability), starting with the environmental characteristics observed. In practice, the correlation existing between the factors affecting production and the actual production is monitored.

The simplest statistical model is one based on the assumption of a linear relationship existing between one or more independent variables and the dependent variable (production). Some variables may be decisive for some agro-environmental systems, and the use of a simple linear regression may therefore be justified. The water regime for dry cultures in dry sub-humid and semi-arid zones comes to mind. However, it is clear that the matter is more complicated in actual fact. Multiple regression equations enable greater articulation, putting the dependent variable, the production, in relation to different independent variables. A well-known example of such an approach is the one De La Rosa et al. (1981) realized for Spain, subsequently incorporated into the MicroLEIS software package (De La Rosa, 2002; De La Rosa et al., 1992, 2004). De La Rosa et al. (1981) proposed a series of equations correlating the yields of various Mediterranean cultures with land properties such as pH, depth, texture and capacity for cationic exchange. Significant statistical relationships have been discovered in Italy between tobacco yields and the soil physical qualities, in particular, air capacity and water content at field capacity (Costantini et al., 1992). By making calculations in the ripening period, Tomasi et al. (1999) more recently found a relationship between sugars in the grape and available water in soil.

Use of linear relationships is not possible in many situations because the relationship between variables and the system's response is not of the linear type. Ordinary transformations can resolve the problem in certain cases. Indeed, there are various ways to infer regression equations of observed data on environmental variables, but the chance of defining one better than the others is unlikely: in fact, this is always an arbitrary process based on a subjective opinion. Non-linear, multivariate statistical techniques may take interaction effects between variables into account, with consequent synergic or antagonistic action among the variables considered. An example of a synergic effect can be that which causes high levels of just one nutritional element to foster a poor response in production, until an analogous level is reached for all the nutritional elements. An example of antagonistic effect is what ensures that a high level of nutrients leads to increased efficiency in water absorption by the plant. In this way, the increased water employed has a less evident response than that obtained when the level of nutritional elements is low.

In view of the numerous advantages of statistical models in land evaluation (which are in the bargain very frequently used), a few drawbacks have also been recorded. Statistical models supply a good description of land behaviour also in statistical terms with regard to a determinate use, but if not based on rigorous experimentation in the field, which is extremely laborious, there may be the disadvantage of excessive subjectivity both in the choice of variables taken into consideration and in processing methods.

Deterministic and Dynamic Models

Dynamic simulation models quantitatively describe physical and biological processes, such as a crop growth, the soil water balance, the leaching of nutrients or pollutants, and soil erosion. In land evaluation, such models are used essentially to quantify a crop production, but also to investigate the effects of drought, the loss of nutritional elements, soil erosion, and other conditions under different crops and management techniques. The use of dynamic models can even be limited to dynamic assessments of particularly significant land qualities. As already mentioned, the soil water regime, a decisive quality for the possibility of agricultural use especially in dry sub-humid and semi-arid regions, may be assessed by dynamic simulation models.

Unlike traditional systems, dynamic models have the advantage (among other things) of considering the temporal variability both of land features and qualities and also of kinds of use. The characteristic of deterministic models, whether of crop growth, of water balance or of erosion, is in general to require a great amount of measured input data. Furthermore, such models must be gauged in the environment in which they are applied and the validation of results is invariably difficult. This often leads to the failed correspondence between the level of detail of the available input data and that required by the model: to compensate for this drawback, it usually becomes necessary to increase the level of detail in the survey, or resort to using the so-called pedotransfer functions (Bouma, 1989), i.e., more or less complex algorithms to estimate the data necessary, from the more easily accessible data. Although these models supply the quantitative yield of a crop as the final output, in many cases, this does not represent the prime purpose of their application. Indeed, they are often employed to investigate physical processes involved in crop production, in different land uses or to support management. WOFOST (Van Diepen et al., 1989) is an example of dynamic simulation model using some land characteristics as input (Rossiter, 1994). Another example is given by the CropSyst model (Cropping Systems Simulation Model, Stockle and Nelson, 1994, 1996). This is a crop growth simulation model, developed with the aim of studying the effects on cropping systems management on productivity and on the environment. Besides productivity, CropSyst simulates soil water balance, nitrate balance, and pesticides' fate. Therefore, it can be used for assessing crop yields, soil erosion, and potential risk for nitrate and pesticide leaching. The model is available on internet (http://www.bsyse.wsu.edu/cropsyst).

Pedotransfer Functions for Estimating Model Input

Among the input data necessary to run simulation models, the physico-hydrological parameters of the soil are the least accessible in routine surveys and among the most laborious to measure.

Therefore, numerous pedotransfer functions for the derivation of physical-hydrological parameters of the soil have been developed over the years, that is to say, retention characteristics and hydraulic conductivity. Although universally applicable relationships for deriving hydraulic properties from the soil do not exist, it is possible to indicate some pedotransfer functions derived from substantial sets of data and validated in the field and in the laboratory. Their choice and consequent use in assessing hydrological parameters is mainly function of factors such as the scale of investigation, availability and kind of data measured and the degree of accuracy required. Analytical formulation of a soil's water retention curve has been formalized by van Genuchten (1980):

$$\theta = \theta_r + (\theta_s - \theta_r)\,[1+\alpha\psi)^n]^{-m} \tag{2}$$

where θ is the water content at a certain tension ψ, θ_s and θ_r (cm³cm⁻³) are the water saturation content and residue, α (cm⁻¹), n (-) and m (-) are empirical parameters determining the form of the curve: α is approximately equal to the inverse of the potential value at the curve's inflection point (air entry potential, h_b, cm), corresponding to the limit between the micropores and the macropores domain, while n determines the curve slope or rather the velocity with which the curve approaches the ordinate axis as the potential diminishes (Fig. 1.2).

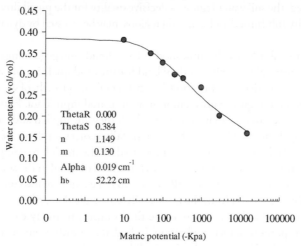

Fig. 1.2. Water retention curve according to van Genuchten's model (dots, experimental values; solid line, fitted model).

Some simulation models of the water balance (e.g. MACRO, Jarvis et al., 1997) use the formulation by Brooks and Corey (1964) for the water retention curve:

$$\theta = \theta_r + (\theta_s - \theta_r)\left(\frac{h}{h_b}\right)^{-\lambda}, \text{ by } h/h_b > 1 \tag{3}$$

$$\theta = \theta s, \text{ by } h/h_b \leq 1 \tag{4}$$

where h is the potential, h_b (cm) the air-entry potential, λ (-) the pore size distribution index that determines the water retention curve slope, θ_r the residual water content (cm³cm⁻³), and θ_s

the one at saturation (cm^3cm^{-3}). In general, the λ values are greater for soils with uniform pore dimension distribution, while values are lower in soils with pores of extremely variable dimension. Residual water content represents the maximum quantity of water not contributing to the flow, on account of absorption on the solid phase or absence of continuity in the pores. It could formally be defined as the water content corresponding to which both $d\theta/dh$ and K conductivity drop to zero for potential increasing h values. Residual water content is an extrapolated parameter and therefore does not necessarily represent the soil's minimum water content possible.

Van Genuchten's parameters are linked to Brooks and Corey's parameters of the following relationships:

$$\alpha = 1/ h_b \tag{5}$$
$$n = \lambda + 1 \tag{6}$$
$$m = \lambda /(\lambda + 1) \tag{7}$$

Both retention functions examined may be combined with water conductivity prevision models, once a functional relationship between the two is known and defined. The numerous approaches formulated to develop continuous functions that are able to describe $K(h)$ and $K(\theta)$ (Gardner, 1958; Kunze et al., 1968; Millington and Quirk, 1961; Russo, 1988) are distinguished by important conceptual differences; still, some of the differences in their descriptive and predictive capacities are due to the fact of the different formulations being calibrated on data sets of scant dimension and relative to distinct textural classes. On account of their simplicity, the use of predictive models for $K(h)$ and $K(\theta)$ is extremely widespread for the numerical solution to Richards' equation for the unsaturated flow. The results obtained show that such models work satisfactorily in the case of coarse and loamy soils, but for previsions of fine and well-structured soils, they are substantially hardly reliable.

An updated presentation in the field of pedotransfer functions can be found in Pachepsky and Rawls (2005).

Decision Support Techniques

There are techniques that go beyond land evaluation procedures in the strict sense, indeed they integrate them, aiming to resolve the problem of allocating resources in the scope of planning, at both individual farm and regional levels. These sometimes use the evaluation procedure results as input parameters in the successive resource allocation stage or, on the contrary, they carry out both the function of resource evaluation instrument and that of backup to the actual decisional stage. Land evaluation and regional planning, though two distinct concepts and two distinct processes, are strictly interdependent. Land evaluation is often used as a study preliminary to planning but, at the same time, elements relative to planning also serve to make a prior selection of the possible uses of the land to be considered in the evaluation process (FAO, 1993, 2007).

Multi-Criteria Analyses

According to the definition given by Roy (1996, in Chakhar and Martel, 2003), multi-criteria analysis is a mathematical decision support tool that will permit alternatives and different scenarios to be compared following different criteria often in competition among themselves. The criteria are characteristics of the territory on which the choice of one scenario over another depends. Thanks to the help provided by the multi-criteria analysis, the planner reaches a decision that at the end turns out to be a compromise, duly taking into account the various pros and cons and the preferences of local subjects (Ceballos-Silva and Lopez-Blanco, 2003).

To do that, it is necessary to reach a compromise between the various alternatives and to rank those alternatives according to their attractiveness. It is multi-criteria analysis that serves to identify this compromise, maximizing benefits for all interested in the evaluation/planning process, and reducing costs and damage to the environment to a minimum. Multi-criteria analysis may be integrated into a GIS environment for constructing support systems to decisions that take into account the spatial dimension (Spatial Decision Support System). An example of commercial software is the GIS Idrisi package (Clark Labs.), which, in the latest versions, has a module for multi-criteria evaluation set in place.

This kind of analysis may turn out useful in approaching land evaluation of the participatory kind (Cools et al., 2003). In this case, participatory approach refers to the direct and joint involvement in the evaluation process of the various operators interested (e.g. the farmers, local technicians, policy makers). Considering that the process involves all people and/or corporations that are or will become affected by the modifications carried out on the territory, this approach raises new problems: different participants often have different objectives and needs and, therefore, planning must take into account all these components and achieve a compromise between them.

An example of participatory approach to planning has recently been produced by the FAO in Bosnia and Herzegovina (FAO, 2004) in a study designed to create an inventory of post-war situation of land resources (http://www.plud.ba/). In this project, carried out in collaboration with Cooperazione Italiana, a model of planning process was tried out that puts the stakeholder at the centre, i.e., any individual, association or institution interested in participating in decisions regarding them. Potential conflict that may be created between the various interests are here resolved through a series of steps that are formalized in their succession and realization: laying out basic information, examining intermediate products, discussing results, and even re-examining them.

Linear Programming

Linear programming is a special case of mathematical programming designed to optimize an objective function subjected to a series of constrictions or limitations. Both objective function and limitations are of the linear type. Inasmuch as they are instruments of optimization, the linear programming techniques are used as a decision support. The various managerial and/or technological decisions often required in environmental planning studies and/or socio-economic studies may be rationalized by means of mathematical algorithms (Reveliotis, 1997).

The MODAM model (*Multi-Objective Decision support tool for Agroecosystem Management*), developed in Germany by ZALF (Zander, 2001), constitutes an example of applying linear programming to planning at a territorial level. In this approach, the economic optimization at the farm level is integrated with environmental restrictions (e.g., risk of soil erosion, risk of pollution, conservation of biodiversity), for the sustainable management of the territory. A recent MODAM application in seven European rural landscapes can be found at http://www.mea-scope.eu/.

1.6.3. The DPSIR Model (with the collaboration of Andrea Vacca[1])

The DPSIR conceptual model (*Driving forces, Pressures, State, Impacts, Response*), devised by the European Environment Agency (EEA, 1999) as a development of the simpler PSR model (*Pressure State Response*) by the OECD (Organization for Economic Cooperation and Development, 1998), is an instrument for describing and analysing interactions between society and the environment. It is used to study factors affecting the environment itself, their interrelation and the efficacy of responses to the effects of such an impact. The DPSIR is, consequently, a generic model rather than an actual methodology, created to organize information concerning the state of the environment; its structure, derived from that originally applied to studies of a social kind, has been amplified and is applied more generally to the organization of environmental indicators and, successively, to sustainable development. The model is based on relationships of cause and effect between social, economic and environmental components acting on the environment, to be precise:

- *Driving forces*: determinants that induce change in the environment (e.g., industrial production), that is, anthropic activities, deriving from social, economic and productive needs, that go on over a territory and create pressure on the environment.
- *Pressures*: pressures on the environment (e.g., drain water) deriving from human activities and behaviours.
- *State*: the state of the environment (e.g., water quality in rivers and lakes), that is, the qualities and characteristics of environmental resources, expressed through chemical, biological, physical, naturalistic and economic parameters, that can be negatively altered by pressure and that must be safeguarded.
- *Impacts*: impacts on the population, economy and ecosystem (e.g., undrinkable water), that is, significant changes of the state of the environment that appear as alterations in ecosystems, in their capacity to sustain life, human health, social and economic performances.
- *Response*: society's response (e.g., watershed management), or rather that of public administration, to any DPSIR component in handling impacts and promoting sustainable development.

Thus, the DPSIR model allows the representation (even visual) of themes and environmental phenomena, correlating them with the policies exercised on them (Fig. 1.3). Therefore, it is easily understood even by non-experts.

Environmental indicators are used to describe the state of a phenomenon, an area or an environment within the DPSIR model; these indicators permit the synthesis of a phenomenon, reducing the number of measures and parameters that would normally be needed to give a complete view of the situation, and also facilitating communication with the user. In fact, a limited number of indicators are selected according to their capacity in interpreting and summarizing a great number of related data; furthermore, they are able to highlight the evolutionary tendency of phenomena and to facilitate correlation between them.

Within the model, the indicators representing the environmental components are connected by a circular logical causality relationship, according to which pressures on the environment modify its state, which, in turn, influences the response to be followed through

---

[1] University of Cagliari, Italy.

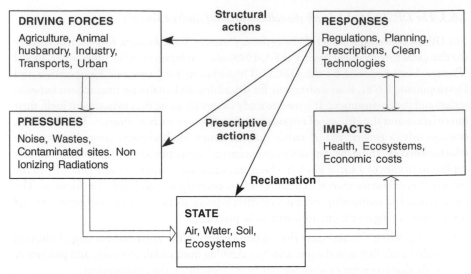

Fig. 1.3. The DPSIR model.

so as to reach the standard desired. Vacca and Marrone (2004) have applied the DPSIR model to the soil, using it as a reference system for the analytical study of cause and effect in degradation processes. Six main types of soil degradation were considered, broken down into their components by means of the application of the DPSIR model.

By way of example, we describe in Fig. 1.4 the application of this model in the case of soil pollution.

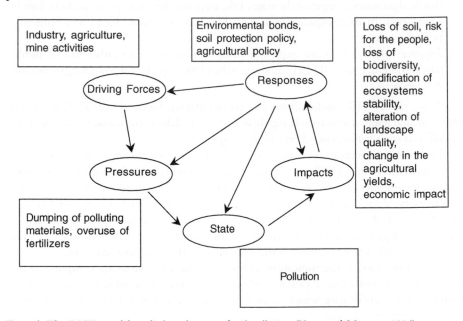

Fig. 1.4. The DPSIR model applied to the case of soil pollution (Vacca and Marrone, 2004).

References

ACUTIS, M., DONATELLI, M. 2003. SOILPAR 2.00: Software to estimate soil hydrological parameters and functions. European Journal of Agronomy 18, 373–377.

BANDIBAS, J.C. 1998. A Land Evaluation System Using Artificial Neural Network Based Expert's Knowledge and GIS. Proceedings of ACRS, 1998. http://www.gisdevelopment.net/aars/acrs/1998/ps1/ps1023a.shtml.

BEEK, K.J. 1978. Land evaluation for agricultural development: some explorations of land use systems analysis with particular reference to Latin America. Publ. 23, ILRI, Wageningen.

BLOEMEN, W. 1980. Calculation of hydraulic conductivities of soils from texture and organic matter content. Z. Pflanzenernähr. Bodenkd. 43, 581–605.

BORSELLI, L. 1995. Fusle: A Computer Program for Soil Loss Risk Analysis by Fuzzy Variables and Possibility Distributions. Quaderni di scienza del suolo, C.N.R., Firenze VI, 81–98.

BOUMA, J. 1989. Using soil survey data for quantitative land evaluation. In: Stewart, B.A. (Ed.), *Advances in Soil Science* 9, Springer and Verlag, Berlin 177–213.

BOUMA, J. 2000. Land evaluation for landscape units. In: Sumner, M.E. (Ed.), Handbook of Soil Science, cap. 7, E393–E412

BRAKENSIEK, D.L., RAWLS, W.J., STEPHENSON, G.R. 1984. Modifying SCS hydrologic soil groups and curve numbers for rangeland soils. ASAE Paper N. PNR-84–302, St. Joseph, Michigan.

BROOKS, R.H., COREY, A.T. 1964. Hydraulic properties of porous media. Hydrology paper no. 3, Colorado State University, Fort Collins.

BUOL, S.W., SANCHEZ, P.A., CATE, R.B., GRANGER, M.A. 1975. Soil fertility capability classification: a technical soil classification system for fertility management. In: Bornemisza, E., Alvarado, A. (Ed.), *Soil Management in Tropical America*, N.C. State Univ., Raleigh, 126–145.

BURROUGH, P.A. 1989. Fuzzy mathematical methods for soil survey and land evaluation. Journal of Soil Science 40, 477–492.

Burrough, P.A., MACMILLAN, R.A., van DEURSEN, W. 1992. Fuzzy classification methods for determining land suitability from soil profile observations and topography. Journal of Soil Science 43, 193–210.

CALZOLARI, C., UNGARO, F., BUSONI, E., SALVADOR, P. 2001a. Metodi indiretti per la stima delle proprietà fisico-idrologica dei suoli. I. Validazione di pedofunzioni da letteratura. Progetto SINA Carta Pedologica, in Aree a Rischio Ambientale, Rapporto n. 9.1, April 2001. http://www.regione.emilia-romagna.it/geologia/suoli/suol_docu_sina.htm.

CALZOLARI, C., UNGARO, F., GUERMANDI, M., LARUCCIA, N. 2001b. Suoli capisaldo della pianura Padano-Veneta: bilanci idrici e capacità protettiva. Progetto SINA Carta Pedologica, In: Aree a Rischio Ambientale, Rapporto n.10.1, aprile 2001. http://www.regione.emilia-romagna.it/ geologia/suoli/suol_docu_sina.htm.

CAMPBELL, G.S. 1985. *Soil Physics with BASIC: Transport Models for Soil-Plant Systems*. 150, Elsevier, New York.

CAMPBELL, G.S., SHIOZAWA, S. 1992. Prediction of hydraulic properties of soils using particle-size distribution and bulk density data. In: Van Genuchten, M.T., Leij, F.J., Lund, L.J. (Ed.), Proceedings of the International Workshop on Indirect Methods for Estimating the Hydraulic Properties of Unsaturated Soils; Riverside, California, October 11–13, 1989, 317–328.

CASTRIGNANO, A., COSTANTINI, E.A.C., BARBETTI, R., SOLLITTO, D. 2009. Accounting for extensive topographic and pedologic secondary information to improve soil mapping, Catena 77, 28–38.

CEBALLOS-SILVA, A., LOPEZ-BLANCO, J. 2003. Delineation of suitable areas for crops using a Multi-Criteria Evaluation approach and land use/cover mapping: a case study in Central Mexico. Agricultural System 77, 117–136.

CHAKHAR, S., MARTEL, J.M. 2003. Enhancing geographical information systems capabilities with multi-criteria evaluation functions. Journal of Geographic Information and Decision Analysis 2, 47–71.

COOLS, N., DE PAUW, E., DECKERS, J. 2003. Towards an integration of conventional land evaluation methods and farmers' soil suitability assessment: a case study in northwestern Syria. Agriculture, Ecosystems and Environment 95, 327–342.

COSBY, B.J., HOMBERGER, G.M., CLAPP, R.B., GLINN, T.R. 1984. A statistical exploration of the relationships of soil moisture characteristics to the physical properties of soils. Water Resources Research 20, 682–690.

COSTANTINI, E.A.C., BARBETTI, R., RIGHINI, G. 2002. Managing the uncertainty in soil mapping and land evaluation in areas of high pedodiversity. Methods and strategies applied in the Province of Siena (Central Italy). Acts of the Int. Sym. on Soils with Med. type of Climate, CIHEAM-IAMB, Bari, 45–56.

COSTANTINI, E.A.C., CASTELLI, F., CASTALDINI, D., RODOLFI, G., NAPOLI, R., PANINI, T., BRAGATO, G., PELLEGRINI, S., ARCARA, P.G., CHERU-BINI, P., SPALLACCI, P., BIDINI, D., SIMONCINI, S. 1992. Valutazione del territorio per la produzione di tabacco di tipo Virginia Bright: uno studio interdisciplinare nel comprensorio veronese (Italia settentrionale). Supp. Annali ISSDS, XX, 157.

COSTANTINI, E.A.C., SULLI, L. 2000. Land evaluation in areas with high environmental sensitivity and qualitative value of the crops: the viticultural and olive-growing zoning of the Siena province. Boll. S.I.S.S. 49 (1–2), 219–234.

DAVIDSON, D.A. 1980. Soils and Land Use Planning. Longman, London, 129.

DE GRUIJTER, 1996. Discussion on the paper by D.G. Rossiter. Geoderma 72, 191–202.

DE LA ROSA, D. 2002. *MicroLEIS: Conceptual framework—Agro-ecological Land Evaluation.* [online] http://leu.irnase.csic.es/ microlei/microlei.htm.

DE LA ROSA, D., ALMORZA, J., CARDONA, F. 1981. Crop yield prediction based on properties of soils in Sevilla, Spain. Geoderma 25, 267–274.

DE LA ROSA, D., MAYOL, F., DIAZ-PEREIRA, E., FERNANDEZ, M., DE LA ROSA JR, D. 2004. A land evaluation decision support system (MicroLEIS DSS) for agricultural soil protection: With special reference to the Mediterranean region. Environmental Modelling and Software 19, 929–942.

DE LA ROSA, D., MAYOL, F., MORENO, J.A., BONSON, T., LOZANO, S. 1999. An expert system/ neural network model for evaluating agricultural soil erosion in Andalucia region, southern Spain. Agriculture, Ecosystems and Environment 73, 211–226.

DE LA ROSA, D., MORENO, J.A., GARCIA, L.V., ALMORZA, J. 1992. MicroLEIS: a microcomputer-based Mediterranean land evaluation information system. Soil Use and Management 8, 89–96.

DENT, D., YOUNG, A. 1981. *Soil Survey and Land Evaluation.* George Allen & Unwin, London.

EEA EUROPEAN ENVIRONMENT AGENCY. 1999. *DPSIR Framework.* http://www.ceroi.net/ reports/ rendal/dpsir.htm.

FAGGIANI, A. 1999. La logica fuzzy nella soluzione dei problemi territoriali multicriteriali. Tesi di laurea di Agostino Nardocci, n.3, DAEST, Venezia.

FAO. 1967. Soil survey interpretation and its use. Soils Bulletin 8, Food and Agriculture Organization of the United Nations, Rome.

FAO. 1976. A framework for land evaluation. Soils Bulletin 32, Food and Agriculture Organization of the United Nations, Rome.

FAO. 1978a. Report on the agro-ecological zones project. Methodology and results for Africa. World Soil Resources Report 48, Vol. 1. FAO, Rome.

FAO. 1978b. Report on the agro-ecological zones project. Results for Southwest Asia. World. Soil Resources Report 48, Vol. 2. FAO, Rome.

FAO. 1980. Report on the agro-ecological zones project. Results for Southeast Asia. World Soil Resources Report 48, Vol. 4. FAO, Rome.

FAO. 1981. Report on the agro-ecological zones project. Results for South and Central America. World Soil Resources Report 48, Vol. 3. FAO, Rome.

FAO. 1983. Guidelines: land evaluation for rainfed agriculture. Soils Bulletin 52, Food and Agriculture Organization of the United Nations, Rome, Italy.

FAO. 1984. Land evaluation for forestry. Forestry paper 48, Food and Agriculture Organization of the United Nations, Rome, Italy.

FAO. 1985. Guidelines: land evaluation for irrigated agriculture. Soils Bulletin 55, Food and Agriculture Organization of the United Nations, Rome.

FAO. 1991. Guidelines: land evaluation for extensive grazing. Soils Bulletin 58, Food and Agriculture Organization of the United Nations, Rome.

FAO. 1978–81. Report on the Agro-ecological Zones project. Vol. 1, Methodology and results for Africa, Vol. 2, Results for southwest Asia, Vol. 3, Methodology and results for South and Central America, Vol. 4, Results for Southeast Asia, FAO World Soil Resources Report 48/1, 4.

FAO. 1993. Guidelines for land-use planning. Development Series no. 1, Food and Agriculture Organization of the United Nations, Rome.

FAO. 2004. Participatory Land Use Development in the Municipalities of Bosnia Herzegovina. Guidelines. Food and Agriculture Organization of the United Nations, Cooperazione Italiana.

FAO. 2007. Land evaluation, towards a revised framework. Food and Agriculture Organization of the United Nations, land and water discussion paper 6, Rome.

GARDNER, W.R. 1958. Some steady state solutions of the unsaturated moisture flow equation with application to evaporation from a water table. Soil Science 85, 228–232.

GIORDANO, A. 1983. Alcune osservazioni sui termini "land", "land evaluation" terra, terreno e territorio. In: atti del convegno "Land Evaluation", Firenze 25–27 maggio 1983, Edizioni del Centro Interregionale, Firenze 162–172.

GOOVAERTS, P. 2001. Geostatistical modeling of uncertainty in soil science. Geoderma 103, 3–26.

HOOSBEEK, M.R., BRYANT, R.B. 1992. Towards a quantitative modelling of pedogenesis: a review. Geoderma 55, 183–210.

IAO. 1999. Land resources and land evaluation of the central valley of Cochabamba (Bolivia). 19th Course professional master "Geomatics and natural resources Evaluation".

JARVIS, N., STENEMO, F. 2002. Guidance document and manual for the use of MACRO_DB V.2.0. Dept. Soil Science, SLU, Uppsala Sweden, 25.

JARVIS, N.J., HOLLIS, J.M., NICHOLLS, P.H., MAYR, T., EVANS, S.P. 1997. MACRO-DB: a decision-support tool for assessing pesticide fate and mobility in soils. Environmental Modelling and Software 12, 251–265.

JRC. 2007. Display of thematic maps, purity maps and confidence-level maps [online]. HTTP://EUSOILS.JRC.IT/ESDB_ARCHIVE/_ESDBV2/FR_THEMA.HTM

KING, D., LE BAS, C. 1996. Towards a European Soil information System: past activities and perspectives of the soil and GIS Support Group. In: *Soil Databases to Support Sustainable Development* JRC, ISPRA, 115–124.

KLINGEBIEL, A.A., MONTGOMERY, P.H. 1961. *Land Capability Classification*. USDA Agricultural Handbook 210, US Government Printing Office, Washington, DC.

KUNZE, R.J., UEHARA, G., GRAHAM, K. 1968. Factors important in the calculation of hydraulic conductivity. Soil Science Society of America Proceedings 32, 760–765.

L'ABATE, G., COSTANTINI, E.A.C. 2001. I suoli e la valorizzazione delle risorse naturali e culturali del territorio. Un prototipo di sistema esperto applicato in Alta Val d'Elsa. Elsanatura, anno 2000, 1, 8–22.

MANCINI, F., COLL., E. 1966. *Carta dei suoli d'Italia*. Comitato per la Carta dei Suoli, Firenze, Italy.

MANCINI, F., RONCHETTI, G. 1968. *Carta della potenzialità dei suoli d'Italia*. Comitato per la Carta dei Suoli, Firenze, Italy.

MILLINGTON, R.J., QUIRK, J.P. 1961. Permeability of porous solids. Transactions of the Faraday Society 57, 1200–1207.

MINISTERO DELL'AGRICOLTURA E DELLE FORESTE—GRUPPO DI LAVORO PER LA VALUTAZIONE DEL TERRITORIO A FINI AGRICOLI E FORESTALI, 1983. Proposta metodologica di classificazione attitudinale del territorio (esemplificazione per la coltura del mais da granella). Suppl. a Annali Istituto Sperimentale Studio e Difesa del Suolo XIV, 55.

NAPOLI, R., COSTANTINI, E.A.C., GARDIN, L. 1999. Un sistema informativo pedologico per le valutazioni agro-ambientali a scala di dettaglio e semi-dettaglio. Agricoltura Ricerca XXI, 159–176.

OECD. 1998. *Joint Working Party of the Committee for Agriculture and the Environment Policy Committee— Proposed Work on Trade, Agriculture and Environment*. Document: COM/AGR/CA/ENV/EPOC (98), 142.

PACHEPSKY, Y., RAWLS, W.J. (Ed.). 2005. Development of pedotransfer functions in soil hydrology. Developments in Soil Science 30, Elsevier.

RAWLS, W.J., BRAKENSIEK, D.L., SAXTON, K.E. 1982. Estimation of soil water properties. Transactions of the ASAE 25, 1316–1320.

RAWLS, W.J., BRAKENSIEK, D.L. 1985. *Prediction of soil water properties for hydrological modelling*. In: Jones, E., Ward, T.J. (Ed.), Proceedings of the ASCE Symposium Watershed Management in the Eighties, Denver, Colorado 30 Apr.–2 May 1985, 293–299.

RENARD, K.G., FOSTER, G.R., WEESIES, G.A., MCCOOL, D.K., YODER, D.C., (COORDINA-TORS). 1997. *Predicting soil erosion by water: a guide to conservation planning with the revised universal soil loss equation (RUSLE).* United States Department of Agriculture, Agriculture Research Service, Agriculture Handbook no. 703, 384.

REVELIOTIS, S.A. 1997. *The "Prototype" Web-based Electronic Text Project: An Introduction to Linear Programming and the Simplex Algorithm.* School of Industrial & Systems Eng., Georgia Institute of Technology, 1997.

ROSSITER, D.G. 1990. A framework for land evaluation using a microcomputer. Soil Use and Management 6, 7–20.

ROSSITER, D.G. 1994. Lecture Notes: "Land Evaluation". http://www.css.cornell.edu/landeval/ le_notes/ lecnot.htm.

ROSSITER, D.G. 1996. A theoretical framework for land evaluation. Geoderma 72, 165–190.

RUSSO, D. 1988. Determining soil hydraulic properties by parameter estimation: on the selection of a model for the hydraulic properties. Water Resources Research 24, 453–459.

SANCHEZ, P.A., COUTO, W., BUOL, S.W. 1982. The fertility capability soil classification system: interpretation, applicability and modification. Geoderma 27(4), 283–309.

SANCHEZ, P.A., PALMA C.A., BUOL S.W. 2003. Fertility capability soil classification: a tool to help assess soil quality in the tropics. Geoderma 114, 157–185.

SAXTON, K.E., RAWLS, W.J., ROMBERGER, J.S., PAPENDICK, R.I. 1986. Estimating generalised soil-water characteristics from texture. Soil Science Society of America Journal 50, 1031–1036.

SCHEINOST, A.C., SINOWSKI, W., AUERSWALD, K. 1997. Regionalization of soil water retention curves in a highly variable soilscape, I. Developing a new pedotransfer function. Geoderma 78, 129–143.

SIMOTA, C., MAYR, T. 1996. Pedotransfer function. In: Loveland, P.J., Rounsevell, M.D.A. (Ed.), *Agro-Climatic Change and European Soil Suitability (EuroACCESS), A spatially distributed, soil, agro-climatic and soil hydrological model to predict the effects of climatic change on land-use within the European Community,* Commission of the European Communities, Directorate-General XII Science, Research and Development, 234.

SOIL SURVEY STAFF. 1999. *Soil Taxonomy—Agricultural Handbook 436.* Washington, USA, 869.

STOCKLE, C.O., Nelson, R.L. 1996. Cropsyst User's manual (Version 2.0). Biological Systems Engineering Dept., Washington State University, Pullman, Washington, USA.

STOCKLE, C.O., Nelson, R.L. 1994. Cropsyst User's manual (Version 1.0). Biological Systems Engineering Dept., Washington State University, Pullman, Washington, USA.

STORIE, R.E. 1933. *An Index for Rating the Agricultural Value of Soils.* University of California, Agricultural Experiment Station.

STORIE, R.E. 1978. *The Storie Index Soil Rating Revised.* University of California, Division of Agricultural Sciences, Special Publication (3203).

TOMASI, D., CALÒ, A., BISCARO, S., VETTORELLO, G., PANERO, L., DI STEFANO, R. 1999. Influence des caractéristiques physiques du sol, sur le développement de la vigne, dans la composition polyphénolique et anthocyanique des raisins et la qualité du vin de Cabernet sauvignon. Bull. O.I.V. 72, 819–820.

UNGARO, F., CALZOLARI, C. 2001. Using existing soil databases for estimating water-retention properties for soils of the Pianura Padano-Veneta region of North Italy. Geoderma 99, 99–121.

UNGARO, F., CALZOLARI, C. 2005. Development of pedotransfer functions using a group method of data handling for the soil of the Pianura Padano-Veneta region of North Italy: water retention properties. Geoderma 124, 293–317.

USBR DEPARTMENT OF THE INTERIOR BUREAU OF RECLAMATION. 1953. *Irrigated Land Use, Part 2: Land Classification.* B.R. Manual, vol. 5, US Government Printing Office, Washington.

VACCA, A., MARRONE, V.A. 2004. *Soil degradation systems in the Italian Regions of the EU Objective 1: application of the DPSIR Model as a reference System for the Typological Systematisation of the Phenomena 2004.* CD-rom Proceedings of Eurosoil 2004 Freiburg, 4–12 September 2004, 10.

VAN DIEPEN, C.A., VAN KEULEN, H., WOLF, J., BERKHOUT, J.A.A. 1991. Land evaluation: from intuition to quantification. In: *Advances in Soil Science,* Stewart, B.A. (Ed.), Springer, New York, 139–204.

VAN DIEPEN, C.A., WOLF, J., VAN KEULEN, H., RAPPOLDT, C. 1989. WOFOST, a simulation model of crop production. Soil Use and Management 5, 16–24.

VAN GENUCHTEN, M.T. 1980. A closed-form equation for predicting the hydraulic conductivity of unsaturated soils. Soil Science Society of America Journal 44, 892–898.

VEREECKEN, H., MAES, J., FEYEN, J., DARIUS, P. 1989. Estimating the soil moisture retention characteristic from texture, bulk density and carbon content. Soil Science 148, 389–403.

VEREECKEN, H., MAES, J., FEYEN, J. 1990. Estimating unsaturated hydraulic conductivity from easily measured soil properties. Soil Science 149, 1–12.

WILLIAMS, J.R. 1975. Sediment routing for agricultural watersheds. Water Resources Bulletin 11(5), 965–974.

WISCHMEIER, W.H., SMITH, D.D. 1965. *Predicting Rainfall Erosion Losses from Cropland East of the Rocky Mountains—Guide for Selection of Practices for Soil and Water Conservation.* Agriculture handbook no. 282, USDA, Washington.

WISCHMEIER, W.H., SMITH, D.D. 1978. *Predicting Rainfall Erosion Losses—A Guide to Conservation Planning.* Agriculture handbook no. 537, USDA, Washington.

YIALOURIS, C., KOLLIAS, V., LORENTZOS, N., KALIVAS, D., SIDERIDIS, A. 1997. An integrated expert geographical information system for soil suitability and soil evaluation. Journal of Geographic Information and Decision Analysis 1(2), 90–100.

ZANDER, P. 2001. Interdisciplinary modelling of agricultural land use: MODAM—Multi-Objective Decision Support Tool for Agroecosystem Management. In: Helming, K. (Ed.), Multidisciplinary Approaches to Soil Conservation Strategies: Proceedings; International Symposium ESSC, DBG, ZALF, Müncheberg, Selbstverlag (ZALF-Berichte, no. 47), 155–160.

PART II

SOIL AND LAND EVALUATION AND PLANNING

2. The Land Capability Classification

Edoardo A.C. Costantini[1]

2.1. Introduction

The Land Capability Classification (LCC) is a method used for the classification of land by a more or less wide range of agro-pastoral or silviculture systems, instead of by specific cultivations or agricultural practices. The original methodology was elaborated by the Soil Conservation Service of the U.S. Department of Agriculture (Klingebiel and Montgomery, 1961) on the basis of detailed soil surveys on a scale ranging from 1:15,000 to 1:20,000. It is important to remember that the U.S. Soil Conservation Service received a formidable impulse from the Soil Conservation and Domestic Allotment Act of 1935. This law was passed after the drastic collapse of agricultural production in the late 1920s, caused by soil erosion in vast areas that were characterized by intensive agricultural activities, based upon mono-succession and completely lacking in soil conservation measures. The realization that this productive collapse was one of the causes of the Great Depression motivated political effort to orient farmers' choices toward a more sustainable agricultural approach, particularly towards conserving fertility and preventing soil erosion.

Through surveys and maps, under the LCC, soils were grouped by their capacity to produce common crops, forage, or timber without deteriorating over the long term. The goal of the LCC map was to provide an easily consultable document for farmers that would subdivide their farmland into areas with different productive potentials, soil erosion risks, and management difficulties for agricultural and woodland activities.

Following the success achieved in the United States, many European and non-European countries developed specific classifications, based upon the characteristics of their own territories. These differed from the original American version in the number and meaning of the classes and subclasses used. For example, while in the United States eight classes and four subclasses are used, in Canada and in England there are seven classes and five subclasses.

The methodology developed by the United States is still by far the most commonly used in Italy also, albeit slightly modified over the years to adapt it to the Italian situation, to ever-growing pedologic knowledge, and to changing purposes. In fact, the LCC is no longer

---

[1] CRA-ABP—Centro di Ricerca per l'Agrobiologia e la Pedologia, Firenze, Italy.

preferred by specialists in soil conservation working at a detailed level, because more advanced systems have been developed for the estimate of soil erosion risks, still based upon studies carried out in the United States.

Instead, the LCC has been used increasingly for territorial programming and planning, therefore upon a wider scale of reference than that of the individual farm. Along these lines there have been many research projects carried out, such as the Italian soil capabilities map by Mancini and Ronchetti (1968), and the regional maps of Piedmont (IPLA, 1982), Emilia-Romagna (Regione Emilia-Romagna, 1981), and Lombardy (Regione Lombardia, 2000). There are also many maps in greater detail, such as those produced in Tuscany (Costantini, 1987; Regione Toscana, 1994) and in Campania (Regione Campania, 2004).

In some cases, the LCC has become a fundamental instrument for the exchange of knowledge between specialists from different fields, particularly pedology, agronomy, architecture, and more generally territorial planners, with a strong impact on public administration decision-making. Along this line can be mentioned the studies carried out at a municipality level in Lombardy (Brenna and Madoi, 2004), in Piedmont (Comune di Carugo, 2004), and in Sardinia (Fantola et al., 1995; Lai et al., 1995).

2.2. Classification Characteristics

The LCC is founded upon a series of guiding concepts.

> The evaluation refers to the set of cultivation activities advisable in the study territory and not to a particular cultivation.
> Socio-economic factors are not included.
> The concept of limitation (subclass) is tied to that of cultivation flexibility, in the sense that an increase of the degree of limitation corresponds to a decrease in the range of possible agro-pastoral or silviculture uses (Fig. 2.1).
> The limitations considered are those that are permanent, not temporary limitations that may be resolved by appropriate improvements such as temporary drainage or fertilization.
> Under the term "management difficulties" are placed all the conservational practices and systems (soil and water management systems) necessary so that the use of the soil does not cause loss of fertility or deterioration.
> The evaluation takes into consideration a medium-high level of management, but all the same it is accessible to the majority of farmers.

The classification uses three categories: class, subclass and unit.

Each land capability class comprises subclasses that have the same degree of limitation or risk. The classes are numbered I to VIII based on the number and severity of their limitations. They are defined as follows:

Arable lands

> Class I: soils with few or no limitations in their agricultural use. Do not require particular conservation practices and allow a wide choice of crops among those practised in the territory.
> Class II: soils with moderate limitations that reduce choice of crops or that require some conservation practices, such as an efficient network of drainage channels.

Class III: soils with strong limitations that reduce the choice of crops or that require accurate and continuous maintenance of the hydraulic, agronomic, and forestry systems. Class IV: soils with severe limitations to their agricultural use. Allow only a limited number of choices.

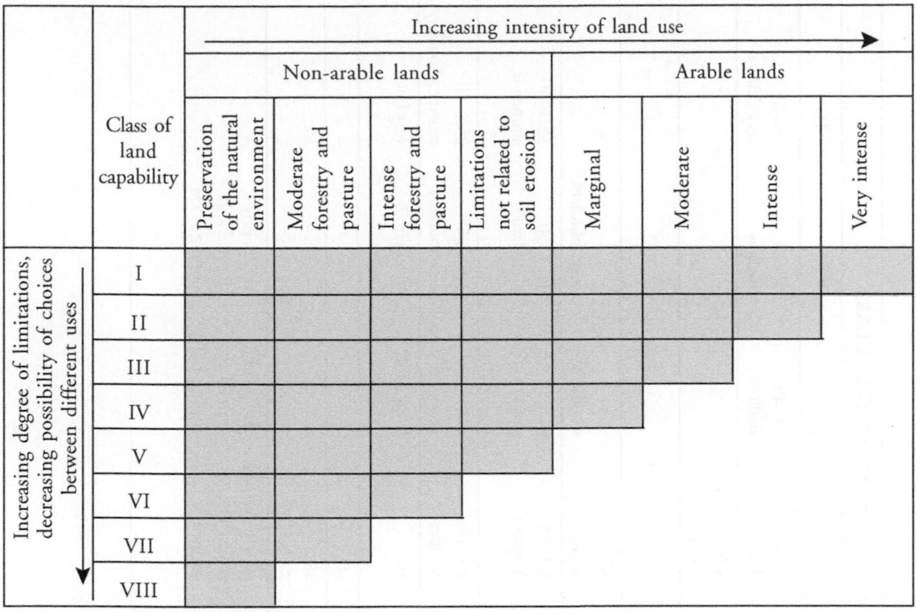

		Increasing intensity of land use →							
	Class of land capability	Non-arable lands				Arable lands			
		Preservation of the natural environment	Moderate forestry and pasture	Intense forestry and pasture	Limitations not related to soil erosion	Marginal	Moderate	Intense	Very intense
Increasing degree of limitations, decreasing possibility of choices between different uses ↓	I								
	II								
	III								
	IV								
	V								
	VI								
	VII								
	VIII								

Fig. 2.1. Conceptual relationships between land capability classes, intensity of soil limitations and risks, and intensity of land use.

Non-arable lands

Class V: soils having little or no erosion phenomena, but with limitations that are not easily removable, which restrict their use to forage production, grazing, or preservation of the natural environment (for example, very stony soils, soils with shallow ground water table).
Class VI: soils with such permanent limitations that their use is restricted to woodland, pasture, or forage production.
Class VII: soils with such permanent limitations that they require conservation practices even for use as woodland or pasture.
Class VIII: soils unsuited for any type of agricultural or forestry use. To be used exclusively for natural reserves or for recreational activities, providing necessary intervention for soil conservation and to aid vegetation (Fig. 2.2).

2.2.1. Matching Table

The following is an original evaluation method obtained by working on the results of the project SINA, *Sistema Informativo Nazionale Ambientale* (National Environmental Information System), and the sub-project "land capability cartography" coordinated by the Region of Emilia Romagna. Several soil conservation services of northern Italian regions (Lombardy,

Table 2.1. Matching table of land capability class.

PROPERTIES	I	II	III	IV	V	VI	VII	VIII
					LAND CAPABILITY CLASS			
Rooting depth (cm)	>100 deep and very deep	>100 deep and very deep	50–100 moderately deep	25–49 shallow	25–49 shallow	25–49 shallow	10–24 very shallow	<10 extremely shallow
AWC: available water at rooting depth (mm)	≥100 from moderate to high	≥100 from moderate to high	51–99 low	≤50 very low	–	–	–	–
USDA surface horizon texture	S, SF, FS, F, FA	L, FL, FAS, FAL, AS, A	AL	–	–	–	–	–
Rock fragments	<5 absent or scarce	5–15 common	16–35 frequent	36–70 abundant	>70 slope < 5%	>70 very abundant	–	–
Medium and large surface stoniness %	<0.3 absent and very scarce	0.3–1 scarce	1.1–3 common	3.1–15 frequent	>15 slope < 5%	15.1–50 abundant	15.1–50 abundant	>50 very abundant and stone outcropping
Rockiness %	0 absent	0 absent	≤2 scarcely bouldery	2.1–10 bouldery	>10 slope < 5%	10.1–25 very bouldery	25.1–50 extremely bouldery	>50 extremely bouldery
Chemical fertility of surface horizon **	good	partly good	moderate	low	from good to low	from good to low	very low	–
Surface horizon salinity mS/cm **	<2	2–4	2.1–8	>8	–	–	–	–
Sub-surface horizon salinity (<1 m) mS/cm ***	<2	2–8	>8	>8	–	–	–	–
Internal drainage	well drained, moderately well drained	well drained, moderately well drained	somewhat poorly drained, sometimes excessively drained	poorly drained, excessively drained	very poorly drained and sloping < 5%	very poorly drained and sloping > 5%	–	–

Flooding risk	none	rare	occasional	occasional	frequent and open water table	–	–	–
Slope %	<13 level or moderately inclined	14–20 highly inclined	21–35 strongly inclined	36–60 very strongly inclined	–	36–60 very strongly inclined	61–90 steeply inclined	>90 very steeply inclined
Erosion	absent	moderately diffuse	strongly diffuse or moderate gully or moderate eolian or solifluction	strong gully or strong eolian	–	mass erosion due to collapse and landslide	–	–
Climate interference ****	absent	slight	moderate	from none to moderate	from none to moderate	strong	very strong	–

* Surface horizon is the layer of soil that limits work operations and trafficability. It corresponds, in arable soils, to the depth of the horizon involved by the main cultivation, for example, ploughing. It may therefore be made up of the Ap1 and Ap2 sub-horizons. This same concept of surface horizon is used in reference to both cultivation and chemical fertility. In this case, in many soils it corresponds to the layer with the most root development of the species of interest. In the case of tree crops with herbaceous cover, though, the thickness penetrated by annual rooting of the tree species may differ from that of the herbaceous cover. In forested lands, the surface horizon refers to the uppermost mineral horizons, extending to the depth corresponding to a hypothetical main rooting activity.

** For functional characteristics of chemical fertility, see Table 2.2.

*** Sub-surface horizon is the layer of soil lying under the surface horizon, where there is more development of the perennating roots of pluri-annual species. Its lower limit in many soils is less than 1 m in depth. The sub-surface horizon may be composed of several horizons in the profile.

**** For climate interference classes, see Table 2.3.

Table 2.2. Functional characteristics of chemical fertility.

Description	LCC Class	pH		BS*		CaCO3 total		CEC		ESP
good	I	6.6–8.4	and	>50	and	<40%	and	>10	and	<8
partly good	II	5.6–6.5	or	35–50	or	>40%	or	5–10	and	<8
moderate	III	4.5–5.5 or >8.4	or	<35	or	any	or	<5	or	<8 and 8–15 within 1m
low	IV	<4.5	and	any	and	any	and	any	or	<15 and any within 1m
from good to low	V	any	and	any	and	any	and	any	and	<8 and any within 1m
from good to low	VI	any	and	any	and	any	and	any	and	<8 and any within 1m
very low	VII	any	and	any	and	any	and	any	and	>15
any	VIII	any	and	any	and	any	and	any	and	any

BS means base saturation.

Fig. 2.2. Example of soils with different land capability classification classes. Class I includes soils of the lower river terraces that are level, deep, without limitations. Because of limitations related to the nature of the soil, river terraces at a higher level are in classes II and III. Class IV includes soils on moderately sloping land that have a high erosion risk. Those of class V have no erosion risk, but have a high risk of frequent flooding, being near watercourses. In class VI there are thin soils on a slope, left for pasture. The soils with the strongest slope and highest risk of erosion are in class VII and require conservative silviculture. The areas found in class VIII are not suited for forestry or agriculture, but are set aside for non-productive purposes (e.g., natural ecosystem, wildlife habitat, water collection, recreation).

Table 2.3. Climate interference classes.

Code	Class	Description
1	Absent	
2	Slight	The climate negatively influences some agricultural crops in certain years (e.g., occasional cold spells on valley floors and fog for olive groves and vineyards)
3	Moderate	The climate negatively influences some agricultural crops most years (e.g., areas with a dry xeric pedoclimate, where there is a high risk of low cereal yield due to heat stress and where the practice of leaving land fallow is widespread)
4	Strong	The climate limits soil use to woodlands or pastures (e.g., mountain areas)
5	Very strong	The climate limits soil use to grazing (e.g., high altitude pastures, above the timberline)

Piedmont, Veneto, Friuli Venezia Giulia) and the CRA-ABP—Centro di Ricerca per l'Agrobiologia e la Pedologia (Firenze) also participated in its creation. It was also tested on soils in other regions of Italy.

2.2.2. Land Capability Subclasses

Within the land capability subclass it is possible to group the soils by their type of limitation on forestry or agricultural use. One or more lowercase letters are placed after the roman numeral indicating the class, from which the user can immediately read whether the limitation, whose intensity determined the class placement, is due to properties of the soil (s), excess water (w), erosion risk (e) or climate (c).

The soil and land properties adopted to evaluate the LCC are grouped as follows:

s **limitations due to nature of soil**
   rooting depth
   texture
   rock fragments
   surface stoniness
   rockiness
   chemical fertility of the surface horizon
   salinity
   excessive internal drainage

w **limitations due to excess water**
   poor internal drainage
   flooding

e **limitations due to risk of erosion and risk of agricultural machinery overturning**
   slope
   surface water erosion
   mass erosion

c **limitations due to climate**
   climatic limitations on soil use and management

Class I has no subclasses, because the soils belonging to this class have few limitations of weak intensity. Class V may have subclasses only indicated by the letters s, w, and c, because the soils in this class are not subject to or very slightly effected by erosion, but have other limitations that restrict use mainly to pasture, forage production, wildlife conservation, silviculture, environment conservation.

2.2.3. Land Capability Unit

If deemed necessary, the land capability unit may be used to identify soils that have the same potential use by forestry or agricultural practices and that present analogous management and resource conservation problems. An Arabic numeral placed after the lowercase letters (e.g., s1) indicates soils that present analogous limitations. This allows the identification of soils similar in terms of behaviour, management difficulties, and specific agro-technical measures. The land capability units are designated as outlined in Table 2.4.

2.2.4. Prime Farmlands and Unique Farmlands

In the United States, the use of the LCC has given rise to the concepts of "prime farmlands" and "unique farmlands". "Prime farmland" has soils that, because of their permanent physical characteristics, represent the most productive part of the national territory, in other words, those on which are based the gross marketable production estimates of the principal food commodities quoted on the stock market. In fact, these soils have been recognized as strategically

Table 2.4. Land capability unit.

Unit	Limitation
1	rooting depth
2	surface horizon texture
3	rock fragments
4	surface stoniness
5	rockiness
6	chemical fertility of the surface horizon
7	salinity
8	poor internal drainage
9	flooding risk
10	slope
11	surface water erosion
12	mass erosion
13	climatic interference

important, not only in furnishing domestic food supply, but also increasingly in the import/export balance and as an instrument of political pressure (Miller, 1979). Generally, these lands belong to the land capability classes I and II, but in some cases land of class III (IIIw–drained) was also included. "Unique farmlands" are specific soils, those with peculiar qualities, of a limited extension. Many highly valued products are in fact produced in lands or ecosystems that could not qualify as prime farmlands under the aforesaid terms. Rice lands are an example. In Mediterranean Europe, unique farmlands are particularly frequent. For instance, soils that produce superior wines sometimes have very poor general agronomic conditions (on thin rocky land on a steep slope), but represent an important economic and eco-systemic legacy of high value agroecosystems.

References

BRENNA, S., MADOI, R. 2004. Informazioni pedologiche e pianificazione territoriale: un esempio dalla Lombardia. Boll. Società Italiana della Scienza del Suolo 1–2, 409–414.

COMUNE DI CARUGO. 2004. *Carta della capacità d'uso*. www.comune.carugo.co.it/DocumentiUfficioTecnico/studioGeologico/Tav2pedologica.pdf.

COSTANTINI, E.A.C. 1987. Cartografia tematica per la valutazione del territorio nell'ambito dei sistemi produttivi. Bacini dei torrenti Vergaia e Borratello: Area rappresentativa dell'ambiente di produzione del vino Vernaccia di San Gimignano (Siena). Ann. Ist. Sper. Studio e Difesa Suolo, Firenze, XVIII, 23–74.

FANTOLA, F., LODDO, S., PUDDU, R., ARU, A. 1995. *Studi pedologici e territoriali per l'adeguamento del Piano Urbanistico Comunale al Piano Territoriale Paesistico nel comune di Assemini*. Atti del Convegno annuale S.I.S.S. "Il Ruolo della Pedologia nella Pianificazione e Gestione del Territorio", Cagliari, pp. 231–246.

IPLA. 1982. *Capacità d'uso dei suoli in Piemonte ai fini agricoli e forestali con carta scala 1:250.000*. IPLA, Torino.

KLINGEBIEL, A.A., MONTGOMERY, P.H. 1961. *Land Capability Classification*. USDA Agricultural Handbook 210, US Government Printing Office, Washington, DC.

LAI, M.R., LODDO, S., PUDDU, R., SERRA, G., ARU, A. 1995. *Lo studio geopedologico nella pianificazione degli interventi di difesa del suolo e di mitigazione della desertificazione. Salvaguardia della risorsa pedologica nel quadro della legge n°.183/89: il bacino del Rio S. Lucia di Capoterra*. Atti del Convegno annuale S.I.S.S, "Il Ruolo della Pedologia nella Pianificazione e Gestione del Territorio", Cagliari, pp. 349–352.

MANCINI, F., RONCHETTI, G. 1968. *Carta della potenzialità dei suoli d'Italia.* Comitato per la Carta dei Suoli, Firenze.

MILLER, F.P. 1979. *Defining, Delineating and Designating Uses for Prime and Unique Agricultural Lands in Planning the Uses and Management of Land.* Monograph of Agronomy, 21.

REGIONE CAMPANIA, 2004. *I Suoli della Piana in Destra Sele.* www.sito.regione.campania.it/ agricoltura/ pubblicazioni/pdf/destra-sele.pdf

REGIONE EMILIA ROMAGNA. 1981. *Capacità d'uso dei suoli della Regione Emilia-Romagna. Servizio Geologico Sismico e dei Suoli.*

3. Land Evaluation for Protecting Soil from Contamination

Costanza Calzolari,[1] Fabrizio Ungaro,[1] Marina Guermandi,[2] Giovanni Aramini,[3] Caterina Colloca[3], Anna Maria Corea,[3] Raffaele Paone,[3] Vincenzo Tamburino,[4] Santo Marcello Zimbone[4] and Serafina Andiloro[4]

3.1. Introduction

One of the primary functions of soil is that of buffering, filtration and transformation of many substances including pollutants. This function is of primary importance when dealing with land evaluation for protecting soil from contamination. Soil protective capacity, or soil attenuation capacity, is the potential of soil to filter polluting agents and to mitigate harmful effects without compromising its functionality. This quality is complex and delicate and can undergo substantial modification even over a brief period. In order for it to be assessed, soil "vulnerability" must be taken into consideration.

Soil vulnerability is the "capability for the soil system to be harmed in one or more of its ecological functions" (Batjes and Bridges, 1993). These functions include: (1) biomass production; (2) the already mentioned function of filter, buffer, storage and transformation; and (3) biological habitat and gene pool. Therefore, static and dynamic properties of soil that control the movement of pollutants also condition its vulnerability.

Soil vulnerability is a more complex concept than water vulnerability, which is normally defined as the potential pollution water bodies may undergo and is often measured in terms of current pollution. Water vulnerability is linked to the protective capacity of the soils at the interface between the hydrogeological reference system and the atmosphere. The concept of

[1] CNR-IRPI—Istituto Ricerca Protezione Idrogeologica, Unità Operativa di Supporto, Pedologia Applicata, Firenze, Italy.

[2] Regione Emilia Romagna—Servizio Geologico, Sismico e dei Suoli, Bologna, Italy.

[3] ARSSA Calabria—Servizio di Agropedologia, Catanzaro Lido, Italy.

[4] Università Mediterranea di Reggio Calabria—Dipartimento di Scienze e Tecnologie Agroforestali e Ambientali, Reggio Calabria, Italy.

water vulnerability even takes on a regulatory sense, inasmuch as European Directive 91/ 676/CEE, better known as the "Nitrates Directive", requires Member States to define as vulnerable zones the areas of land of their territory draining to water polluted by nitrates from agricultural sources. To this aim, in addition to water monitoring data, adequate knowledge/ information on areas vulnerable to the risk of contamination and pressures from agriculture are essential elements for supporting the designation process.

In order to apply environmental regulations, another concept has also been used over time, that of suitability for application of biowaste of various origins (e.g., sewage sludge, slurries). In this case, maximum loads must be evaluated, together with the ways in which organic material is applied, thus increasing the amount of organic matter in the soil and supplying nutritional elements, without, however, causing problems by polluting the soil itself or the waters.

In assessing soil in terms of its protective capacity, one must remember that it is not necessarily the soil that is to be protected, but rather its function of filtering, storage and transformation. Furthermore, the assessment goal must be clear, because soil with a good protective capacity regarding water infiltration does not necessarily possess analogous capacity towards the crops it supports or towards the food chain they will join.

Regarding what has been stated above, it is important to underline that there does not exist a soil protective capacity that is valid for all intents and purposes and for all pollutants, and that even very protective soils for water tables, for example, can accumulate pollutants with a possible "time bomb" effect, i.e., the risk that a change in any factor of state or equilibrium might lead to release of the accumulated pollutant load.

For all the above reasons, it is necessary to protect soils from contamination, currently one of the primary degradation risks to which European soils are subject. This has been recognized at European level, with the Thematic Strategy for Soil Protection, recently adopted by the European Commission.

3.2. Thematic Strategy for Soil Protection

In September 2006, the European Commission adopted the Thematic Strategy for Soil Protection. It consists of a Communication from the Commission to other European Institutions, (COM(2006)231), a proposal for a Directive from the European Parliament establishing a framework for the protection of soil (COM(2006)232), and an Impact Assessment (SEC(2006)620).

The strategy lays down some basic principles. Because of its low rate of formation, soil is in reality a non-renewable resource, which performs several functions crucial to ecosystems and human activities. Soil degradation has very high costs, mainly handled by society as a whole, while phenomena of soil degradation are increasing all over Europe, where there are several examples of permanent loss of the resource. The current instruments for soil protection are considered inadequate for ensuring a coherent and comprehensive approach at EU level, with some risk of internal market distortion mechanisms.

The communication identifies the main threats for European soils (erosion, compaction, organic matter decline, salinization, loss of biodiversity, contamination), lays down the political objectives of soil policy, and addresses research, with special regard for conserving soil biodiversity and raising public awareness about the soil. Moreover, it establishes the next actions to be taken at a European level.

In the proposal for a directive, a frame is defined for soil protection and for the maintenance of its main functions, namely: production of biomass; storage, filtering and transformation of nutrients and water; pool of habitat, species and genetic biodiversity; physical and cultural environment of human activities; raw material source; organic carbon storage; and archive of the geological and cultural heritage.

The directive, specifically addresses some threats to the soil that are distinguished between those linked to the agricultural use of land, i.e., erosion, decline in organic matter, compaction and salinization, and the others, i.e., soil contamination and sealing. It then defines different approaches for addressing the two groups. In the first case, the Member States must identify risk areas according to their own perception of the risk, set targets to be achieved and accordingly adopt measures.

In the case of soil sealing, the Member States are required to take measures to limit or mitigate the effects of sealing, i.e., by rehabilitating brownfield sites and limiting the use of greenfields, and facilitating the use of construction techniques that allow the maintenance of as many soil functions as possible.

In the communication, soil contamination caused by point sources is differentiated from that caused by diffuse sources. In the directive proposal, however, only point contamination is considered, adopting the prevention principle for the non-point, on the basis of which Member States are required to take measures to limit the introduction of dangerous substances into the soil.

A specific procedure is envisaged for contamination from point sources, which is perceived as one of the main threats to soil integrity. In the impact assessment accompanying the strategy, it is estimated that there are 3.5 million potentially polluted sites for Europe 25, with 0.5 million that require a remediation strategy.

First, a common definition is given for soil contamination. A contaminated site is defined as a site "where there is a confirmed presence, caused by man, of dangerous substances of such a level that Member States consider they pose a significant risk to human health or the environment [...]." On the basis of this definition, Member States are required to establish an inventory of contaminated sites and adopt a national remediation strategy. This strategy must indicate the goals to be achieved and the priorities, and must include a mechanism to fund remediation for the sites, so-called orphan sites, for which a clear responsibility of polluters cannot be found. A report on soil status is also envisaged for all the contaminated sites, in which the site history and characteristics are recorded. Such reports, to be compiled at the time of site property transactions, will be one of the tools to feed national inventories of contaminated sites.

3.3. Current Community Legislation

Several instruments in the environmental legislation developed by Europe over recent decades concern soil. Once transferred into national legislation, these build a complex framework. A non-exhaustive overview is given of the main tools in the following paragraphs.

Council Directive 91/676/EEC, also known as the Nitrates Directive, has the objective of reducing and preventing water pollution due to nitrates from agricultural sources. In order to achieve this objective, the directive provides for several actions. Member States are requested to designate and delimit the nitrate-vulnerable zones, defined as all areas draining into nitrate-polluted or potentially nitrate-polluted waters. For identification of vulnerable zones (and their periodical revision), appropriate monitoring networks should be implemented. Codes of Good Agricultural Practices must be established, mandatorily in vulnerable zones and elsewhere on a voluntary basis. These codes should consider the periods in which land application of fertilizer is inappropriate, the definition of fertilizer application in sloping areas, and the establishment of buffer areas near water courses. Measures such as crop rotation, use of cover crops and practices for limiting runoff may be included as well. Member States are also required to periodically review nitrate-vulnerable zones, on the basis of the results of water monitoring.

In the Nitrates Directive report by the Commission to the Council of Europe and the European Parliament on its implementation (COM(2007) 120 final), it was confirmed that agriculture made a significant contribution to nitrate pollution of ground water and surface water and to eutrophication. Even though progress in implementing this directive has been reported, there is still a delay in the designation of nitrate-vulnerable zones, which between 1999 and 2003 increased from 35.5% of EU-15 territory up to 44%, especially in the southern European Member States. A delay in implementing action programmes is also reported.

Even if soil is not considered directly, the control of polluting sources is addressed in the Water Framework Directive (GWD, 2000/60/CE), with the Groundwater Daughter Directive 2006/118/EC, developed in response to the requirements of Article 17 of the GWD. The purpose of this directive is to establish a framework for the protection of surface and groundwater "inter alia, through specific measures for the progressive reduction of discharges, emissions and losses of priority substances and the cessation or phasing-out of discharges, emissions and losses of the priority hazardous substances". Member States are required to take appropriate measures through river basin management plans.

In 2006, the European Commission adopted a strategy aimed at improving the way pesticides are used across the EU, the so-called Thematic Strategy on the Sustainable Use of Pesticides (COM(2006) 373). The strategy consists of a Communication, accompanied by a proposal for a Framework Directive, which sets out common objectives and requirements, while allowing for flexibility in correspondence with the geographic, agricultural and climatic situation of each Member State. Accompanied by a proposal for a revision of Directive 91/414/EEC concerning the placing on the market of plant protection products (see the relative paragraph), it thus complements and integrates the existing EU legislation controlling which pesticides may actually be made available on the market, and the related procedures.

Other European legislation initiatives attend, more or less directly, to the issue of soil pollution. This is the case of the Waste Framework Directive (75/442/EEC), which, among other things, requires the disposal of waste without endangering the environment, and the Sewage Sludge Directive (86/278/EEC), which regulates the use of sewage sludge in agriculture in such a way as to prevent harmful effects on soil. The European Commission is at present (mid 2009) assessing whether the current Directive should be reviewed, and to which extent.

The Common Agriculture Policy reform (Regulation (EC) No 1782/2003) contains explicit and mandatory requirements for maintaining the agricultural and environmental performances of land. In order to receive payments, farmers should be able to demonstrate that they have adopted measures to maintain the Good Agricultural and Environmental Condition of agricultural land according to what has been established by Member States.

3.4. Pedological Basis for Evaluation of Soil and Water Protection from Contamination

The problem of protecting soil and water from pollution is extremely complex, presenting normative aspects and aspects linked to the soil as such, to the soil-water-plant-atmosphere system, and to the different substances and molecules considered.

Mainly driven by existing environmental legislation, several assessment methodologies have been produced, mostly in regions with a higher environmental pressure (intensive agriculture and animal husbandry areas). Concerning Italian environmental conditions and priorities, research and legislative actions have focussed on slurries from intensive animal husbandry and, more recently, on olive oil mill waste water and sludges.

3.4.1. Soil Attenuation Capacity

Soil attenuation capacity, or the capacity of soil to protect water, is influenced by properties of the soil and of the pollutants, by the quantity and quality of water going through the soil, and by practices of soil management. Furthermore, since the processes involved are very different, it is necessary to consider the different properties involved in protecting the ground water and the risk of contaminating surface runoff waters. The risk of pollution of runoff water is linked to soil surface properties, as well as to the properties of the pollutants, the quantity of water reaching the soil, and land use and management practices.

The functional aspects of the soil related to the protection of runoff water are, therefore, its infiltration capacity, in turn linked to its water retention capacity and to its overall hydraulic conductivity and that of surface horizons, and the tendency to form superficial crusts (risk of incrustation or soil crustability); that is, all the properties connected to runoff generation. Soil protective capacity towards ground water is also associated with soil properties, pollutant properties, the hydraulic load on the soil, and land use and management practices.

Soil with properties that allow the rapid transmission of pollutant substances to the underlying ground water is defined as being scarcely protective, and the area in which it persists is defined as *sensitive*, i.e., subject to property change. However, that does not necessarily mean that the presence of a scarcely protective soil signifies pollution risk, given that management practices could nullify the risk. On the other hand, inappropriate management practices could produce the risk of groundwater pollution, even in non-sensitive areas.

The main functional aspects of soil related to its protective capacity toward ground water are the mechanic filter effect, physico-chemical buffer capacity, resistance to molecule mobility, microbiological buffer capacity, and the buffer effect on pH. These functions are in turn linked to: hydraulic conductivity or permeability, and groundwater conditions, which together determine the leaching potential; organic matter content; the soil structure; and the content and typology of clayey material that affect the adsorption capacity. The hydraulic load also influences the quantity of percolation towards ground water at the bottom of the profile. It is obvious that when the quantity of water infiltrating into the soil is superior to the evapotranspiration needs of the crops or the vegetation present and the storage capacity of the soil, percolation will take place at the bottom of the profile, with the leaching of any pollutant elements.

The outlines developed over time for evaluating the soil attenuation capacity all take into consideration the properties briefly illustrated above. The number of properties considered and the weight attributed to them in the overall evaluation generally depend on the variability found locally and the experts' perception of risk regarding the local context.

To counteract excessive weight of the local expert's subjective evaluation, the use of mathematical models has become relatively frequent for describing and forecasting soil leaching of various kinds of pollutants in farmland conditions and for simulating hypothetical management scenarios. These models vary widely in their conceptual approach and in their complexity, as they are heavily influenced by the environment in which they were developed and their more or less applicative purpose.

Empirical Approach for Assessment of Soil Attenuation Capacity

Relatively easy to handle, and usually easily understandable by regulators and end-users, empirical approaches have been widely used for assessing the soil protection capacity, or attenuation capacity, particularly in the late 1990s. These models are typically conceived as schemes in which several classes of soil behaviour are considered according to the soil properties and environmental conditions.

In Table 3.1, an example is given that regards soil capability for storing potential pollutants at rooting depth for a period sufficient for complete degradation (Brenna and Rasio, 2000).

An empirical approach was developed in the United Kingdom based on the properties of soils directly linked to catchment hydrology. The HOST conceptual model, Hydrology of Soil Types (Boorman et al., 1995), was developed as a collaborative project by the Institute of Hydrology (now Centre for Ecology and Hydrology), the Soil Survey and Land Research Centre (now Cranfield University National Soil Resources Institute), the Macaulay Land Use Research Institute, and the Department of Agriculture in Northern Ireland. HOST is based

Table 3.1. Example of an empirical scheme (ERSAF, 1997a, b).

Protective capacity	Permeability	Water table depth (cm)	Granulometric class
HIGH	Low (< 0.36 cm/h)	> 100 with low permeability	Fine, very fine, fine silty, fine loamy, coarse silty, loamy, skeletal clay, and highly contrasting classes including those in which the first term is fine, very fine or fine silty.
MODERATE	Moderate (0.36–3.6 cm/h)	50–100 with low permeability > 100 with high permeability	Loamy coarse, loamy skeletal, and all the other classes on sandy, sandy skeletal and fragmental.
LOW	High (> 3.6 cm/h)	≤ 50 with low permeability ≤ 100 with moderate permeability	Sandy, sandy skeletal and fragmental, and highly contrasting classes including those in which the first term is sandy, sandy skeletal or fragmental.

Chemical modifiers: pH and cationic exchange capacity.
pH ≤ 5.5 and cationic exchange capacity < 10 cmol kg⁻¹ lead to decrease by one class.

on the re-interpretation of existing soil surveys and a classification of UK soils has been derived that can be applied via soil maps to aid hydrological studies and analyses.

This system consists of conceptual models of the processes carried out in the soil or substratum. Such models are based on three kinds of hydrological conditions: soils on permeable substrata with ground water deeper than 2 m; soils on a permeable substratum with superficial ground water; soils with an impermeable or scarcely permeable horizon within 1 m. Within these three situations there are subdivisions based on different soil properties and on the presence of organic or hydromorphic horizons. From these stem 11 groups, which in turn are ranked in 29 classes according to substratum properties. At the moment, the model is maintained by Macaulay Land Use Research Institute (http:// www.macaulay.ac.uk/host/).

Based on this approach, modified for local conditions, a classification for soil attenuation capacity has been applied in Italy in Piedmont (IPLA, 2007). The scheme adopted is detailed in Table 3.2.

Table 3.2. Example of an empirical scheme (IPLA, 2007).

HIGH protective capacity	Soils without or with a scarce presence of rock fragments, clay loam to clay or silty, without cracks, and with no hydromorphic features within the first 150 cm.
MODERATELY HIGH protective capacity	Soils with one or more of the following characteristics: 36–60% rock fragments; sandy loam texture; cracks in the topsoil; gley features between 50 and 100 cm deep.
MODERATELY LOW protective capacity	Soils with one or more of the following characteristics: rock fragments > 60%; loamy sand or sand; cracks also in the subsoil; gley features in the top 50 cm.
LOW protective capacity	Soils with one or more of the following characteristics: rock fragments between 16 and 35%; loam, silty loam, sandy clay loam, sandy clay; reversible cracks in the topsoil, horizons permanently reduced, hydromorphic features between 100 and 150 cm.

Model Application for Soil Attenuation Capacity Assessment at Regional Level

As an alternative to the empirical outlines briefly described, the soil protective capacity towards water may be assessed by using mathematical models physically based on the soil's functional response to external input. In this approach, the risk of contamination from diffuse sources such as agriculture is evaluated through simulation of soil-plant-atmosphere systems typical of certain agro-environmental contexts.

Over recent years, numerous mathematical models have been developed for the description and modelling of pollutants leaching from the soil in field conditions and for simulating hypothetical management scenarios (Wosten et al., 2001). There are several examples of the application of physically based models at a general scale (from small catchments to regions) (Vassiljev et al., 2004; de Paz and Ramos, 2004; Grizzetti et al., 2005; Kersebaum et al., 2003; Almasri and Kaluarachchi, 2007), some of them implemented in Italy (Ungaro and Calzolari, 2000; Severi et al., 2002; Ragazzi et al., 2004; Balderacchi et al., 2007).

In Italy, a recent example of model application is given by Calzolari et al. (2000). In this application, the MACRO water balance model (Jarvis, 1994), coupled with the SOILN nitrogen balance model (Eckersten et al., 1996; Larsson, 1999; Larsson and Jarvis, 1999a, b), has been applied to a number of benchmark soils in Northern Italy for representative agricultural management scenarios. The assessment of soil protection capacity, expressed in classes, is based on both water and nitrogen fluxes from the soil. Through similitude criteria and with the aid of pedotransfer functions for assessing the parameters necessary for the MACRO model, in the end it is possible to confidently apply the evaluation to soils that have not been specifically surveyed, so as to extend the results of the procedure geographically.

Once validated in controlled test sites (Marchetti et al., 2000, 2004; Mantovi et al., 2005), the model was used, together with SOILN, on 51 soil benchmark profiles of the Pianura Padano-Veneta region in different crop scenarios. By means of statistical analysis, the simulation results were used to define the soil protective capacity classes as in Table 3.3.

Table 3.3. Soil protective capacity classes.

Protective capacity	Relative flux %	Relative loss of NO$_3^-$ %
Low (L)	> 40%	> 20%
Moderately low (ML)	29–40%	11–20%
Moderately high (MH)	12–28%	5–10%
High (H)	< 12%	< 5%

Classes are defined according to water and nitrate flows at the bottom of the simulation profile expressed in "relative" terms, i.e., normalized to the inputs in terms of precipitation (and irrigation) and nitrogen.

The simulation results were used to derive regression equations between the flux at the bottom of the profile and the associated loss of nitrates. The following relation was calculated for a monosuccession of grain maize:

$$y = 0.1337 * x^{1.2875} \quad [r^2 = 0.87] \tag{1}$$

where the dependent variable is represented by the nitrate losses relative to the nitrogenous input, and the independent variable by the relative flux at the bottom of the profile, both expressed in percentages. The relation is shown in Fig. 3.1 (Calzolari et al., 2001).

The approach briefly described was used by some regional bodies in the application of the Nitrates Directive. In Veneto, the Regional Agency for the Protection of the Environment has produced a map of the soil protective capacity of the Venice lagoon watershed (ARPAV - Regione del Veneto, 2004, Fig. 3.2), subsequently extended to the soils of the whole Veneto plain through the use of the MACRO model, coupled with a GIS containing nitrogen loads assessed on the basis of municipalities and the 1:250000 soil map.

In order to accomplish the dictates of the Nitrates Directive in Italy, a new version of the vulnerability map of Emilia Romagna has been released, based on this approach (Severi et al., 2002). What is evaluated in this work is not so much the soil attenuation capacity as the aptitude of agro-ecosystems to influence the passage of nitrates of agricultural origin into surface and ground water. Thus is introduced the concept that the protective capacity is a characteristic of agro-ecosystems rather than of the soil. The outcome is a map with four flux classes flowing out of the system as the contribution of the soil-climate-crop system to potential reloading of ground water. Finally, this map is then cross-checked with the map of the upper limits of coarse deposits from the soil surface in order to obtain the new regional map of aquifer vulnerability (Severi et al., 2002), as in Table 3.4.

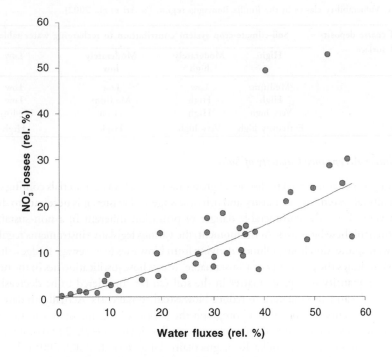

Fig. 3.1. Water fluxes at the bottom of the soil profile vs. leached nitrates.

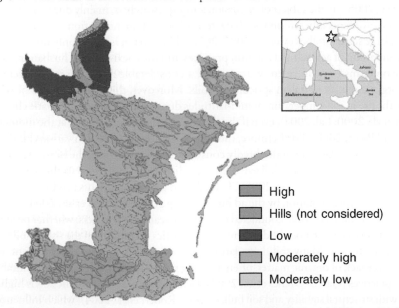

High

Hills (not considered)

Low

Moderately high

Moderately low

Fig. 3.2. Venice lagoon watershed: soil protective capacity (ARPAV-Regione del Veneto, 2004).

Table 3.4. Vulnerability classes in the Emilia Romagna region (Severi et al., 2002).

Depth of coarse deposits from soil surface	Soil-climate-crop system contribution to recharging water tables			
	High	Moderately high	Moderately low	Low
> 10 m	Medium	Low	Low	Low
5–10 m	High	High	Medium	Low
0–5 m	Very high	High	High	Medium
Outcrop	Extremely high	Very high	High	High

3.4.2. Slurry Acceptance Capacity of Soils

In view of the increasing need to dispose of great quantities of waste materials coming out of the agricultural system, agro-industry and urban sewage and wastes, it is necessary to defend soil and water from the considerable danger of pollution inherent in a non-sustainable management of these biowastes. At the moment, the various legislative instruments regulating soil and water protection from pollution deriving from biowastes (e.g., sewage sludges, slurries, compost) indicates the possibility of a "sustainable" use, whose justification lies in the need to increase the quantity of organic matter in the soil (so far considered to be decreasing in European soils) and also the need to reduce biowaste disposal costs. Actually, to date there exist few documents that precisely photograph the supply of organic matter in the soil in detail. Studies have been produced only on a very small scale (Jones et al., 2004) or for limited areas, such as in Italy for the Emilia-Romagna plain (Ungaro et al., 2003, 2005). Temporal trends are even less known, and relatively recent studies published for England and Wales (Bellamy et al., 2005) are the subject of discussion among researchers, mainly due to analytical and sampling problems inherent in the characterization and quantification of soil organic matter on a large scale (Stolbovoy et al., 2005, 2007). However, it is commonly understood that the abandoning of agricultural rotation practices and the coexistence of husbandry and agriculture on farms have, in recent years, caused a considerable reduction in the average supply of organic matter in Italian agricultural soils. Moreover, this trend could be heavily affected by the rise in soil temperature at mid-latitudes due to the influence of climatic change (Davidson et al., 2000; Lal, 2004), even if the picture is not sufficiently clear at the moment (Giardina and Ryan, 2000). Furthermore, the organic matter contained in various kinds of waste material and sludges can undoubtedly constitute a resource, once the biowastes have been appropriately treated and stabilized. This is explicitly acknowledged in the Thematic Strategy for Soil Protection (COM(2006)231 final) and in the documents produced during the process of its implementation by the ad hoc working group (Marmo et al., 2004).

Numerous studies have been conducted over recent years on the effects, whether positive or negative, of organic matter on the different characteristics of soil and on the risk of soil and surface water pollution. Many soil degradation processes in a semi-arid environment are connected to the lack of organic matter, often deriving from intensive soil use and unsuitable agronomic practices (Garcia-Orenes et al., 2005). In fact, organic matter in the soil is highly correlated with structural stability and soil bulk density (Robert et al., 2004), which influences the hydrological behaviour of the upper soil layers, and ultimately its infiltration capacity, resistance to erosion (De Ploey and Poesen, 1985) and water retention capacity, linked in its turn to the high retention capacity of organic matter and its hydrophobicity. It is logical, therefore, that great importance is given to the possibility of using exogenous organic matter (Marmo et al., 2004) deriving from different kinds of biowastes, to improve the physical characteristics of the soil. Much research has been carried out in recent years demonstrating the positive effects of sludges on the tendency to form crusts (Pagliai et al., 1983), on structural stability and bulk density (Garcia-Orenes et al., 2005), and on soil porosity (Sort and Alcaniz, 1999; Pagliai et al., 1983), both in quantitative and qualitative terms (Pagliai and Vittori Antisari, 1993). The importance of the appropriate use of biowastes for fighting erosion, for the combined effect on numerous properties linked to erosive processes, is then discussed and highlighted in the documents accompanying the Thematic Strategy on Soil Protection at the European level (Crescimanno et al., 2004).

Obviously, the possibility of using the nutrients contained to a varying degree and typology in biowastes has also been the subject of considerable attention (Robert et al., 2004; Hernandez et al., 2002; Dauden and Quilez, 2004). However, large amounts of biowastes are needed to obtain an increase in the content of nutrients, especially of nitrogen (Stamatiadis et al., 1999). The use of biowastes to improve the soil, as well as directly influencing the available source of nutritive elements, strongly affects the biological activity of the soil (Dick, 1992; Banerjee et al., 1997).

Many of the aforesaid studies, and indeed many others, underline that the multiple and interconnected influences exerted by the application of exogenous organic matter to the soil are not always positive. Alongside the positive effects reported, very often there are also negative ones, or at least cumulative effects that would advocate caution in using sludges, especially as concerns heavy metals (e.g., Cu and Zn, as in Mantovi and Baldoni, 2004) and persistent organic pollutants (Brändli et al., 2005). Excessive hydrophobicity may be the cause of an increased risk of runoff; chemical characteristics of the sludge used may cause salinity to rise or, instead, have a depressive effect on pH values; it was also noted that an increase in soil biological activity may cause a reduction in biodiversity. If the intention is to preserve the soil health, exogenous organic matter must be added only after careful consideration (Doran and Zeiss, 2000; Marmo et al., 2004).

Once it has been established that there is a high level of organic matter in biowastes from various sources, and that it can, in principle, be recycled in the soil for all the positive effects just mentioned and in general for a correct balance of biological cycles, the biowastes must obviously be correctly and usefully applied in a way that is also sustainable. Therefore, a risk assessment connected to biowaste use is also needed (Lulli, 1997). For a correct risk evaluation, even though expressed in terms of "soil suitability to application", the following points must necessarily be made clear:

the typology of the biowaste (e.g., animal slurries, urban or industrial depuration sludge, waste from oil mills);
the characteristics of the biowaste (chemical, physical and biological characteristics and type of maturation process undergone);
pedological and environmental characteristics; and
the legislative reference framework.

Properties that Condition Suitability of Soil and Environment for Spreading of Animal Slurries
(Marina Guermandi)

Initial studies carried out in the 1980s on the use of animal slurries (Lulli, 1997; Lulli et al., 1980; Bigi et al., 1989) highlighted the importance of considering every environmental aspect of any given study area when assessing territorial suitability, i.e., not just the soil, but also morphological and hydrogeological conditions. Although these analyses focus principally on soil and on some of its properties, it is nonetheless imperative to draw on in-depth knowledge of the type of geological unit (for example, permeability, drainage density), the morphology (type of basin, slope) and the aquifer. We must consider the soil in relation to the underlying lithology, the climate, the crop, and farming practices.

Appreciation of the complexity of the translocation, degradation and fixation processes that occur in soil, and that implicate interaction between numerous factors, subsequently led to the identification of complex soil properties. For example, the "hydrological behaviour" (Guermandi et al., 1991) of soil was proposed in order to describe its capacity to condition movement of rain water or other liquids towards aquifers and the surface water network. To define this, the main parameters taken into consideration are permeability, oxygen availability, drainage conditions and slope, which together define the soil's capacity to accept rain, or rather the land's suitability to accept rain water and therefore also liquid manure, without surface runoff and/or rapid deep percolation.

Another factor affecting hydrological behaviour is the infiltration capacity, namely the ability of the soil surface to allow water to penetrate. This property is particularly important when considering agricultural soils. The infiltration capacity of these soils varies over the course of the year, depending on crop sequence and land management practices. High after ploughing, it tends to decrease progressively, falling to minimum levels before harvest. Soils with significantly different infiltration capacity consequently have different susceptibility to favour water loss through surface runoff and/or surface ponding.

High correlation was found in Emilia Romagna between soils with a high crusting index, the presence of crusting, and low, steady-state experimental infiltration rates. Another soil quality that conditions hydrological behaviour is water retention capacity, i.e., the amount of water soil is able to retain, thus preventing water from percolating down below the level explored by the root systems of crops. For this purpose a thickness of 150 cm is generally used as a standard.

Other evaluation models used for assessing hydrological behaviour focus principally on soil parameters collected during routine monitoring carried out for the compilation of soil maps at different scales. The majority of models adopted by soil survey technical units consider the following soil properties: texture, rock fragments in the soil, vertic features, and the coefficient of linear extensibility. Other properties sometimes observed include drainage, depth of a shallow water table and the presence of organic horizons (ERSAL, 1996). Although these models differ from one another and may appear unconnected, the majority of them focus on soil properties that control the risk of degradation of surface and subsurface water resources. One may find that soil and environmental properties that are important in one model are far less so in another, and this can inevitably lead to classifications that are effective only in the area in which they were calibrated. For example, the low incidence of coarse-textured soils may lead to their being included in a single hydrological behaviour class, whereas in areas with a higher incidence of such soils, we may find that different behaviours exist depending on the granulometric composition of sand, the type of gravel or the presence of textural discontinuities. In the case of sandy-loam or loam soil textures, with high percentages of very fine sand, drainage can also be significantly impeded (Petrella, 1997).

The use of different assessment models has frequently made it difficult to draw comparisons, even between similar or adjacent territories. More detailed, comprehensive methods have proved complicated to implement outside the original territorial environment; on the other hand, some simpler assessment models have proved inadequate. And what has become evident is that there is a pressing need for a general methodology that might prevent conflicting assessments of similar environments and soils (Ragazzi and Vinci, 1997).

This is now partly possible with the implementation of mathematical models that are used to describe the water flow from the soil system and, consequently, to estimate the soil's protective capacity in relation to water. Although this has facilitated comparison of assessments, we must underline the importance of verifying results obtained from simulation models at a local level through the use of local experimental data or assessments made by local experts. Although the risk of soil pollution as a result of fertilization with animal effluents is virtually negligible, especially as compared to the risk of nitrate contamination of ground water, some assessment models do take this aspect into account. The soil can in effect be described and classified not just in terms of its hydrological behaviour, which is relatively simple to describe, but also by its capacity to store, filter and transform substances. This capacity or behaviour, known as attenuation capacity, highlights soil's role as a natural laboratory where chemical and biological reactions take place; these lead to the degradation of organic matter, the release of nutritional elements readily assimilable by crops and the storing of molecules not naturally present in the soil system. The following elements can be used to describe this capacity: soil rooting depth, content of rock fragments, reaction, and cationic exchange capacity.

The processes that the slurry undergoes within the soil largely take place within the layer explored by the root systems of crops. The available rooting depth is therefore a good indicator of the thickness of the "active layer" of soil capable of carrying out these processes. On the other hand, the volume occupied by the rock fragments in the soil (particles and fragments of rock with a diameter greater than 2 mm) can be considered virtually "inactive" as regards adsorption and degradation of slurries, which take place in the soil matrix. A good capacity to transform the applied organic matter maximizes fertilizing efficiency and reduces leaching of contaminants.

Soil reaction also influences microbial activity and therefore the processes of alteration and degradation of organic matter. Low pH values coincide with significant decrease of this activity. A soil reaction as close as possible to neutrality (pH values from 6.5 to 7.5) is important for immobilization of heavy metals and phosphates present in animal slurries, and for the cationic exchange capacity, which highlights the soil's capacity to adsorb such potentially polluting substances.

Interpretative models for assessment of soil suitability for the spreading of animal slurry, based on the soil properties that describe its hydrological behaviour and, less frequently, its purifying capacity, enable us to divide the different types of soil examined into various capacity classes, generally four, according to the Land Suitability Classification (FAO, 1976), which range from highly suitable to unsuitable.

The assessment of soil suitability for the spreading of animal slurry helps identify areas characterized by high environmental fragility, where the soil is unable to exercise an adequate protective function with regard to water and the crops themselves. An informed understanding of the behaviour of soil in relation to the spreading of animal effluents can also guide agronomic soil management (volume, timing and method of distribution) to reduce associated pollution risks, within Regional Action Plans for Nitrate Vulnerable Zones, as designated by national and European legislation.

In commonly used methods of assessment of aquifer vulnerability, although agriculture is considered mainly responsible for groundwater pollution, soil is generally attributed marginal importance. It is therefore imperative that we pool our resources into improving our understanding, survey and description of soil behaviour to facilitate better management in agronomic terms (better fertilization efficiency), which, as a result, implies fewer nitrates leaching into the environment.

The identification of soil functional types enables us to check or better describe the suitability classes currently proposed, providing estimation, completed by data measured in the field regarding permeability, water retention and infiltration capacity, and assessment models of the vulnerability of aquifers and surface water to pollution.

3.4.3 *Soil Suitability for Vegetation Water Application* (Giovanni Aramini, Caterina Colloca, Anna Maria Corea, Raffaele Paone, Vincenzo Tamburino, Santo Marcello Zimbone, Serafina Andiloro)

Normative Aspects

As in the case of other biowastes, the normative issue is a major one also in the case of olive mill waste waters. In the following paragraphs, reference is made to an Italian case study. In Italy, the management of olive mill waste waters is regulated by a complex normative framework. According to the present laws (and in particular Law 574/96), olive mill waste water should be considered a resource rather than a waste product. Indeed, this law provides for its agronomic use, subject to the need to limit its environmental impact. The law imposes quantitative limits, specifies modes of storage and use, and excludes certain soil typology. The hydraulic load is set at 50 m³/ha/year for oil mills with a traditional discontinuous pressing system and 80 m³/ha/year for oil mills with a continuous-cycle centrifugation system.

Vegetation water (VW) storage is allowed for no more than 30 d in silos, tanks and basins, subject to prior communication to the mayor of the municipality in which the containers are located. The application method must ensure proper distribution and incorporation of VW in the soil, preventing surface runoff. Application of VW is forbidden in the following cases:

soils located within 300 m of drinking water catchment protection areas;
soils located within 200 m of inhabited areas;
soils with water tables within 10 m depth;
icy, snowy, waterlogged or flooded soils.

The agronomical use of VW is subject to prior communication to the mayor of the municipality in which the soils are located. The communication must be accompanied by a technical report covering pedo-geomorphological and hydrogeological aspects, as well as the general characteristics of the receiving environment.

The competent authority may request further verification or directly order inspections and controls. In the case of any effective risk of damage to water, soil, subsoil or other environmental resources, the mayor may decide to stop or reduce VW application.

Article 7 of Law 574/96 is particularly interesting as it enables the regional authorities to draw up a specific plan for VW application, based on the evaluation of the characteristics of the different soils present in their region. This plan, a concrete instrument to plan interventions in the field, must be based on the specific pedological characteristics of the receiving environment (pedoenvironment) and must apply to homogeneous areas in terms of olive production and milling plants. A copy of the plan must be submitted to the Ministry of Agricultural and Forestry Policy.

Effects of Olive Mill Effluents on Soils and Waters

In the last few decades, much scientific research has explored the agronomic use of olive VW in order to establish its effects on the soil's physical and chemical properties, growing crops and, last but not least, water tables. Looking at the extensive bibliography available, the results can be summarized as follows:

> VW application has a fertilizing effect, particularly due to its high potassium content (Saviozzi et al., 1993; Pagliai et al., 2001);
> Temporary acidification following VW application is normally neutralized within a few weeks or months (Alianello, 2001);
> After an initial increase, the soil phenolic content tends to return to normal values in the medium term;
> VW application temporarily reduces water infiltration speed and the hydraulic conductivity of the soil. This effect is directly proportional to the amount of VW applied and varies according to its composition. It is generally ascribed to the absorption, by the superficial layers of the soil, of hydrorepellent fat compounds present in VW. In summer months, with the mineralization of organic matter, the phenomenon tends to disappear (Pagliai et al., 2001);
> After application of the effluent, microbiological activity ceases (probably hindered by the presence of bacterial-synthetic or bactericidal substances), but in a few weeks it returns to normal (the hindrance effect lasts longer in fine-textured soils);
> Regarding the risk of water table pollution, research shows that the soil has a filtering role, which, however, varies with different parameters, particularly texture; fine-textured soils with clayey horizons are preferable in this respect (De Simone, 2001);
> VW application on tree crops does not seem to have specific contraindications (Briccoli Bati et al., 1991);
> VW supply to growing herbaceous crops may produce phytotoxic effects resulting in a generalized productivity decrease (Bonari et al., 2001; Bonari and Ceccarini, 1993).

Overall examination of the research conducted in this field reveals two critical points:

1) failure to pay due attention to the soil's vertical variability (researchers often took into account only the superficial horizon) and to compare soils appreciably different in genetic aspects, horizon organization, and physical, chemical and hydraulic properties;

2) limited monitoring of the treated soils, with data extending to the medium and long term only in very few cases.

Soil Suitability for VW Application

Evaluation of soil suitability for VW application serves to determine, in the light of our present knowledge, the fundamental criteria that must be followed to conduct it in the most suitable way possible for every type of pedoenvironment (the soil or "receiving environment" mentioned in Article 7 of Law No. 574/96). To achieve this goal it is necessary to provide a frame of reference for the interpretation of the relevant soil properties. Such evaluation must take into account:

the information provided by research and experimentation;
legal obligations and limits; and
the working schedule.

In the evaluating process we must be cautious so to counterbalance the shortcomings of currently available research results.

Soil suitability evaluation, while not legally mandatory, can be extremely useful as a general reference framework for the VW Application Plan that the regional authorities are required to draw up under Article 7 of Law No. 574/96.

In Calabria, the ARSSA (Regional Agency for Development and Services in Agriculture) has drawn up a soil suitability map at a scale of 1:250,000. This map, the first to divide the region's farmland into soil mapping units grouped on the basis of their suitability for olive mill effluent use, will, it is hoped, be the starting point for more detailed work. It also aims to stimulate monitoring, in the medium and long term, of soils belonging to different suitability classes.

Soil Suitability Map: the ARSSA Method—Calabria

ARSSA evaluation of soil suitability for olive mill effluents is based on the FAO Land Suitability Classification (1976). To determine the parameters affecting the relationship between VW and soil (intended in its broader sense of "land"), researchers took into account those properties that render VW potentially phytotoxic, such as high COD, BOD, and salinity content as well as the presence of biotoxic compounds (polyphenols). Land characteristics affecting suitability include pedological, hydrological and morphometrical features (Marrara et al., 2004; Pagliai et al., 2001; Favi et al., 1993). Classification took into account:

soil capacity to accept olive mill effluents without surface runoff;
soil wastewater treatment and protective capacity;
water table depth; and
altitude and slope.

The necessary basic information has been largely inferred from the database of the Calabria soil map (ARSSA 2004), at a scale of 1:250,000. Morphometry is derived from a Digital Terrain Model (DTM) with a 40 m resolution, while the information relating to the water table depth is derived from further analysis of the data from the piezometric and water quality maps drawn up for the Special Project No. 26 of the CASMEZ (*Cassa per il Mezzogiorno,* Southern Italy Development Fund) in 1970.

Since the ARSSA soil suitability map is at a scale of 1:250,000 (ARSSA, 2005), it is not possible to highlight graphically some of the discriminating factors envisioned by Law 574/96, namely:

soils located within 300 m of drinking water catchment protection areas;
soils located within 200 m of inhabited areas.

In consideration of the sources used, researchers used the same basic evaluation units as the mapping units of the above-mentioned Calabria soil map. Normally the FAO Land Suitability Classification system is applied to the three classical hierarchical levels: orders, classes and subclasses. There are two orders: suitable and non-suitable land. Suitable land is, in turn, subdivided into classes with different degrees of suitability. In the ARSSA classification, these classes are: S1, soils with no limitations; S2, soils with moderate limitations; S3, soils with major limitations; and S4, lands with severe limitations. Within each class, subclasses specify the nature of the limitations involved.

The evaluation process is based on the concept of sustainable use of natural resources, attainable only if VW application fulfils the following objectives:

Prevent soil degradation; in order to do so, the physical and chemical properties of the soil (oxygen availability, pH, active limestone, electrical conductivity) must favour the degradation of the components supplied by VW, thus preventing their accumulation.
Prevent groundwater pollution that might originate from VW's deep percolation through layers of soils and subsoils without adequate treating capacity.
Prevent pollution of water bodies due to surface runoff.
Prevent crop damage.
Exploit a by-product of undoubted fertilizing effect, due particularly to its high potassium content.

Soil Capacity to Receive Olive Mill Effluent without Surface Runoff

Vegetation water runoff is one of the major risks against which the law seeks to protect surface water bodies, even those located far from the treated area. Assessment of the runoff risk is based on the slope and the soil infiltration capacity. As can be inferred from Table 3.5, the capacity to receive olive mill effluents without surface runoff risk is directly proportional to infiltration capacity and inversely proportional to the slope of the soil.

The different infiltration capacity classes were determined on the basis of soil hydraulic conductivity at saturation point and its crust formation risk, as reported in Table 3.6.

Both the hydraulic conductivity at saturation and the crust formation risk were inferred from the information in the Calabria soil map database. Crust formation risk has been evaluated by the following:

$$Ic = (1.5 \, Zf + 0.5 \, Zc)/(C + 10 \, OM) \tag{2}$$

where Zf = fine silt %; Zc = coarse silt %; C = clay %; and OM = organic matter %.

Vegetation Water Treatment and Protective Capacity of Soils (Modified by Unsaturated Zone's Permeability)

Among its functions, soil acts as a protective barrier against potential pollutants. In the specific case of olive mill effluent, it fulfils this action by first retaining and then degrading, through biological oxidation, the effluent's organic compounds (Fig. 3.3).

Table 3.5. Classification of soil capacity to receive olive mill effluents without surface runoff.

SLOPE %	INFILTRATION CAPACITY					
	Very high	High	Moderately high	Moderately low	Low	Very low
Concave	Very high	Very high	Very high	Very high	Very high	Very high
<1	Very high	Very high	High	High	Moderate	Moderate
1–5	Very high	High	High	Moderate	Moderate	Low
6–12	High	High	Moderate	Moderate	Low	Low
13–20	High	Moderate	Moderate	Low	Low	Very low

Table 3.6. Infiltration capacity classification.

Hydraulic conductivity at saturation Ks (cm/ha)	Incrustation risk (Ic)		
	< 1.2	1.2–1.6	> 1.6
> 35	Very high	Very high	High
3.5–35	Very high	High	Moderately high
0.35–3.5	High	Moderately high	Moderately low
0.035–0.35	Moderately high	Moderately low	Low
0.0035–0.035	Moderately low	Low	Very low
< 0.0035	Low	Very low	Very low

The VW treatment and protective capacity of the soil varies with its physical, chemical and biological properties. A rapidly draining soil, for example, has a good capacity to "metabolize" the organic compounds it receives but, at the same time, also easily leaches potential pollutants. Inversely, a soil with hindered drainage is a very effective barrier against groundwater pollution, but its low oxygen availability slows down degradation of the VW organic compounds.

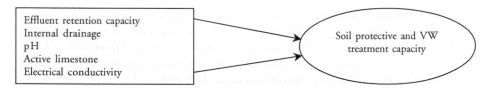

Fig. 3.3. Evaluation of soil protective and VW treatment capacity.

Therefore, in order to evaluate a soil's suitability for VW application, the two characteristics of its protective and treatment capacity must both be taken into account, as only a good combination of them results in a suitable soil. ARSSA researchers have evaluated the following soil properties: effluent retention capacity, pH, internal drainage, active limestone, and electrical conductivity. Some of these parameters, particularly electrical conductivity, are also predictive of the soil's capacity to receive oil mill effluent without undergoing agronomic degradation.

Soil Capacity to Retain Olive Mill Effluent

The risk of effluent percolation is minimized by a good retention capacity. Evaluation of such capacity is extremely important given that VW is generally applied in a period with a positive water balance, i.e. with surplus water. Therefore, in the case of soils with a low water retention capacity, there is a real risk of deep percolation. The capacity to retain VW varies with the texture and the skeletal structure of the soil's different horizons (which condition the water retention capacity per unit of volume) and with soil depth.

In our case, the USDA Soil Taxonomy's "textural families", based on texture, coarse fragments content and horizon succession, appeared the most functional. The "textural families" of Calabria's soils were taken from the aforementioned Calabria soil map. VW retention capacity is directly proportional to a soil's depth and inversely proportional to its content of coarse fragments, as can be seen in Table 3.7.

Also, soils subject to deep water percolation may protect water tables by removing solutes from the percolating solution through various physical means (e.g., retention through mechanical filtration, adsorption). This action, difficult to describe and predict theoretically, is always correlated to the parameters affecting soil retention capacity of the effluent.

pH and Active Limestone

Since pH conditions the mobility of elements, it is a parameter normally used to evaluate a soil's protective capacity. In the case of effluent application, pH values directly affect microbiological activity by influencing the chemical reactions responsible for organic matter degradation. For most soil micro-organisms the ideal pH values are around neutrality. VW has an acid pH value, between 4 and 5.5, that in most cases causes acidification of the treated soil, even though it lasts only a few months. In agreement with the models for assessing soil suitability proposed among the references, in the ARSSA work, pH was considered as a discriminating factor, calling for extreme caution in applying VW when dealing with acid or subacid soils.

The presence of active limestone guarantees a buffer action against the acidifying reaction of VW and, since it favours soil structuring processes, enhances both water infiltration and oxygenation of the soil surface layers.

Table 3.7. Classification of soil capacity to retain olive mill effluents.

Soil depth (cm)	Textural family		
	Very fine, fine, fine silty, fine loamy, coarse silty, coarse loamy, loamy, clayey-skeletal, medial	Coarse loamy, loamy skeletal plus all strongly contrasting classes (including over-sandy, sandy-skeletal and fragmental classes) in which the first term is *fine, very fine* or *fine silty*	Sandy, sandy-skeletal and fragmental, plus all the remaining over-sandy, sandy-skeletal and fragmental classes, as well as all strongly contrasting classes in which the first term is *sandy, sandy-skeletal* or *fragmental**
> 100	High	High	Moderate
100–50	High	Moderate	Low
50–25	Moderate	Low	Low
< 25	Very low	Very low	Very low

* For further information see Keys to Soil Taxonomy, USDA (2006) http://soils.usda.gov/technical/classification/tax_keys/keys.pdf

Internal Drainage

Internal drainage affects the length and frequency of the periods in which the soil is partly or totally waterlogged. This parameter conditions soil oxygen availability and hence its microbiological activity, characterized by oxidation processes. While good drainage ensures good oxygen availability, excessively fast drainage increases the risk that undegraded surface organic matter may percolate deep into the soil. Therefore, soils in the ARSSA classification with difficult or excessively fast drainage were considered unsuitable for VW application.

Electrical Conductivity

Because of the high salinity of oil mill effluents, electrical conductivity, which expresses soluble salts content in soil, was considered a crucial parameter in the evaluation process. VW use in soils that have poorly leachable salt might have negative effects on their physico-chemical properties, in particular causing de-structuring processes.

Soil Classification According to Its VW Treatment and Protective Capacity

The soil properties we have reviewed (effluent retention capacity, pH, active limestone, internal drainage, electrical conductivity) were used to classify it according to its VW treatment and protective capacity, as shown in Table 3.8.

    In the case of soils that are poorly protective due to their low or very low water retention capacity or their excessively fast drainage, we must consider the attenuation capacity of the unsaturated zone. Indeed, if the soil is the first "barrier" against the risk of water body pollution, the unsaturated zone, the part of the subsoil between the soil surface and the water table, also has a protective function. Therefore, as indicated in Table 3.8, the results of the evaluation of a soil's VW treatment and protective capacity must be adjusted to account for the unsaturated zone's permeability. For this purpose, ARSSA researchers took into account the information provided by the regional map of land vulnerability to agriculturally derived nitrates (ARSSA, 2002). This work divides the different lithologies into two classes, non-permeable and permeable, according to a simplified version of the Sintacs model scheme (Civita and De Maio, 2001) reported in Table 3.9.

Water Table Depth

To evaluate the aquifer's presence and depth, whether it be free or connected to the surface, researchers used the data contained in the piezometric and water quality maps drawn up for

Table 3.8. Soil classification according to its VW treatment and protective capacity.

VW treatment and protective capacity	Effluents retention capacity	pH	Active limestone %	Internal drainage	Electrical conductivity (mS/cm)
High	4—High	> 6.4	Present	Poor to good	< 0.5
Moderate	3—Moderately high	6.4–5.6	Absent	Poor to good	0.5–1
Low*	2—Low	6.4–5.6	Absent	Slow, fast, very slow	1–2
Very low**	1—Very low	< 5.6	Absent	Hindered	> 2

* This class becomes moderate in the case of impermeability of the unsaturated zone.
** This class becomes low in the case of impermeability of the unsaturated zone. This consideration holds only when the soil has been assigned to a low or very low class for problems related to its water retention, organic matter retention-degradation or draining capacity.

Table 3.9. Mitigating action of rocks composing the unsaturated zone and relative score—Modified Sintacs Model (Civita and De Maio, 2001).

Unsaturated zone characteristics	non-permeable					permeable				
	1	2	3	4	5	6	7	8	9	10
Coarse alluvial deposit								X	X	
Karstified limestone									X	X
Fractured limestone						X	X	X	X	X
Fissured dolomite		X	X	X						
Medium-fine alluvial deposit			X	X						
Sand complex						X	X			
Sandstone, conglomerate						X	X	X		
Fissure plutonic rock			X	X						
Turbidic sequence (Flysch)		X	X	X						
Marl, claystone	X									
Clay, silt, peat	X									
Pyroclastic rock		X	X	X						
Fissured metamorphic rock		X	X	X						

the Special Project No. 26 of the *Cassa per il Mezzogiorno* (Southern Italy Development Fund). This work, conducted between 1970 and 1980, is a complete study of Calabria's water resources. Much more recent detailed data show that in recent decades there has been a general lowering of water table depth (and this confirms the need to follow a cautious approach, as has been done in this work).

Law 574/96 forbids VW application on soils that contain water tables within 10 m depth. While ARSSA researchers complied with this regulation for this study, according to many authors it is an exceedingly cautious limit, which excludes VW from being applied on most alluvial plains.

Altitude and Slope

Given the widespread practice of olive farming in Calabria, in the evaluation process ARSSA researchers set the limit of 800 m a.s.l., thus excluding upland areas which, being very distant from oil mills, in any case are not concerned with VW application. The slope limit of 20% derives from the difficulty of distributing VW by ordinary means on steeper soils and from the risk of surface runoff.

Soil Suitability for Olive Mill Effluent Application

For each of the considered parameters (effluent retention capacity, pH, active limestone, internal drainage, electrical conductivity), researchers have elaborated the relative thematic map, managed in a GIS environment.

From joint interpretation of the above parameters, according to the evaluation scheme in Table 3.10, the different suitability classes were defined and the corresponding map was drawn up at a scale of 1:250,000. The distribution of potentially suitable soils is largely coincident with the main olive-growing areas of Calabria. It is extremely interesting to compare

Table 3.10. Evaluation scheme.

Suitability class	Effluents' acceptance capacity	Protective and VW treatment capacity, modified by the unsaturated zone permeability	Water table depth (m)	Altitude (m)	Slope (%)
Soils with no limitations (S1)	High Very high	High	> 10	< 800	< 20
Soils with moderate limitations (S2)	Moderate	Moderate	> 10	< 800	< 20
Soils with major limitations (S3)	Low	Low	> 10	< 800	< 20
Soils with severe limitations (S4)	Very low	Very low	> 10	< 800	< 20
Non-suitable soils (N)			< 10	> 800	> 20

Table 3.11. Ratio between annual volume of VW and areas suitable for application.

Suitability class	Area (ha)	Available VW (m³)	Available VW/ Suitable soils (m³/ha)	VW/soils S1 + S2 (m³/ha)
Non-suitable soils (N)	963,826	902,226		
Suitable soils				
S1	6,124	902,266	1.66	3.72
S2	236,302			
S3	198,851			
S4	99,610			

the yearly volume of VW produced and the suitable surface area. In Calabria this amounts to about 1.5 m³/ha (Table 3.11), which shows that VW management can be relatively simple within the region.

Regulations

The evaluation system used allows subdivision of the regional territory into different application suitability orders, classes and subclasses. Within a given order, the classes express the different degrees of limitations. Such limitations are absent in class S1, moderate in S2, major in S3 and severe in S4. Growing limitations imply a growing risk of environmental (particularly soil and water) degradation. At the same time, careful consideration of the type and degree of limitation can allow optimization of effluent application strategies and/or the management of the soils to be treated. Indeed, in many cases, if specific techniques or monitoring actions are adopted, the limitations of the physical environment can be overcome. For every type and degree of limitation, ARSSA researchers defined a series of possible "regulations", summarized in Table 3.12.

In the case of limitations caused by surface runoff, the adoption of working techniques that follow the slope's contours and the use of instruments that prevent excessive refining of the soil and maximize its roughness may permit VW to be used with no particular environmental risks. Alternatively, to favour water infiltration, VW volume should be restricted to 40 m³/ha for each application, with administration intervals longer than 15 d. Proper soil plant cover with wild plants and/or pruning residues significantly attenuates runoff risks. The combined adoption of all these techniques may permit effluent to be spread even where there is a high runoff risk (class S3).

Table 3.12. Regulations for different suitability classes and types of limitations.

Type of limitation	Prescriptions		
	S2—Moderate limitations	S3—Major limitations	S4—Severe limitations
r capacity to receive olive mill effluents without surface runoff	Maximum dose of 40 m³/ha per administration, with an interval longer than 15 d **or** mechanical operations following the slope's contours and conducted with instruments that maximize soil ruggedness **or** adequate vegetation cover (>50%) with wild plants and/or pruning residues.	Maximum dose of 40 m³/ha per administration, with an interval longer than 15 d and mechanical operations following the slope's contours and conducted with instruments that maximize soil ruggedness **or** the above-mentioned maximum doses and adequate vegetation cover (>50%) with wild plants and/or pruning residues.	This subclass is not present.
t Soil capacity to retain olive mill effluents	Maximum dose of 40 m³/ha per administration, with an interval longer than 15 d.	Maximum dose of 40 m³/ha per administration, with an interval longer than 30 d.	
d Internal drainage	This subclass is not present.	Fractioned VW supply with single doses lower than 40 m³/ ha, administered at intervals longer than 1 month and agronomical interventions enhancing drainage.	This subclass is not present.
p pH	Maximum dose of 40 m³/ha per administration, with an interval longer than 15 d and monitoring of the main chemical parameters of the surface layer of the soil at least every 3 years. Alternatively VW neutralization with CaO.	This subclass is not present.	Maximum dose of 20 m³/ ha per administration, with an interval longer than 30 d and monitoring of the main chemical parameters of the surface layer of the soil at least every 2 years. Alternatively VW neutralization with CaO.
c active limestone	Monitoring of the main chemical parameters of the surface layer of the soil at least every 3 years.	Monitoring of the main chemical parameters of the surface layer of the soil at least every 3 years.	Monitoring of the main chemical parameters of the surface layer of the soil at least every 3 years.

The soil's effluent retention capacity depends on texture and on soil depth; coarse textural classes and low depth negatively affect application suitability. In this case it is necessary to reduce the volume and frequency of application, in extreme cases limiting VW administration to 20 m³/ha at intervals longer than 30 d.

The same regulations apply to limits linked to drainage, which affects the soil's capacity to "metabolize" the organic components of VW. In this case, however, it is necessary to control oxygen availability as well, favouring the soil's aeration with agronomic interventions.

In the case of limits resulting from low pH, constant monitoring of the soil's main chemical parameters may reveal degradation processes and possibly suggest a halt to VW application. To our knowledge, it is always advisable to limit the doses according to pH values. The pH problem does not exist when VW is neutralized prior to application.

Lastly, the presence of active limestone guarantees a buffer action against the acidity of VW. In the case of non-calcareous soils, regular monitoring of chemical parameters is a valid precautionary measure.

In all cases, it must be highlighted that VW incorporation into soil is a valid solution to the initial soil surface impermeabilization that follows application and that is caused by the hydrophobic action of fats and by the clogging of pores by solid particles in suspension.

3.5. Pedological Basis for Assessment of Vulnerability to Pesticides

Agricultural activities are a major source of groundwater and surface water contamination because of the widespread use of chemicals ranging from fertilizers to phytopharmaceuticals, which, if applied without due precautions and unaccompanied by a profound knowledge of the means by which solutes are transported through the soil, may give rise to phenomena of toxicity and the pollution of water resources.

The fate of these compounds in the environment is influenced by various factors linked to:

characteristics of the compound;
characteristics of the soil;
climatic conditions;
agricultural management practices;
interaction between the above factors.

Because of the complexity of the problem, mathematical models in recent decades have increasingly been resorted to as evaluation instruments to estimate water susceptibility to contamination from pollutants and xenobiotics used in agriculture. At the legislative level in this context, Community Directive 91/414/EEC is brought to bear, obliging all Member States to legislate with regard to the registration of phytopharmaceuticals, ensuring that registration proposals are accompanied by appropriate studies on the environmental impact of the various substances. In particular, the directive underlines the importance of using "validated models" to calculate Predicted Environmental Concentrations (PECs) as the basis for defining the environmental risk and evaluating the effects on health. The results reached in terms of PECs must be used in accordance with the directive both to evaluate the necessity for further experimental checks and to guide decisions in management terms and in the safeguarding of land resources. Faced with the lack of shared criteria and common methodologies, the European Commission, Directorate General Environment has instituted a forum for coordinating models for the fate of pesticides and their use, also known as FOCUS (FOrum for the Co-ordination of Pesticide Fate Models and Their USe, 1995, 1996, 1997, 2000, 2001). FOCUS, therefore, saw to the definition of practical guidelines for the use of simulation models in the ambit of Community registration procedures and for the development of calibration and validation procedures for using the models in standardized scenarios relative to the different environmental, climatic, pedological and hydrological conditions encountered in the EU. Two distinct stages of the business of calibration and validation were followed through: the first for ground waters and the second for surface waters. Regarding ground waters, the scenarios put in place by FOCUS in order to give a primary evaluation approximation of the leaching potential are defined by a set of standard combinations of climatic, pedological and cultural data (five crops common to all scenarios and 20 crops present in at least one scenario), concerning nine sites: Châteaudum (France), Hamburg (Germany), Jokioinen (Finland), Kremsmünster (Austria), Okehampton (United Kingdom), Piacenza (Italy), Oporto (Portugal), Seville (Spain) and Thiva (Greece). Depending on which scenarios are implemented, input data for four simulation models were defined: MACRO 4.2, PEARL 1.1, PELMO 3.2 and PRZM 3.2.

The scenarios identified do not represent specific field conditions, nor do they try to represent the locality or state where they are situated, but they are "realistic" examples of worst-case scenarios, useful in the definition of leaching risks, adopting 0.1 mg/L as the threshold limit value of concentration. Simulations have a duration of 20 years in the case of pesticides applied every year; the length of duration rises to 40 and 60 years, respectively, in the case of products applied every two or every three years. Each simulation is preceded by an initialization period of six years, the results of which are ignored for purposes of the evaluation of the leaching potential. Simulation results are given as average annual concentrations at a defined depth, given by the ratio between the integral of the solute flux (the total quantity of the active substance or its metabolites that reach the specified depth throughout the year) and the integral of the water flux at the same depth in the same time span (annual recharge flux); in the years in which the net recharge flux at the specified depth is zero, the concentration of the solute and its metabolites are placed at zero level. In equation form, the average annual concentration at a certain depth is calculated as follows:

$$C_i = (\Sigma_{i,i+j} J_s) / (\Sigma_{i,i+j} J_w) \tag{3}$$

where C_i is the average concentration at the specified depth (mg/L) for a period of time that goes from day i to day j (365 or 366 d for 20-year scenarios, 730 or 731 d for 40-year scenarios; 1095 or 1096 d for 60-year scenarios), J_s is the daily flux of solute (mg/m²/d), and J_w is the daily water flux (L/m²/d). The reference depth is fixed at 1 m for all models.

The simulation output, i.e., the 20 average annual values, are placed in increasing order; the 17th value (the fourth highest one) is used to represent the value of the 80th percentile associated with the climatic guide variables adopted for specific simulation conditions, while the concentrations relative to the 90th percentile represent the vulnerability associated with pedological and climatic conditions. For a correct interpretation of the results, as well as the average concentration at a depth of 1 m, there should also be listed the parameters necessary to calculate the water and mass balances, the annual precipitation (and any possible irrigation additions), evapotranspiration, runoff, percolation at 1 m, and variation of the water stored in the soil. In the case of mass balances, the following items should be expressed on an annual basis: the quantity applied (or produced, in the case of metabolites); losses from erosion and runoff; crop uptake; degradation; losses from volatilization; leaching at depths of more than 1 m; and storage within 1 m.

The models adopted by FOCUS have different conceptual architectures, just like the environmental and hydrological conditions that they can handle, as well as the processes they can simulate. PELMO and PRZM are capacitative models of the functional kind, while MACRO and PEARL are more complex mechanistic models.

MACRO (Jarvis, 1994) is a numerical model for the simulation of water balance and solute transport in the soil. The model simulates fluxes in conditions of both unsaturated and saturated soils, in the presence or not of crops, of ground water (whether permanent or fluctuating), and of the drainage system. A peculiarity of the model is the subdivision of the soil into two flux domains, microporous and macroporous, each characterized by its own water content, with its own flux intensity and solute concentration. Richards' equation and the convection-dispersion equation are used to describe the fluxes in the micropore domain, while fluxes in the macropores are of the laminar-gravitative type, in accordance with Darcy's law. The maximum water content in the two domains, corresponding to the field capacity in the case of the micropores and to the difference between total saturation and field capacity in the case of macropores, is derived from examining experimental retention curves. The two flux domains are in dynamic equilibrium with each other and the exchanges are calculated depending on the soil type and degree of aggregation. The form, size and degree of aggregation as described in the field are used to calculate the parameter that regulates the flux and solute exchange between the two porosity domains. Further assumptions of the model include: (1) kinetics of the first order for pesticide degradation in each of the four pools present in the soil (micropores and macropores, solid/liquid stages; the degradation rates are determined on the basis of reference values modified for soil moisture and temperature conditions); (2) instantaneous equilibrium between the stage in solution and the adsorbed one, according to Freundlich's isotherm; and (3) adsorption sites divided between micropores and macropores.

PEARL (*Pesticide Emission Assessment at Regional and Local Scales*) is a model developed by two Dutch research centres (RIVNM and Alterra Green World Research), based on two pre-existing models PESTLA (*PESTicide Leaching and Accumulation*; version 1: Boesten and Van der Linden, 1991; version 3.4: Van den Berg and Boesten, 1999) and PESTRAS (*PESticide TRansport ASsessment*; Tiktak et al., 1994; Freijer et al., 1996), the latter being a modified version of the first version of PESTLA. PEARL is based on: (1) convection-dispersion equation, including the diffusion in the gaseous stage with Henry's coefficient related to the temperature; (2) Freundlich's two-site adsorption model (one in equilibrium and the other dynamic); (3) transformation rates dependent upon the water content, temperature and the soil depth; and (4) the crop uptake. The model also deals with the formation and processes of the metabolites and describes the lateral transport to the drainage system. Preferential fluxes or those in macropores are not handled. The water and thermic balance of the soil as such are not simulated by PEARL, but are derived from another model, SWAP (*Soil WaterAtmosphere Plant*; version 2.0: Van Dam et al., 1997). Water fluxes in SWAP are described through reference to Richards' equation and the model can handle a number of different boundary conditions in terms of water fluxes. Evaporation from the soil and crop evapotranspiration are calculated by applying suitable multiplicative coefficients to the potential reference evapotranspiration, one for the soil and one for the crop. The model can simulate the presence of ground water at the bottom of the soil profile and its oscillations as affected by the rainfall regime.

PELMO (Klein, 1991) is a monodimensional simulation model of pesticide movement in the soil, in accordance with the simplified approach of the chromatographic column; the water balance is simulated in conformity with an approach of the capacitative type, based on field capacity. PELMO was constructed on the American PRZM 1 model of the Environment Protection Agency (EPA) (Carsel et al., 1984), but it has been considerably modified and improved in order to satisfy the requests of German control authorities. Routines were inserted into version 2.01 (1995) to calculate runoff based on the Curve Number method and for the volatilization of the active principles. PELMO 2.01 was validated in a combined project of the Industrieverband Agrar (IVA), the German EPA, and the Fraunhofer-Institut für Umweltchemie und Ökotoxikologie of Schmallenberg (Klein et al., 1997). In 1998, another module was added to handle the products of pesticide degradation and to include further metabolization until CO_2 is released (PELMO 3.0; Jene, 1998). The possibility of simulating lysimeters and experimental parcels was also added recently (Fent et al., 1998). The version implemented by FOCUS is 3.2, developed in 1999. In the last version, the model can manage up to eight degradation products, as well as the kind applied.

PRZM is a monodimensional simulation model for the vertical movement of pesticides in the soil according to the simplified chromatographic column approach; the water balance is simulated according to an approach of the capacitative kind. The first official version dates back to 1984 (Carsel et al., 1984); the second updated version, PRZM2, was released as part of the RUSTIC model (Dean et al., 1989a, 1989b) and successively as a stand-alone model. In the latest versions (PRZM 3.1 and 3.2), routines for pesticide degradation have been introduced in relation to the soil temperature and moisture content and the possibility of using Freundlich's isotherm to describe adsorption processes inserted; the model handles up to two metabolites simultaneously, as well as the kind applied. Also in the case of PRZM, analogously to PELMO, the generation of runoff is modelled using the Curve Numbers approach.

MACRO has recently been applied in Italy to produce pesticide vulnerability maps of the Marche region at a 1:50,000 scale. A stand-alone software has been developed, FitoMarche (Balderacchi et al., 2007), linking a GIS application to the model, and based on the regional soil and land use maps. In assessing the fate of pesticides, the software can use a probabilistic approach.

MACRO is still used as a decision-supporting tool in the multiscale approach developed under the European FP6 project FOOTPRINT (http://www.eu-footprint.org/home.html), aimed to further develop the FOCUS approach in the scope of the revision of Directive 91/414/EEC, within the future Thematic Strategy on the Sustainable Use of Pesticides.

References

ALIANELLO, F. 2001. Effetti della somministrazione di acque reflue di frantoi oleari sulle caratteristiche chimiche e biochimiche del suolo. Progetto Panda MIPAF (Ed.), L'informatore Agrario, Verona.

ALMASRI, M.M., KALUARACHCHI, J.J. 2007. Modeling nitrate contamination of groundwater in agricultural watersheds. Journal of Hydrology, 343, 211–229.

ARPAV-REGIONE DEL VENETO. 2004. *Carta dei Suoli del bacino scolante in laguna di Venezia.* Ponzano (TV), 399.

ARSSA. 2002. Utilizzazione agronomica delle acque reflue olearie. Progetto POM Misura 2.

ARSSA. 2004. I suoli della Calabria-Carta dei suoli della regione Calabria in scala 1:250.000.

ARSSA. 2005. Carta di attitudine dei suoli allo spargimento delle acque di vegetazione della Regione Calabria in scala 1:250.000.

BALDERACCHI, M., BOCELLI, R., TREVISAN, M. 2007. *Tools to Assess Pesticide Environmental Fate.* La Goliardica Pavese, Pavia, 142 p.

BANERJEE, M.R., BURTON, D.L., DEPOE, S. 1997. Impact of sewage sludge application on soil biological characteristics. Agriculture, Ecosystems and Environment, 66, 241–249.

BATJES, N.H., BRIDGES, E.M. 1993. Soil vulnerability to pollution in Europe. Soil Use and Management, 9, 25–29.

BELLAMY, P.H., LOVELAND, P.J., BRADLEY, R.I., LARK, R.M., KIRK, G.J.D. 2005. Carbon losses from all soils across England and Wales 1978–2003. Nature, 473, 245–248

BIGI, L., BOCCIAI, P., FAVI, E., GALANTI, E., GONNELLI, I., MAIANI, S., ROSSI, L., RUSTICI, L., TEGLIA, P., VINCI, A. 1989. Val di Chiana Senese. Valutazione dell'attitudine dei suoli allo spandimento dei liquami suini. Regione Toscana, Dipartimento Agricoltura e Foreste, 86 p.

BOESTEN, J.J.T.I., VAN DER LINDEN, A.M.A. 1991. Modelling the influence of sorption and transformation on pesticide leaching and persistence. Journal of Environmental Quality, 20, 425–435.

BONARI, E., CECCARINI, L. 1993. Sugli effetti dello spargimento delle acque di vegetazione sul terreno agrario: risultati di una ricerca sperimentale. Genio Rurale no. 5, Edagricole, Bologna.

BONARI, E., GIANNINI, C., CECCARINI, L., SILVESTRI, N., TONINI, M., SABBATICi, T. 2001. Spargimento delle acque di vegetazione dei frantoi oleari sul terreno agrario. Inf. Agr., suppl. 50, 8–12.

BOORMAN, D.B., HOLLIS, J.M., LILLY, A. 1995. *Hydrology of soil types: a hydrologically-based classification of the soils of the United Kingdom.* Institute of Hydrology Report no. 126, Institute of Hydrology, Wallingford.

BRANDLI, R.C., BUCHELI, T.D., KUPPER T., FURRER, R., STADELMANN, F., TARRADELLAS, J. 2005. Persistent organic pollutants in source-separated compost and its feedstock materials—a review of field studies. Journal of Environmental Quality, 34, 735–760.

BRENNA, S., RASIO, R. 2000. Valutazione della funzione protettiva dei suoli a supporto delle decisioni nella Regione Lombardia. Bollettino della Società Italiana della Scienza del Suolo, 48(1–2), 247–252.

BRICCOLI BATI, C., GRANATA, R., LOMBARDO, N. 1991. Valutazione del pericolo di inquinamento delle falde acquifere in seguito allo spandimento di acque di vegetazione su terreno agrario. Patron Editore, Bologna.

CALZOLARI, C, UNGARO, F., BUSONI, E., FILIPPI, N., GUERMANDI, M., TAROCCO, P., BRENNA, S., MICHELUTTI, G., PIAZZI, M., VINCI, I. 2000. The SINA project in the Padano-Veneto basin. Proceedings of the International Congress on "Vulnerabilità e Sensibilità dei Suoli", Firenze 18–21 ottobre 1999, 287–307.

CALZOLARI, C., UNGARO, F., GUERMANDI, M, LARUCCIA, N. 2001. *Suoli capisaldo della pianura padano-veneta: bilanci idrici e capacità protettiva.* Rapporto SINA 10.1, April 2001.

CARSEL, R.F., SMITH, C.N., MULKEY, L.A., J.D. DEAN, JOWISE, P. 1984. *User's manual for the pesticide root zone model (PRZM): Release 1.* EPA-600/3-84–109, US EPA, Athens, Georgia.

CASSA PER IL MEZZOGIORNO. 1970–1980. Carte Piezometriche e della qualità della Acque. Progetto Speciale, 26.

CIVITA, M., DE MAIO, M. 2001. Valutazione e cartografia automatica della vulnerabilità degli acquiferi all'inquinamento con il sistema parametrico SINTACS R5. Pitagora (Ed.), Bologna, 248 p.

CRESCIMANNO, G., LANE, M., OWENS, P.N., RYDEL, B., JACOBSEN, O.H., DÜWEL, O., BÖKEN, H., BERÉNYI-ÜVEGES, J., CASTILLO, V., IMESON, A. 2004. Soil Erosion. Task Group 5 on Links with Organic Matter and Contamination, Working Group and Secondary Soil Threats. Vol.2, Soil Erosion, European Commission, Brussels, Belgium.

DAUDÉN, A., QUILEZ, D. 2004. Pig slurry versus mineral fertilization on corn yield and nitrate leaching in a Mediterranean irrigated environment. European Journal of Agronomy 21, 7–19.

DAVIDSON, E.A., TRUMBORE, S.E., AMUNDSON, R. 2000. Soil warming and organic carbon content. Nature, 408, 789–790.

DE PAZ, J.M., RAMOS, C. 2004. Simulation of nitrate leaching for different nitrogen fertilization rates in a region of Valencia (Spain) using a GIS-GLEAMS system. Agriculture, Ecosystems and Environment, 103, 59–73.

DE PLOEY, J., POESEN, J. 1985. Aggregate stability, runoff generation and interrill erosion. In: Richards, K.S., Arnett, R.R., Ellis, S. (Ed.), *Geomorphology and Soils,* Allen and Unwin, London, pp. 99–120.

DE SIMONE, C. 2001. Effetti fitotossici dei reflui oleari. Progetto Panda MIPAF (Ed.), L'informatore Agrario, Verona.

DEAN, J.D., HUYAKORN, P.S., DONIGIAN, A.S., VOOS, K.A., SCHANZ, R.W., MEEKS, Y.J., CARSEL, R.F. 1989a. Risk of Unsaturated/Saturated Transport and Transformation of Chemical Concentrations (RUSTIC). Vol. 1: Theory and Code Verification. EPA/600/3-89/048a, U.S. EPA Environmental Research Laboratory, Athens, Georgia.

DEAN, J.D., HUYAKORN, P.S., DONIGIAN, A.S., VOOS, K.A., SCHANZ, R.W., CARSEL, R.F. 1989b. Risk of Unsaturated/Saturated Transport and Transformation of Chemical Concentrations (RUSTIC). Vol. 2: User's Guide. EPA/600/3-89/048b, U.S. EPA Environmental Research Laboratory, Athens, Georgia.

DICK, R.P. 1992. A review: long-term effects of agricultural systems on soil biochemical and microbial parameters. Agriculture, Ecosystems and Environment, 40, 1–4, 25–36.

DORAN, J.W., ZEISS, M.R. 2000. Soil health and sustainability: managing the biotic component of soil quality. Applied Soil Ecology, 15, 3–11.

ECKERSTEN, H., JANSSON, P.E., JOHNSSON, H. 1996. *SOILN model (ver 9.1), User's Manual,* 3d ed. Sveriges Lantbruksuniversitet, Uppsala, 92.

ERSAF. 1997a. *Manuale per la compilazione delle schede delle unità cartografiche, ver. 3.1.* April 1997.

ERSAF. 1997b. Norme generali per il rilevamento e la compilazione della Carta Pedologica della Lombardia alla scala 1:50.000. Progetto carta pedologica, ERSAL, Milano.

ERSAL. 1988. Progetto "Carta pedologica". I suoli della bassa pianura bresciana fra i fiumi Mella e Chiese. SSR 1, Milano, 112 p.

ERSAL. 1996. Progetto "Carta pedologica". I suoli del Trevigliese. S.S.R.18, Milano, 56–59.

FAO. 1976. A framework for land evaluation. Soil Bulletin, no. 32, Rome.

FAVI, E., FOSSI, F., GIOVANNELLI, P. 1993. Lo spargimento delle AA.VV.: indicazioni agronomiche e caratteristiche pedologiche. Genio Rurale no. 5, 1993, Edagricole, Bologna.

FENT, G., JENE, B., KUBIAK, R. 1998. Performance of the Pesticide Leaching Model PELMO 2.01 to predict the leaching of bromide and 14C-Benazolin in a sandy soil in comparison to results of a lysimeter- and field study. Staatliche Lehr und Forschungsanstalt für Landwirtschaft, Weinbau und Gartenbau (SLFA) Neustadt, Poster Abstract 6B-030, IUPAC Congress Book of Abstracts, London.

FOCUS. 1995. Leaching Models and EU Registration. European Commission Document 4952/VI/95.

FOCUS. 1996. Soil Persistence Models and EU Registration. European Commission Document 7617/VI/96.

FOCUS. 1997. Surface Water Models and EU Registration of Plant Protection Products. European Commission Document 6476/VI/96.

FOCUS. 2000. FOCUS groundwater scenarios in the EU review of active substances—The report of the work of the Groundwater Scenarios Workgroup of FOCUS (Forum for the Co-ordination of pesticide fate models and their USe), Version 1 of November 2000. EC Document Reference Sanco/321/2000 rev. 2, 202 p.

FOCUS. 2001. *FOCUS Surface Water Scenarios in the EU Evaluation Process under 91/414/EEC.* Report of the FOCUS Working Group on Surface Water Scenarios, EC Document Reference SANCO/4802/2001-rev.2, 245 p.

FREIJER, J.I., TIKTAK, A., HASSANIZADEH, S.M., VAN DER LINDEN, A.M.A. 1996. PESTRAS v3.1.: a one dimensional model for assessing leaching, accumulation and volatilisation of pesticides in soil. RIVM report 715501007, Bilthoven, the Netherlands.

GARCÌA-ORENES, F., GUERRERO, C., MATAIX-SOLERA, J., NAVARRO-PEDREN, J., GOMEZ, I., MATAIX-BENEYTO, J. 2005. Factors controlling the aggregate stability and bulk density in two different degraded soils amended with biosolids. Soil and Tillage Research, 82, 65–76.

GIARDINA C.P., RYAN, M.G. 2000. Evidence that decomposition rates of organic carbon in mineral soil do not vary with temperature. Nature, 404, 858–861.

GRIZZETTI, B., BOURAOUI, F., DE MARSILY, G. 2005. Modelling nitrogen pressure in river basins: A comparison between a statistical approach and the physically-based SWAT model. Physics and Chemistry of the Earth, 30, 508–517.

GUERMANDI, M., DI GENNARO, A., FILIPPI, N. 1991. Evaluation of the pollution risk to surface waters and groundwaters arising from the application of potential pollutants to the soil. Mem. Descr. Carta Gel. d'It., XLVI, 89–101.

HERNANDEZ, T., MORAL, B., PEREZ-ESPINOSA, A., MORENO-CASELLES, J., PEREZ-MURCIA, M.D., GARCIA, C. 2002. Nitrogen mineralisation potential in calcareous soils amended with sewage sludge. Bioresource Technology, 83, 213–219.

IPLA. 2007. Carta della Capacità Protettiva dei Suoli nei confronti delle acque sotterranee, scala 1:250.000. Regione Piemonte, http://www.regione.piemonte.it/agri/suoli_terreni/suoli1_250/atlante_carto.htm

JARVIS, N. 1994. *The MACRO model (version 3.1). Technical Description and Sample Simulations.* Department of Soil Sciences, Swedish University of Agricultural Sciences, Reports and Dissertations, 19, Uppsala.

JENE, B. 1998. *PELMO 3.00 Manual extension.* Staatliche Lehr- und Forschungsanstalt für Landwirtschaft, Weinbau und Gartenbau, D-67435 Neustadt/Wstr.

JONES, R.J.A., YLI-HALLA, M., DEMETRIADES, A., LEIFELD, J., ROBERT, M. 2004. Task Group 2 status and distribution of soil organic matter in Europe. In: Van-Camp, L., Bujarrabal, B., Gentile, A.R., Jones, R.J.A., Montanarella, L., Olazabal, C., Selavaradjou, S.K., Reports of the Technical Working Groups Established under the Thematic Strategy for Soil Protection. EUR 21319 EN/3, 827, Office for Official Publications of the European Communities, Luxembourg, Vol. 3, Organic Matter and Biodiversity.

KERSEBAUM, K.C., STEIDL, J., BAUER, O., PIORR, H.-P. 2003. Modelling scenarios to assess the effects of different agricultural management and land use options to reduce diffuse nitrogen pollution into the river Elbe. Physics and Chemistry of the Earth, 28, 537–545.

KLEIN, M. 1991. PELMO: Pesticide Leaching Model. Fraunhofer-Institut für Umweltchemie und Ökotoxikogie, D57392 Schmallenberg.

KLEIN, M., MÜLLER, M., DUST, M., GÖRLITZ, G., GOTTESBÜREN, B., HASSINK, J., KLOSKOWSKI, R., KUBIAK, R., RESSELER, H., SCHÄFER, H., STEIN, B., VEREECKEN, H. 1997. Validation of the pesticide leaching model PELMO using lysimeter studies performed for registration. Chemosphere 35(11), 2563–2587.

LAL, R. 2004. Soil carbon sequestration to mitigate climate change. Geoderma 123, 1–22.

LARSSON, M.H. 1999. Quantifying macropore flow effects on nitrate and pesticide leaching in a structured clay soil. Field experiments and modeling with the MACRO and Clay Soil. Acta Universitatis Agriculturae Sueciae, Agraria, 164, 34.

LARSSON, M.H., JARVIS, N. 1999b. A dual porosity model to quantify macropore flow effect on nitrate leaching. Journal of Environmental Quality, 28, 1298–1307.

LARSSON, M.H., JARVIS, N.J. 1999a Evaluation of a dual-porosity model to predict field-scale solute transport in a macroporous soil. Journal of Hydrology, 215, 153–171.

LULLI, L. 1997. Esperienze sul problema dei reflui zootecnici. criteri e logiche. Seminario AIP, "L'utilizzazione agronomica dei reflui zootecnici: il contributo della pedologia." Bologna, October 1997.

LULLI, L., LORENZONI, P., LEONI, G., DELOGU, F. 1980. Un esempio di cartografia pedologica applicata alla pianificazione territoriale. I suoli, la loro capacità d'uso, la loro attitudine alla edificabilità e la zonizzazione per l'allevamento suinicolo del comune di Sorbolo (Parma). Annali Ist. Sper. Studio Difesa Suolo, Vol. XI, 97–120.

MANTOVI, P., FUMAGALLI, F., BERETTA, G.P., GUERMANDI, M. 2005. Nitrate leaching through the unsaturated zone following pig slurry applications. Journal of Hydrology, 316, 195–212.

MANTOVI, P., BALDONI, G. 2004. Effects of 12 years use of sewage sludge on the plant-soil system. In: Bernal, M.P., Moral, R., Clemente, R., Paredes, C. (Eds.), Proceedings of the 11th International Conference of the FAO ESCORENA Network on the Recycling of Agricultural, Municipal and Industrial Residues in Agriculture. Murcia, Spain, 6–9 October 2004.

MARCHETTI, R., PONZONI, G., SPALLACCI, P., CEOTTO, E., UNGARO F., CALZOLARI, C. 2000. Simulating water flow in areas at environmental risk with the MACRO model. Model evaluation with data from lysimeter studies. Proceedings of the International Conference "Soil Vulnerability and Sensitivity", Florence, 1999 October 18–21, pp. 323–330.

MARCHETTI, R., SPALLACCI, P., PONZONI, G. 2004. Simulating nitrogen dynamics in agricultural soils fertilized with pig slurry and urea. Journal of Environmental Quality, 33, 1217–1229.

MARMO, L., FEIX, I., BOURMEAU, E., AMLINGER, F., BANNICK, C.G., DE NEVE, S., FAVOINO, E., GENDEBIEN, A., GILBERT, J., GIVELET, M., LEIFERT, I., MORRIS, R., RODRIGUEZ CRUZ, A., RÜCK, F., SIEBERT, S., TITTARELLI, F. 2004. Task Group 4 Exogenous organic matter. In: Van-Camp, L., Bujarrabal, B., Gentile, A.R., Jones, R.J.A., Montanarella, L., Olazabal, C., Selavaradjou, S.K., Reports of the Technical Working Groups Established under the Thematic Strategy for Soil Protection. EUR 21319 EN/3, Office for Official Publications of the European Communities, Luxembourg, Vol. 3. Organic Matter and Biodiversity.

MARRARA, G., TAMBURINO, V., ZIMBONE, S.M. 2004. "Prove di accumulo di acque reflue olearie". Valorizzazione di acque reflue olearie e sottoprodotti dell'industria agrumaria e olearia. Laruffa Editore, Reggio Calabria.

PAGLIAI, M., VITTORI ANTISARI, L. 1993. Influence of waste organic matter on soil micro and macrostructure. Bioresource Technology, 43.

PAGLIAI, M., BISDOM, E.B.A., LEDIN, S. 1983. Changes in surface structure (crusting) after application of sludge and pig slurry to cultivated agricultural soils in northern Italy. Geoderma, 30, 35–53.

PAGLIAI, M., PELLEGRINI, S., VIGNOZZI, N., PAPINI, R., MIRABELLA, A., PIOVANELLI, C., GAMBA, C., MICLAUS, N., CASTALDINI, M., DE SIMONE, C., PINI, R., PEZZAROSSA, B., SPARVOLI, E. 2001. Influenza dei reflui oleari sulla qualità del suolo. Informatore Agrario, Supplemento 50/2001, Verona.

PETRELLA, F. 1997. Il contributo della pedologia nell'assistenza tecnica alla regione Piemonte per la predisposizione di criteri e norme tecniche per lo spandimento dei liquami zootecnici. Seminario AIP, "L'utilizzazione agronomica dei reflui zootecnici: il contributo della Pedologia." Bologna, October 1997.

RAGAZZI, F., VINCI, I. 1997. L'attitudine dei suoli allo spargimento dei liquami zootecnici: aspetti normativi e confronti tra diverse metodologie su alcuni suoli della bassa pianura veneta. Seminario AIP "L'utilizzazione agronomica dei reflui zootecnici: il contributo della pedologia." Bologna, October 1997.

RAGAZZI, F., VINCI, I., GARLATO, A., GIANDON, P., MOZZI, P. 2004. *Carta dei suoli del bacino scolante in laguna di Venezia*. ARPAV - Osservatorio Regionale Suolo, Castelfranco Veneto (TV).

ROBERT, M., NORTCLIFF, S., YLI-HALLA, M., PALLIÈRE, C., BARITZ, R., LEIFELD, J., GERHARD BANNICK, C., CHENU, C. 2004. Task Group 1 Functions, roles and changes in SOM. In: Van-Camp, L., Bujarrabal, B., Gentile, A.R., Jones, R.J.A., Montanarella, L., Olazabal, C., Selavaradjou, S.K., Reports of the Technical Working Groups Established under the Thematic Strategy for Soil Protection. EUR 21319 EN/3, pagina, 827. Office for Official Publications of the European Communities, Luxembourg, Vol. 3, Organic Matter and Biodiversity.

SAVIOZZI, A., LEVI MINZI, R., RIFFALDI, R. 1993. Effetto dello spargimento delle acque reflue dei frantoi oleari su alcune proprietà del terreno agrario. Genio Rurale no. 5, 1993, Edagricole, Bologna.

SEVERI, P., BERRÈ, F., CARNEVALI, G., CENTINEO, M.C., DI DIO, G., FARINA, M., FRASSINETI, G., GIAPPONESI, A., GUERMANDI, M., LARUCCIA, N., SARNO, G., TRAGHETTI, T. 2002. Nuova carta regionale della vulnerabilità: aspetti metodologici. 4th European Congress on Regional Geoscientific Cartography and Information Systems, Bologna, Maggio 2003.

SORT, X., ALCANIZ, J.M. 1999. Modification of soil porosity after application of sewage sludge. Soil and Tillage Research, 49, 337–345.

STAMATIADIS, S., DORAN, J.W., KETTLER, T. 1999. Field and laboratory evaluation of soil quality changes resulting from injection of liquid sewage sludge. Applied Soil Ecology, 12, 263–272.

STOLBOVOY, V., MONTANARELLA, L., FILIPPI, N., SELVARADJOU, S., PANAGOS, P., GALLEGO, J. 2005. Soil Sampling Protocol to Certify the Changes of Organic Carbon Stock in Mineral Soils of European Union. EUR 21576 EN, Office for Official Publications of the European Communities, Luxembourg. 12 p.

STOLBOVOY, V., MONTANARELLA, L., FILIPPI, N., JONES, A., GALLEGO, J., GRASSI, G. 2007. Soil sampling protocol to certify the changes of organic carbon stock in mineral soil of the European Union. Version 2. EUR 21576 EN/2. 56 pp. Office for Official Publications of the European Communities, Luxembourg. ISBN: 978-92-79-05379-5.

TIKTAK, A., VAN DER LINDEN, A.M.A., SWARTJES, F.A. 1994. PESTRAS: A one dimensional model for assessing leaching and accumulation of pesticides in soil. RIVM report 715501003, Bilthoven, the Netherlands.

UNGARO, F., CALZOLARI, C., TAROCCO, P., GIAPPONESI, A., SARNO, G. 2005. Quantifying spatial uncertainty of soil organic matter indicators using conditional sequential simulations: a case study in Emilia Romagna plain (Northern Italy). Canadian Journal of Soil Science, 85, 499-510

UNGARO, F., CALZOLARI, C. 2000. Integration of GIS derived soil information with geostatistical estimation of pedotransfer functions inputs for soil modeling applications. In Proceedings of the 4th International Symposium on Spatial Accuracy Assessment in Natural Resources and Environmental Sciences, Amsterdam, 12–14 July 2000, pp. 663–670.

UNGARO, F., CALZOLARI, C., TAROCCO, P., GIAPPONESI, A., SARNO, G. 2003. Soil organic matter in the soils of the Emilia-Romagna Plain (Northern Italy): knowledge and management policies. In: Smith, C.A.S. (Ed.), Soil Organic Carbon and Agriculture: Developing Indicators for Policy Analyses. Proceedings of an OECD expert meeting, Ottawa, Canada. Agriculture and Agri-Food Canada, Ottawa and Organisation for Economic Co-operation and Development, Paris, 113–126.

VAN DAM, J.C., HUYGEN, J., WESSELING, J.G., FEDDES, R.A., KABAT, P., VAN WALSUM, P.E.V., GROENENDIJK, P., VAN DIEPEN, C.A. 1997. Theory of SWAP version 2.0. Simulation of water flow, solute transport and plant growth in the soil-water-atmosphere-plant environment. Technical Document 45, DLO Winand Staring Centre, Wageningen, The Netherlands.

VAN DEN BERG, F., BOESTEN, J.J.T.I. 1999. Pesticide Leaching and Accumulation Model (PESTLA) version 3.4. Description and User's Guide. Technical Document 43, DLO Winand Staring Centre, Wageningen, The Netherlands, 150 p.

VASSILJEV, A., GRIMVALL, A., LARSSON, M. 2004. A dual-porosity model for nitrogen leaching from a watershed. Hydrological Sciences Journal, 49, 313–322.

WOSTEN, J.H.M., PACHEPSKY, Y.A., RAWLS, W.J. 2001. Pedotransfer functions: bridging the gap between available basic soil data and missing soil hydraulic characteristics. Journal of Hydrology, 123–150.

4. Soil Suitability for Irrigation

Paolo Baldaccini[1] and Andrea Vacca[2]

4.1. Introduction

The evaluation of land suitability for irrigation generally follows the basic principles of land evaluation, though differing slightly in some specific features. When determining the suitability of land for irrigation, special attention should be paid to the following fundamental land characteristics:

    topography (e.g., slope, microrelief);
    soil;
    substrata characteristics (e.g., impermeable layers, saltwater lenses);
    quantity and quality of irrigation water, water table fluctuations;
    general drainage conditions; and
    other characteristics such as infrastructure, market accessibility, socio-economic conditions that, although not specific, may strongly influence irrigation development and management.

It is therefore important to highlight the very close relationship between the physical and socio-economic aspects in the process of land evaluation for irrigation. The physical evaluation makes it possible to interpret environmental data and provides useful information for a balanced comparison between the benefits and disadvantages of an irrigation project in different areas of land. The socio-economic evaluation makes it possible to analyse all the variables, such as business organization, amortization periods, interest rates, that in some cases may be far more important than the differences in physical land suitability and may significantly affect the final outcome of land suitability evaluation, because of the major technical and

---

[1] Formerly Università di Sassari, Italy.
[2] Università di Cagliari—Dipartimento di Scienze della Terra, Cagliari, Italy.

financial commitments required of an irrigation plan. In light of the significance of and complexities involved in the process, a multidisciplinary team is required for a complete quantitative study and interpretation. The scale and intensity of development should also be taken into account. Though the general principles may remain the same, the criteria for land suitability evaluation for irrigation may differ depending on whether the evaluation is for identification of suitable irrigable areas, for a feasibility study, for project appraisal or for project implementation.

Since land characteristics may change to different extents over time, evaluation and planning need to be periodically or constantly updated. Hence, the procedures described should not be regarded as fixed schemes but as providing the basis for possible alternatives for the proper evaluation of water and land resources and the most appropriate critical site-specific limits for the suitability classes.

Following this assumption, the subsequent sections provide information on the evaluation of soil suitability for irrigation.

4.2. General Principles

Most of the general principles discussed in this paper are taken from the FAO (1979a, b, 1985) and from Sys et al. (1991). The land use type considered here concerns the irrigation of crops normally cultivated under rainfed conditions that need irrigation in environments with seasonal water deficit. Particular irrigated crops, for instance, rice, are not considered here. For these crops the reader is referred to the pertinent literature (Booker Agriculture International Limited, 1979; Early et al., 1979; De Datta, 1981; Sys et al., 1991, 1993).

The main methods for soil suitability evaluation for irrigation (USBR, 1951; FAO, 1985) consider land and soil criteria to be directly correlated with the specific use and focus special attention on the soil-water relationship. An evaluation of this type clearly has to be supplemented with a specific evaluation for the crop type considered. The relationship between the FAO soil suitability classes for irrigation and the USBR classes is shown in Table 4.1.

Although a quantitative evaluation is desirable, it may be useful to propose criteria for a qualitative evaluation of the physical environment as well. In particular, these criteria could be used in the pre-project phases for reconnaissance and in the detailed survey, providing additional information for selecting the most suitable lands for irrigation, which are then economically evaluated during the feasibility study.

Table 4.1. Relationship between FAO soil suitability classes for irrigation and USBR classes (FAO, 1979a, modified).

FAO	Class definition	USBR
S1	Highly suitable	Class 1
S2	Moderately suitable	Class 2
S3	Marginally suitable	Class 3
Sc	Special use	Class 4
N1	Not suitable at present but potentially suitable	Class 5
N2	Presently and potentially unsuitable	Class 6

The evaluation of the physical environment can be done in terms of land characteristics and land qualities. Figure 4.1 shows these characteristics and qualities and the relationships between them.

It is worthwhile recalling that soil survey investigations for irrigation are among the most complex soil studies for specific uses. The high costs of the technical project, its social implications and the resulting alterations to water and agricultural systems of the land in question require a detailed and careful evaluation of soil suitability for irrigation. Three main aspects distinguish soil survey investigations for irrigation from general investigations. First, relief features need to be examined in greater detail because of their influence on the water regime and on the consequent economic implications of irrigation development. Second, the investigation should also consider the physical properties, such as the ability of the soil to absorb, transmit or retain water and to drain surplus water in a relatively short time. Finally, it should also take into account that land evaluation is also based on economic parameters, such as the potential increase in crop production and hence the possibility of repaying initial investments. These three aspects are strongly influenced by the type of irrigation system chosen and by aridity conditions. Surface irrigation systems are more strongly affected by topography than sprinkler or drip irrigation systems. In arid areas, one of the main purposes of soil survey investigation is to assess salinity and sodicity and the measures required to keep them under control. Moreover, a detailed knowledge of soil characteristics is essential for the design of the irrigation scheme itself (e.g., design and dimension of the distribution network, drainage scheme, soil conservation measures, choice of crops). Finally, a higher survey intensity level, soil description to greater depths (3 m or more), and field and laboratory tests on soil-water relationships (including water retention capacity, water movement in the soil) are required for soil survey investigations for irrigation.

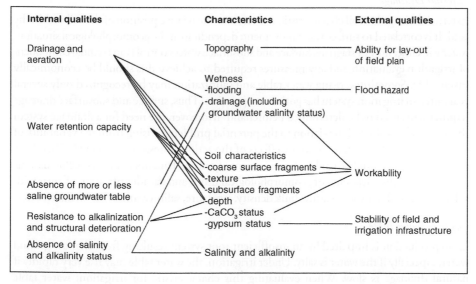

Fig. 4.1. Relationship between land characteristics and land qualities in the suitability evaluation for irrigation (Sys et al., 1991, modified).

4.3. Functional Factors

4.3.1. Topographic Factors

Slope

The slope (field) gradient is the inclination of the soil surface from the horizontal, expressed as a percentage. It strongly influences the feasibility of an irrigation project. Indeed, project costs increase with slope. Slope is also an important factor in the choice of the most suitable type of irrigation system for the site and in the choice of crops. Moreover, the erosion potential due to excessive surface irrigation and the potential runoff from sprinkler systems increase with increasing slope.

Microrelief

Microrelief is defined as the minor surface irregularities and undulations of the land surface, with differences in height between crest and trough ranging from a few centimetres to a few metres. In the evaluation process, the importance of microrelief is related with levelling requirements, with effects on both the economic and soil parameters of the evaluation.

Erosion Hazard

The erodibility of a soil needs to be considered when designing any irrigation system. The rate and method at which water is applied should be controlled so that it does not cause excessive runoff and erosion. Obviously, this is related to slope, soil type, and type of irrigation system.

4.3.2. Hydrological Factors

External Drainage

External drainage is a soil characteristic and refers to the relative position of the soil within the land. It is correlated to surface water movement; depending on the geomorphological situation, water will either flow or stagnate. Surface and subsurface water control is an essential component of irrigation evaluation and any measures required to achieve this should be economically sustainable. Sometimes, rising water table and salinization may be recognized only several years after an irrigation system has gone into operation. Thus, surface and subsurface drainage requirements need to be determined beforehand. Moreover, the need for a drainage system should also be evaluated in relation to the potential problems arising from the variability of deep layers and the related economic effects of the actions foreseen or required. Therefore, special attention should be paid to the assessment of soil and land drainability, the fluctuations in water table, the variability of geomorphological units, and the identification of areas with substantially different soil texture, bulk density, structure, salinity, and other characteristics.

Depth to Water Table

Crop production is impaired by an insufficient root system resulting from shallow ground water, especially if the water is salty. Under irrigation, the water table may rise very rapidly if natural drainage is slow. When evaluating this characteristic for irrigation, water table fluctuations due to rainfall, the natural drainage network, infiltration rate, and the future irrigation design should be taken into account. The highest level to which the water table should be permitted to rise during the irrigation season is an important factor in estimating the cost of a subsurface drainage system. The permissible depth depends on several factors such as capillary conductivity of the soil, evapotranspiration conditions, rooting depth, soil aeration, rainfall intensity, amount and frequency, quality of irrigation water, and crop salt tolerance.

Flood Hazard

Since many irrigable areas are sited in alluvial plains, lands located in areas susceptible to flooding should be evaluated for possible flood damage and feasible protective measures, which will affect the economic parameters of the evaluation. Important information to be collected in flood hazard zones include location and extent of flooding, frequency, duration and seasonality, type of flood damage, local evidence of the effect of flooding on irrigated farming and on crop choice and yield.

4.3.3. Physical Soil Factors

Rock Outcrops

Rock outcrops reduce the available surface area for irrigation and are therefore an important limiting factor. Their removal, feasible in some cases, will negatively affect the economic parameters of the evaluation.

Coarse Surface Fragments

Coarse surface fragments affect soil workability and are therefore also to be taken into account in the evaluation. The cost of removing excessive amounts of coarse rock fragments will negatively affect the economic parameters of the evaluation.

Soil Depth

Soil depth is defined as the thickness of the loose soil above solid rock, limiting root penetration or water percolation. All other conditions being equal, deep soils hold more water available for plant uptake than shallow soils. Therefore, soil depth affects the time interval between irrigations or net precipitation, when plants are not subjected to water stress.

Presence and Depth of Limiting Horizons

Horizons that limit root penetration or water percolation, such as cemented horizons or clayey layers, or horizons that limit water retention capacity, such as gravelly or pebbly horizons, may be present in soils. For the same reasons indicated for soil depth, the presence of these horizons and their depth in the soil profile affect their irrigation suitability. However, the presence of limiting horizons in the subsoil may in some cases, after irrigation or rainfall, increase the available water capacity of the topsoil. Indeed, in these cases the water retained by the limiting horizon could potentially be used, even partly, by the plants.

Texture of Topsoil

The texture of the topsoil (the plough layer) is the numerical proportion by weight (g/kg) of the less than 2 mm soil sand, silt and clay particles. Texture is determined in relation to its influence on water retention capacity, infiltration rate, and hydraulic conductivity.

Fine-textured soils generally hold more water than coarse-textured ones. Soils with intermediate texture have more water available for plant uptake than clayey soils.

It is difficult to establish satisfactory textural diagnostic criteria for identifying the different suitability classes for irrigation. Indeed, because of the large variability of textural classes, only an empirical approach and comparison with all the other soil properties is likely to produce really useful information.

Texture of Subsoil

As regards texture of the subsoil (the horizon underlying the plough layer), the same considerations discussed above apply.

Fragment Content in Topsoil

The water retention capacity and workability of soil are affected by the rock fragment content of the topsoil. Soils containing abundant rock fragments may have very low available water content and may be difficult to work. Since fragments have an effect on the wear and tear of agricultural machinery, not only the amount but also the kind (lithological nature), size and degree of weathering of fragments must be determined. In coarse-textured soils, greater rock fragment content will result in greater macroporosity and, therefore, lower soil water tension for retaining water. In such cases, water will mainly be lost by percolation. In fine-textured soils, the volume occupied by the fragments does not play a role in soil water tension. Therefore, here again the soil will have a lower water content. However, these processes are affected by the nature and state of weathering of the rock fragments. Indeed, some lithologies are more porous than others (e.g., marls, tuffs) and, under the same lithological conditions, the degree of weathering of the fragments may differently affect their ability to partake in soil hydrological processes.

Fragment Content in Subsoil

In general, the same considerations as above apply to the fragment content in the subsoil. Rock fragment content in the subsoil strongly affects soil permeability. Excessive fragments in the subsoil rooting volume may accelerate soil drainage, reducing the amount of water available for plant uptake.

Structure

Soil structure refers to the nature and degree of aggregation of soil particles into discrete soil units of various type and size. The type of structure affects porosity and consequently water and air movement, rooting and nutrients content. Structureless soils may be single-grained, for example, loose sand, where water percolates rapidly, or massive, like many clayey soils, where water moves very slowly. More favourable soil-water relationships exist in soils that have subangular blocky or granular structure. Some aggregates are quite stable upon wetting, while others disperse readily. Aggregate stability is therefore very important in irrigation, especially when flood or sprinkler irrigation systems are envisaged. Many cultural practices improve soil structure and aggregate stability (e.g., conservation tillage, liming, use of grasses in crop rotation, addition of organic matter). On the contrary, other practices, for instance, tilling with heavy equipment, overworking, high intensity sprinkler irrigation and the use of saline water can damage soil aggregates. Moreover, variations in structure along the soil profile should also be taken into account. For instance, the presence of a massive horizon or of a poorly aggregated horizon at a depth of 60 cm does not create problems in dry conditions, but under irrigation can prevent or slow down water percolation.

Consistency

Soil consistency is the degree and kind of cohesion and adhesion between the soil particles and/or the resistance of a soil to deformation or rupture under stress. It depends strongly on texture, organic matter and moisture content of the soil. Consistency is important in evaluating irrigable soils in relation to agricultural practices such as tillage, compaction caused by heavy equipment, or irrigation systems.

Infiltration Rate

The infiltration rate is the speed at which water penetrates the soil. It is a very important factor in the choice of suitable irrigation method and system design and management techniques. For instance, soils with high basic infiltration rates may be unsuitable for flood or furrow irrigation and drip or sprinkler irrigation may be preferable.

Permeability (Hydraulic Conductivity)

Hydraulic conductivity quantifies the ability of a soil to transmit water. This ability depends on pore geometry (texture and structure) and on the fluid properties (viscosity and density). The average hydraulic conductivity of a soil profile is used to determine subsurface drainage and to verify the occurrence of a perched water table, which may damage crop roots. No universally acceptable minimum values can be established for hydraulic conductivity. These values depend on the depth at which the slowly permeable zone occurs and on the crops to be grown. However, the minimum hydraulic conductivity should be adequate to avoid saturated condition persisting in the root zone. Moreover, owing to the many factors affecting this parameter, it is not always easy to obtain truly representative values of hydraulic conductivity that can be confidently used for evaluating soil suitability for irrigation, particularly when measurements have to be made in dry soils rather than in saturated soils or in soils with abundance of coarse fragments. In this sense, it would be desirable that field descriptions place special emphasis on the morphological features influencing or reflecting hydraulic conductivity (texture, structure and structural stability, consistency, fragment content, colour and mottles, presence or absence of insoluble carbonates, porosity, depth to impermeable layers). Good soil descriptions provide the basis for deciding whether measurements of hydraulic conductivity are likely to be representative or not.

Available Water Capacity

Available water capacity (AWC) is the term traditionally used to express the amount of water held in the soil available for use by most plants. Conventionally, the AWC has been regarded as the difference between the soil moisture content at "field capacity" and "permanent wilting point". It is generally expressed in g/kg of dry soil. It depends on crop rooting depth and several soil characteristics (texture, structure, bulk density, salinity, sodicity, mineralogy, soil chemistry, organic matter content). Of these, texture is the predominant factor in mineral soils. The AWC is one of the main soil factors in irrigation scheduling, especially for the choice of irrigation frequency or for determining water volumes. Indeed, only a partial depletion of the AWC should be allowed. The level of this depletion is a management decision based on the type of crop grown, stage of crop growth, total AWC of the soil profile, rainfall patterns and the availability of pumped or delivered water.

4.3.4. Chemical Soil Factors

Soil Reaction (pH)

The degree of acidity or alkalinity of a soil is generally expressed as a pH value. Soil reaction is significant for crop production because of the effect on solubility and availability of nutrients. Soil pH measurements chiefly serve the purposes of irrigation suitability surveys by providing a general indication of whether soil acidity, alkalinity or sodicity prevails. In many cases they can be used to appraise remediable soil deficiencies in relation to economic correlation, such as the need for soil amendments (lime for acid soils and gypsum for sodic soils) and optimum land management including cropping practices.

Organic Matter

Organic matter content by itself is very seldom a proper criterion for grouping soils into classes of varying suitability for irrigation. However, its determination allows important correlations with other physical and chemical soil properties (e.g., structure, permeability, AWC, cation exchange capacity). In general, irrigation favours a more balanced distribution of organic matter within the soil profile. Soils rich in organic matter (peat soils, Histosols) present major problems for irrigation because of their instability and require special management techniques, such as subirrigation.

Exchange Complex

The soil cation exchange characteristics that are important for evaluating irrigation suitability include cation exchange capacity (CEC), the nature of exchangeable cations, and the sodium adsorption ratio.

Cation Exchange Capacity

In the evaluation of soils for irrigation, the CEC is correlated to potential productivity, in terms of the capacity of the soil to retain and supply plant nutrients, and can be considered an indication of the amount and nature of the organic matter and clay minerals present in the soil. Moreover, the CEC affects some physical and structural soil characteristics. Usually, CEC values exceeding 8–10 cmol/kg in the upper 30 cm can be considered sufficient for satisfactory production under irrigation if all the other factors are favourable. Lower values in strongly weathered soils can be considered suitable only for special crops (e.g., rice).

Nature of Exchangeable Cations

Soils with medium or high productivity usually have an exchange pool dominated by calcium and magnesium, containing only minor amounts of potassium and sodium. In very acid soils, a large proportion of the exchange sites is occupied by exchangeable hydrogen or aluminium ions rather than by nutrient bases. Highly productive soils are usually at least 50% base saturated. Unsaturated or very acid soils may need dressings of lime to correct calcium deficiency or aluminium toxicity and their cost is an important factor in the evaluation of irrigation suitability. Cation imbalance may also have detrimental effects on the soil structure and hence on infiltration capacity, permeability, soil tillage, and susceptibility to erosion. Excessively large amounts of exchangeable sodium or of sodium plus magnesium are very important in irrigation suitability assessment.

Sodium Adsorption Ratio

The sodium adsorption ratio (SAR) refers to the degree of Na saturation of the soil exchange complex and is calculated from the concentration of sodium, calcium, and magnesium ions in the soil-water extract using the following formula:

$$SAR = [Na^+]/(0.5[Ca^{2+}] + 0.5[Mg^{2+}])^{1/2}$$

The SAR is a good indicator of the structural stability of a soil and of the physical response that may be expected when water is applied. In general, the physical properties become increasingly unfavourable with increasing levels of exchangeable sodium, and this is much more evident in soils with expanding clay minerals. It is important to bear in mind that the value of SAR in a soil in equilibrium with the irrigation water has greater significance than the SAR of the soil prior to irrigation, especially as far as drainage conditions and irrigation water quality are concerned.

Mineralogy of Clay Fraction

Many of the physical and chemical characteristics of a soil are determined by the kinds of clay minerals contained therein. However, the nature of the clay itself is not a suitable criterion for judging irrigation suitability. This is partly because the clay fraction of a soil is usually composed of a mixture of clay minerals and partly because the influence of the clay type is very closely integrated with other soil characteristics. For this reason, a knowledge of the exact proportions of clay minerals is less important than an understanding of the general nature of the clay. Field observations and the comparison of total clay content with CEC can provide sufficient indications as to whether 1:1 or 2:1 clay mineral types predominate. Nevertheless, a knowledge of the clay minerals is important in the event that physical and chemical soil characteristics need to be improved.

Calcium Carbonate

Calcium carbonate may be present in soils formed from limestone and may accumulate in soils developed in arid and semiarid climates. These accumulations may be dispersed throughout the soil profile in the form of soft aggregates (concretions) or may be concentrated in horizons of varying hardness (Bk, Bkm, Ck, Ckm) and at varying depths below the surface. The amount of carbonates present, their distribution within the soil profile, and the depth to the lime-rich horizons are all important factors in judging soil suitability for irrigation. Indeed, the moisture characteristic curves of highly calcareous soils are similar to those of coarse-textured non-calcareous soils. This behaviour seems to be related to the formation of different structural units rather than to a lower water retention capacity. It is also important to bear in mind that large amounts of carbonates reduce the water holding capacity (especially at high tensions) and this will affect irrigation scheduling. Surface crusting can sometimes be a serious problem in irrigated calcareous soils, especially in poor soils. Crusts not only affect infiltration and soil aeration but also impede or prevent seedling emergence. The physical characteristics of calcareous soils often change when they are irrigated: soils may become more coherent and resistant to root penetration and stable dry soil aggregates may disintegrate when wet, with consequent negative effects on ploughing operations and soil erosion. Finally, a knowledge of the depth to horizons rich in calcium carbonate, whether hardened or not, is of the utmost importance in evaluating soil suitability for irrigation because of the technical and economic effects of their presence.

Gypsum

Soils containing gypsum are widespread in arid and semiarid areas. The same considerations apply for gypsum content as for calcium carbonates.

Salinity

Excess soluble salts, measured in terms of electrical conductivity, constitute one of the soil properties detrimental to crop growth in arid and semiarid areas. Owing to their solubility, these salts can be removed by leaching with excess water under satisfactory drainage condition. An excess of soluble salts in the soil is often associated with an excess of exchangeable sodium, the latter reducing the possibilities for leaching with consequent negative effects on the economic parameters of the evaluation. Salinity and sodicity levels are unstable soil properties and can both be altered by irrigation. Therefore, important factors in irrigation evaluation are soil and subsoil drainage conditions, possible rise of the saltwater table under irrigation, irrigation water quality, infiltration rate, levelling needed to create a suitable surface for leaching, soil amendments, and crop selection (reclamation costs).

4.4. General Criteria for Determining Soil Suitability Classes for Irrigation

The general criteria for determining soil suitability classes for irrigation are shown in Table 4.2, which should be considered only as a reference table. The critical ranges that specify for each individual class-determining factor the different suitability levels may vary depending on climate, local geomorphological and pedological conditions, type and intensity of agriculture, socio-economic conditions, design standards, and other factors. For instance, the critical limits for slope may be higher for the first suitability classes when drip irrigation is envisaged or for moderately erodible soils and where soil conservation measures exist or are foreseen. A further example is soil depth. In general, deeper soils are preferable, but in some cases shallow soils with excellent characteristics such as nearly level surface, good drainage, high fertility status, favourable water-air-soil relationships, very favourable climate, and accessibility to markets may belong to the first suitability classes for irrigation (e.g., soils in some areas of Salento and "Murge Baresi", in the Apulia region).

As an example of the variability of criteria and limits for determining the different classes of suitability for irrigation, we report here the one proposed by Southard (2001) for California (Table 4.3).

Table 4.2. General criteria for determining soil suitability classes for irrigation. *This table is to be considered only as a reference.* (#) s = stones, g = gravel.

	S1	S2	S3	N1	N2
Topography					
slope (%)	< 4	4–10	10–20	20–30	> 30
microrelief	none	slight	moderate	strong	very strong
erosion hazard	none	slight	moderate	severe	extreme
Hydrology					
external drainage	well drained	moderately well drained	somewhat poorly drained or somewhat excessively drained	poorly drained or excessively drained	very poorly drained or very excessively drained
depth to freshwater table (m)	> 2	2–1.20	1.20–0.75	0.75–0.50	< 0.50
depth to saltwater table (m)	> 3	3–2	2–1.20	1.20–1	< 1
flood hazard	none	very rare	rare	occasional	frequent and very frequent
Soil					
rock outcrops (%)	0	< 2	2–10	10–25	> 25
coarse surface fragments (%)	< 0.1	0.1–3	3–15	15–50	> 50
soil depth (m)	> 1	1–0.75	0.75–0.50	0.50–0.25	< 0.25
depth to limiting horizons (m)	> 1	1–0.75	0.75–0.50	0.50–0.25	< 0.25
texture of topsoil	from SL to CL	from SIL to permeable C	from LS to moderately permeable C	from S to slowly permeable C	from S to impermeable C
texture of subsoil	from SL to CL	from LS to permeable C	from LS to poorly permeable C	from S to very slowly permeable C	from S to impermeable C
fragment content in topsoil (%)#	< 3s; <5g	3–5s; 5–10g	5–10s; 10–20g	10–35s; 20–40g	> 35s; > 40g
fragment content in subsoil (%)	< 15	15–35	35–55	35–55	> 55
structure	granular and fine and medium subangular blocky	coarse sub-angular blocky and fine and medium angular blocky	coarse angular blocky and fine, medium, and coarse prismatic	angular blocky and very coarse prismatic	massive and single grain
infiltration rate (cm/h)	0.8–3.5	0.2–0.8/ 3.5–11	0.1–0.2/ 11.0–12.5	0.1–0.2/ 11.0–12.5	< 0.1/> 12.5
permeability	moderate	moderately slow or moderately rapid	slow or rapid	very slow or very rapid	impermeable or very rapid
AWC (g/kg)	> 150	115–150	75–115	< 75	< 50
calcium carbonate (%)	< 15	15–25	25–40	40–75	> 75
gypsum (%)	< 5	5–15	15–30	15–30	> 30
salinity(mmhos/cm)	< 4	4–8	8–16	16–30	> 30
sodicity (SAR)	< 4	4–13	13–22	22–38	> 38

Table 4.3. Criteria and limits proposed for different irrigation systems in California (Southard, 2001, modified)

Restricting factors	Limits				Feature
	Sprinkler irrigation	Drip or trickle irrigation	Furrow irrigation	Basin or paddy irrigation	
Texture (topsoil)	COS, S, LCOS, LS ≥ 60% clay	not used	COS, S, FS, VFS, LCOS, LS, LFS, LVFS ≥ 40% clay	COS, S, FS, VFS, LCOS, LS, LFS, LVFS, SL, FSL, L not used	Fast intake Slow intake
Slope	6–15% (0–1)> 15% (1)	not used	> 2%	> 2%	Slope
Fragments (> 75 mm)	> 25%	not used	> 10%	not used	Large stones
Ponding	present	present	present	not used	Ponding
Depth to high water table (during growing season)	60–90 cm (0–1) < 60 cm (1)	< 60 cm	60–90 cm (0–1) < 60 cm (1)	not used	Wetness
AWC(100 cm in depth)	10–15 cm (0–1) < 10 cm (1)	not used	10–15 cm (0–1) < 10 cm (1)	10–15 cm (0–1) < 10 cm (1)	Droughty
Permeability	< 0.5 cm/h	< 0.5 cm/h	< 0.5 cm/h > 15 cm/h	not used > 3 cm/h	Slow permeability Rapid permeability
Depth to bedrock	75–125 cm (0–1) < 75 cm (1)	< 50 cm	75125 cm (0–1) < 75 cm (1)	75–125 cm (0–1) < 75 cm (1)	Depth to rock
Depth to cemented pan	< 100 cm	50–100 cm (0–1) < 50 cm (1)	< 100 cm	< 100 cm	Cemented pan
K factor (USLE)	> 0.32	not used	not used	not used	Erodes easily
Flooding (during growing season)	frequent or very frequent	frequent or very frequent	frequent or very frequent	frequent or very frequent	Flooding
EC (mmhos/cm) (100 cm in depth)	4–8 (0–1) > 8 (1)	4–8 (0–1) > 8 (1)	4–8 (0–1) > 8 (1)	4–8 (0–1) > 8 (1)	Excess salt
SAR	0.5–10 (0–1) > 10 (1)	0.5–10 (0–1) > 10 (1)	0.5–10 (0–1) > 10 (1)	0.5–10 (0–1) > 10 (1)	Excess sodium
Sulphidic materials*	"sulfa" taxonomic great groups	"sulfa" taxonomic great groups	"sulfa" taxonomic great groups	"sulfa" taxonomic great groups	Excess sulphur

* *Sensu Soil Survey Staff (1999). (0–1) = moderate limitation. (1) = Severe limitation. When only one value is written, the limitation has to be considered severe for that restricting factor.*

4.5. Examples of Application in Italy

4.5.1. Implementations

In Italy, reported or known implementations of soil suitability evaluation for irrigation are based on the general principles drawn up by the United States Bureau of Reclamation (USBR, 1951). In particular, the evaluation procedure has been applied in Sicily (Fierotti et al., 1990a, 1990b, 1990c; Fierotti and Dazzi, 1990, 1993), in Sardinia (Arangino et al., 1986a, 1986b), in Apulia (Aru et al., 1979) and in Lombardy (ERSAL, 1990). A table showing the restricting factors and critical limits for determining soil suitability classes for irrigation has been published for Sardinia (Table 4.4). This table has been used, within the *Piano Generale delle Acque della Regione Autonoma della Sardegna* (Water Plan of the Sardinian Region), to evaluate the suitability of soils in the irrigable areas of Sardinia.

Most of the irrigation plans and projects implemented in southern Italy within the activities of the Cassa per il Mezzogiorno (Southern Italy Development Fund) and of the Consorzi di

Table 4.4. Criteria and limits proposed for Sardinia for determining soil suitability classes for irrigation (Arangino et al., 1986a, modified).

	Class 1	Class 2	Class 3	Class 4
Soil				
texture	L; CL; SICL; SCL; SL; well-structured C	from SC to C and S with medium structure	from C to S with faint structure	like class 3
soil depth (cm)	80	80–50	50–35	< 35
coarse surface fragments (%)	0–0.1	0.1–3	3–15	> 15
rock outcrops (%)	0	0–2	2–10	10–20
permeability or internal drainage	medium	slow	very slow or fast	impermeable or very fast
degree of weathering of minerals	little weathered	moderately weathered	weathered	very weathered
salinity	0	0	moderately saline	possible medium to high salinity
carbonates (%)	3–25	25–50	50	50
Topography				
slope (%)	< 10	10–20	20–30	30–40
erosion hazard	slight or moderate	moderate	severe	from severe to extreme
Drainage				
soil and topography	conditions are such that there is no need for drainage	conditions are such that low cost drainage interventions are required	conditions are such that high cost drainage interventions are required	like class 3
drainage class	well-drained soils	from somewhat poorly drained to somewhat excessively drained soils	poorly drained or excessively drained soils	like class 3

Bonifica (land reclamation consortia) envisaged a soil study and the relative soil suitability evaluation for irrigation. These plans and projects are generally made available by the authorities concerned. However, whether they can be freely consulted should be verified case by case.

4.5.2. Small-scale Studies and Evaluations (edited by Rosario Napoli[1])

In the scope of activities carried out by the Istituto Nazionale di Economia Agraria (INEA) for the POM project (1998–2002), with the title "Water Resource Irrigation Use: study on irrigated crops and their profitability", and in particular with Action 1b of Measure 3 of this project, an Information System for Water Resource Management in Agriculture (SIGRIA) has been set up (Bonati et al., 2000). Some of the thematic information produced regards the soil suitability to irrigation data base in the territories of the 66 Reclaim and Irrigation Consortia in Regions of Southern Italy (ex Obiettivo 1).

This evaluation concerned the whole area at present managed by the Consortia, which amounts to 8,130,246 ha, with particular reference to areas equipped with both internal and external irrigation in districts of the Consortia, for the total extent of about 1,600,000 ha. The Research Centre for Agrobiology and Pedology (previously called Istituto Sperimentale per lo Studio e la Difesa del Suolo or ISSDS) treated the aspects of acquisition of pedological data from surveys and past cartography in the whole area under study, as well as the construction and application of evaluation matrices.

Soil Characters and Qualities Useful in Evaluating Irrigation

In accordance with the methodology and purposes envisaged by the Project, a set of features was taken into consideration that are necessary for evaluation for irrigation purposes, referring to both the soil site and the profile, starting with experiences undergone by the United States Bureau of Reclamation (USBR, 1951), successively modified and gauged expressly for the Project by the CRA-ABP.

The following characteristics were taken into consideration with regard to habitat: surface stoniness and rockiness (expressed in percentages) and surface erosion; for inside environments, the fine fraction texture of the topsoil and subsoil, suitable depth,[2] drainage, reaction, total carbonates and salinity (of the topsoil and subsoil).

Adjustment of the Matching Table in Relation to Irrigation Typologies

It has been necessary to introduce the variable "typology of irrigation system" among the features to be considered for the evaluation. Indeed, evaluation of the slope classes assumes a different value depending on the irrigation typology used. In particular, three large, simplified typologies have been distinguished and introduced in the matching table: flooding, submersion and lateral irrigation (furrows), sprinkler (rain-fed) and localized irrigation (drip or pierced

---

[1] CRA-ABP: Centro di Ricerca per l'Agrobiologia e la Pedologia, Firenze, Italy.
[2] Defined as the maximum depth in cm, given by the sum of horizons that may be penetrated by roots. A horizon impenetrable to roots is assumed to present a possibility of rooting below 30%. Rooting possibility is the percentage of soil volume that can be reached by the roots and may be assessed by the profile characteristics and the root distribution (if present) in the soil. Impenetrable horizons or horizons hard to penetrate could be rock, consolidated sediments, densipan, fragipan, duripan and petrocalcic, petrogipsic, petroferric, and placic horizons, or horizons with a permanent water table.

hose), on the basis of data registered by INEA on about 70% of the total area equipped by the consortia, regarding distribution of the principal irrigation techniques used in the equipped areas of the districts in the eight regions involved (Table 4.5).

The final evaluation table (Table 4.6) details the above-mentioned characteristics of the soil site and profile and refers to the irrigation system in relation to the corresponding items and classes in the CRA-ABP (ISSDS) handbook (Gardin et al., 2002) used for acquiring and harmonizing pedological data.

The four evaluation classes are detailed in the suitability cartography with the following correspondence to soils in the cartographic units:

good class: suitable soils
moderate class: fairly suitable soils
poor class: marginally suitable soils
no class: non-suitable soils

Evaluation of Information Bases

The information layers enabling this work come from various sources. Data inherent to pedology are based on available national and regional data sources, as well as on experimental data (detailed cartography produced by research institutes and corporations). On the other hand, with regard to information relative to land use and irrigation techniques concerned, we made use of information deriving from the CASI 3 project, managed directly by INEA (INEA, 2001a). Land suitability to irrigation classification was achieved through evaluation of data from pedological surveys, set according to other physical factors (slope, height); economic and social factors were not considered.

The soils in each cartographical unit are classified according to the four classes presented in the evaluation table (Table 4.6), following the criterion of the most limiting characteristic (the worst condition determines the evaluation class). Each class has been attributed with one or more subclass limitation codes with regard to the principal limiting characteristics responsible for their collocation in the specific class, according to the chart in Table 4.7.

Table 4.5. Distribution of main irrigation techniques in equipped areas included in the Consortia districts of the eight Southern Italian regions.*

Administrative region	Irrigation technique (ha)					
	Flooding	Submersion	Lateral infiltration (furrows)	Sprinkler irrigation	Drip irrigation	Total
Campania	66	65	5,514	7,115	362	13,122
Puglia	506	0	0	37,944	16,920	55,370
Sicily	3,645	200	500	11,407	25,703	41,475
Sardinia	20	4,118	23	39,731	15,637	59,529
Calabria	1,993	730	0	6,902	681	10,306
Abruzzo	15,551	11	0	44,215	89	59,866
Molise	0	0	0	1,615	311	1,925
Basilicata	2,735	0	2,527	13,532	15,230	34,024
Total	24,516	5,124	8,564	162,461	74,932	275,617

* *Extracted data from the Land use Database CASI 3—INEA (INEA, 2001a).*

Table 4.6. Evaluation outline of Soil Suitability to irrigation in the POM 1998–2002 Project "Water resource irrigation use: study on irrigated crops and their profitability" (from the United States Bureau of Reclamation, modified by ISSDS).

	CLASSES OF IRRIGABILITY							
	1-good		2-moderate		3-poor		4-none	
Feature	values	CRA-ABP classes	values	CRA-ABP classes	values	CRA-ABP classes	values	CRA-ABP classes
Topsoil texture USDA Class	FA or FLA or FL or FSA		AL or F or FSA or L or A or AS or FS		SF		S	
Subsoil texture USDA Class	FA or FLA or FL or FSA		AL or F or FSA or L or A or AS or FS		SF		S	
Depth exploitable by roots, cm	≥100	5 or 4	<100 e ≥ 50	3	–	–	<50	1 or 2
Stoniness, %	≤0.1	0 or 1	>0.1 and ≤3	2	> 3% and ≤15	3	>15	4 or 5
Hard crusts,* %	≤0.1	0 or 1	>0.1 and ≤3	2	> 3% and ≤15	3	>15	4 or 5
Rockiness, %	0	0	–	–	> 0 and <2	1	>2	2–5
Drainage	good	3	moderate	4	somewhat excessive or poor	2 or 5	excessive, slow or very slow	1 or 6 or 7
Topsoil reaction, pH	≤9	<9	–		–		>9	9
Subsoil reaction, pH	≤9	<9	–		–		>9	9
Topsoil carbonates, %	1–20	3 or 4 or 5	<1 or 20–40	1 or 2 or 6	>40	7	–	
Subsoil carbonates, %	1–20	3 or 4 or 5	<1 or 20–40	1 or 2 or 6	>40	7	–	
Topsoil salinity, mmhos/cm	<2	0 or 1	–	–	2–4	2	>4	>2
Subsoil salinity, mmhos/cm	<2	0 or 1	–	–	2–4	2	>4	>2
Surface erosion	absent	0	moderately diffuse	1	moderately canalized	2 or 4	strong	3 or ≥5
Localized irrigation slope, %	0–13	1–2	13–21	3	21–35	4	>35	5–6
Sprinkler irrigation slope, %	≤5		5–13	–	–	–	>13	
Irrigation slope for flooding, submersion and lateral infiltration, %	1		3				>3	

* *The same legend of stoniness class is used to estimate covered area percentages.*

Table 4.7. Subclass codes for the main limitations.

CODE*	LIMITATION
PE	slope
D	drainage
PR	suitable depth
S	salinity
SK	surface stoniness
CR	calcrete
R	rockiness
T	texture
E	erosion
C	carbonates
N-n	no limitation
ND	not determined**

* Composite symbols may be used to indicate a limitation in surface (topsoil) or depth horizons (subsoil), for example, cs= carbonates in the subsoil (where such a determination can be made).
** Item retrieved from the bibliography without the possibility of applying the matching table and minimal information even for an expert's assessment.

Cartography of soil suitability for irrigation with respect to any typology of irrigation system is quoted in "Atlante dell'Irrigazione nelle Regioni Meridionali" (INEA, 2001b).

Application of Evaluation Methodologies to Different Data Sources and Determination of Evaluation Confidence

On the basis of the elements below, a degree of trust in the information added to the data base and used for evaluation was assigned to characters necessary for evaluation for irrigation purposes and for use capacity:

some data came from various surveys, different in scale and purpose;
some data was expressly surveyed for the purposes required and had varying accuracy levels;
some characteristics described in past soil chart legends could have been expressed with different classes (thus attributed afresh with a probable loss of information) or could have been missing in that form (in this case, it was accordingly necessary to translate, derive or re-interpret the information missing from other features);
some features could be missing entirely (it was necessary to assess whether to omit the characteristic or arrange for a targeted survey for acquisition of such a characteristic).

Each feature therefore has a degree of trust expressed qualitatively by three classes—high, medium and low—and their application, together with the evaluation, has given general results summarized in Tables 4.8 and 4.9. The criteria used to attribute confidence classes were:

Table 4.8. Pedological information confidence classes (% of the total consortia surface area).

Region	Abruzzo	Basilicata	Calabria	Campani	Molise	Puglia	Sardinia	Sicily
Class								
High	0.04	6.61	0.002	0.00	1.75	0.00	33.10	0.00
Medium	0.07	0.95	92.43	10.48	4.52	81.95	5.59	0.00
Low	73.99	48.76	0.034	68.46	40.69	0.00	0.00	66.32
Total	74.10	56.32	92.47	78.94	46.96	81.95	38.69	66.32
Na *	25.90	43.68	7.53	21.06	53.04	18.05	61.31	33.68

* Areas not assessed for exclusion or for lack of sufficient data.

Table 4.9. Summary of assessment results of soil attitude towards irrigation in Consortilium areas in Southern Italian regions. Source: CRA-ABP (ISSDS) up-dated to 2002

Objective regions 1	Suitability 1		Suitability 2		Suitability 3	
	ha	%	ha	%	ha	%
Suitable	1,890,655	23.3	3,774,544	46.4	4,148,215	51.0
Not suitable	2,957,232	36.3	1,073,343	13.2	699,672	8.6
Not assessed	3,282,359	40.4	3,282,359	40.4	3,282,359	40.4
Overall total	8,130,246	100.0	8,130,246	100.0	8,130,246	100.0

Suitability legend for different irrigation techniques: Suitability 1 = flooding, submersion and lateral infiltration; Suitability 2 = sprinkler; Suitability 3 = localized.

1. geographical criterion: the quantity and the distribution of observations made over the land are satisfactory, middling or poor in relation to the scale and density objective of information required by the project;
2. the criterion of quality of the punctual data: the presence (total, partial or absent) of laboratory-analysed measurements carried out with standard suitable methodologies, and precise soil profile descriptions with up-to-date handbooks and pedogenetic and diagnostic horizon reconnaissance.

References

ARANGINO, F., ARU, A., BALDACCINI, P., VACCA, S. 1986a. I suoli delle aree irrigabili della Sardegna. Piano Generale delle Acque, Ente Autonomo del Flumendosa, Assessorato della Programmazione, Bilancio ed Assetto del Territorio, Regione Autonoma della Sardegna, 133 p.

ARANGINO, F., ARU, A., BALDACCINI, P., VACCA, S. 1986b. Carta dei suoli delle aree irrigabili della Sardegna. Piano Generale delle Acque, Ente Autonomo del Flumendosa, Assessorato della Programmazione, Bilancio ed Assetto del Territorio, Regione Autonoma della Sardegna, 25 sheets 1:100,000, 1 legend.

ARU, A., BALDACCINI, P., FIEROTTI, G. 1979. Studi pedologici per scopi irrigui in tre aree campione della Puglia. Progetto speciale n. 14, Cassa per il Mezzogiorno, Rome, 119 p.

BONATI, G., FAIS, A., NINO, P., RAIMONDI, G. 2000. *Il SIGRIA-Sistema Informativo per la Gestione delle Risorse Idriche in Agricoltura.* MondoGIS no. 23, Rome.

BOOKER AGRICULTURE INTERNATIONAL LIMITED. 1979. Sugarcane as an irrigated crop. In: Land evaluation criteria for irrigation, FAO World Soil Resources Report n. 50, Rome, 103–113.

DE DATTA, S.K. 1981. *Principles and Practices of Rice Production.* International Rice Research Institute, Los Baños, Philippines, 618 p.

EARLY, A.C., SINGH, V.P., TABBAL, D.F., WICKHAM, T.H. 1979. Land evaluation criteria for irrigated lowland rice. In: Land evaluation criteria for irrigation, FAO World Soil Resources Report n. 50, Rome, 114–144.

ERSAL. 1990. *I suoli dell'Isola bergamasca (a cura di Bonalumi G., Roncalli W., Vitali G., Assi I.).* Progetto Carta Pedologica, Milano, 296, 1 map 1:25,000, 4 maps 1:50,000.

FAO. 1979a. Land evaluation criteria for irrigation. FAO World Soil Resources Report No. 50, Rome, 219 p.

FAO. 1979b. Soil survey investigations for irrigation. FAO Soils Bulletin No. 42, Rome, 188 pp.

FAO. 1985. Guidelines: land evaluation for irrigated agriculture. FAO Soils Bulletin No. 55, Rome, 231 p.

FIEROTTI, G., DAZZI, C. 1990. Caratteristiche ambientali e possibilità irrigue nel perimetro consortile del Salso inferiore. In: Il Consorzio di Bonifica del Salso inferiore, 1939-1989, Ed. APE, pp. 95–132.

FIEROTTI, G., DAZZI, C. 1993. Il quadro pedologico del comprensorio del Consorzio di Bonifica dell'Alto e Medio Belice ed esempi di valutazione dei suoli all'irrigazione. Ricerca pilota a valenza interregionale sulla riconversione colturale ed ammodernamento delle strutture nel comprensorio dell'Alto e Medio Belice. In: Il Consorzio di Bonifica dell'Alto e Medio Belice, Ed. ARBOR, Palermo, pp. 109–178.

FIEROTTI, G., SARNO, R., CIRRITO, V., DAZZI, C., RAIMONDI, S. 1990a. Utilizzazione a scopi irrigui delle acque reflue depurate dei comuni di Caltanissetta e S. Cataldo: studio pedo-agronomico di alcune aree irrigabili in territorio di Caltanissetta. In: Il Consorzio di Bonifica del Salso inferiore, 1939–1989, Ed. APE, pp. 163–198.

FIEROTTI, G., DAZZI, C., LOMBARDO, V., OLIVERI, G., RAIMONDI, S. 1990b. Studio pedo-agronomico dei suoli potenzialmente irrigabili con le acque del serbatoio "Laura". In: Il Consorzio di Bonifica del Salso inferiore, 1939-1989, Ed. APE, pp. 199–250.

FIEROTTI, G., DAZZI, C., LOMBARDO, V., OLIVERI, G., RAIMONDI, S. 1990c. Studio pedo-agronomico dei suoli potenzialmente irrigabili con le acque del serbatoio "Gibbesi". In: Il Consorzio di Bonifica del Salso inferiore, 1939–1989, Ed. APE, pp. 251–298.

GARDIN, L., COSTANTINI, E.A.C., NAPOLI, R. 2002. Guida alla descrizione dei suoli in campagna e alla definizione delle loro qualità. http://www.soilmaps.it/ita/downloads.html

INEA. 2001a. *Il Progetto CASI: Guida tecnica e presentazione dei risultati.* Collana "rapporto Irrigazione", Ed. INEA, MiPAF, Min. Infrastrutture e Trasporti e UE, Fondo Europeo di Sviluppo Regionale, Rome.

INEA. 2001b. *Atlante dell'Irrigazione nelle Regioni Meridionali—Capitolo 5 Elementi Fisici Pedologia: Irrigabilità dei suoli regioni Obiettivo 1.* Napoli, R. (Ed.) INEA, MiPAF, Min. Infrastrutture e Trasporti e UE, Fondo Europeo di Sviluppo Regionale, Rome.

SOIL SURVEY STAFF. 1999. *Soil Taxonomy. A Basic System of Soil Classification for Making and Interpreting Soil Surveys.* 2nd ed. Agriculture Handbook No. 436, USDA-NRCS. US Govt. Printing Office, Washington, DC.

SOUTHARD, S.B. 2001. Improved irrigation interpretations using a National Soils Information System. Extended Abstracts of the 7th International Meeting on Soils with Mediterranean Type of Climate "Preserving the Mediterranean Soils in the Third Millennium", Valenzano (Ba), Italy, 23–28 September 2001, pp. 121–122.

SYS, C., VAN RANST, E., DEBAVEYE, J. 1991. *Land Evaluation. Part II. Methods in Land Evaluation.* Agricultural publications no. 7, General Administration for Development Cooperation, Brussels, Belgium, 247 p.

SYS, C., VAN RANST, E., DEBAVEYE, J., BEERNAERT, F. 1993. *Land Evaluation. Part III. Crop Requirements.* Agricultural publications no. 7, General Administration for Development Cooperation, Brussels, Belgium, 199 p.

USBR. 1951. *Bureau of Reclamation Manual. Vol. V. Irrigated Land Use. Part 2. Land Classification.* Bureau of Reclamation, Dept. of Interior, Denver Federal Center, Denver, Colorado, USA.

5. Soil Erodibility Assessment for Applications at Watershed Scale

Lorenzo Borselli,[1] Paola Cassi[1] and Pilar Salvador Sanchis[1]

5.1. Introduction

The term "soil erodibility" indicates the aptitude of a soil, based on its properties, to be eroded by the following processes and exogenous agents: rainfall, runoff, mass movements and wind.

The concept of erodibility gained in importance during the last 50 years in the field of soil erosion modelling and applications of soil conservation. However, erodibility is a concept borrowed from geomorphological literature that was developed and adopted up to the beginning of the 20th century.

In this context, the concept of erodibility was often used to give a qualitative assessment of the effectiveness of various forms of erosion caused by exogenous agents such as water, ice and wind (Davis, 1909). It was used mainly by geologists and geographers for a long time and related to the processes most effective in the characterization of landform dynamics (Taylor and Eggleton, 2001; Turkington et al., 2005). It is easy to associate varying rates of erodibility to compactness of igneous and metamorphic rock masses compared to marls and clay shale, deeply eroded by gullies. Indeed, the effectiveness of different processes and geomorphic agents is directly linked to the characteristics of the bedrock in its state of weathering.

In the late 20th century, the great development of studies on soil erosion and experimentation in the field of soil conservation led to an extension of the erodibility concept also in soil science. The term "soil erodibility" was for the first time used by Middleton (1930).

---

[1]CNR-IRPI—Istituto di Ricerca per la Protezione Idrogeologica, Unità Operativa di Supporto, Pedologia Applicata, Firenze, Italy.

An important development in the concept of soil erodibility came about from the study and implementation of the extensive modelling of USLE (Universal Soil Loss Equation) (Wischmeier and Smith, 1978; Lafflen and Moldenhauer, 2003) and from other relevant works (Foster et al., 1981, 2002; Renard et al., 1997; Kinnell and Risse, 1998), which are implementations of the same basic conceptual modelling, which we name "USLE-TYPE".

The original qualitative concept of erodibility used by geologists and geomorphologists assumed a quantitative definition in soil science, through the parameterization of one or more soil characteristics (Bryan, 2000). In USLE-TYPE models, the erodibility mainly assumed an agronomic and soil conservation value. In a broad overview of the theory and modelling of the processes of soil erosion, the term "erodibility" has a major importance in the fields of geomorphology, hydraulics, agriculture and hydrology (Bryan, 2000).

Soil erodibility is not a physical quantity or measurable property like length, mass, or strength. It can be expressed qualitatively (high, medium, low) or quantitatively (a magnitude expressed numerically). In the first case, we proceed through comparison of the results of field or laboratory experiments on different soil exposed to the same erosion agent. In the second case, the soil erodibility can be expressed as an index, number or parameter defined as a coefficient of proportionality between a force, pressure, energy, linked to one or more specific processes or erosive agents. In this case, the numerical value that expresses the erodibility is closely related to a model to estimate the rate of soil erosion.

The use of a numerical parameter to express the soil erodibility is useful, especially in the modelling, but it is also limited because it is strictly linked to the model chosen to estimate erosion. Hence, a universal erodibility index does not exist.

There are various factors influencing soil erodibility. In the case of water erosion processes, soil erodibility is linked to the following:

1. The conditions of the state of aggregation and stability of aggregates during the first wetting (resistance to slaking)
2. Dispersivity of the clay fraction (chemical dispersion)
3. Resistance to surface shear stress (resistance to detachment operated by the rain drops and surface runoff)
4. The infiltration capacity and production of surface runoff
5. The tendency to formation of surface crusts (sealing)

According to Torri and Poesen (1997), the soil erodibility grows with increase in the characteristics listed in points 2 and 5 and may decrease with the characteristics listed in points 1, 3 and 4.

The soil response to the processes listed above is not constant but varies over time. For example, the initial soil moisture at the beginning of the rainfall has an important influence on the hydrological response and on the possibility of a process of slaking (disintegration and weakening of dry soil aggregates due to the matrix tensions developed with the first wetting) active in the early stages of the event. The components of erodibility are therefore highly dynamic over time (Torri and Borselli, 2000). For this reason, the time interval in which the erodibility is evaluated is crucial. Some processes described above operate on different time scales. Soil erodibility evaluated yearly has a different degree of complexity from erodibility evaluated on the basis of a single rainfall event.

The aims of this chapter are to provide the reader with a choice of criteria for assessing soil erodibility in a watershed context, to give an overview of the soil properties that mostly affect erodibility, and at the same time to provide some general guidelines on the various existing modelling approaches.

5.2. Models and Criteria for Soil Erodibility Assessment

The quantitative assessment of erodibility is necessarily linked to a modelling where this parameter is used to assess, within a predefined space-timescale, a potential rate of soil erosion through diffuse or concentrated erosion by water.

More than 80% of current soil erosion models (Doe et al., 1999; Merritt et al., 2003) use an explicit parameterization of erodibility through a coefficient of erodibility, sometimes with more than one (Foster et al., 1995; Misra and Rose, 1996; Rose et al., 1997; Laflen et al., 1997; Morgan et al., 1998).

The diversity of modelling techniques used, space-time steps, type of models (physical, conceptual, distributed, event-based, annual) makes uncertain a single soil erodibility characterization. Some models of soil erosion estimation cannot be applied at watershed scale. In this chapter, we will analyse the erodibility assessment criteria for models that have a higher number of applications, as for USLE-TYPE models.

5.2.1. The K Factor in USLE- TYPE Models

In USLE-TYPE models, the soil erodibility is derived from the study of the combined effect of rainfall and runoff.

According Wischmeier and Smith (1978) and Lafflen and Moldenhauer (2003), erodibility K factor is the rate of soil loss for unitary rainfall erosion energy as measured on a unit plot. The unit plot is 22.18 m long, has a 9% slope, and is continuously maintained in a clean fallow condition with tillage performed upslope and downslope.

In the international system (SI), K is expressed in Mg ha/h/MJ/mm (Foster et al., 1981) and the K values represent an integrated average of annual values of soil and soil profile reaction to a large number of erosion and hydrological processes. The data are collected over a long term (20 years).

The K values are identified using Wischmeier's nomograph (Wischmeier, 1971), then modified by Foster et al. (1981) and Rosewell (1993), in order to express K in SI units (Mg ha/h/MJ/mm). The nomograph in (SI units) is based on the following formula:

$$K=2.77\times10^{-7}(12-OM)M^{1.14}+4.28\times10^{-3}(s-2)+3.29\times10^{-3}(p-3) \qquad (1)$$

where

$$M = T_f(100\text{-}C) \qquad (2)$$

and T_f is the soil fraction in percentage by weight between 0.002 and 0.1 mm (silt + very fine sand), OM is the organic matter (%), C is the clay fraction (%) (< 0.002 mm), p is a code indicating the class of permeability (Table 5.1), and s is a code for structure size, type and grade (Table 5.2) (Soil Survey Staff, USDA, 1951). The formula is applicable for a silt content not exceeding 70%. Note that the use of eq. 1 necessarily entails the use of classification and textural subdivisions into classes according to the USDA classification.

Table 5.1. Soil permeability classes (Soil Survey Staff, USDA, 1951).

Soil permeability classes	Permeability rates for the entire soil profile
1	Rapid, > 130 mm/h
2	Moderately rapid, 60–130 mm/h
3	Moderate, 20–60 mm/h
4	Moderately slow, 5–20 mm/h
5	Slow, 1–5 mm/h
6	Very slow, < 1 mm/h

Table 5.2. Soil structure classes (Soil Survey Staff, USDA, 1951).

• Size	• Type	• Class
• Fine	• Platy, prismatic, • angular blocky, • subangular blocky,	• 4
• Medium	• Platy, prismatic, • angular blocky, • subangular blocky,	• 4
• Coarse	• Platy, prismatic, • angular blocky, • subangular blocky	• 4
• Very Coarse	• Platy, prismatic, • angular blocky, • subangular blocky	• 4
• Fine	• Granular	• 1
• Medium	• Granular	• 2
• Coarse	• Granular	• 3
• Very coarse	• Granular	• 3

The soil permeability class value can be predicted through a qualitative assessment following the original criteria of the USDA manual (Soil Survey Staff, USDA, 1951: 168-170), here represented in Table 5.1. Note that the classes of permeability in subsequent editions of the USDA manuals are changed. However, the nomograph was developed using as reference the 1951 edition.

To use eq. 1, we need to know the fraction between 0.002 and 0.1 mm. If the fraction less than 0.1 mm is not measured directly, it must be obtained through interpolation of the size distribution curve (Nemes et al., 1999). To overcome this element of uncertainty, some authors (Loch and Rosewell, 1992; Rosewell, 1993; Loch et al., 1998), suggest the following relationship, based on several tests on Australian soils for the estimation of M in eq. 1, be used to extend this relationship to all soils as well as the well-aggregated soils and soils with high clay content:

$$M \, (\%) < 0.125 \text{ mm} \qquad (3)$$

The authors also suggest that, in addition to eq. 3 for the assessment of M, the following modification be considered to estimate the *K* factor:

$$K_m = \frac{K}{d_s - 1} \tag{4}$$

where K_m is the modified *K* values, d_s is the wet density of sediments expressed in Mg/m³, which is also assessed by the following relationship (Loch and Rosewell, 1992; Loch et al., 1998).

$$d_s = 1.462 + 0.048(1.03259^x) \tag{5}$$

where *x* is the percentage of soil fraction greater than 0.02 mm.

According to Loch and Rosewell (1992) and Loch et al. (1998), eqs. 3 and 4 should be used as an alternative to the classical methodology (eqs. 1, 2) and always in the case of soils with high clay content, or strong aggregation (e.g., vertisols or soils with high structural stability).

Equations 1 and 2 have a broad range of application, especially with the proposed changes (eqs. 3, 4, 5), but there are also other pedotransfer functions and pedoalgorithms for *K* assessment.

5.2.2. Other Pedotransfer Functions and Pedoalgorithms for Calculating K Factor (Applicable to Models USLE-TYPE)

The Revised Universal Soil Loss Equation model (RUSLE, Renard et al., 1997) uses the *nomograph* expressed by eq. 1. To get a more general application, however, other authors have proposed some changes in the calculation of *K*. The RUSLE model manual (Renard et al., 1997) contains an alternative equation for estimating *K* based on a global set of experimental data (255 soils). Only soils with skeleton less than 10% were considered. The value of *K* expressed in international units (Mg ha/h/MJ/mm) is estimated using the following equation:

$$K = 0.0034 + 0.0405 e^{\left[-0.5 \left(\frac{\log_{10} dg + 1.659}{0.7101} \right)^2 \right]} \tag{6}$$

where *dg* is the geometric mean particle diameter (mm), calculated as indicated by Shirazi and Boersma (1984):

$$dg = e^{[0.01 \sum f_i \ln m_i]} \tag{7}$$

where f_i is the particle size fraction as a percentage and m_i is their average diameter (mm).

This equation is presented in RUSLE for general use as an alternative to eq. 1. The eq. 6 is useful when the data for estimating *K* factor by the Wischmeier-Foster approach are not available. Note that the evaluation of *K* factor obtained by eq. 6 does not take into account the possible effects of organic matter content. Furthermore, the high value of the coefficient of determination obtained by the authors for eq. 6 is due to the fact that eq. 6 is derived by means of a non-linear regression between the average values of K for different values dg. In practice, considering the actual dispersion of original data set, we obtain a real value of the coefficient of determination that is extremely low.

Trying to develop a more integrated and comprehensive approach, Torri et al. (1997, 2002) proposed an alternative method, which requires also the organic matter content, derived from a large data set on 240 global soils with a skeleton fraction less than 10%.

The alternative wording proposed for the calculation of the K factor, expressed in international units (Mg ha/h/MJ/mm), is as follows:

$$K = 0.0293\left(0.65 - Dg + 0.24Dg^2\right)e^{\left\{-0.021\frac{OM}{C} - 0.00037\left(\frac{OM}{C}\right)^2 - 4.02C + 1.72C^2\right\}} \tag{8}$$

where C is the fraction of total clay content, OM is organic matter as a percentage, and D_g is logarithm of the geometric mean of the particle size distribution, which can be directly calculated according to Shirazi et al. (1988):

$$Dg = \sum f_i \; \log_{10}\left(\sqrt{d_i d_{i-1}}\right) \tag{9}$$

where f_i is the mass fraction of particles in the class with range of diameters d_i and d_{i-1} (in mm). If only three main textural components of soil (sand, loam and clay) are available, the argument of square root in eq. 9 can be calculated as in Table 5.3.

Table 5.3. Calculation of parameter Dg in the case of three basic textural components.

Textural component	d_i (mm)	d_{i-1} (mm)	$d_i d_{i-1}$	$\log_{10}\left(\sqrt{d_i d_{i-1}}\right)$
clay	0.002	0.00005*	0.0000001	−3.5
loam	0.05	0.002	0.0001	−2
sand	2	0.05	0.1	−0.5

Conventionally for the lower limit of clay particles is equal to 0.00005 (mm) (Shirazi et al., 1988).

When only the data of the textural classes as a percentage (sand (S), loam (L) and clay (C)) are available, D_g can be calculated by the following simplified formula:

$$Dg = \frac{-3.5C - 2.0L - 0.5S}{100} \tag{10}$$

or directly estimated by the diagram in Fig. 5.1.

A comparative study of pedotransfer functions (6, 8) conducted over a large database of 190 Italian soil profiles (Francaviglia et al., 2003) indicated that these two pedotransfer functions are almost equivalent and have the same areas of uncertainty in the estimation of K.

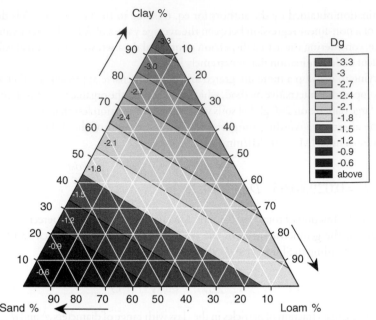

Fig. 5.1. Ternary diagram for the estimation of the parameter D_g (logarithm to the base 10 of the geometric mean of the particle size distribution) used in eq. 8.

The same study also involved a test of the procedure based on a pedoalgorithm based on fuzzy logic implemented in software FUZKBAS (Torri et al., 1997; http://www.fi.cnr.it/irpi/software.htm). The latter procedure, as revealed by Francaviglia et al. (2003), shows more interesting results and the best way to discriminate and differentiate the K values in the soil database analysed.

In the software FUZKBAS (Torri et al., 1997), the K value is evaluated with the same input parameters of pedotransfer function in eq. 8 and the output is a distribution of values of K around a reference value, hence not a single value of K. K value is given as a fuzzy variable for a given soil type (Torri et al., 1997). The analysis with fuzzy techniques of the global database values of K is certainly an improvement over a traditional statistical analysis, but does not provide a definitive result.

Salvador Sanchis et al. (2008) extended the analysis of soil erodibility global databases to soils with rock content (skeleton + rockiness) > 10% in volume, increasing the dataset to a total of 334 soils.

Salvador Sanchis et al. (2008), examining the new enlarged dataset, noted that the climate and rock content have the greatest influence on the K factor. Figure 5.2 shows the main groups into which the database is subdivided and highlights the difference of various populations based on the climate classification: cool continental climate (Df, Cf) or warm climate/tropical/Mediterranean (A, Cs,) following the Koppen-Geiger classification.

Note that this diversification based on climate is present in both soils with rock content < 10% and soils with rock content > 10%.

The major evidence is that the soils developed in continental climate have an erodibility that is double that of the soils developed in warm climate.

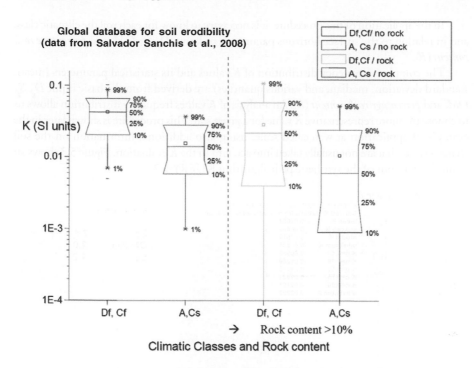

Fig. 5.2. Box and Wiskers plot of soil erodibility values distributions measured for various groups of climatic class following Koppen-Geiger climate classification and rock content %: left < 10%, right > 10%.

A new technique and a specific software are developed to generalize the non-deterministic method for estimating K. The new software KUERY1.2 (http://www.irpi.fi.cnr.it/software.htm) (2005) makes it possible to query the database and obtain a statistical distribution of values of K, corresponding to given values of soil properties. The input data of KUERY1.2 are climate classification, percentage of total rock content (R), percentage of organic matter content, logarithm of geometric mean of the particle size distribution (D_G) (Shirazi et al., 1988), the logarithm of geometric standard deviation of geometric mean of the particle size distribution (S_g).

In the new procedure, the S_g parameter which is the logarithm of geometric standard deviation, already defined by Shirazi and Boersma (1984) and Shirazi et al. (1988), is calculated with the following equation:

$$S_g = \sqrt{\sum f_i \left[\log_{10}\left(\sqrt{d_i d_{i-1}}\right) - D_G\right]^2} \tag{11}$$

If only the data of the textural classes in percentage (sand (S), loam (L), clay (C)) are available and D_G is calculated by eqs. 9–10, S_g can be calculated using the following equation:

$$Sg = \sqrt{\frac{\left[C(-3.5 - Dg)^2 + L(-2.0 - Dg)^2 + S(-0.5 - Dg)^2\right]}{100}} \tag{12}$$

In the application of this procedure, it is necessary to know for each soil the climatic class, and in relation to the surface horizons parameters to know D_g, S_g, *OM*, and *percentage rock content* (*R*).

The complete frequency distribution of *K* values and its statistical parameters (mean, standard deviation, median, and various quantiles) are derived from the basic input D_g, S_g, *OM*, and *percentage rock content* (*R*). An analysis of *K* values frequency distribution allows us to choose the more representative *K* value for a given soil. This procedure is shown later in the example of application at watershed scale, using also additional information on the soil characteristics that are not usually taken into account in the *K* evaluation. Figure 5.3 shows an example of output of the synthetic pedoalgorithm KUERY 1.2.

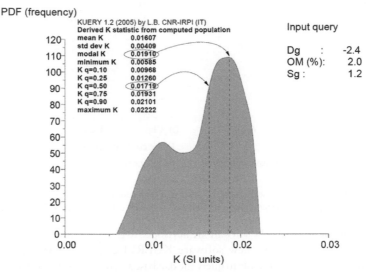

Fig. 5.3. Distribution of possible values of *K* for a soil located in the warm Mediterranean climate group. The probability density distribution (PDF) refers to a soil with rock content less than 10%, $D_g = -2.4$; OM = 2%, $S_g = 1.2$ in a warm and/or Mediterranean climate.

5.2.3 Seasonal Variation of Soil Erodibility

The seasonal variability of *K* value within the year is also considered in USLE and RUSLE models. This variability can be seen on a monthly basis (Rhernad et al., 1997), calculating up to 12 different values depending on the soil moisture content and freezing and thawing cycles.

Usually a corrective coefficient to the mean annual *K* value is calculated by the following equation:

$$K_{mr} = \frac{K_m}{K} \tag{13}$$

where K_{mr} is the ratio between the monthly erodibility K_m and the mean annual *K* value.

The change in erodibility is taken into account using seasonal correction factors, which are strongly influenced by the thermo-pluviometric behaviour of the weather station considered

(Zanchi, 1983; Mutchler et al., 1983). Zanchi (1983, 1988), examining a data set of 169 rainfall events between December 1977 and August 1983, found a very different trend for each month of the year. The best interpolation of experimental data was obtained with the following equation:

$$K_{mr} = 1 + 0.9\cos(0.532t) \tag{14}$$

where t is the month, which is expressed by a number (January = 0.5, February = 1.5, etc.), and the argument of cosine is expressed in radians.

A new, more general approach has been developed from Salvador Sanchis et al. (2008) using a dataset from three locations: Minnesota (USA), Mississippi (USA), Tuscany (Italy) and four different soils.

The new method performs calculation of an index based on the average cumulative temperature of the month starting with the lowest average temperature.

$$T_U = \frac{1}{T_{max} - T_{min}} \sum_{M_0}^{M} T_J \ , T_J > 0 \tag{15}$$

where:

T_U is a dimensionless index related to the average cumulative air temperature;

T_{max} (°C) is the highest mean monthly temperature;

T_{min} (°C) is the lowest mean monthly temperature;

M_0 is the numerical integer index of the month with the lowest value of average air temperature;

M is the numerical integer index of the month for which T_U is calculated (it varies from 1 to 12);

J is a numerical integer index of the month considered, which varies between M_0 and M ;

T_J (°C) is the average temperature of the month J.

Using this definition, K_{mr} can be calculated using the following equation:

$$K_{mr} = 4.75\cos[0.373\ln(3.223T_U)] + \frac{24.1}{14.0 - T_U} - 4.83 \tag{16}$$

Please note that if calculated values of K_{mr} are $< 10^{-3}$, K_{mr} must be equal to 0.

5.2.4. General Consideration on Model for Soil Erodibility Assessment

The uncertainty is evident in all the K factor assessment processes, because there are model and parametric uncertainties at the same time. The last type of uncertainty in particular is mainly due to the level of spatial-temporal representativeness of the soil parameters assumed locally for K evaluation.

The choice of the model used for K estimation has key importance. The application of a single pedotransfer function gives an incomplete evaluation of K values, because there is no equation that can interpolate trends, as the global database shows. Pedoalgorithms are valid alternatives because, using basic soil parameters, they estimate the complete frequency distribution of possible *K* values. The use of pedoalgorithm KUERY 1.2 will be shown in the examples of application. The next section will describe the additional soil features that could be a complement and help in K assessment.

5.3. Assessment of Soil Erodibility: Integrations from Erosion Features and Soil Additional Characteristics

All the pedotransfer functions previously presented allow an immediate estimate of soil erodibility for modelling USLE-TYPE on an annual basis, and other developments make it possible to achieve a dynamic assessment of seasonal average values for *K*.

However, the pedotransfer function for estimating the *K* factor from the basic characteristics is unable to consider many other important characteristics that have strong influence on soil erodibility.

At field scale, we can observe processes of degradation of the soil characteristics that are directly related to erodibility: sealing, crusting, dispersivity. All these phenomena are always linked to a low soil structural stability (Valentin and Bresson, 1992) and can have a strong impact on soil erodibility (Robinson and Phillps, 2001). The *K* values global database, used by several authors to derive different pedotransfer functions, refers to very different soil type, geographic area and climate and the information required for *K* assessment is limited. For example, it does not require information on the potential structural stability and/or dispersivity. Nevertheless, these properties are readily observable in field: qualitative characteristics of soil erosion and re-deposition, and texture and thickness of the depositional structure.

The dispersion of the soil particles is closely related to the chemical characteristics of the circulating solutions and clay mineral constituents (Sumner, 1993; Agassi et al., 1994). There are many tests to assess the dispersion potential (dispersivity) (Vacher et al., 2004). One of the most rapid and well-known is the "crumb test" (Emerson, 1967).

The test can assess the degree of dispersivity of some soil aggregates placed on a Petri-plate with deionized water. A visual analysis of the behaviour of clods after wetting with deionized water is sufficient to give an assessment of the potential of soil dispersivity.

The crumb test can be performed as follow (Fig. 5.4, DNR, 2005):

- Place five air-dried soil crumbs, each about 5 mm diameter, in a squat beaker or glass jar containing 100 ml deionized water.
- Add the soil to the water, not the water to the soil.

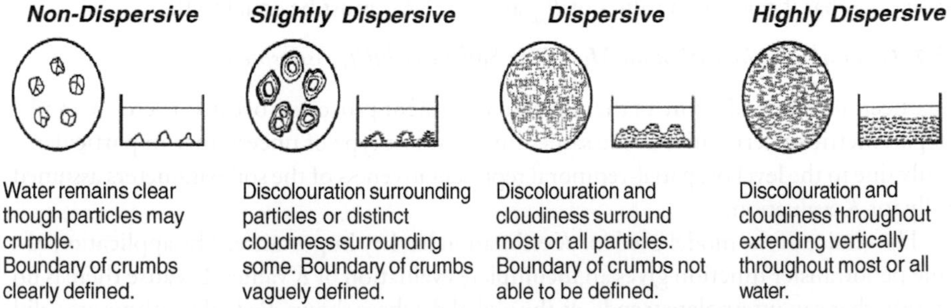

Non-Dispersive	**Slightly Dispersive**	**Dispersive**	**Highly Dispersive**
Water remains clear though particles may crumble. Boundary of crumbs clearly defined.	Discolouration surrounding particles or distinct cloudiness surrounding some. Boundary of crumbs vaguely defined.	Discolouration and cloudiness surround most or all particles. Boundary of crumbs not able to be defined.	Discolouration and cloudiness throughout extending vertically throughout most or all water.

Fig. 5.4. Crumb test: visual appraisal and identification of class of soil dispersion potential (from DNR, 2005).

- Let stand without shaking for at least 1 h.
- Note the turbidity of the solution. If the solution is clear, then little trouble from dispersion can be expected.
- Examine the interface between the soil sample and the water. If there is a cloud at this interface, it is indicative of a dispersive soil.

A criterion to assess the soil dispersion potential is the concentration of the major cations of soluble salts in the soil (Na, K, Ca and Mg), and derived or associated parameters: ESP (exchangeable sodium percentage), SAR (Sodium adsorption ratio), EC (electrical conductivity) (Sherard et al., 1976; USDA-SCS, 1991; Sumner, 1993; Bell and Walke, 2000). One of the most interesting parameters related to the soil dispersion potential is the ESP. Many authors suggest conditions of soil dispersion with ESP values higher than 4% (Sherard et al., 1976; USDA-SCS, 1991; Sumner, 1993; Bell and Walker, 2000), and sometimes with values even lower (Barre et al., 2004; Sotelo, 2005). However, the value of ESP is not sufficient by itself to express the soil dispersion potential (Sumner, 1993; Bell and Walker 2000; Vacher et al., 2004). A promising method to identify soils with dispersion potential is one proposed by Barre et al. (2004), which uses ESP and the *Loveday-Pyle* index (LPI), derived from the dispersion test of Emerson (2002) (see Fig. 5.5). This diagram shows five classes with decreasing dispersion from 1 to 5. Classes 4 and 5 do not have dispersive behaviour. In class 4, in cases of high ESP values, the high content of carbonates prevents dispersion.

Soil aggregate stability is also checked by analysing the response to mechanical and/or electrochemical stress. This method is widely used to determine the potential ability of soil to change its state of aggregation (structural collapse) and the consequent promote mobility of fine sediments, which is strictly related to soil erodibility (Le Bissonais, 1996; Le Bissonais and Le, 1996; Le Bissonnais and Arrouays, 1997; Barthès and Roose, 2002; Legout et al., 2005). Several tests have been proposed for soil aggregate stability assessment (Imeson and Vis, 1984;

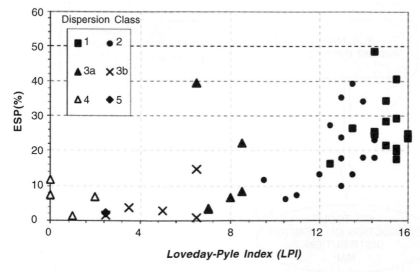

Fig. 5.5. Loveday-Pyle Index (LPI) vs ESP indicating the various classes of dispersion (from Barre et al., 2004).

Kemper and Rosenau, 1986; Le Bissonais, 1996). In all these works, the authors clearly highlight the importance of the stability test to estimate the potential aptitude of soil to be eroded. An example of how the various indices of aggregates stability are linked to soil erodibility, as derived from systematic observations in the field, can be found in Barthès and Roose (2002).

If additional features are known, for crusting potential, dispersivity, and low structural stability, a reliable *K* factor can be better estimated.

These soil characteristics are extremely useful to evaluate erodibility at watershed scale from basic information derived from a soil database integrated with field observations. The process of assessing the K value following the pedoalgorithm KUERY 1.2 can be used to define an expected frequency distribution of K values, obtained through a statistical analysis of *K* values calculated for soils with similar characteristics. For a practical application, however, it is necessary to use a single value of K. With additional information on the aggregate structural stability or dispersion susceptibility, the *K* value to be used within the expected frequency distribution of K values can be selected (Fig. 5.3). The choice can be made with the knowledge of additional properties. For example, in case of clear conditions of structural instability or dispersion, the choice can be addressed more to the 75th or 90th percentile than to the median or modal values. The choice of the highest percentiles of K values distribution takes into account the possibility of high erodibility values in the presence of an additional factor that indicate a major aptitude to erosion (lower aggregate stability, high dispersion potential). An example of this procedure is provided in the next section.

5.4. Soil Erodibility Maps: an Application

The elements for the production of soil erodibility maps at watershed scale are illustrated in Fig. 5.7. It is necessary to produce a reliable soil erodibility map, which requires a detailed soil

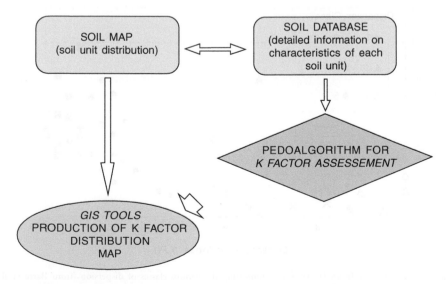

Fig. 5.6. Elements for the production of maps of soil erodibility at watershed scale.

map (at least 1:25,000) and databases with all the relevant properties of each soil unit mapped.

Two other key elements are related to elaborate a soil erodibility map: a pedoalgorithm for estimating the *K* factor for each soil map unit and GIS tools for spatial distribution of K values within the basin according to existing soil units (Fig. 5.6).

In this process, the bond soil unit/reference profile/K value is the key point of the whole system. This bond is a strength point when for each soil unit there is a sufficient number of observations that define the fundamental properties of the unit. The absence of an adequate number of observations, in proportion to the scale of the map, is a weak point of the process of associating the properties of a soil profile to a soil cartographical unit.

The following is an example of the process of mapping the *K* factor at watershed scale.

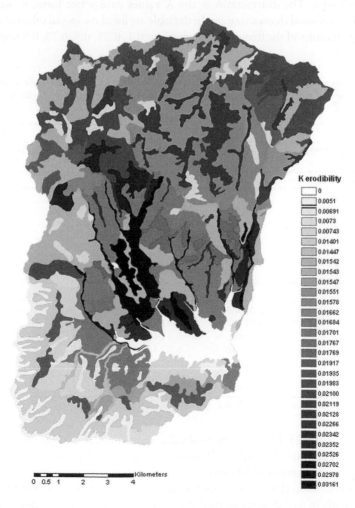

Fig. 5.7. Distribution of soil erodibility values in the Bilancino Basin.

The map was developed in 14 months from 2003 to 2004 with the objective of studying the primary source areas of sediments in the Bilancino basin (Mugello, Tuscany, BABI project, Borselli et al., 2004). In the absence of a soil map at the scale needed, a soil map at scale 1:25,000, functional to estimate the K factor for each soil unit, was produced.

The pedoalgorithm KUERY 1.2 was subsequently applied using the soil unit information and new analytical data. Table 5.4 shows the important parameters for the K assessment for four different soil units. Units 23 and 25 refer to soil with rock content lower than 10%, while units 13 and 19 refer to soil with rock content higher than 10%. The climate group is temperate/hot/Mediterranean for all the units (Cs).

All data in Table 5.4 refer to the A and Ap horizons of the unit reference profiles.

The pedoalgorithm KUERY 1.2 provides the frequency distributions (PDF) of the K values for each unit. The distributions of the K values have a free form, so we used a non-parametric statistical characterization. In the table are listed the modal values of K and of the following quantiles of the frequency distribution: 0.1, 0.25, 0.5, 0.75, 0.9 respectively (see also Fig. 5.3)

Table 5.4. Parameters relevant to the application of pedoalgorithm KUERY 1.2 for the assessment of erodibility of some soil units. The figures refer to the A or Ap horizons of reference profile identified for each unit.

Unit no.	23	25	13	19
Classification (WRB98)	Profondic Luvisols	Gley-Dystric Cambisols	Eutri-Skeletic Cambisol	Gley-Dystric Regosols
Textural class(USDA)	FS	F	FL	AL
Clay %	18	24	23	40
Loam %	31	32	60	45
Sandy %	51	44	17	15
Rock content %	< 10	< 10	50	26
OM %	1.5	1.35	3.5	2.5
D_g	−1.505	−1.7	−2.09	−2.375
S_g	1.3	1.2	0.944	1.047
K (SI units)	*K* values			
K (modal)	0.02967	0.02292	0.00384	0.01693
$K_{0.10}$	0.01297	0.015	0.00286	0.00667
$K_{0.25}$	0.0188	0.01911	0.00327	0.01221
$K_{0.50}$	0.02616	0.02352	0.0051	0.01547
$K_{0.75}$	0.03006	0.02776	0.01144	0.01842
$K_{0.90}$	0.03161	0.02979	0.0144	0.01977
K (used)	0.03161*	0.02352	0.0051	0.01547

*We use the 0.9 quantile because horizon A and Ap has ESP > 4.

In the case of unit 25, 13 and 19, the final value used is the median (quantile 0.5). In the case of unit 23, since the value of the ESP was 6%, we used the 0.9 quantile because of the probable dispersive behaviour of A horizon.

For the whole basin (150 km^2), in most of the units we used the median value of the K distribution. Only in 6 of 29 units we used the 0.9 quantile due to the assumed dispersive behaviour (ESP > 4) of A and Ap horizon.

Figure 5.7 is a map of the distribution of *K* values in the entire watershed estimated from benchmark soil parameters associated with each soil unit. The darker portions show higher soil erodibility values and the white portion shows the Bilancino reservoir surface.

The benefits of a detailed soil map are evident from comparing the previous existing GIS representations for the distribution of soils of the Mugello and their estimated erodibility.

In a previous work on the same watershed (Cassi, 2002), due to lack of soil maps, the erodibility was estimated by extrapolating data from lithological characteristics of the substratum and on two *K* values calculated experimentally in two soils in the area of Mugello (Zanchi, 1988).

The comparison of the two erodibility maps, the Cassi (2002) map and the map derived with the KUERY 1.2 methodology (Borselli et al., 2004), both developed in a raster resolution 10x10 m, evidences the obvious limits of the methodology for estimating the *K* factor on the basis of lithology alone. The distribution of soil units obtained of the soil map in the BABI project has a low correspondence to the division into lithological units, consequently leading to a different distribution of *K* values.

The *K* values derived from lithological map underestimate the soil erodibility up to 80% in northern and central areas near the ridge and central valley areas surrounding the Bilancino reservoir (mainly in units associated with different soil orders of terraces), and overestimate up to 250% in areas of the southern ridge of the watershed (Fig. 5.8).

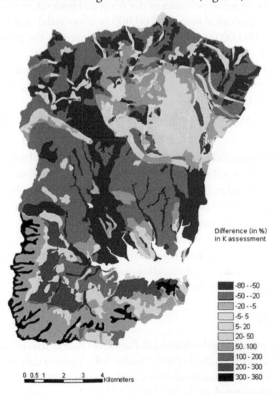

Fig. 5.8. Map of the difference between *K* derived from lithologic map and *K* derived from soil map.

These data were obtained through a calculation of the difference in percentage between the *K* values derived from lithological map (Cassi, 2002) and the *K* derived from soil map (Borselli et al., 2004).

These results give clear evidence how low is the correspondence between data derived from lithology and those derived from soil, and how great may be the error in the extrapolation of erodibility values, bound to the physical and chemical properties of soil (texture, organic matter, rock content), from lithological characteristics that provide none of these data.

The algorithms and rules that make it possible to estimate soil erodibility from physical and chemical characteristics of various soil horizons are consolidated techniques of soil science and of soil conservation modelling. The erodibility values of a particular lithology, on the other hand, are largely subjective and not supported by any database of measured values, as are the values of soil erodibility (Torri et al., 1997).

5.5. Guidelines for Assessing Soil Erodibility

From the application at watershed scale we understand the importance of soil map cartography and a soil database appropriate for the purposes of charting a soil erodibility map that is reliable and based on current scientific knowledge. Estimation based only on lithological information or geological map generally has poor reliability and, as the example above indicates, could lead to large over- or underestimates of the *K* value. However, maps of soil erodibility from inadequate maps (geological map at 1:100,000 or lower scale) or from unsuitable scale (soil map of Italy 1:1,000,000) have been widely produced in the past. These practices should be discouraged.

An acceptable compromise between cost and benefits is a soil map at 1:25,000 scale. The knowledge gained about the soils in a given region, the sensitivity and the experience may, however, represent key elements of the estimate of *K*.

The guidelines for assessing erodibility at watershed scale can be summarized in the following points:

- Availability of adequate soil mapping and an organized soil database.
- For each soil unit in the soil database, selection of the physical-chemical soil parameters necessary for the selected pedotransfer function or pedoalgorithms for *K* factor estimate.
- Where available, use of additional information on phenomena related to structural instability and dispersivity potential of the soil.
- The lack of information does not justify the use of alternative map type (geological maps or insufficiently detailed soil maps) from a technical and scientific point of view.
- The use of lithological characteristics alone for the estimation of soil erodibility has no scientific basis and therefore should be rejected.

Acknowledgements

This research was partly funded by the Autorità di Bacino del Fiume Arno (Arno River Watershed Authority)-ITALY; BABI project (2003–2007) "Studio della dinamica delle aree sorgenti primarie di sedimento nell'area pilota del bacino di Bilancino". The research described in this paper was also conducted in the framework of the EC-DG RTD-6th Framework Research Programme (sub-priority 1.1.6.3)—Research on Desertification—project DESIRE (037046): Desertification Mitigation and Remediation of Land—a global approach for local solutions.

References

AGASSI, M., BLOEM, D., BEN-HUR, M. 1994. Effect of drop energy and soil and water chemistry on infiltration and erosion. Water Resources Research 30, 1187–1193.

BARRE, D.A., BIGGS, A.J.W., SHARP, G.C. 2004. Sodicity and dispersion relations in some Darling Downs subsoils. SuperSoil 2004: 3rd Australian New Zealand Soils Conference, 5–9 December 2004, University of Sydney, Australia. Published on CD-ROM. Website: www.regional.org.au/au/asssi/

BARTHÈS, B., ROOSE, E. 2002. Aggregate stability as an indicator of soil susceptibility to runoff and erosion; validation at several levels. Catena 47, 133–149.

BELL, F.G., WALKER, D.J.H. 2000. A further examination of the nature of dispersive soils in Natal, South Africa. Quarterly Journal of Engineering Geology and Hydrogeology 33, 187–199.

BORSELLI, L., CASSI, P., SALVADOR SANCHIS, P., UNGARO, F. 2004. Studio della dinamica delle aree sorgenti primarie di sedimento nell'area pilota del Bacino di Bilancino: PROGETTO (BABI) - RAPPORTO FINALE. CNR-IRPI, Consiglio Nazionale delle Ricerche, Istituto di Ricerca per la Protezione Idrogeologica, Firenze, 104 p.

BRYAN, R.B. 2000. Soil erodibility and processes of water erosion on hillslope. Geomorphology 32, 385–415.

BRYAN, R.B., GOVERS, G., POESEN, J. 1989, The concept of soil erodibility and some problems of assessment and application. Catena 16, 393–412.

CASSI, P. 2002. Ricerca Bibliografica e Verifica Metodologico-Applicativa Dell'equazione Usle Nel Sottobacino Del Bilancino (Bacino Dell'arno). università degli studi di padova-Master in Difesa e Manutenzione del Territorio.

CERDA, A. 1996. Soil aggregate stability in three Mediterranean environments. Soil Technology 9, 133–140.

DAVIS, W.M. 1909. Geographical Essays. Ginn, Boston.

DNR. 2005. Dispersive soil. Test for dam construction. Department of Natural Resources, Australia. Fact sheet. http://www.dnr.qld.gov.au

DOE, W.W., JONES, D.S., WARREN, S.D. 1999. The Soil Erosion Model Guide for Military Land Managers: Analysis of Erosion Models for Natural and Cultural Resources Applications. U.S. Army Engineer Waterways Experimental Station. Vicksburg. 129 p.

EMERSON, W.W. 2002. Emerson Dispersion Test. In: McKenzie, N., Coughlan, K., Cresswell, H. (Eds.), Soil Physical Measurement and Interpretation for Land Evaluation. CSIRO Publishing, Collingwood.

EMERSON, W.W. 1967. A classification of soil aggregates based on their coherence in water. Australian Journal of Soil Research 5, 47–57.

FOSTER, G.R., D.C. YODER, D.C., WEESIES, G.A., MCCOOL, D.K., MCGREGOR, K.C., BINGNER, R.L. 2002. User's guide—Revised Universal Soil Loss Equation Version 2 (RUSLE 2). USDA-Agricultural Research Service, Washington, DC.

FOSTER, G.R., FLANAGAN, D.C., NEARING, M.A., LANE, L.J., RISSE, L.M., FINKNER, S.C. 1995. Ch. 11 Hillslope erosion component. In: Flanagan, D.C., Nearing, M.A. (Ed.), USDA-Water Erosion Prediction Project: Hillslope Profile and Watershed Model Documentation. NSERL Rep. No. 10. USDA-ARS Nat. Soil Erosion Research Laboratory, West Lafayette, Indiana, pp. 11.1–11.13.

FOSTER, G.R., MCCOOL, D.K., RENARD, K.G., MOLDENHAUER, W.C. 1981. Conversion of the universal soil loss equation to SI metric units. Journal of Soil Water Conservation 36, 355–359.

FRANCAVIGLIA, R., MECELLA, G., SCANDELLA, P., MARCHETTI, A. 2003. Confronto di metodi per la stima del fattore d ierodibilità K del suolo. Bollettino della Società Italiana di Scienza del Suolo 52(1–2), 537–546.

IMESON, A.C., VIS, M. 1984. Assessing soil aggregate stability by water-drop impact and ultrasonic dispersion. Geoderma D. Reidel Publishing, Dordrecht, 34, 185–200.

KEMPER, W.D., ROSENAU, R.C. 1986. Aggregate stability and size distribution. In: Klute, A. (Ed.), Methods of soil analysis Part I—Physical and Mineralogical Methods, 2nd ed. American Society of Agronomy, Madison, Wisconsin, pp. 425–442.

KINNELL, P.I.A., RISSE, L.M. 1998. USLE-M: Empirical modelling rainfall erosion through runoff and sediment concentration. Soil Science Society of America Journal 62, 1667–1672.

LADO, M., BEN-HUR, M. 2004. Soil mineralogy effects on seal formation, runoff and soil loss. Applied Clay Science 24, 209– 224.

LAFLEN, J.M., MOLDENHAUER, W.C. 2003. Pioneering soil erosion prediction: the USLE story. WASWC special publication 11. World Association of Soil and Water Cosnervation. Thailand. ISBN 974-91310-3-7.

LAFLEN, J.M., ELLIOT, W.J., FLANAGAN, D.C., MEYER, C.R., NEARING, M.A. 1997. WEPP— Predicting water erosion using a process-based model. Journal of Soil Water Conservation 52, 96–102.

LE BISSONNAIS, Y. 1996. Aggregate stability and assessment of soil crustability and erodibility: I. Theory and methodology. European Journal of Soil Science, 47, 425–437.

LE BISSONNAIS, Y., ARROUAYS, D. 1997. Aggregate stability and assessment of soil crustability and erodibility: II. Application to humic loamy soils with various organic carbon contents. European Journal of Soil Science 48, 39–48.

LE BISSONAIS, Y. 1996. Soil characteristic and aggregate stability. In: Agassi, M. (Ed.), Soil Erosion, Conservation, and Rehabilitation. Marcel Dekker, USA, pp. 41–60.

LEGOUT, C., LEGUEDOIS, S., LE BISSONNAIS, Y. 2005. Aggregate breakdown dynamics under rainfall compared with aggregate stability measurements. European Journal of Soil Science, 56, 225–237.

LOCH, R.J., SLATER, B.K., DEVOIL, C. 1998 Soil erodibility (Km) values for some Australian soils. Australian Journal of Soil Research, 36(6), 1045–1056.

LOCH, R.J., ROSEWELL, C.J. 1992. Laboratory methods for measurement of soil erodibilities (K Factors) for the universal soil loss equation, Australian Journal of Soil Research 30, 233–248.

LOCH, R.J., POCKNEE, C. 1995. Effects of aggregation on soil erodibility: Australian experience. Journal of Soil Water Conservation 50, 504–506.

MERRITT, W.S., LETCHER, R.A., JAKEMAN, A.J. 2003. A review of erosion and sediment transport models. Environmental Modelling and Software 18, 761–799.

MIDDLETON, H.E. 1930. Properties of soils which influence soil erosion. U.S. Dep. Agric. Tech., 178.

MISRA, R.K., ROSE, C.W. 1996. Application and sensitivity analysis of process-based erosion model GUEST. European Journal of Soil Science 47, 593–604.

MORGAN, R.P.C., QUINTON, J.N., SMITH, R.J., GOVERS, G., POESEN, J.W.A., AUERSWALD, K., CHISCHI, G., TORRI, D., STYCZEN, M.E. 1998. The European Soil Erosion Model EUROSEM: a dynamic approach for predicting sediment transport from fields and small catchments. Earth Surface Processes Landforms 23, 527–544.

MUTCHLER, C.K., CARTER, C.E. 1983. Soil erodibility variation during the year. Transactions of the ASAE 1102–1104.

NEMES, A., WOSTEN, J.H., LILLY, A., OUDE VOSHAAR, J.H. 1999. Evaluation of different procedures to interpolate particle-size distribution to achieve compatibility within soil databases. Geoderma 90, 187–202.

RENARD, K.G., FOSTER, G.R., WEESIES, G.A., MCCOOL, D.A., YODER, D.C. 1997. Predicting soil erosion by water: a guide to conservation planning with the Revised Universal Soil Loss Equation RUSLE. Agric. Hbk. 703, USDA, WA.

ROBINSON, D.A, PHILLIPS, C.P. 2001. Crust development in relation to vegetation and agricultural practice on erosion susceptible, dispersive clay soils, from central and southern Italy. Soil Tillage and Research 60, 1–9.

ROSE, C.W., COUGHLAN, K.J., CIESIOLKA, C.A.A., FENTIE, B. 1997. Program GUEST (Griffith University Erosion System Template). In: Coughlan, K.J., Rose, C.W. (Ed.), A new soil conservation methodology and application to cropping systems in tropical steeplands. ACIAR Tech. Report, No. 40, Canberra. Australian Centre for International Agriculutural Research, Canberra, pp. 34–58.

ROSEWELL, C.J. 1993. SOILOSS—A program to assist in the selection of management practices to reduce erosion (SOILOSS Handbook). Technical Handbook No. 11 (2nd ed.), Soil Conservation Service, Sydney.

SALVADOR SANCHIS, M.P., TORRI, D., BORSELLI, L., POESEN, J. 2008. Climate effects on soil erodibility. Earth Surface Processes and Landforms 33, 1082–1097.
SEYBOLD, C.A., HERRICK, J.E. 2001. Aggregate stability kit for soil quality assessments. Catena 44, 37–45.
SHERARD, J.L., DUNNIGAN, L.P., DECKER, R.S. 1976. Identification and nature of dispersive soils. Journal of Geotechnical Engineering Division, ASCE 102, 287–301.
SHIRAZI, M.A., BOERSMA, L. 1984. A unifying quantitative analysis of soil texture. Soil Science Society of America Journal 48, 142–147.
SHIRAZI, M.A., BOERSMA, L., HART, W. 1988. A unifying analysis of soil texture: improvement of precision and extension of scale. Soil Science Society of America Journal 52, 181–190.
SINGER, MICHAEL J., LE BISSONNAIS, Y. 1998. Importance of surface sealing in the erosion of some soils from a mediterranean climate. Geomorphology 24, 79–85.
SOIL SURVEY STAFF. 1951. Soil survey manual. United States Department of Agriculture, Soil Conservation Service, Agricultural Handbook No. 18. U.S. Government Printing Office, Washington, DC.
SOTELO, R.R. 2005. Identificación de Arcillas Erodibles Dispersivas Utilizando Ensayos Agronómicos de Suelos. http://www1.unne.edu.ar/cyt/tecnologicas/t-051.pdf
SUMNER, M.E. 1993. Sodic soils: New perspectives. Australian Journal of Soil Research, 31, 683–750.
TAYLOR, G., EGGLETON, R.A. 2001. Regolith Geology and Geomorphology. Wiley and Sons, Chichester, 325 p.
TORRI D., BORSELLI, L. 2000. Water erosion. In: Sumner, M.E. (Ed.), Manual of Soil Science. CRC Publications, New York, pp. G171–G194.
TORRI, D., POESEN, J. 1997. Erodibilità. In: Pagliai, M. (Ed.), "Metodi di Analisi Fisica del Suolo", sezione VII, Ministero delle politiche agricole e Forestali-Societa Italiana di Scienza del Suolo, Rome.
TORRI, D., POESEN, J., BORSELLI, L. 1997. Predictability and uncertainty of the soil erodibility factor using a global dataset. Catena 31, 1–22.
TORRI, D., POESEN, J., BORSELLI, L. 1998. Erratum to predictability and uncertainty of the soil erodibility factor using a global dataset. Catena 32, 307–308.
TORRI, D., POESEN, J., BORSELLI, L. 2002. Corrigendum to "Predictability and uncertainty of the soil erodibility factor using a global dataset" [Catena 31 (1997) 1–22] and to "Erratum to Predictability and uncertainty of the soil erodibility factor using a global dataset" [Catena 32 (1998), 307–308]. 46(4), 309–310.
TURKINGTON, A.V., PHILLIPS, J.D., CAMPBELL, S.W. 2005. Weathering and landscape evolution. Geomorphology 67, 1–6.
USDA–SCS. 1991. Dispersive clays. Soil Mechanics note no. 13, 32 p.
VACHER, C.A., LOCH, R.J., RAINE, S.R. 2004. Identification and management of dispersive mine spoils. Final Report. Australian Centre for Mining Environmental Research. Landloch Pty. Ltd. 84 p.
VALENTIN, C., BRESSON, L.M. 1992. Morphology, genesis and classification of surface crusts in loamy and sandy soils. Geoderma 55, 225–245.
VAN DE GRAAFF, R., PATTERSON, R.A. 2001. Explaining the Mysteries of Salinity, Sodicity, SAR and ESP in On-site Practice. In: Patterson, R.A., Jones, M.J. (Eds.), Proceedings of On-site '01 Conference: Advancing On-site Wastewater Systems. Lanfax Laboratories, Armidale.
WILLIAMS, J.R. 1975. Sediment-yield prediction with universal equation using runoff energy factor. In: Present and Prospective Technology for Predicting Sediment Yield and Sources ARS, S-40, USDA, WA.
WISCHMEIER, W.H., SMITH, D. 1978. Predicting rainfall erosion losses: a guide to conservation planning. USDA-ARS Agriculture Handbook No. 537, Washington DC. 58 p.
WISCHMEIER, W.H., JOHNSON, C.B., CROSS, B.V. 1971. A soil erodibility nomograph for farmland and construction sties. Journal of Soil and Water Conservation 26(5), 189–192.
YU, B., ROSE, C.W. 1998. Application of a physically based soil erosion model, GUEST, in the absence of data on runoff rates I. Theory and methodology. Australian Journal of Soil Research, 37(1), 1–12.
ZANCHI, C. 1983. Primi risultati sperimentali sull'influenza di differenti colture (frumento, mais, prato) nei confronti del ruscellamento superficiale e dell'erosione. Annali Istituto Sperimentale per lo Studio e la Difesa del Suolo, Firenze, XIV, 277–288.
ZANCHI, C. 1988. Soil loss and seasonal variation of erodibility in two soils with different texture in the Mugello valley in Central Italy. Catena Supplement, 12(1), Catena Verlag, Germany, 167–174.

6. Evaluation of the Hydrologic Soil Group (HSG) with the Procedure SCS Curve Number

Devis Bartolini[1] and *Lorenzo Borselli*[1]

6.1. Introduction

The process of runoff generation during a rainfall event is affected by various soil properties, soil surface physical characteristics and land use. The relationship between precipitation and runoff is a process related to many complex interactions. A physical model of the infiltration into soil was mainly developed in the 20th century (Green and Ampt, 1911; Richards, 1931; Horton, 1940; Philip, 1957; Smith and Parlange, 1978; Morel-Seytoux and Verdin, 1981; Swartzendruber, 1987; Morin and Kosovsky, 1995). The physical and chemical properties of soil affect the evolution of the capacity of infiltration, which itself is not a constant but varies continuously during a rainfall event. It is well known that even the conditions of antecedent soil moisture influence the hydrological response during a rainfall event. One of the most interesting behaviours of the rainfall/runoff process is the strong non-linearity that can be experimentally observed between the amount of precipitation and runoff. The non-linearity is accentuated in the early stages of a rain. Given the complexity of the phenomenon, many researchers developed in the 20th century a series of simplified and empirical approaches to estimate the potential runoff volume from a given rainfall volume and soil characteristics (e.g., permeability, texture) that are easily assessed and usable for a rapid application (a constraint in a period when personal computers were not available) (Hudson, 1986).

The most famous model of rainfall-runoff is the Soil Conservation Service Curve Number model (SCS-CN USDA, 1969; USDA, 1972, 1985; NRCS, 2007). With this model, the volume of runoff produced from each precipitation is estimated through a single parameter that summarizes the influence of both the superficial aspects and deep soil, linked to the characteristics of different horizons, such as the saturated hydraulic conductivity, in addition to the characteristics and use of plant cover, soil and humidity before the precipitation event. In essence, the SCS-CN model is a generalization of the note rational formula in which the volume of outflow is estimated through a multiplication factor of between 0 and 1 drop of rain. Since the description of the model SCS-CN is beyond the goals of this chapter, the reader is referred to other works (Mishra and Singh, 2003; NCRS, 2007) for technical details. It is,

[1]CNR-IRPI—Istituto di Ricerca per la Protezione Idrogeologica, Unità Operativa di Supporto, Pedologia Applicata, Firenze, Italy.

however, important to note that in the past forty years the SCS-CN model has been a basis for many other more complex hydrological models at various operative scales (Singh, 1995; Mishra and Singh, 2003), where it was necessary to evaluate the production of outflow depending on the complex interactions between soil, topography and plant cover; hence the extreme importance of procedures for determining the parameters of the model related to intrinsic characteristics of the soil.

In practice, to apply the SCS-CN model, we need to classify a soil in one of four groups, each identifying a different potential runoff generation (A, B, C, D): each is a hydrologic soil group (HSG: SCS-CN USDA, 1969; NRCS 2007) and the potential runoff generation increases from group A to group D.

The classification of a soil into one of these groups is in fact an essential condition for applying the SCS-CN model to any scale. In the case of applications at basin scale, knowledge of the spatial distribution of land is crucial.

In this chapter, we describe the official definition of the various HSG and give a set of guidelines for their assignment, according to standard practice or by following alternative methods. Finally, an example of application is provided as a complement to the previous guidelines.

6.2. Definition of Hydrologic Soil Group (HSG)

The definition of the HGS provided by the United States Department of Agriculture (SCS-CN USDA, 1969; USDA 1972, 1985; Mishra and Singh, 2003) has been recently revised and updated (NRCS, 2007) and is determined by the water-transmitting soil layer with the lowest saturated hydraulic conductivity (in the first 50 cm) and depth to any layer that is more or less impermeable or water depth to a water table (if present). The least transmissive layer can be any soil horizon that transmits water at a slower rate than those horizons above or below it.

The saturated hydraulic conductivity of an impermeable or nearly impermeable layer may start from essentially less than 3.6 mm/h. For simplicity, either case is considered impermeable for HSG purposes (NRCS, 2007).

The HSG are defined by NRCS (2007) as follows:

- *Group A*: Soils have low runoff potential when thoroughly wet. Water is transmitted freely through the soil. Group A soils typically have less than 10% clay and more than 90% sand or gravel and have gravel or sand textures. Some soils having loamy sand, sandy loam, loam or silt loam textures may be placed in this group if they are well aggregated, have low bulk density, or contain more than 35% rock fragments.
- *Group B*: Soils have moderately low runoff potential when thoroughly wet. Water transmission through the soil is unimpeded. Group B soils typically have 10% to 20% clay and 50% to 90% sand and have loamy sand or sandy loam textures. Some soils having loam, silt loam, silt, or sandy clay loam textures may be placed in this group if they are well aggregated, have low bulk density, or contain more than 35% rock fragments.
- *Group C*: Soils have moderately high runoff potential when thoroughly wet. Water transmission through the soil is somewhat restricted. Group C soils typically have 20% to 40% clay and less than 50% sand and have loam, silt loam, sandy clay loam, clay loam, and silty clay loam textures. Some soils having clay, silty clay, or sandy clay textures may be placed in this group if they are well aggregated, have low bulk density, or contain more than 35% rock fragments.

• *Group D*: Soils have high runoff potential when thoroughly wet. Water movement through the soil is restricted or very restricted. Group D soils typically have more than 40% clay and less than 50% sand and have clayey textures. In some areas, they also have high shrink-swell potential. All soils with a depth to a water-impermeable layer less than 50 cm (20 in.) and all soils with a water table within 60 cm (24 in.) of the surface are in this group, although some may have a dual classification.

6.3. Procedures for Evaluation

There are various methods for assigning the HSG. In the old description of the procedure (SCS-CN USDA, 1969), the HSG code represents the synthesis of certain elements characteristic of the soil, particularly texture, structure, degree of permeability, depth to which permeability is reduced, and extent of the reduction. This procedure for assignment was based on sub-factor scoring and their successive summation in order to obtain a global score. The global score obtained was compared with a reference table for a direct assignment of final HSG class (SCS-CN USDA, 1969).

Recent changes (NRCS, 2007) introduce a new procedure by the use of a table with diagnostic physical characteristics (Tables 6.1, 6.2 and 6.3).

Table 6.1 summarizes the diagnostic physical characteristics of the various groups, while Tables 6.2 and 6.3 show the decision matrix used to determine the soil HSG following the last updated procedure of NRCS (2007).

Table 6.1. Diagnostic physical characteristics of soil HSG (NRCS, 2007, modified).

Hydrologic soil group	Diagnostic physical characteristics
A	The saturated hydraulic conductivity of all soil layers exceeds 144 mm/h. The depth to any water-impermeable layer is greater than 50 cm (20 in.). The depth to the water table is greater than 60 cm (24 in.). Soils deeper than 100 cm (40 in.) to a water-impermeable layer are in group A if the saturated hydraulic conductivity of all soil layers within 100 cm (40 in.) of the surface exceeds 36 mm/h.
B	The saturated hydraulic conductivity in the least transmissive layer between the surface and 50 cm (20 in.) ranges from 10.0 µm/sec (1.42 in./h; 36 mm/h) to 40.0 µm/sec (5.67 in./h; 144 mm/h). The depth to any water-impermeable layer is greater than 50 cm (20 in.). The depth to the water table is greater than 60 cm (24 in.). Soils deeper than 100 cm (40 in.) to a water-impermeable layer or water table are in group B if the saturated hydraulic conductivity of all soil layers within 100 cm (40 in.) of the surface exceeds 14.40 mm/h but is less than 36 mm/h.
C	The saturated hydraulic conductivity in the least transmissive layer between the surface and 50 cm (20 in.) is between 1.0 µm/sec (0.14 in./h; 3.6 mm/h) and 10.0 µm/sec (1.42 in./h; 36 mm/h). The depth to any water-impermeable layer is greater than 50 cm (20 in.). The depth to the water table is greater than 60 cm (24 in.). Soils deeper than 100 cm (40 in.) to a restriction or water table are in group C if the saturated hydraulic conductivity of all soil layers within 100 cm (40 in.) of the surface exceeds 1.44 mm/h but is less than 4.0 µm/sec 14.40 mm/h).
D	For soils with a water-impermeable layer at a depth between 50 and 100 cm (20 and 40 in.), the saturated hydraulic conductivity in the least transmissive soil layer is less than or equal to 3.6 mm/h. For soils deeper than 100 cm (40 in.) to a restriction or water table, the saturated hydraulic conductivity of all soil layers within 100 cm (40 in.) of the surface is less than or equal to 1.44 mm/h.

Sometimes a soil with a favourable saturated hydraulic conductivity is classified in Group D on the basis of the presence of a water table within 60 cm (24 in.) of the surface. Dual HSGs (A/D, B/D and C/D) have been originated to classify these soils when they are properly

Table 6.2. Criteria for assignment of HSG when a water-impermeable layer exists at a depth between 50 and 100 cm (20 and 40 in.) (NRCS, 2007, modified).

Soil property	Group A	Group B	Group C	Group D
Saturated hydraulic conductivity of the least transmissible layer	> 144.0 mm/h	≤ 144.00 to > 36.0 mm/h	≤ 36.0 to > 3.60 mm/h	≤ 3.6 mm/h
	and	and	and	and/or
Depth to water-impermeable layer	50 to 100 cm	50 to 100 cm	50 to 100 cm	< 50 cm
	and	and	and	and/or
Depth to high water table	60 to 100 cm	60 to 100 cm	60 to 100 cm	< 60 cm

Table 6.3. Criteria for assignment of HSG when any water-impermeable layer exists at a depth greater than 100 cm (40 in.) (NRCS, 2007, modified).

Soil property	Group A	Group B	Group C	Group D
Saturated hydraulic conductivity of the least transmissive layer	> 36.0 mm/h	≤ 36.0 to > 14.40 mm/h	≤ 14.40 to > 1.44 mm/h	≤ 1.44 mm/h
	and	and	and	and/or
Depth to water-impermeable layer	> 100 cm	> 100 cm	> 100 cm	> 100 cm
	and	and	and	and/or
Depth to high water table	> 100 cm	> 100 cm	> 100 cm	> 100 cm

drained. The first letter of a dual HSG applies to the drained condition, according to the saturated hydraulic conductivity and water table depth when drained, and the second to the undrained condition.

As a result of construction and other disturbances, the soil profile can be altered from its natural state and the listed group assignments generally no longer apply, nor can any supposition based on the natural soil be made that will accurately describe the hydrological properties of the disturbed soil. In these circumstances, an on-site investigation should be made to determine the HSG (NRCS, 2007).

In the case of topsoil characteristics, for example, a tendency for cracks to form on desiccation, dichotomous behaviour on a seasonal basis can be indicated with an assignment of dual HSG (e.g., A/D, B/D): the second HSG indicated (D in these examples) is that resulting from the model, while the first is arbitrarily defined by the operator and indicates the temporal phase in which the soil cracks. It should be noted that for soils with a tendency to form surface crusts and sealing, due to low stability of the structure, the group resulting from the hydrological model can be underestimated.

Other changes were intended to improve the accuracy and consistency of the HSG assessment, reducing the subjectivity, but leaving unchanged the relative simplicity of use and application. An example comes from Langan and Lammers (1991). The authors compared the characteristics and properties of soils defined in the USDA soil taxonomy with the definition of HSG. They examined the detail of the definition of HSG, using diagnostic characters of soil classification in the USDA soil taxonomy. Some of the keys they use to identify HSG are, in particular, the presence of specific horizons or contacts that improve or restrict the movement of water in the soil or the development of roots, the soil moisture regime, as well as the slope and the interactions between these and other parameters. The result of this study is a guide to the use of the soil taxonomy to assign the right class of HSG to a particular soil.

McCuen (1982) gives evaluation of the HSG based on the value of the minimum rate of infiltration of the soil (Table 6.4). In this case, the minimum infiltration rate should be considered the infiltration rate at saturation (steady state infiltration rate during a rainfall event).

Table 6.4. HSG identification following McCuen (1982).

HSG	Minimum infiltration rate (mm/h)
A	< 7.6
B	7.6–3.8
C	3.8–1.3
D	1.3–0

6.4. Example of Application

In the following example, the new procedure (NRCS, 2007) is applied to a soil (source: Regione Emilia Romagna, Geological, Seismic and Soil Survey: hilly and montainous Soil Database). Using the basic soil physical characteristics of each horizon, we estimated saturated conductivity of each horizon. We also simulated the change of the depth of the less conductive soil layer that will affect the final HSG classification.

There are many pedotransfer functions available in the literature. In the following example, the Brakensiek et al. (1984) equation is used to estimate the saturated hydraulic conductivity (mm/h) using inputs such as soil texture and bulk density characteristics. The computation was done using the PEDON SEI interface (http://www.irpi.fi.cnr.it/pedone/Pedonintrod.html).

Table 6.5 shows data on a soil profile extract from the database of soils of Emilia Romagna. The value of bulk density, required for the application of the Brakensiek et al. (1984) pedotransfer function, is the average in the area in which the same type of soil is present (see "technical notes" at http://www.irpi.fi.cnr.it/pedone/Pedonintrod.html).

Table 6.5. Data from a soil profile extracted from the soil database of region Emilia Romagna (1996).

Horizon	Upper limit (cm)	Lower limit (cm)	Structure size and type	Grade of Structure	Bulk density Mg/m³	Sand tot. %	Loam tot. %	Clay %	Textural class	Ks (Brakensiek et al., 1984) (mm/h)
Ap	0	45 (55 or 105)	33	7	1.42	54	29	17	sandy loam	24.23
2 B	45 (55 or 105)	170	34	6	1.42	30	38	32	clay loam	1.48
3Cg	170	270	34	6	1.42	49	33	18	loam	15.34

The value of Ks in the upper horizon classifies this soil as Group C (3.6 mm/h < Ks < 36 mm/h, Table 6.2), but the depth of soil with lower water conductivity is at 45 cm, so this condition bring the classification to Group D (Table 6.2).

If the depth of horizon 2B is 55 cm (instead of 45 cm), the same soil is classified as Group C (3.6 mm/h < Ks < 36 mm/h, depth of water-impermeable > 50 cm).

If the depth of horizon 2B is assumed to be 105 cm, we can classify the soil as Group B (14.40 mm/h < Ks < 36 mm/h and depth of water-impermeable layer > 100 cm).

The codes used in Table 6.5 related to soil structure characteristics are defined in Tables 6.6 and 6.7.

Table 6.6. Codes used in the description of soil structure (size and type) (Regione Emilia Romagna, 1995).

Stucture size	Stucture type				
	Platy	Prismatic	Blocky angular	Blocky sub-angular	Granular
Fine	11 1–2	12 10–20	13 5–10	14 5–10	15 1–2
Medium	21 2–5	22 20–50	23 10–20	24 10–20	25 2–5
Coarse	31 5–10	32 50–100	33 20–50	34 20–50	35 5–10
Very coarse	41 > 10	42 > 100	43 > 50	44 > 50	45 > 10

Table 6.7. Codes used in the description of the soil structure grade (from Regione Emilia Romagna, 1995).

Grade	
1	Stuctureless
2	Massive
3	Weakly developed
6	Moderately developed

References

BRAKENSIEK, D.L., RAWLS, W.J., STEPHENSON, G.R. 1984. Modifying SCS hydrologic soil groups and curve numbers for rangeland soils. ASAE Paper No. PNR-84-302, St. Joseph, Michigan.

GREEN, W.H., AMPT, G.A. 1911. Studies on soil physics. 1. The flow of air and water through soils. Journal of Agricultural Science 4, 1–12.

HORTON, R.E. 1940. An approach towards a physical interpretation of infiltration capacity. Soil Science Society A. Proceedings 5, 399–417.

HUDSON, N. 1986. Soil conservation. B T Batsford Limited, London, 324 p.

LANGAN, L.N., LAMMERS, D.A. 1991. Definitive criteria for Hydrologic Soil Groups. Fall (1991)—Soil Survey Horizons. 32(3), 69–73.

MCCUEN, R.H. 1982. *A Guide to Hydrologic Analysis Using SCS Methods*. Prentice Hall, Englewood Cliffs, New Jersey.

MISHRA, S.K., SINGH, P. 2003. *Soil Conservation Service Curve Number (SCS-CN) methodology*. Water Science and Technology Library, vol. 42, 513.

MOREL-SEYTOUX, H.J., VERDIN, J.P. 1981. *Extension of the Soil Conservation Service Rainfall-Runoff Methodology for Ungaged Watersheed*. Federal Highway Administration, Environmental Division, Washington, DC, Report no. 81–10.

MORIN, J., KOSOVSKY, A. 1995. The surface infiltration model. Journal of Soil and Water Conservation 50(5), 470–476.

NRCS. 2007. *National Engineering Handbook—Part 630 Hydrology*. Chapter 7, Hydrologic Soil Groups (ver. Amended and reprinted, 05/15/2007). http://directives.sc.egov.usda.gov/OpenNonWebContent.aspx?content=22526.wba

PHILIP, J.R. 1957. The theory of infiltration. 4. Sorptivity and algebraic infiltration equation. Soil Science 84, 257–264.

REGIONE EMILIA ROMAGNA. 1995. *Carta dei Suoli Regionale scala 1:50.000. Normativa Tecnica Regionale*. Ufficio Pedologico, Servizio Cartografico, 117 p.

RICHARDS, L.A. 1931. Capillary conduction of liquids through porous media. Physics 1, 318–333.

SCS-CN USDA. 1969. *Soil Conservation Service National Engineering Handbook, section 4, Hydrology*. Washington, DC.

SINGH, V.P. 1995. *Computer Models of Watershed Hydrology*. Water Resources Publications, Colorado, USA, 1130 p.

SMITH, R.E., PARLANGE, J.Y. 1978. A parameter-efficient hydrologic infiltration model. Water Resource Research 14(3), 533–538.

SWARTZENDRUBER, D. 1987. A quasi solution of Richards' equation for the downward infiltration of water into soil. Water Resource Research 23, 809–817.

USDA. 1972. *Soil Conservation Service National Engineering Handbook, section 4, Hydrology*. Washington, DC.

USDA. 1985. *SCS National Engineering Handbook, Section 4: Hydrology*. Washington, DC.

7. Land Evaluation for Developing Countries
Luca Ongaro[1] and Paolo Sarfatti[1]

It is commonly acknowledged internationally that sound scientific data on natural resources are needed for their intelligent and sustainable use to ensure food security, eradicate poverty and preserve the environment for the benefit of future generations. A huge quantity of geographical data, such as low, medium and high resolution satellite images and digital elevation models, exist, but their use for natural resources evaluation applications is far below the potential.

Several methodologies are involved in the process of land evaluation: these methodologies constitute a reliable tool to provide a good overall knowledge about available resources in a given region, and to analyse alternative development scenarios.

Land evaluation is particularly important in developing countries during cooperation projects, when the introduction of new technologies (e.g., irrigation) or new crops is foreseen or changes in land and water management are planned, or, more generally, whenever land use planning is needed.

During the past 20 years, the Istituto Agronomico per l'Oltremare (IAO) of Florence has developed its own methodology, through a long series of case studies, in the framework of both development projects (Algeria, Tunisia) and training courses (Bolivia, China, Eritrea, Morocco, Senegal, Tunisia). This methodology involves in a first phase the preparation of a land unit map, followed by the application of different evaluation schemes through consolidated models or by setting up ad hoc procedures. In this framework, the Master's Degree in Geomatics and Natural Resource Evaluation, which the IAO has organized annually since 1974, is a "laboratory" in which this methodology is developed.

7.1. Land Unit Mapping for Land Evaluation: the IAO Approach (*Luca Ongaro*)

To deal with international development cooperation, it is fundamental that project preparation and management should be based on a sound knowledge of the available environmental resources. In most cooperative projects in developing countries, however, scientific as well as

---

[1]IAO—Istituto Agronomico per l'Oltremare, Firenze, Italy.

organizational, logistical and economic problems arise, in a completely different way than expected when operating in more developed countries. In fact, often the environmental data do not exist or are scattered, outdated or are at a scale that is too small to be really useful. Sometimes time is a constraint, or the field check must be carried out in a single mission, without the possibility of visiting the study area several times. Moreover, project sites often have harsh climates, with poor road networks and inadequate logistics. Thus, an operational methodology to carry out inventory and evaluation of the natural resources was developed that had to be functional in such a context, scientifically coherent, and internationally acceptable. Also, multidisciplinary integration is an essential tool wherever the development project must start from scratch.

From the beginning, the chosen framework was the holistic approach to land unit systems, which largely rely upon the experiences developed by foreign and international institutions such as FAO, CSIRO (Australia), LRDC (UK), and ITC (Netherlands). The IAO's goals were mainly to create an operational assembly of remote sensing along with GIS techniques, and to set up field survey procedures as rapid, accurate and effective as possible. Special attention is paid to the evaluation (or "pragmatical classification") of the natural resources, following the FAO framework (FAO, 1976), which in many aspects is the most suitable for this type of activity.

From the scientific point of view, the inventory and evaluation of natural resources is a very complex problem, as many different disciplines are involved, such as vegetation and soil sciences, climatology, and geology. Moreover, one should consider the large number of variables involved, for example, in the production of biomass, or all the possible interrelationships among environmental and management factors to see how an operationally exhaustive approach is far from being available. It is here that the possibility of using simplified methods becomes very important, in a technically and scientifically sound framework that allows the description of the reality with a degree of approximation that is acceptable for practical application.

In the IAO methodology, everything is centred on the land unit map, which is the synthetic document that fully describes the environment and its resources. "Land" is defined as "an area of the earth's surface, the characteristics of which embrace all reasonable stable, or predictable cyclic, attributes of the biosphere vertically over and under this area, including those of the atmosphere, the soil and the underlying geology, the hydrology, the plant and animal populations, and the result of past and present human activity, to such an extent, that those attributes exert a significant influence on present and future uses of the land by man" (FAO, 1976).

Based on this definition, the landscape is the visual aspect of the land, so the landscape unit is the object to be identified and analysed, first on the remotely sensed data, then on the terrain. Once described in all its various components, the landscape unit becomes a land unit, i.e., a portion of territory that is homogeneous, at the working reference scale, for a set of environmental factors whose properties have been measured or described, along with their interrelationships. In practice, what we try to apply is a holistic approach to a system of land units: "holism" is a philosophical concept used to signify the tendency of nature to create objects, by creative evolution, that are more than the sum of their parts. So a holistic land survey takes into consideration, or uses, the principle that the landscape is something more than the mere sum of its observable attributes and measurable parameters taken one by one. The term, in effect, is almost synonymous with "integrated" (Zonneveld, 1979). The land units, as practical expressions of holistic concepts, in reality represent ecosystems, and as such they provide an excellent framework for traditional disciplines. In operational practice, then, we have efficient tools to study, describe and evaluate the land units.

The first step is the classification of the land, i.e., the grouping of similar elements into hierarchical subdivisions. The IAO methodology provides, pragmatically, three "hierarchical" levels of land classification, in decreasing scale order: the site, the facet and the system.

The **site** is the smallest holistic entity: it is homogeneous in all or almost all its attributes and corresponds to the location chosen for the point-based observations during the fieldwork.

The **facet** is a portion of territory formed by a population of sites that are related both spatially and in terms of their attributes, at least for one of them, typically landform or land cover. Generally the land facet is also homogeneous in terms of soil, and especially from the management point of view, which in fact is one of the main diagnostic criteria during the fieldwork. The land facet is a fundamental object for land evaluation.

One of the main characteristics of the IAO methodology is the adoption of an open system of classifiers for the cataloguing of facets that greatly simplifies the fieldwork, the land classification and the setup of a legend system.

The **land system is** "a combination of land facets together, forming one convenient mapping unit on a reconnaissance scale" (Zonneveld, 1972). Thus, a land system is mainly constituted by geographically associated facets that form recurrent patterns, whose limits coincide with those of the main geological and geomorphologic characteristic and/or processes (Ongaro, 1998).

While the site is a real portion of a territory, both the facet and the land system are pure cartographical abstractions produced by a process of generalization, whose limits and extent largely rely upon the surveyor's experience and interpretation of the landscape evidence.

The land unit, as delineated on the map, can contain any aggregation of facets, pure or mixed, and/or land systems, without any relation with the different hierarchical levels, according to exclusively the mapping scale and the characteristics of the study area. The mapping scale, in fact, determines the size of the minimum mapping unit, i.e., the smallest portion of territory that can be meaningfully delineated on the final map. The legend of a land unit map, therefore, is always composed of two tables, one being a facet legend and the other the land unit legend, linked through a system of weights and relative compositions. An analogue system of relationships is used also to correlate the field observations with the facet to which they refer, and therefore ultimately with the mapping units. The Geographic Information System so created is scalable (i.e., the study area can be extended, the mapping scale can be changed) and is easy to update, allowing the user at any time to insert new field observations and change the legend. Moreover, a GIS based on the land unit, as described here, is exactly the opposite of the traditional GIS technique of adding different information layers to derive, applying specific algorithms, a synthesis or a thematic map (Fig. 7.1).

With the land unit approach, the starting point *is* the synthesis, from which the different thematic maps are derived by successive analysis. The advantages of this reversed approach are many: apart from avoiding any problem of thematic or geometric coherence between the different layers, it ensures at the beginning the integration of various types of information and hence the correct analysis of the interrelationships among the different environmental factors.

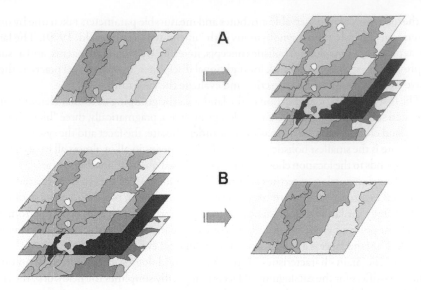

Fig. 7.1. Schematic difference between a land unit-based approach (A) and a traditional GIS (B).

A holistic approach based on the land unit can therefore be used as an excellent basis for a wide range of evaluation procedures, not only for production purposes, but also for environmental impact assessment.

7.2. Land Evaluation in Developing Countries: Examples of Applications (*Paolo Sarfatti and Luca Ongaro*)

During the past 20 years, IAO has tested many different applications of land evaluation:

- General land capability evaluation
- Land suitability for specific crops, differentiating between traditional (e.g., terracing, animal traction, organic manure) and modern farming systems (e.g., mechanization, chemical fertilizers)
- Suitability for irrigation, using different techniques
- Suitability evaluation for specific forest species, timber and fuel wood production, soil conservation or social and recreational purposes
- Naturalistic value and eco-tourism evaluation
- Erosion risk and erosion assessment
- Suitability for urban development

For each application, ad hoc indexes and tables have been developed, adopting parametric methods, both qualitative and quantitative. We report here three examples developed in the framework of development projects or training courses carried out by the IAO in Algeria, Bolivia and Tunisia (Sarfatti, 1995; IAO, 1999, 2004).

7.2.1. Land Evaluation for Durum Wheat Cultivation (Algeria, 1995)

The applied research project on durum wheat and lentil in the region of Tiaret (Algeria) was carried out from 1993 to 1997. In the framework of this project a land unit map at 1:200,000 scale was produced for four *Wilayas*, for a total surface of about 30.000 km², in one of the most important areas of the country for production of durum wheat, an essential staple crop.

The suitability evaluation was carried out for winter wheat without irrigation, taking into account edaphic, climatic and topographic information. The crop cycle in the study area goes from November to June, so for water availability the total rainfall of this period was considered. The climatic data from the existing stations were interpolated using krieging techniques.

Table 7.1. Evaluation of workability according to texture and structure (from Sys, 1985, modified).

Code	Description	Evaluation
Cm	Clay, massive	S3
SiCm	Silty clay, massive	S3
C+60,v	Fine clay, prismatic	S1
C+60,s	Fine clay, poliedric	S1
C-60,v	Clay, prismatic	S1
C-60,s	Clay, poliedric	S1
SiCs	Silty clay, poliedric	S1
Co	Clay, very weak poliedric	S1
SiCL	Silty clay loam	S1
CL	Silty loam	S1
Si	Silty	S1
SiL	Silty loam	S1
SC	Sandy clay	S1
L	Loam	S1
SCL	Sandy clay loam	S1
SL	Sandy loam	S2
LfS	Fins sandy loam	S3
LS	Loamy sand	S3
LcS	Coarse sandy loam	N
fS	Fine sandy	N
S	Sandy	N
cS	Coarse sandy	N

Concerning pedological data, the most important land qualities were considered. Quantitative morphological, hydrological and pedological observations such as runoff, permeability, mottles, and colour were used to assess the internal and external soil drainage, in order to estimate oxygen availability for the roots. For nutrients, the organic matter content of the topsoil and the cation exchange capacity (weighted average of the top 50 cm) were considered. Regarding rooting conditions, one of the more problematic qualities in the study area, the soil depth explored by the roots was considered, taking as lower limit the bedrock, the water table or the presence of horizons that impede root penetration, such as hard calcareous crusts or pans. For germination conditions, the average air temperature of the corresponding month was considered; it was estimated through multiple regression analysis, using the latitude, the distance from the sea coast and the elevation, obtained from a digital elevation model. Flowering and ripening conditions were evaluated using rainfall and temperature data for the corresponding months, obtained as described above. Salinity and sodicity were used to evaluate excess salt, and active $CaCO_3$ to evaluate toxicity. Workability was estimated through the sequence of an index that combines texture and structure (Table 7.1).

Potential for mechanization was assessed by combining slope, surface stoniness and rock outcrops. Finally, a very important element for Algeria, wind and water erosion risk, was estimated in the field through a qualitative assessment of the erosion type, its degree of intensity and the percentage of area affected (Table 7.2).

Table 7.2. Matrix for wind and water erosion risk evaluation (based on field observation).

Area affected	Intensity		
	Weak	Moderate	Strong
Minimal	S1	S1	S2
Localized	S1	S2	S3
Diffuse	S2	S3	N

Table 7.3 summarizes the complete rating table adopted. In many cases the thresholds found in the literature were adapted to local conditions. The rigid adoption of existing criteria would have led almost all the units to be classified as unsuitable, while during the research it emerged that in some areas yields are quite high for Algerian conditions (> 16 Q/ha), where unsuitable units have average yields less than 4 Q/ha.

The crop requirements were estimated in strict collaboration with the project's agronomists and breeders, local scientists and extension technicians, focusing on the most diffuse varieties and genotypes, and taking into account the real field conditions.

The results of this evaluation were used by the extension services to identify and select farms for seed multiplication and commercial seed production. Moreover, the analysis of the limiting factors served to better define the ideotypes of durum wheat for the different zones, in order to set up specific breeding programmes for the different geographical macro-zones.

7.2.2. *Land Evaluation for Urban Development (Bolivia, 1999)*

One of the main concerns for land use planning is the selection of the best-suited sites for specific land use types. The following examples deal with the urban development of Cochabamba, Bolivia, a town in rapid and disordered expansion. A site was selected on the basis of several criteria. Different characteristics, for example, distance from the urban centre, slope, development costs, environmental hazards, have their own influence on suitability for urban development. Moreover, the site could be highly suitable for other uses that compete with the urban development, e.g., agricultural use.

The methodology is based upon the assignment of a set of weights to the different criteria in order to estimate the potential for urban development. It is worthwhile to outline that any evaluation implies a certain degree of subjectivity, so the results should be considered only as a support to decision making.

Table 7.3. Rating table for land suitability evaluation for durum wheat in Algeria.

Land quality	Characteristic	Unit	Highly suitable S1	Moderately suitable S2	Marginally suitable S3	Unsuitable N
1 Water availability	Average rainfall during growing season	mm	350–1250	250–1500	200–1750	Any
2 Drainage conditions	Soil drainage class		Good to moderate	Imperfect	Slow	Very slow
3 Nutrient availability	Availability in deep horizons	pH	6.6–6.8	6.3–7	6–7.3	Any
4 Nutrient retention capability	Topsoil organic matter	%	> 1 (> 0.4)	> 0.5 (any)	Any	Any
	Cation exchange capacity of deep horizons	Meq/ 100 g	> 20	12–20	12–20	< 12
	Base saturation of deep horizons	%	> 50	> 35	Any	Any
5 Rooting conditions	Soil depth	cm	> 60	60–40	40–25	< 25
6 Conditions affecting germination	Average air temperature during germination	°C	10–20	6–25	2–37	Any
7 Conditions for ripening	Average air temperature during flowering	°C	12–26	10–32	8–36	Any
	Average air temperature during ripening	°C	14–30	12–36	10–42	Any
	Rainfall during flowering	mm	30–120	> 15	> 10	Any
	Rainfall during ripening	mm	30–120	10–150	Any	Any
8 Flood hazards	Frequency of flooding		Absent or rare	Reduced or moderate	High	Very high > 5 times a year
9 Climatic hazards	Frequency of frost during spring		Absent	Rare	Frequent	Very frequent
	Frequency of *sirocco* (hot wind)		Absent	Rare	Frequent	Very frequent
10 Excess of salts	Salinity (electric conductivity)	mmhos/ cm	< 8	< 12	< 16	Any
	Sodicity (ESP)	%	< 25	< 35	< 45	Any
11 Soil toxicities	CaCO₃ in the soil profile	%	< 30	< 40	< 60	Any
12 Soil workability	Texture/structure	Sys's seque- nce	from C-60, v to L	from C+60, v to SCL	from C+60, v to LfS	from Cm to Cs
13 Potential for mechanization	Slope	%	< 8	< 16	< 30	Any
	Rockiness	%	< 2	< 10	< 10	Any
	Stoniness	%	< 1	< 5	< 15	Any
14 Erosion hazard	Water erosion		Nil	Low	Moderate	High
	Wind erosion		Nil	Low	Moderate	High

The information was processed into a dataset of raster layers, duly correlated by a multi-criteria analysis, using the appropriate module from the IDRISI software (Eastman, 1997). Two distinct types of evaluation criteria were considered: factors and constraints.

The factors, generally known as decision variables or structural variables, are continuous variables (e.g., distance from roads, slope gradient), while the constraints are Boolean and are intended to exclude a given area from the evaluation.

The following factors were considered:

- distance from main roads
- distance from the central business district (the administrative centre of the town)
- slope
- landslide and other mass movement hazards (based on slope, land cover, geology, presence of clay in the topsoil and other factors)
- natural sensitive areas

The factors must have the same scaling system. The simplest mode is to perform a linear scaling. In this case, their values were re-scaled between 0 and 255, using a fuzzy function. Once normalized, the values were weighted according to their importance for the urban development. To define the constraints, we took into account the following:

- flooding hazards (based on climatological, topographic and hydrological parameters)
- distance from point sources of pollution (e.g., airport, refinery, wastewater treatment)

Factors and constraints were combined using multi-criteria analysis to obtain a final map of general aptitude for urban development. The selected method was a Weighted Linear Combination (WLC). This method starts by multiplying each factor by its *factor/tradeoff weight*,[1] then summing the results (Fig. 7.2); the constraints are applied through subsequent *zero out* multiplications, to exclude unsuitable areas.

Then, weights were assigned to each factor according to a hierarchical scale of factor relevance. Normally, weight assignment should be made following priorities defined by local communities or by administrators. In our case, we decided to attribute a larger weight to the environmental hazard factors (Fig. 7.3).

In the final phase, a land suitability map for urban development was produced, following the classification presented in Table 7.4.

Finally, to highlight the conflict between different potential uses, a comparison was made between land suitability for urban development and land capability, which enhances the best agricultural areas. The final product is a map of "use priorities" that shows the priority areas for agriculture and for urban development, the current urban areas, the water bodies and the areas that are not prioritized for either of the two kinds of use considered.

7.2.3. Land Evaluation for Nature Conservation (Tunisia, 2004)

This application was developed in a study area (Oued R'mel) encompassing some zones of high naturalistic interest, such as the high Djebel Zaghouan mountain, another mountain

---

[1] The factor/tradeoff is the degree at which a factor can be replaced by another; the way this compensation happens is controlled by a set of tradeoff weights.

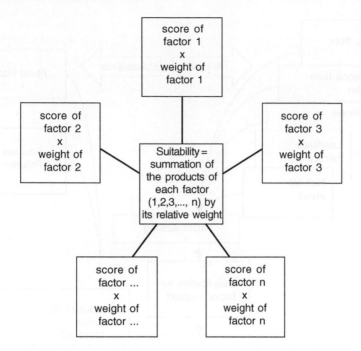

Fig. 7.2. Weighted Linear Combination (WLC): multiplication of each factor's score by the respective weight and their summation.

Table 7.4. Classification system for the land suitability map for urban development.

Suitability class
Suitable
Moderately suitable
Marginally suitable
Unsuitable
Excluded areas
Industrial areas
Flooded areas
Water bodies

ridge with extensive formation of *Tetraclinis articulata* (a tree included in the IUCN's red list of species at risk of extinction), and some humid areas, important because of their location along the bird migration routes between Europe and Africa. Starting from the land unit map, several GIS layers were produced:

- vegetation composition (floristic and structural-physiognomic)
- range conditions (e.g., richness of species, grazing pressure, vegetation cover)
- vegetation frequency (estimated using the frequency of each vegetation type in the whole land unit map)
- natural areas (a combined index derived from the three layers above)

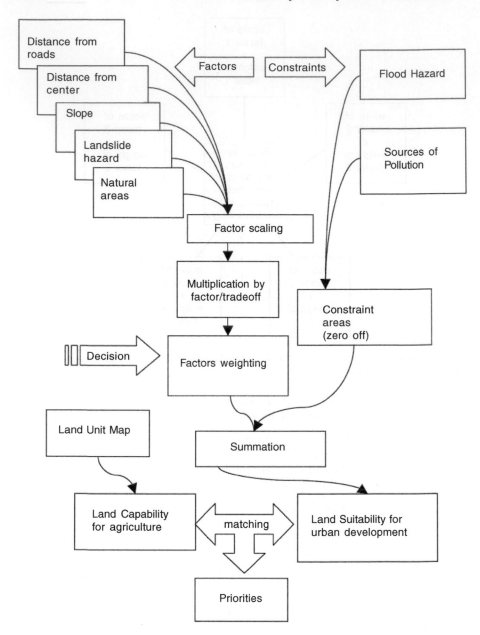

Fig. 7.3. Flux diagram for land suitability evaluation for urban development (valley of Cochabamba, Bolivia).

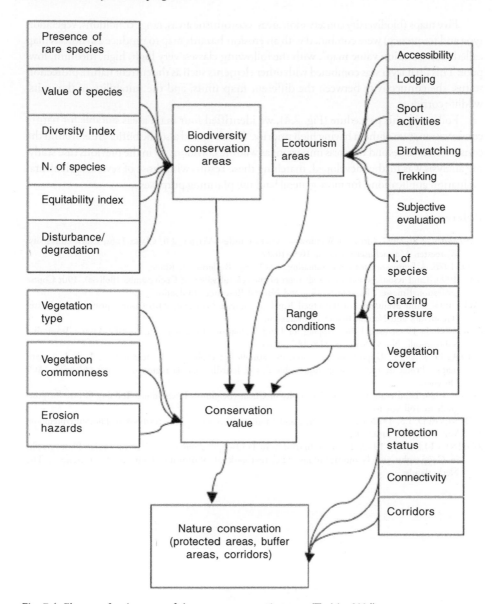

Fig. 7.4. Elements for the setup of the nature conservation map (Tunisia, 2004).

- biodiversity conservation value (presence of rare species, value of species, total number of species, Shannon's diversity index, equitability index, degree of disturbance and/or degradation)
- ecotouristic areas (11 areas chosen according to factors such as accessibility, lodging infrastructures, possibility of sporting activities, bird watching, and trekking, along with a subjective evaluation from a panel of expert witnesses)

Five maps (biodiversity conservation areas, ecotouristic areas, range conditions, vegetation type and frequency) were combined with an erosion hazards map to produce a synthesis map called "conservation value map", with the following classes: very high, high, medium, low, poor. This final layer was combined with other elements such as the current nature protection status, the connectivity between the different map units, and the suitability of the units' wildlife corridors.

Following this procedure (Fig. 7.4), we identified four main areas suitable for nature conservation, two of them having high priority. We identified also the buffer areas around the conservation zones and the areas important as wildlife corridors. As in the previous case study, the analysis was further developed, matching these results with those of several agricultural evaluation applications, for more general land use planning purposes.

References

EASTMAN, J.R. 1997. Idrisi for Windows— User's Guide—Version 2.0. Clarks Labs, Clark University, Worcester, Massachusetts (USA), 10.9–10.22.

FAO. 1976. A framework for land evaluation. FAO Soils Bulletin 32, Rome.

IAO. 1999. Land resources and land evaluation of the central valley of Cochabamba (Bolivia). 19th Course professional master "Remote Sensing and Natural Resources Evaluation".

IAO. 2004. Land evaluation in the Oued Rmel catchment (Tunisia). 24th Course professional master "Geomatics and Natural Resources Evaluation".

ONGARO, L. 1998. Land Unit Mapping for Land Evaluation. Relazioni e monografie Agrarie Tropicali e Subtropicali, Nuova Serie, n. 115, IAO, Firenze.

SARFATTI, P. 1995. Rapport sur les ressources naturelles et évaluation des terres (Vols. 1, 2, 3, 4, with maps). Projet de recherche appliquée: blé dur et lentille dans la région de Tiaret (Algerie), IAO, Florence.

SYS, C. 1985. Land evaluation. State University of Ghent (Belgium), International Training Centre for post-graduate soil scientists.

ZONNEVELD, I.S. 1972. Land Evaluation and Land(scape) Science. ITC Textbook of Photointerpretation, Vol. VII, ITC, Enschede.

ZONNEVELD, I.S. 1979. Land evaluation and landscape science. In: Use of Aerial Photographs in Geography and Geomorphology, ITC textbook of photointerpretation, 27, Enschede, The Netherlands.

8. Soil Restoration: Remediation and Valorization of Contaminated Soils

Claudio Bini[1]

8.1. Introduction

Remediation of contaminated soils is one of the most important environmental issues. Chemical soil degradation affects 12% of all degraded soils in the world, totalling 2 Mld ha (Adriano et al., 1995). Soil contamination is not only a social and sanitary issue, but also an economic concern, since it involves major costs related to decreasing productivity, product quality and monetary evaluation of the contaminated sites. Moreover, costs related to remediation of contaminated soils (particularly with heavy metals) are very high. Therefore, few developed countries (USA, UK, the Netherlands, Germany, Australia) have started remediation actions. Many developing countries have not yet started remediation projects, although they are affected by high environmental hazards (e.g., As in soils and ground water in Bangladesh and U in soils of Bosnia, as a consequence of the recent civil war).

    In the USA, the remediation of the sites listed in the National Priority List in 1986 (40% of the whole) would account for US$7 billion (Salt et al., 1995), and remediation of the over 1,000 sites that have been identified as hazardous would account for more than US$35 billion.

    In Switzerland, 10,000 ha of arable land have Zn concentration above the target value, and 300,000 ha present high levels of Cd, Pb and Cu (Vollmer et al., 1995).

    Research carried out in five European Union countries (Table 8.1) allowed identification of more than 22,000 contaminated industrial sites that are in critical condition (totally 0.2% of the land), for which an immediate intervention is required to safeguard public health, or

[1] Università di Venezia-Dipartimento di Scienze Ambientali, Venezia, Italy.

that have severe limitations in their use, and more than 50,000 sites need further investigation in order to assess their actual hazard.

Table 8.1. Number of contaminated sites in selected European Union countries (Adriano et al., 1995).

Country	Contaminated sites(total)	Sites in critical condition
Germany	32,000	10,000
Belgium	8,300	2,000
Italy	5,600	2,600
Netherland	5,000	4,000
Denmark	3,600	3,600

The Italian Environmental Agency estimated in 2005 that there are over 10,000 contaminated sites that need remediation. Of these, areas previously occupied by highly contaminating factories (e.g., chemicals at Porto Marghera, Venice; metallurgy at Bagnoli, Naples; tannery factories at Arzignano, Vicenza and S. Croce, Pisa) present very high contamination levels by organic as well as inorganic substances.

Many of the organic substances (e.g., PCB, PAH) contribute to contamination of ecosystems and are highly poisonous to living organisms and harmful to human health. Correspondingly, many metals, when present at high concentrations in the environment, are critical or toxic to plants and animals (Salomons, 1995) and may enter the food chain and therefore affect humans. The risk assessment for human health, therefore, is assuming increasing importance in the solution of problems connected with soil remediation. Indeed, the risk assessment criteria are applied to identify and classify the various sites on the basis of intervention priority, establish objectives and standard of decontamination, and select the most appropriate and site-specific technology.

In areas affected by high contamination, direct and indirect health hazards require urgent restoration and acceptable costs, regardless of the remediation technology selected for the site. In other cases, such as land with non-hazardous contaminant levels, or excessive costs compared to the expected benefits, remediation may eliminate or reduce the environmental hazard and contribute to the improvement of green areas, public services, and arable land otherwise not usable.

Decision makers should evaluate the selection of the remediation technologies also in relation to the effects that it may have on the soil quality. Many processes, indeed, determine significant changes in soil characteristics (e.g., pH variation, red-ox conditions, fertility, structure loosening, sterilization and decline of biological activity). Action for restoration of degraded areas, therefore, should take care of costs for remediation as well as management of the site, the hazards arising from the site, and the benefits derived from restoration.

Metal contamination persistence and limited knowledge of mechanisms regulating the soil-metal interaction and the absorption of contaminants by living organisms make soil remediation particularly difficult and expensive. Any of the current technologies are actually effective and applicable at wide scale. The most frequently implemented technical solutions are clearly inadequate for cleaning large areas of moderately contaminated land, where soft and environment-friendly technologies are needed to restore soil fertility in such a way that the land can be used for agriculture or public/residential green areas. Therefore, in recent years the interest of both public authorities and private companies (e.g., Dupont, Monsanto) in innovative methodologies for decontamination and restoration of contaminated sites is increasing.

8.2. Background and Legislative Soil Reference Values

A soil is contaminated[1] when its concentration of contaminants exceeds the background level. Background level corresponds to the total content of metals in soils not affected by human activities. These values are available in a number of publications (Alloway, 1995; Adriano, 2001).

Background values may vary as a function of the locality from which a given soil is sampled. For example, metal concentrations in serpentine-derived soils can be highly toxic to animals and plants as a result of the naturally elevated metal contents of the parent rock from which the soil is derived. Metal concentrations in soils are known to be affected by the clay content of soils and increase almost linearly as a function of it. Because of the spatial variability of metals in soils, background values will not serve as good reference values for legislative purposes. Therefore, it is not possible to arrive at a single background value for any of the metals. In an effort to expedite remediation of hazardous waste sites in the absence of a national soil clean-up standard, many national agencies have developed their own clean-up standards. In general, clean-up levels promulgated for industrial sites tend to be up to one order of magnitude less stringent than those for residential sites. On the other hand, soil clean-up levels established to protect groundwater quality tend to be more stringent than those based on direct human exposure to contaminated soils. In addition, carcinogens tend to be assigned more stringent levels than non-carcinogens. Although no US federal levels have been developed for hazardous constituents in soil, soil screening levels based on health risk were drafted by the US-EPA in the autumn of 1993 (Bryda and Sellman, 1994). These levels are used to help assess contaminated soils at sites that pose potential concern, as well as screen out those soils that do not require additional actions. Clean-up levels developed for metals in the USA are based on average background concentrations found in soils or standard risk assessment methods (Bryda and Sellman, 1994).

Some other countries, notably Canada, the UK, Belgium and the Netherlands, have progressed further in setting soil standards for soil remediation. In the Netherlands, the discovery of a contaminated residential area created a major public outcry that led to a legislative mandate for soil restoration in that country. Metals, inorganics, and a wide range of organic compounds were involved.

For each contaminant, three different values were initially adopted:

1. Mean reference value.
2. Threshold value for pollution, above which no biological or ecological damage is yet observed; soils with this level of pollution, however, should be further monitored.
3. Threshold value above which restoration is recommended.

---

[1] Contamination occurs when soil composition deviates from the normal composition. In their natural state, contaminants may not be classified as pollutants unless they have detrimental effects on organisms. Pollution occurs when a substance is present in greater than natural concentration as a result of human activities and has a net detrimental effect on the environment and its components (Adriano et al.,1995).

These criteria were recently revised. Table 8.2 presents the values for metals adopted by the Dutch legislature. The intervention values for soil remediation will be used to assess whether contaminated land poses a serious threat to public health. These values indicate the concentration levels of the metals in soil above which the functionality of the soil for human, plant, and/or animal life is seriously compromised or impaired. Concentrations in excess of the intervention values correspond to serious contamination. The intervention values replace the old C values in the soil protection guidelines.

Table 8.2. Dutch target values (also referred to as A-value or reference value) and intervention values (also referred to as C-value) for selected metals for soil (mg/kg dry matter).

Metal	Target value	Intervention value
Arsenic	29	55
Barium	200	625
Cadmium	0.8	12
Chromium	100	380
Cobalt	20	240
Copper	36	190
Mercury	0.3	10
Lead	85	530
Molybdenum	10	200
Nickel	35	210
Zinc	140	720

Source: Dutch Ministry of Housing, Spatial Planning and Environment, the Hague, the Netherlands.

The present values

- are based not only on considerations of the natural concentrations of the contaminants that indicate the degree of contamination and its possible effects but also on the local circumstances, which are important with regard to the extent of and scope for spreading or contact;
- are related to spatial parameters. The soil is regarded as being seriously contaminated if the metal mean concentration in at least 25 cu m of soil volume exceeds the intervention values;
- are dependent on soil type, since they are related to the content of organic matter and clay in the soil.

The target values (Table 8.2) are important for remedial as well as for preventive policy. They indicate the ultimate target soil quality levels for a given use. These values are derived from the analysis of field data from relatively pollution-free rural areas regarded as non-contaminated and take into account both human toxicological and ecotoxicological considerations.

In other European countries, the public has asked for similar legislation for soil restoration. Tolerable metal concentrations for agriculture and horticulture were published in Germany

(Kloke, 1980) and in Switzerland (Vollmer et al., 1995). In the UK, soil use after restoration was proposed as a criterion for determining the threshold value for redevelopment of contaminated sites (guidance 59/83, Department of the Environment, London, 1987). In Germany, Eikmann and Kloke (1993) introduced threshold values for playgrounds, parks, parking areas, and industrial sites. In agricultural and horticultural soils, lower threshold values were proposed when growing leafy vegetables than for fruit production or for the cultivation of grain or ornamental plants (Eikmann and Kloke, 1995). A similar approach has been proposed in Poland where agricultural and horticultural uses vary according to the severity of soil contamination (Kabata-Pendias, 1997).

In recent environmental legislation in Belgium, the threshold values for restoration that correspond somewhat to the intervention values vary with the intended land use for the remediated site. These threshold values were defined using the Human Exposure to Soil Pollution Model by Stringer (1990), which estimates the transfer of contaminants from soil to humans by different pathways (i.e., inhalation, ingestion, drinking water, animal or plant food). It was recently improved and several other models are proposed to assess the human risk of soil pollution.

In Italy, Legislation Act n° 471/99 proposes criteria for identifying contaminated sites, suggesting soil remediation and environmental restoration, determining the threshold value and the possible intervention to clean up a contaminated site permanently. Practically, a site is contaminated when the concentration of just one of the contaminants surpasses the threshold values reported in the national contaminant list for green areas, and residential and industrial sites (Table 8.3).

8.3. Soil Remediation and Risk Assessment

The risks associated with polluted soils vary from site to site according to scientific database, public perception, political perception, national priority, and other factors. While severely contaminated soils may require some form of remediation, there may be instances where remediation is not desirable (Adriano et al., 1995), as in the following cases:

1. Them cost of clean-up far exceeds the expected benefits of clean-up in terms of human health and ecological sustainability.
2. The contaminated soil is not being used and is unlikely to be used in the future.
3. There are inexpensive substitutes for the contaminated soil in question.
4. The site will not be used after remediation because potential users have other alternatives.
5. The contamination does not degrade soil and/or water quality to an unsafe or unhealthy level (NRC, 1993).

In the USA, a systematic procedure for remedial action comprises the following steps:

1. reporting and identification;
2. selection of response action;
3. preliminary assessment/site investigation;
4. remedial investigation/feasibility study;
5. remedial design/remedial action;
6. operation and maintenance/post-closure monitoring.

Table 8.3. Maximum concentration values recordable in soil and subsoil of contaminated sites, with reference to specific land use (D.M. 471/99, annex 1).

	Green and residential areas mg/kg d.m.	Commercial and industrial areas mg/kg d.m.
Inorganic compounds		
Antimony	10	30
Arsenic	20	50
Berillium	2	10
Cadmium	2	15
Cobalt	20	250
Chromium (total)	150	800
Chromium VI	2	15
Mercury	1	5
Nickel	120	500
Lead	100	1,000
Copper	120	600
Selenium	3	15
Tin	1	350
Thallium	1	10
Vanadium	90	250
Zinc	150	1,500
Cyanides	1	100
Fluorides	100	2,000
Organic compounds		
Benzene	0.1	2
Ethylbenzene	0.5	50
Styrene	0.5	50
Toluene	0.5	50
Xylene	0.5	50
Benzo(a)anthracene	0.5	10
Benzo(a)pyrene	0.1	10
Benzo(b)fluorantene	0.5	10
Benzo(k)fluorantene	0.5	10
Crisene	0.5	50
Dibenzo(a)pyrene	0.1	10
Dibenzo(a,h)anthracene	0.1	10
Indenopyrene	0.1	50
Pyrene	5	10

In arriving at a remedial decision, there are three categories of criteria that must be considered according to the National Contingency Plan (Grasso, 1993):

1. Threshold criteria:
    Overall protection of human health and the environment
    Compliance with applicable or relevant and appropriate requirements

2.   Primary balancing criteria:
       Long-term effectiveness and permanence
       Reduction of toxicity, mobility, or volume through treatment
       Short-term effectiveness
       Implementability
       Cost
3.   Modifying criteria:
       State acceptance
       Public consensus

Risk assessment is playing an increasingly important role in the remediation of hazardous waste sites. Regulations call for the use of risk assessment techniques to help in identifying and prioritizing sites requiring remediation, developing remedial objectives and clean-up standards, and selecting the most appropriate remedy for a particular location. Protection of human health may not ensure adequate environmental protection. Thus, there has been an emphasis on the development of ecological risk assessment methods. The ultimate choice of remedial technology largely depends on the nature and degree of contamination, the intended function or use of the remediated site, and the availability of innovative and cost-effective techniques. The choice is further complicated by environmental, legal, geographical, and social factors. More often the choice is site-specific. For example, home gardens and agricultural fields in large rural areas that are contaminated may require a remedial approach different from that for smaller but heavily contaminated areas. Similarly, large areas around old mining and smelter sites need an approach that differs from that of a heavily polluted spot.

8.3.1. Methods of Soil Remediation

The methods and techniques for remediating contaminated soils may be subdivided into two strategies: confinement and treatment.

Confinement technologies include (civil) engineering techniques that have the objective of removing or isolating the source of contamination, or of modifying migration pathways or percourses. Such techniques comprise the following:

* excavation and landfilling both inside and outside the site;
* barriers created in the contaminated soil;
* soil encapsulation; soil solidification;
* hydraulic intervention (pumping, washing).

Important factors driving the selection of such techniques are:

* large space available within the contaminated land;
* available geological and hydrogeological background studies;
* availability of natural/seminatural materials (geomembranes, geotextiles) to dress the excavated materials and to cover the contaminated material;
* the possible impact derived from excavation and/or disturbance;
* sterilization of the whole area devoted to infrastructure and buildings.

None of these techniques are entirely satisfactory (Exner, 1995). Landfilling is a temporary solution that delays remediation. Furthermore, it has been discontinued in most countries.

Encapsulation and solidification do not remove the contaminant from the soil, thereby greatly limiting the value of the soil. Soil washing and flushing have been used extensively in Europe but have had limited use in the USA. The process entails excavation of the contaminated soil, mechanical screening to remove various oversized materials, separation processes to generate coarse- and fine-grained fractions, treatment of those fractions, and management of the residuals generated. Soil washing performance is closely tied to three key physical soil characteristics: particle size distribution, contaminant distribution among the particles of different sizes, and strength of the soil-contaminant bond. In general, soil washing is most appropriate for soils that contain at least 50% sand and gravel, such as coastal sandy soils and soils with glacial deposits (Westinghouse Hanford Co., 1994).

Treatment technologies are based on processes designed to remove, stabilize or destroy contaminants.

- Contaminants may be removed by mobilization and/or accumulation processes (leaching, sorption), concentration and recovery processes (physical separation), or a combination of processes (accumulator plants).
- *In situ* stabilization consists of immobilization of the contaminant by a combination of physical, chemical and biological processes, thereby making it less toxic.
- Contaminants can be destroyed by physical, chemical or biological degradation (e.g., thermic or microbiological treatments).

Treatment processes may be defined according to their application, namely:

- *ex situ*, when carried out in the area of the contaminated site;
- *in situ*, when carried out without removing the contaminated soil;
- on-site, when carried out in the area of the contaminated site by moving and removing the contaminated material;
- off-site, when contaminated material is moved from the site and transported to treatment plants or to a landfill. <start

Chemical treatments involve destruction or removal of contaminants by the following means:

- Oxidation (change to higher chemical valence): many organic compounds, for instance, are oxidized to CO_2.
- Reduction (change to lower chemical valence): CrVI may be reduced to CrIII, which is less mobile and less toxic than CrVI.
- Immobilization: contaminant mobility is reduced through precipitation as an insoluble complex, or adsorption on solid matrices, for example.
- Extraction: contaminant is extracted from soil material by application of different extractants (e.g., organic compounds, acids, tensioactives); percolating liquid may be collected and treated for more degradation, or sent to a landfill.
- Substitution: some chemical groups of the contaminant may be replaced by other groups that make the contaminant less toxic (e.g., dealogenation of chloride solvents).

Physical treatments are aimed at separating contaminants from the soil matrix, taking into account the differences between contaminant and soil characteristics (e.g., volatility, magnetic properties, density). They include electrokinetics, electrolysis, electroosmosis, electrophoresis, stripping extraction, and other processes.

Thermic treatments use elevated temperatures to prime physical and chemical processes such as volatilization, ash flying, or pyrolysis, thus allowing contaminant removal or destruction, or immobilization in the soil matrix. Two main thermic treatments are available:

- desorption (working temperature is in the range 100–800°C), and
- incineration (working temperature is in the range 800–2500°C; contaminant is destroyed).

When contaminants are volatilized from the soil, they may be successively removed from the gas-phase through condensation or combustion. High-temperature incineration is the most common method of dealing with metal-contaminated soils because it has been proven to be the most reliable way to destroy the broadest range of wastes (Lee et al., 1990). However, it is costly and is often difficult to obtain a legal permit for it.

Biological treatments are known also as "bioremediation", i.e., "utilization of living organisms to reduce or eliminate environmental contaminants" (Adriano et al., 1999). Biological treatments include one or more of the following processes:

- Degradation: biochemical degradation of contaminant by soil microorganisms (bacteria, fungi, actinomycetes).
- Transformation: biochemical conversion of contaminant to make it less toxic and/or less mobile.
- Accumulation: organic and inorganic contaminants may accumulate in tissues of living organisms (particularly plants).
- Mobilization: a contaminant-bearing solution may be separated form the contaminated soil.

Microorganisms can detoxify several kinds of contaminated sites and bring them back to the original state.

Higher plants are used to stabilize, decrease or remove contaminants (heavy metals, organics, radionuclides, explosives) from soil and water. This technology, known as phytoremediation, is potentially minimally destructive, environment friendly and cost effective and can be applied to large areas of contaminated land to attenuate the risks related to soil contamination (Barbafieri, 2001; Li et al., 2002). It is based on several natural processes that involve plants:

- direct sorption of metals or moderately hydrophobic organic compounds;
- accumulation or transformation of chemicals by lignification, metabolization, volatilization;
- catalysis and degradation of organic compounds by enzymes released by plants;
- release of exudates in the rhizosphere, with pH modification, carbon and microbial activity increase, and contaminant degradation.

In the following paragraphs are described some techniques used in phytoremediation for removing inorganic contaminants from soils based on the above processes, techniques that the clean-up industry is already using or seriously considering.

Phytostabilization is a process in which plants tolerant to contaminant metals are used to reduce the mobility of contaminant metals, thereby reducing the risk that they will cause further environmental degradation by leaching into the ground water or spreading through the air (Smith and Bradshaw, 1979; Losi et al., 1994; Vangronsveld et al., 1995a). In *in situ* metal stabilization, soil amendments can be combined with the use of metal-tolerant plants to enhance plant growth and stabilizing effect of the treatment (Vangronsveld et al., 1995a, b, 1996). Metal-tolerant plants immobilize contaminants at the root-soil interface by absorption, precipitation or complexation, thus reducing mobility and migration to ground water, or to the food chain.

Contaminants can be stabilized in the following places:

- In the root zone: proteins are released (by roots) in the rhizosphere, and determine precipitation of contaminants.
- On cell walls: proteins associated with root cell walls may stabilize contaminants outside the root cell, thus preventing the contaminant from being transported inside the plant (barrier effect).
- In root cells: proteins present on root membranes may enhance contaminant transport inside the cell, where it is sequestered in vacuoles (Fig. 8.1).

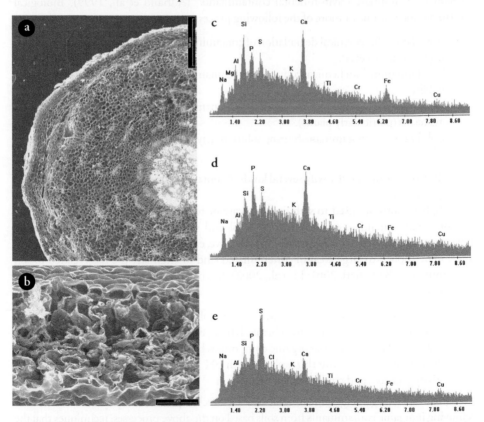

Fig. 8.1. SEM-EDS observations of wild plants of *Taraxacum officinale* from the tannery district in Italy (adapted from Bini et al., 2007). (a) Root cross-section of *Taraxacum*; (b) leaf cross-section of *Taraxacum*; (c, d) elemental spectra of *Taraxacum* roots; (e) elemental spectrum of *Taraxacum* leaf.

Arboreal and shrubby plants seem to be more prone than herbaceous ones to act as phytostabilizers, owing to different means of sorption and metal mobility in the plant organs. Some ornamentals, such as bay (*Laurus nobilis*), pitosphor (*Pitosphorum tobira*), or oleander (*Nerium oleander*), proved effective in metal stabilization (Carratù et al., 2001). Among the herbaceous plants, monocotyledons, both native (e.g., *Lolium perenne*) and cultivated (e.g., *Zea mays, Avena sativa*), are more suitable than dicotyledons (Argese et al., 2001).

Phytostabilization is particularly suitable at sites in which it is important to keep metals in non-mobile form, in order to impede dispersion, as happens for chromium (Bini et al., 2000a, 2007). Moreover, when metal concentration is very high, phytoextraction would require a long time to achieve the objectives of land restoration.

Rhizofiltration is a process by which plant roots absorb, precipitate and concentrate heavy metals from polluted wastewater streams (Dushenkov et al., 1995). It proved particularly effective in taking up As from wetland areas (Sette et al., 2001).

Phytostimulation involves stimulation of microbial or fungal degradation by means of exudates and enzymes in the rhizosphere (Schnoor et al., 1995). Root exudates (organic acids, alcohols, sugars) have a stimulating effect on microbial activity, thus increasing the biodegradation capacity of bacteria and fungi, as observed by Mehmannavaz et al. (2001) with *Sinorhizobium meliloti* on PCB in the rhizosphere.

Phytostimulation is a symbiotic relation between plants and soil microorganisms. The former provide nutrients to microorganisms, and the latter determine soil decontamination and favour root development.

Phytoextraction makes use of several plants that show a marked ability to accumulate contaminants (heavy metals, radioactives) in their aerial parts, and for this reason are known as "hyperaccumulator plants" (Wenzel et al., 1993; Baker et al., 2000). At the end of the phenological cycle (or when the metal sorption is concluded), the plant may be harvested and the contaminant recovered by distillation. Hyperaccumulator plants are able to accumulate metals at a rate up to 500 times the metal concentration in non-accumulator plants (Lasat, 2002), with metal concentration in leaves even higher than 5% dry weight (McGrath, 1998). Hyperaccumulator plants may accumulate more than 10 mg/kg Hg, 100 mg/kg Cd, 1,000 mg/kg Cu, Cr, Pb, and 10,000 mg/kg Zn. More than 400 species of hyperaccumulator plants are known at present (Baker et al., 2000), in 45 families, of which the most widely represented is *Brassicaceae* (Marchiol et al., 2004). Not all metals are accumulated in plants in the same way and to the same extent: some plants absorb only one metal, others more metals; for some metals, such as thallium, there are not yet known accumulator plants; for others, such as arsenic, some species, for example the fern *Pterix vittata*, have been discovered recently (Ma et al., 2001; Tu et al., 2004).

As a general rule, metals such as Cu, Cd, Zn, Ni, Pb, Se are easily accumulated, while As, Co, Cr, Mn, Fe, U are difficult to accumulate.

Phytoextraction efficiency depends upon several factors:

- the nature and concentration of contaminant (i.e., red-ox status, binding, bioavailability);
- soil/sediment chemical-physical characteristics (e.g., pH, texture, CEC);
- morphological and physiological plant characteristics (e.g, root pattern, absorption capacity, metal synergism or antagonism).

A coefficient currently used to evaluate phytoextraction efficiency is the biological absorption coefficient (BAC = $(metal)_{plant}/(metal)_{soil}$; Ferguson, 1992). The total amount of removed contaminants results from the harvested plant biomass multiplied by the metal concentration in plant.

As already stated, once the absorption and translocation of metal from roots to the aerial parts has ceased (i.e., the phenological cycle is ended, the maximum metal concentration allowed is achieved, more metal is unavailable), plants may be harvested and transported to a landfill; the contaminated site may be subjected to repeated cultivation cycles in order to decrease metal concentration to acceptable levels, or at best to eliminate it. Researchers have also explored the opportunity to recover metals absorbed by plants by distillation (Cunningham and Berti, 1997). This technique proved interesting in an economic perspective: in the USA, Canada, Australia and possibly other countries, some private companies have been established to economically use these low-cost "ore outcrops" (Chaney et al., 1997). One of the plants most frequently experimented on is the known Zn-Cd hyperaccumulator *Thlaspi caerulescens*. However, the reduced plant biomass (which is common to many hyperaccumulator plants) does not allow recovery of significant metal amounts. To overcome this important limiting factor, which would diminish the economic relevance of this technique, research in genetic engineering is in progress (Raskin, 1996; McNair, 1997; Prasad, 2003) to produce plants with higher biomass and therefore with major ability to remove metal from soil.

Assisted Phytoremediation addresses one of the major limits in soil decontamination with plants, i.e., the small biomass of most plants, especially those that grow in cold to temperate climates. A second factor limiting phytoextraction effectiveness is the soil metal bioavailability, i.e., the ability of plants to take up a metal from the soil. Bioavailability depends upon chemical and physical conditions, bacteria, fungi and plants that may influence such conditions (Ernst, 1996), organic complexes formed, precipitation of inorganic compounds, and other factors. Assisted phytoextraction increases metal bioavailability by applying to soil synthetic chelatants such as EDTA, HEDT, or DTPA. The most effective in increasing Ni and Pb extraction by several plants is EDTA (Kramer, 1996; Blaylock et al., 1997). This technique, however, has a high environmental impact, since chelatants may be leached to subsoil and ground water. Moreover, chelatant treatment may influence plant growth, if the phytotoxicity threshold is surpassed (McGrath, 1998). Alternatives to chelatant application are increasing microflora in the rhizosphere with microorganisms (Whiting, 2001) and application of ameliorants (e.g., zeolites: Zaccheo et al., 2001) that make the soil more suitable, thus enhancing plant growth.

8.3.2. Current Status and Perspectives in Phytoremediation

Phytoremediation is an emerging technique to clean up contaminated sites. It allows both organic and inorganic contaminants to be removed from soil and water, and physical stabilization of contaminated soils. It presents numerous advantages over other treatment techniques:

- it improves chemical, physical and biological soil properties;
- it reduces erosion rate;
- it reduces waste production;

- it allows potential metal recovery;
- it improves the aesthetics of the land;
- it is highly acceptable to the public;
- it is cheaper than other remediation technologies (40% reduction with respect to other *in situ* technologies, and up to 90% reduction with respect to *ex situ* technologies).

There are, however, some disadvantages:

- it is strongly dependent on environmental and climatic conditions;
- it is strongly dependent on contaminant concentration and bioavailability;
- it depends on the extent and depth of the contamination area (depth should be less than 5 m);
- it requires a longer time for land restoration than other technologies.

Although current literature on phytoremediation is abundant, the physiological mechanisms that control and regulate the above processes are not yet elucidated, partly because many of these are site-specific. The specificity depends first of all on the kind of contaminant, its composition and possible transformation (evolution), and, consequently, on the plant species to be used for remediating the site. Further investigation, therefore, is needed in order to elucidate both the mechanisms involved and selection of plants, possibly applying genetic engineering to increase plant biomass.

Phytoremediation is particularly suitable at sites where contamination is rather low and diffused over large areas, its depth is limited to the rhizosphere or the root zone, and there are no time constraints for the intervention. It has been calculated, indeed, that with present accumulator plants, at least 3–5 years are needed to achieve appreciable results in cleaning up a moderately contaminated soil (McGrath, 1998). McGrath (1995) calculated that nine harvests of *Thlaspi caerulescens* are necessary to decrease Zn concentration in soil from 444 to 300 mg/kg; conversely, a non-accumulator plant, such as radish, needs 2046 harvesting cycles to attain similar results. *Thlaspi caerulescens* and the hyperaccumulator *Cardaminopsis halleri* may remove, within just one harvest, Cd accumulated over decades of phosphate fertilizer application, with a removal rate of 150 g/ha/y and 34 g/ha/y, respectively (McGrath and Dunham, 1997).

One of the peculiar properties of phytoremediation is the economic aspect, since it is much cheaper than the other technologies. Black (1995) estimated costs of only US$80/m^3 of soil cleaned with phytoextraction, and US$250/m^3 with soil washing. According to Cunningham and Berti (1997), costs for phytopurification would be US$100,000/ha, and excavation and landfilling would amount to US$500,000/ha, while costs for traditional technologies of *in situ* treatment would range between US$500,000 and 1,000,000/ha. However, actual costs for phytoremediation are still highly variable and will be better estimated when this innovative technology is found to be really effective. Presently, indeed, the paucity of large-scale application makes it difficult to obtain reliable information on remediation time and costs. A comparison of costs for different remediation technologies is reported in Table 8.4. It is noteworthy that many *ex situ* treatments require the combined use of different items reported in the table, and therefore the whole cost would be higher than the single item.

Table 8.4. Comparative costs of different remediation technologies per soil unit.

In situ treatment	Costs (US$/m³)
Soil flushing	50–80
Bioremediation	50–100
Phytoremediation	10–35
Ex situ treatment	
Excavation and transport to landfill	30–50
Disposal in sanitary landfill	100–500
Incineration or pyrolysis	200–1,500
Soil washing	150–200
Bioremediation	150–500
Solidification	100–150
Vetrification	Up to 250

8.4. Applications

The suitability of phytoremediation for remediating contaminated sites has been continually assessed since the 1980s and the requisite items for its application have been identified, namely:

- laboratory research aimed at a better knowledge of processes regulating metal sorption by plants;
- field and laboratory research for new plant species with high metal sorption capacity, as well as genetic engineering techniques;
- laboratory trials and pilot-scale experiments to find the best conditions for application of different phytoremediation technologies;
- large-scale experiments to assess the concrete possibility of applying such technologies to contaminated soil remediation at environmentally and economically sustainable rates.

Metal bioavailability is the metal availability to a specific living organism (plant or animal) in specific environmental conditions (Adriano et al., 1995); therefore, soil characteristics and behaviour of the plant (living organism) control the actual metal availability. The mechanisms involved in such control are not yet fully elucidated, and often, in bioavailability evaluation, the physiological factors, including metal transport through cell membranes, are neglected. Generally, metal concentration is much higher in roots than in the aerial parts, but in some instances the opposite happens. Metal accumulation occurs as a consequence of compartmentalization and vacuole complexation, as was observed in vacuoles isolated from protoplasts of tobacco cells accumulating elevated levels of Cd and Zn (Barbafieri, 2001).

Metal assimilation, both essential and critical or toxic, is a typical and diffused plant feature (Streit and Strumm, 1993). According to Baker (1981), plants may be subdivided into three categories:

- *Excluder plants*: these species may limit metal sorption or translocation to the aerial parts, irrespective of the metal concentration in the substrate, until toxicity symptoms appear. In many cases, the metal is concentrated in roots that create a barrier effect against translocation.

- *Indicator plants*: in these species, metal concentration in aerial tissues is proportional to that in soil.
- *Accumulator plants*: these species may absorb and actively concentrate metals in the aerial parts at levels higher than in soil.

Among the plants that tolerate high metal concentration, those that present a natural tendency to accumulate very high metal concentrations (up to 500 times the normal concentration) are considered hyperaccumulator plants (Baker and Brooks, 1989; Baker et al., 1994). More than 400 species are presently known as accumulator plants, and a large quantity are indicator plants, useful in environmental quality evaluation. In current literature, information on new indicator/accumulator plants, in particular species with high biomass, for example, *Fragmites australis* (Massacci et al., 2001) and *Cannabis sativa* (Kos and Lestan, 2004), is published almost daily, thus enhancing the potential of this green technology.

A provisional list of accumulator plants, with indication of the metal(s) accumulated, is reported in Table 8.5.

8.4.1. Phytoextraction of Zinc and Cadmium

Zinc and Cd are chemical elements that present high affinity; it is argued, therefore, that they could be accumulated by the same plant species, although with different mechanisms, considering that Zn is an essential micronutrient, while Cd is toxic to plants and animals (Brooks, 1998). Many plants are known to be able to accumulate Zn; most of them belong to the genus *Thlaspi*, with *T. calaminare* accumulating Zn up to 3.96% dry weight. Also, *Cardaminopsis halleri*, *Viola calaminari*, and *Haumaniastrum katangense* present Zn concentration higher than 10,000 mg/kg, while *Thlaspi caerulescens* is considered a Cd hyperaccumulator plant, since it may accumulate over 100 mg/kg dry weight (McGrath, 1998). Other plants, like *Brassica juncea*, proved effective to remove Cd from the soil, although not at hyperaccumulator level (Kumar et al., 1995). Many other plants, however, are likely to accumulate Cd; better knowledge and more research would enhance identification of new Cd accumulator species.

Thlaspi caerulescens requires a minimum concentration of micronutrient Zn much higher than the mean Zn concentration in non-accumulator plants, presumably 1,000 mg/kg in the soil (McGrath, 1998). Mean Zn concentration in tissues of *Thlaspi caerulescens* is much higher than in "normal" plants (300–500 mg/kg vs 15–20 mg/kg). It is likely that the mechanisms responsible for the high metal tolerance are effective even at low metal concentration in the substrate; this may impede or limit the casual diffusion of this species in the surroundings of the contaminated area, ensuring in the meantime adequate Zn level for growing "normal" plants.

8.4.2. Phytoextraction of Lead

Lead phytoextraction is difficult because of the strong binding of the metal to the organic matter and the soil mineral fraction, which limits translocation to the aerial parts. At present, moreover, a few species are known to be able to accumulate Pb; among them, *Thlaspi rotundifolium* may accumulate up to 8,200 mg/kg d.w. High Pb levels in plant tissues could also derive from foliar adsorption of aerial input, a common pathway related to vehicular traffic until the 1990s; this Pb source, of course, should not be accounted for in terms of phytoextraction.

Table 8.5. Provisional list of the main metal accumulator plants.

Plant species	Metal accumulated	Reference
Aeollanthus biformifolius	Co, Cu	8
Agrostis capillaris, A. tenuis	Pb, Cr	7, 8, 12
Altermathera sessilis	Al	1
Alyssum bertoloni, A. murale	Ni	2, 3, 14
Arenaria patula	Zn	8
Armeria maritima	Cu	2, 7
Astragalus sp.	Se	2, 8, 11
Atriplex sp.	Se	8, 11
Becium homblei	Cu	2, 7
Berkeyia coddii	Ni	8
Bonmullera sp.	Ni	8
Brassica jucea, B. napus	Pb, Se	11
Buxus sp.	Ni	8
Calendula arvensis, C. officinalis	Cr	15
Cardaminopsis halleri	Zn	8, 11
Eichhornia crassipes	Cr, As, Cd, Hg, Pb	1, 8
Elodea canadensis	As, Co, Cu, Ni	8
Equisetum arvense	Cu, Zn	8
Festuca arundinacea, F. ovina	Se, Cu, Zn, Cr, Ni, Pb	7, 11, 13
Haumaniastrum robertii	Co, Cu	8
Hordeum vulgaris	Hg	5
Hybanthus floribundus	Ni	8
Ipomea alpina, I. carnea	Cr, Zn, Cu, Pb, Cd	1, 8
Lemna minor	Cr, Cd, Cu, Ni, Pb, Se	1, 12
Lolium perenne	Cu, Zn, Cr, Ni, Pb	13
Minuartia verna	Cu, Co	2, 7, 8
Myriophyllum verticillatum	As, Co, Cu, Mn	8
Phyllanthus sp.	Ni	8
Picea abies	U, Hg	4, 9
Pinus nigra, P. silvestris	U, Hg (Pb, Zn)	4, 9, 10
Plantago lanceolata	Cu, Cd, Pb, Zn	16
Populus nigra	Pb, Zn, Cu, Cr, Cd	4
Potamogeton richardsonii	Co, Cu, Pb, Zn	8
Raphanus sativus	Cd, Zn	14
Salix viminalis	Cd, Zn	14
Sambucus nigra	U	4, 9
Saxifraga sp.	Ni	8
Senecio coronatus	Ni	8
Silene vulgaris, S. cobalticola	Cu, Co	2, 7
Taraxacum officinale	Cu, Cd, Pb, Cr	17
Thlapsi alpestre	Cu, Co, Zn	8
Thlaspi caerulescens	Zn, Cd	3, 8, 11, 14
Thlaspi calaminare	Zn	3,8

Thlapsi rotundifolium	Zn, Pb	7
Typha angustata	Cr	1
Trifolium pratense	Cd, Zn	6
Viola calaminaria	Zn	3, 8
Viscaria alpina	Cu	2
Zea mays	Cd, Zn	14

1. Mhatre and Pankhurst (1997). 2. Pandolfini et al. (1997). 3. Baker and Brooks (1989). 4. Wagner (1993). 5. Panda et al. (1992). 6. Kabata-Pendias et al. (1993). 7. Ernst (1996). 8. Brooks (1998). 9. Steubing and Haneke (1993). 10. Bargagli (1993). 11. McGrath (1998). 12. Zayed et al. (1998). 13. Pichtel and Salt (1998). 14. Felix (1998). 15. Bini et al. (2000a). 16. Zupan et al. (1995). 17. Bini et al. (2000b).

To achieve effective decontamination in a relatively short time, plants having Pb concentration up to 10,000 mg/kg and a biomass higher than 20 t/ha/y (d.w.) would be necessary (McGrath, 1998). The highest Pb levels (up to 3.5% d.w.) have been found by Kumar et al. (1995) in an experimental trial on different species of Brassicaceae, with *Brassica juncea* producing 18 t/ha biomass. These results suggest that Brassicaceae are likely to remove Pb from contaminated soils at a rate of 630 kg/ha/y. Phytoextraction coefficients at full scale, however, are less likely than those obtained with controlled conditions, as shown by Huang and Cunningham (1996) and reported in Table 8.6.

The most promising method of Pb phytoextraction is to use synthetic chelatants to enhance metal assimilation by plants (assisted phytoextraction). Huang and Cunningham (1996) experimented with different chelatants on several plants, and in their experiment EDTA proved the most effective. Phytoextraction coefficients up to 1,000 times those found in the absence of chelatants were assessed, and this was also true of native non-accumulator plants. In such case, plants present early toxic symptoms, and therefore they should be harvested before senescence or death. According to the same authors, two harvests of *Zea mays* yielding 25 t/ha/y, with Pb concentration up to 10,500 mg/kg, would decrease Pb level in soil to 600 mg/kg in just seven years.

In case of chelatant application, the decontamination project should pay attention to percolating waters, in order to avoid groundwater contamination. Moreover, foliar fertilization with phosphate would be necessary to preserve nutritional levels.

Table 8.6. Lead concentration (μg/g) in shoots of various species, both native and cultivated (Huang and Cunningham, 1996).

Species (cultivar)	Hydroponic culture	Soil culture
Zea mays	375	225
Brassica juncea (211,000)	347	129
Brassica juncea (531,268)	241	97
Thlaspi rotundifolium	226	79
Triticum aestivum	139	12
Ambrosia artemisiifolia	96	75
Brassica juncea Cern.	65	45
Thlaspi caerulescens	64	58

8.4.3 Phytoextraction of Copper and Cobalt

A few examples of plants accumulating copper are known at present, all coming from Cu-Co mine areas of Zaire (Brooks, 1998). In the mineralized area, indeed, many endemic species, characterized by very high tolerance towards these two metals, do exist, and accumulate up to 1% of the two metals. Twenty-six plants are Co-hyperaccumulators, and 24 are Cu-hyperaccumulators; of these, 9 hyperaccumulate both Co and Cu (Table 8.7).

The highest Cu concentration in a phanerogam (1.37% d.w.) was recorded in *Aeollanthus biformifolius* and in *A. subacaulis* var. *linearis*. However, evidence for a possible use of such plants in soil clean-up is lacking (McGrath, 1998). Brooks and Robinson (1998) carried out a study on the feasibility of using hyperaccumulator plants as phytomining and estimated a recovery potential of 0.015 t/ha Cu and Co with *H. katangense*, assuming a biomass of 7.5 t/ha after fertilization.

8.5. Case Studies

Numerous experiments have been carried out, in laboratory, in batches and on a large scale, aimed at assessing the effectiveness of phytoremediation and implementing its potential, with the perspective of gaining economic advantage in addition to the environmental restoration of contaminated land. Blaylock et al. (1997) developed a technology that combines the ability of a Pb-accumulator plant, *Brassica juncea*, with fertilization of a large, densely inhabited district, where several cases of Pb poisoning in children were recorded. After the first cultivation cycle, the most contaminated zone (originally up to 1,000 mg/kg Pb) was reduced by 25%, and Pb concentration was 800 mg/kg.

Similar results are reported by Shen et al. (2001) after EDTA application to a mine soil cultivated with *Brassica napus*. Among herbaceous native plants, *Plantago lanceolata* and *Taraxacum officinale* have long been tested as metal indicator/accumulator plants, in soils with different contamination levels (Zupan et al., 1995, 2003; Bini et al., 2000b, 2007).

Some plants, such as mint (*Mentha acquatica*), fern (*Pteris vittata*) and cane (*Phragmites australis*), owing to their high biomass, proved particularly effective in heavy metal rhizofiltration, in particular of As, in hydromorphic conditions (wetland areas, submerged soils). Sette et al. (2001) carried out a research project in a wetland area close to an abandoned

Table 8.7. Plant species hyperaccumulating Cu and Co (µg/g) (Brooks, 1998).

Species	Cu	Co
Aeollanthus biformifolius	3,920	2,820
Anisopappus davyi	2,889	2,650
Buchnera henriquesii	3,520	2,435
Bulbostylis mucronata	7,783	2,130
Gutembergia cupricola	5,095	2,309
Haumaniastrum katangense	8,356	2,240
Haumaniastrum robertii	2,070	10,200
Lindernia perennis	9,322	2,300
Pandiaka metallorum	6,260	2,139

Sb-As mine and found that *Mentha aquatica* may accumulate in roots up to 8,900 mg/kg As; the rate of translocation to shoots and leaves is very low (13 and 177 mg/kg, respectively). Fern too, in particular *Pteris vittata*, proved effective (Table 8.8) in accumulating high amounts of As (up to 4,300 mg/kg) (Blaylock et al, 2003; Caille et al., 2003), with high transfer coefficient (As plant/As soil = 9).

Table 8.8. As accumulation in three species of fern (after Blaylock, 2003).

Species	As concentration mg/kg	Yield kg/ha	Recovery mg/kg
Pteris vittata	900	13,050	5.9
Pteris mayii	2,013	6,100	6.1
Pteris parerii	1,416	5,050	3.6

Phragmites australis proved to be an effective metal accumulator in numerous experimental trials (Yez et al., 1997; Massacci et al., 2001), besides being currently used in domestic and industrial wastewater phytopurification plants, for removing nitrate and phosphate. A peculiar character of this plant is the genetic similarity to cereal plants currently cultivated, such as maize and wheat, which could be used in phytoremediation.

Experimental research on accumulator plants with cultivated species is by far less frequent than expected, although it is increasing. One of the most frequently investigated plants, together with maize, is sunflower (*Helianthus annuus*). This plant, grown on a highly metal-contaminated soil and treated with EDTA (assisted phytoextraction), showed noteworthy ability to accumulate metals in its tissues (Quartacci et al., 2001). Thirty days after seedling, it accumulated 2,500 µg Zn, 500 µg Cu, Cd and Pb, and 100 µg Cr. Most of the metals (over 50%) were concentrated in roots: only Zn proved good metal transfer capacity to aerial parts (50% in shoots, 25% in leaves). Further investigations with sunflower (Sacchi et al., 2001), however, showed important limitations in assisted phytoextraction of Pb: the application of the maximum rate of chelatant determined Pb increases not proportional to the lower rate, and a concentration of 600 mg/kg Pb in leaves. According to the authors, given a biomass yield of 10 t/ha, Pb would be removed from the contaminated soil, at a rate of 6 kg/ha, over more than 100 years, not a cost-effective time period; moreover, chelatant application would be too expensive and pose serious environmental hazard because of the possible leaching of chelatant to subsoil and ground water.

Arboreal plants, until now, received little attention in comparison to the herbaceous ones, due to their different physiology; however, their high biomass would constitute an important advantage in their use in phytoremediation. In the past few years, indeed, experimental research on arboreal plants has progressed quickly, and interesting results have been achieved, especially with short-term coppices. One of the most experimented plants is willow (*Salix viminalis, S. caprea, S. rubens, S. fragilis*). Greger and Landberg (2003) recorded a strong correlation between *Salix viminalis* biomass and metal uptake from soil (Fig. 8.2), and Wieshammer et al. (2003) found high metal concentrations (326 mg/kg Cd, 2,413 mg/kg Zn, 70 mg/kg Pb) in willow leaves, with increments up to 54% Zn, 39% Cd, 21% Pb, in mycorrhized plants inoculated with metal-tolerant bacteria.

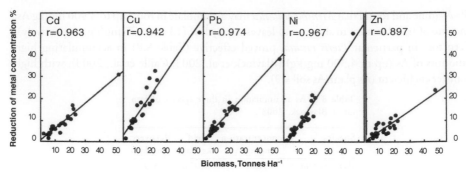

Fig. 8.2. Relationships between *Salix viminalis* biomass and metal removal from soil (adapted from Greger and Landberg, 2003).

A survey carried out in an urban park in the UK (Lepp and Dickinson, 2003) showed that 50 years after birch (*Betula* sp.), maple (*Acer pseudoplatanus*) and willow (*Salix caprea*) plantation on a strongly degraded and contaminated area, the vegetation cover had been naturally reconstructed, suggesting that the potential resilience of natural vegetation was effective in remediation of contaminated soils.

8.6. Summary and Conclusions

Phytoremediation is an emerging technology that holds great potential in cleaning up contaminants that: (1) are near the surface, (2) are relatively non-leachable, (3) pose little imminent risk to human health or the environment, and (4) cover large surface areas. Moreover, it is cost-effective in comparison to current technologies, and environment-friendly.

However, phytoremediation is not yet ready for large-scale application, despite favourable initial cost projections, which indicate that the clean-up market is likely to expand in coming years (Table 8.9).

Most of the available data, until now, has come from microcosm experiments; large-scale experiments could help in assessing the feasibility of phytoremediation and its effective contribution to clean up contaminated soils. Research should be addressed to find out new highly efficient accumulator plants, and related cultivation technologies. Moreover, basic research is necessary to advance our scientific understanding of physical, biological, and chemical processes important for natural accelerated remediation. More research is needed in microbiology, molecular biology, geochemistry, hydrology, transport processes and many

Table 8.9. Phytoremediation market projection for past and future years in the USA (millions US$) (adapted from Glass et al., 1997).

Site categories	2000	2005	2010
Ground water polluted with organic compounds	2–3	10–15	20–45
Soils polluted with heavy metal	1–2	15–22	40–80
Soils polluted with radionuclide	0.1–0.5	2–5	25–30
Waste water polluted with heavy metals	0.1–1.5	1–3	3–5
Other sources	0–1	2–5	12–20
Total US phytoremediation market	3–8	30–50	100–180

other disciplines. Finally, basic research should be focused on the behaviour of complex systems that include mixtures of contaminants and multiple organisms, and this research must account for the natural spatial and temporal variability of such systems.

References

ADRIANO, D.C. 2001. Trace Elements in the Terrestrial Environment, 2nd ed. Springer-Verlag, New York, 866 p.

ADRIANO, D.C., BOLLAG, J.M., FRANKEMBERGER, W.T., SIMS, R.C. 1999. Biodegradation of contaminated soils. Agronomy Monograph 372, Soil Science Society of America, 772 p.

ADRIANO, D.C., CHLOPECKA, A., KAPLAND, D.I., CLIJSTERS, H., VANGROSVELT, J. 1995. Soil contamination and remediation philosophy, science and technology. In: Prost, R. (Ed.), Contaminated Soils. INRA, Paris, pp. 466–504.

ALLOWAY, B.J. 1995. Heavy Metals in Soil. Chapman and Hall, London.

ARGESE, E., DELAUNEY, E., AGNOLI, F., FARAON, F., SORGANO, A., CACCO, G. 2001. Variazione di parametri morfo-fisiologici di piante di orzo ed avena allevate in microcosmi trattati con metalli pesanti. Boll. Soc. It. Sci. Suolo 50(3), 709–722.

BAKER, A., MCGRATH, S., REEVES, R., SMITH, J. 2000. Metal hyperaccumulator plants: a review of the ecology and physiology of a biological resource for phytoremediation of metal-polluted soils. In: Terry, N., Banuelos, G. (Eds.), Phytoremediation of Contaminated Soils. Lewis Publisher, London, pp. 85–107.

BAKER, A., MCGRATH, S.P., SIDOLI, C.M.D., REEVES, R.D. 1994. The possibility of in-situ heavy metal decontamination of polluted soils using crops of metal-accumulating plants. Resource Conservation and Recycling 11, 41–91.

BAKER, A.J.M. 1981. Accumulators and excluders strategies in the response of plants to heavy metals: Journal of Plant Nutrition 3, 643–654.

BAKER, A.M.J., BROOKS, R.R. 1989. Terrestrial higher plants which hyperaccumulate metallic elements —A review of their distribution, ecology and phytochemistry. Biorecovery 1, 81–126.

BARBAFIERI, M. 2001. Applicabilità e limiti della fitodepurazione. In: Siti contaminati: indagini, analisi di rischio e tecniche di bonifica (a cura di L. Bonomo). Politecnico di Milano, pp. 699–717.

BARGAGLI, R. 1993. Plant leaves and lichens as biomonitors of naturals or anthropogenic emissions of mercury. In: Markert, B. (Ed.), Plants as Biomonitors. VCH, Weinheim, pp. 468–484.

BINI, C., CASARIL, S., PAVONI, B. 2000b. Fertility gain and heavy metal accumulation in plants and soils. Toxic Environmental Chemistry 77, 131–142.

BINI C., MALECI, L., ROMANIN, A. (2007). The chromium issue in soils of the leather tannery district in Italy. Jour. of Geoch. Explor., 96

BINI, C., MALECI, L., GABBRIELLI, L., PAOLILLO, A. 2000a. Biological perspectives in soil remediation with reference to chromium. In: Wise, D. (Ed.), Bioremediation of Contaminated Soils Marcel Dekker Inc., New York, pp. 663–675.

BLACK, H. 1995. Absorbing possibilites: phytoremediation. Environmental Health Perspectives, 103(12), 1–6.

BLAYLOCK, M.J., ELLES, M.P., NUTTAL, C.Y., ZDIMAL, K.L., LEE, C.R. 2003. Treatment of As contaminated soil and water using Pteris vittata. Proc. VI ICOBTE, Uppsala, Sweden.

BLAYLOCK, M.J., SALT, D.E., DUSHENKOV, S., ZAKHAROVA, O., GUSSMAN, C., KAPULNOK, Y., ENSLEY, B.D., RASKIN, I. 1997. Enhanced accumulation of Pb in Indian Mustard by soil-applied chelating agents. Environmental Science and Technology 31(3), 860–865.

BROOKS, R.R. 1998. Phytochemistry of hyperaccumulator. In: Brooks, R. (Ed.), Plants that Hyperaccumulate Heavy Metals: Their Role in Phytoremediation, Microbiology, Archaeology, Mineral Exploration and Phytomining. CAB International, UK, pp. 15–53.

BROOKS, R.R., ROBINSON, B.H. 1998. The potential use of hyperaccumulators and other plants for phytomining. In: Brooks, R. (Ed.), Plants that Hyperaccumulate Heavy Metals: Their Role in Phytoremediation, Microbiology, Archaeology, Mineral Exploration and Phytomining. CAB International, UK, pp. 327–356.

BRYDA, L.K., SELLMAN, C.L. 1994. Recent developments in cleanup technology. Remediation (Autumn), 475–489.

CAILLE, N., SWANWICK, S., ZHAO, F.J., MCGRATH, S. 2003. Assessment of As phytoestraction potential of two ferns by field and pot experiments. Proc. VII ICOBTE, Uppsala, Sweden.

CARRATÙ, G., CARAFA, A.M., APRILE, G.G. 2001. Valutazione della capacità di Pb-decontaminante di due specie ornamentali. Boll. Soc. It. Sci. Suolo, 50(3), 733–738.

CHANEY, R.L., LI, Y.M., ANGLE, J.S., BAKER, A. 1997. Converting metal hyperaccumulator wild plants into commercial phytoremediation systems: approaches and progress. Proc. IV ICOBTE, Berkeley, California, pp. 623–624.

CUNNINGHAM, S.D., BERTI, W.R. 1993. Remediation of contaminated soils with green plants: An overview. In Vitro Cellular Devevelopment Biology 29P, 207–212.

CUNNINGHAM, S.D., BERTI, W.R. 1997. Phytoestraction or in-place inactivation: technical, economic and regulatory consideration on the soil-lead issue. Proc. IV ICOBTE Berkeley, California, 627–628.

DUSHENKOV, V., DUMAR, P., MOTTO, H., RASKIN, I. 1995. Rhizofiltration—The use of plants to remove heavy metals from aqueous streams. Environmental Science and Technology 29, 1239–1245.

DUTCH MINISTRY OF HOUSING, SPATIAL PLANNING AND ENVIRONMENT. 1995. Soil protection act. (Parliamentary Paper I, 1993/94, 21556, No. 266. The Hague, Netherlands, 19 pp.

EIKMANN, T., KLOKE, A. 1993. Ableitungskriterien fur die Eikmann-Kloke-Werte. In: Kreysa, G., Wiesner, J. (Eds.), Beurteilung von Schwermetallen in Boden von Ballungsgebieten: Arsen, Blei und Cadmium. Internationale Expertenbeitrage und Resumee der DECHEMA Abeitsgruppe, pp. 469–500.

EIKMANN, T., KLOKE, A. 1995. Nutzungs- und Schutzgutbezogene Orienterungswerte fur (Schad) Stoffe in Boden. In: Rosenkranz, D., Einsele, G., Harress, H.M. (Eds.), Bodenschtz. Erganzbares Handbuchder Massnahmen und Empefehlungen fur Schutz, Pflege und Sanierung von Boden, Landschaft und Grundwasser 1, 1–7.

ERNST, W.H.O. 1996. Bioavailability of heavy metals and decontamination of soils by plants. Applied Geochemistry 11, 163–167.

EXNER, J.H. 1995. Alternatives to incineration in remediation of soils and sediments assessed. Remediation, 5, 1–18.

FELIX, H. 1998. Field trials for in-situ decontamination of heavy metal polluted soils using crops of metal accumulating plants. Z. Pflanzernhar. Bodenk 160, 525–529.

FERGUSON, J.E. 1992. The Heavy Elements. Chemistry, Environmental Impact and Health Effects. Pergamon Press, pp. 383–397.

GLASS, D. & ASSOCIATES, INC. ESTIMATES. 1997. Estimated U.S. markets for phytoremediation 1997–2005. Explorer-Internet.

GRASSO, D. 1993. Hazardous Waste Site Remediation: Source Control. Lewis Publishers, CRC Press Inc., Boca Raton, Florida.

GREGER, M., LANDBERG, T. 2003. Improving removal of metals from soil by salix. Proc. VII ICOBTE, Uppsala, Sweden.

HUANG, J.W., CUNNINGHAM, S.D. 1996. Lead phytoextraction: species variation in lead uptake and translocation. New Phytology, 134, 75–84.

KABATA-PENDIAS, A., PIOTROWSKA, M., DUDKA, S. 1993. Trace metals in legumes and monocotyledons and their suitability for the assessment of soil contamination. In: Markert, B. (Ed.), Plants as Biomonitors. VCH, Weinheim, pp. 485–494.

KABATA-PENDIAS, A. 1997. Trace metal balance in soil—a current problem in agriculture. In: Adriano, D.C., Chen, Z.S., Yang, S.S. (Eds.), Biogeochemistry of Trace Elements. Advanced Environmental Science, Applied Science Publishers, Northwood, England.

KLOKE, A., SAUERBECK, D., VETTER, H. 1980. The contamination of plants and soils with heavy metals and the transport of metals in terrestrial food chains. In: Nriagu J.O. (Ed.), Changing Mmetal Cycles and Human Health. Life Science Research Report 28, 113–141. Springer, Berlin.

KOS, B., LESTAN, D. 2004. Soil washing of Pb, Zn, and Cd using biodegradable chelator and permeable barriers and induced phytoestraction by Cannabis sativa. Plant and Soil 63(1–2), 43–51.

KRAMER, U. 1996. Free histidine as a metal chelator in plants that accumulate nickel. Nature 379, 635–638.

KUMAR, P., DUSHENKOV, V., MOTTO, H., RASKIN, I. 1995. Phytoextration—The use of plants to remove heavy metals from soils. Environmental Science Technology 29, 1232–1238.

LASAT, M.M. 2002. Phytoextraction of toxic metals: a review of biological mechanisms. Journal of Environmental Quality 31, 109–120.

LEE, C.C., HUFFMAN, G.L., SASSEVILLE, S.M. 1990. Incinerability ranking systems for RCRA hazardous constituents. Hazardous Waste and Hazardous Materials 7, 385–394.

LEPP, N.W., DICKSON, N.M. 2003. Natural bioremediation of metal polluted soils. A case history from the U.K. Proc. VII ICOBTE, Uppsala, Sweden.

LI, H., SHENG, G., SHENG, W., XU, O. 2002. Uptake of trifluralin and lindane from water by ryegrass. Chemosphere 48, 335–341.

LOSI, M.E., AMRHEIN, C., FRANKENBERGER, W.T. JR. 1994. Bioremediation of chromate-contaminated groundwater by reduction and precipitation in surface soils. Journal of Environmental Quality 23, 1141–1150.

MA, L.Q., KOMAR, K.M., TU, C., ZHANG, W., CAI, Y., KENNELLY, E.D. 2001. A fern that hyperaccumulates arsenic. Nature 409, 579.

MARCHIOL, L., SACCO, P., ASSOLARI, S., ZERBI, G. 2004. Reclamation of polluted soils: phytoremediation potential of crop-related BRASSICA species. Water, Air and Soil Pollution 158(1), 345–356.

MASSACCI, A., IANNELLI, M.A., PIETRINI, F. 2001. Il fitorimedio: organismi vegetali come potenziali agenti disinquinanti. Bol. Soc. It. Sci Suolo 50(3), 581–588.

MCGRATH, S. 1995. Behaviour of trace elements in terrestrial ecosystems. In: Prost, R. (Ed.), Contaminated Soils. INRA, Paris, pp. 35–54.

MCGRATH, S. 1998. Phytoextraction for soil remediation. In: Brooks, R. (Ed.), Plants that Hyperaccumulate Heavy Metals: Their Role in Phytoremediation, Microbiology, Archaeology, Mineral Exploration and Phytomining. CAB International, UK, pp. 261–287.

MCGRATH, S., DUNHAM, S.J. 1997. Potential phytoextraction of zinc and cadmium from soil using hyperaccumulator plants. Proc. IV ICOBTE, Berkeley, California, pp. 625–626.

MCNAIR, M.R. 1997. The genetics of metal tolerance and accumulation by plants. Proc. IV ICOBTE 647–648.

MEHMANNAVAZ, R., PRASHER, S.O., AHMAD, D. 2001. Rhizospheric effects of alfalfa on biotransformation of polychlorinated biphenyls in a contaminated soil augmented with *Sinorhizobium meliloti*. Process Biochemistry.

MHATRE, G.N., PANKHURST, C.E. 1997. Bioindicators to detect contamination of soils with special reference to heavy metals. In: Pankhrust, C.E. et al., Biological Indicators of Soil Health. CAB Int., Wallingford, UK, pp. 349–369.

NATIONAL RESEARCH COUNCIL (NRC). 1993. Soil and water quality: an agenda for agriculture. National Academy Press, Washington DC, 516 p.

PANDA, K.K., LENKA, M., PANDA, B.B. 1992. Monitoring and assesment of mercury pollution in the vicinity of a chloralkali plant. Environmental Pollution 76, 33–42.

PANDOLFINI, T., GREMIGNI, P., GABBRIELLI, R. 1997. Biomonitoring of soil health by plants. In: Pankhrust C.E. et al., Biological Indicators of Soil Health. CAB Int, Wallingford, UK, pp. 325–348.

PICHTEL, J., SALT, C.A. 1998. Vegetative growth and trace metal accumulation on metalliferous wastes. Journal of Environmental Quality 27, 618–624.

PRASAD, M.N. 2003. Phytoextraction of metals: role of natural hyperaccumulators, transgenic plants and soil amending chelators. Proc. VI ICOBTE, Uppsala, Sweden.

QUARTACCI, M.F., SGHERRI, C.L.M., NAVARI-IZZO, F. 2001. Fitoestrazione da un suolo contaminato da più metalli: accumulo e tolleranza. Bol. Soc. It . Sci. Suolo 50, 3, 649–660.

RASKIN, I. 1996. Plant genetic engineering may help with environmental cleanup (commentary). Proceedings of the National Academy of Science USA 93, 3164–3166.

REBEDEA, I., EDWARDS, R., LEPP, N.W., LOVELL, A.J. 1995. An investigation into the use of synthetic zeolites for in situ land reclamation. Third International Conference of the Biogeochemistry of Trace Elements. Book of abstracts. Paris.

SACCHI, G.A., RIVETTA, A., COCUCCI, M. 1999. Assorbimento radicale e bio acuumulo di metalli pesanti nelle piante: problematiche e prospettive. In Impatto ambientale di metalli pesanti ed elementi in traccie. Pitagora editrice Bologna, pp. 65–76.

SALOMONS, W. 1995. Environmental impact of metals derived from mining activities: processes, predictions, prevention. Journal of Geochemical Exploration 52, 5–23.

SALT, D.E., BLAYLOCK, M., KUMAR, N., DUSHENKOV, V., ENSLEY, B.D., CHET, I., RASKIN, I. 1995. Phytoremediation: A novel strategy for the removal of toxic metals from the environment using plants. Biotechnology 13, 468–474.

SCHNOOR, J.L. 1995. Phytoremediation of organic and nutrient contaminants. Environmental Science and Technology 29(7).

SETTE, B., BOSCAGLI, A., RICCOBONO, F. 2001. Mentha aquatica L. as an arsenic accumulator. Proc. 6th ICOBTE, Guelph, Ontario, Canada, p. 395.

SHEN, Z.G., LI, X.D., CHEN, H.M., WANG, C.C., CHUA, H. 2001. Phytoextraction of lead from a contaminated soil using high biomass species of plants. Proc. 6th ICOBTE, Guelph, Ontario, Canada, p. 133.

SMITH, R.A.H., BRADSHAW, A.D. 1979. The use of metal tolerant plant population for the reclamation of metalliferous wastes. Journal of Applied Ecology 16, 595–612.

STEUBING, L., HANEKE, J. 1993. Higher plants as indicators of uranium occurrence in soil. In: Markert, B. (Ed.), Plants as Biomonitors. VCH, Weinheim, pp. 155–165.

STREIT, B., STUMM, W. 1993. Chemical properties of metals and the process of bioaccumulation in terrestrial plants. In: Markert, B. (Ed.), Plants as Biomonitors, VCH, Weinheim, pp. 31–62.

STRINGER, D.A. 1990. Hazard assessment of chemical contaminants in soil. ECETOC Technical. Rep. No. 40; Aneme Louise 250, Brussels, Belgium.

TU, S., MA, L., LUONGO, T. 2004. Root exudates and arsenic accumulation in As hyperaccumulating Pteris vittata and non hyperaccumulating Nephrolepis exaltata. Plant and Soil 258(1), 9–19.

VANGRONSVELD, J., STERCKX, J., VAN ASSCHE, F., CLIJSTERS, H. 1995a. Rehabilitation studies on an old non-ferrous waste dumping ground: effects of revegetation and metal immobilization by beringite. Journal of Geochemical Exploration 52, 221–229.

VANGRONSVELD, J., VAN ASSCHE, F., CLIJSTERS, H. 1995b. Reclamation of a bare industrial area contaminated by non-ferrous metals: in situ metal immobilization and revegetations. Environmental Pollution 87, 51–59.

VANGRONSVELD, J., COLPAERT, J., VAN TICHELEN, K. 1996. Reclamation of a bare industrial area contaminated by non-ferrous metals: physico-chemical and biological evaluation of the durability of soil treatment and revegetation. Environmental Pollution.

VOLLMER, M.K., GUPTA, S.K., KREBS, R. 1995. New standards on contaminated soils in Switzerland— Comparison with Dutch and German quality criteria. In: Prost, R. (Ed.), Contaminated Soils. INRA, Paris, pp. 445–459.

WAGNER, G. 1993. Large scale screening of heavy metals burdens in higer plants. In Markert, B. (Ed.) Plants as Biomonitors. VCH, Weinheim, pp. 425–434.

WENZEL, W.W., SATTLER, H. JOCKWER, F. 1993. Metal hyperaccumulator plants: a survey on species to be potentially used for soil remediation. Agronomy Abstracts, p. 52.

WESTINGHOUSE HANFORD COMPANY. 1994. Soil washing: bench-scale tests on 116-F-4 Plupo Crib soil. WHC-SD-EN-TI-268. Westinghouse Hanford Co., Richland, Washington.

WHITING, S.N. 2001. Rhizosphere bacteria mobilize Zn for hyperaccumulation by Thlaspi caerulescens. Environmental Sciency and Technology 35, 3144–3150.

WIESHAMMER, G., SOMMER, P., GORFER, M., STRAUSS, J., WENZEL, W.W. 2003. Bioavailable contaminant stripping of Cd and Zn using willows. Proc. VII ICOBTE , Uppsala, Sweden.

YETZ, H., BAKER, A., WONG, M.H., WILLIS, A.J. 1997. Zinc, lead and cadmium tolerance, uptake and accumulation by Fragmites australis. Annals of Botany 80, 363–370.

ZACCHEO, P., CRIPPA, L., GIGLIOTTI, C. 2001. Studio dell'efficienza del mais nella phytoremediation di un suolo contaminato da metalli pesanti. Boll. Soc. It. Sci. Suolo 50(3), 685–692.

ZAYED, A., GOWTHAMAN, S., TERRY, N. 1998. Phytoaccumulation of trace elements by wetland plants: duckweed. Journal of Environmental Quality 27, 715–721.

ZUPAN, M., HUDNIK, V., LOBNIK, F., KADUNC, V. 1995. Accumulation of Pb, Cd, Zn from contaminated soil to various plants and evaluation of soil remediation with indicator plant (Plantago lanceolata) In: Prost, R. (Ed.), Contaminated Soils. INRA, Paris, pp. 325–335.

ZUPAN, M., KRALJ, T., GRCMAN, H., HUDNIK, V., LOBNIK, F. 2003. The accumulation of Cd, Zn, Pb in Taraxacum officinale and Plantago lanceolata from contaminated soils. Proc. VII ICOBTE, Uppsala, Sweden.

PART III

LAND SUITABILITY AND LAND ZONING

Land Suitability for row crops

9. Bread Wheat (*Triticum aestivum* L.)

Giancarlo Chisci[1] and Letizia Venuti[2]

9.1. Crop Characteristics

Bread wheat (*Triticum aestivum* L.) is originally from the Middle East but is now grown in many regions of the world and in very different environmental conditions, because of its adaptability and the use of different varieties.

9.2. Cultural Practices

9.2.1. Crop Rotation

Crops are rotated in order to increase the level of nitrogen and water storage in the soil and to control weeds and pathogens. The practice of re-sowing wheat on the same parcel causes yield reduction and serious pest attacks. The optimal position for bread wheat within a crop rotation is after a row crop (e.g., maize, potato, sunflower, beet, hemp), which clears weeds from the soil and leaves it in good chemical and physical condition because of the previous deep soil tillage and fertilization. Forage crops are also a good choice before wheat cropping, even if the residual fertility of the soil is better used by a row crop. In dry areas (with precipitation of 300–400 mm/year), the rotation wheat-fallow is very convenient, because it allows both water storage and fertility to be restored (Baldoni and Giardini, 1981). The fallow field can be bare or used for grazing. In these areas, pulse crops (lentil in dry areas and broad bean in less dry areas) are usually included in the crop rotation because they leave a high

---

[1] Università di Firenze, Firenze, Italy.

[2] CNR-IRPI—Istituto di Ricerca per la Protezione Idrogeologica, Unità Operativa di Supporto, Pedologia Applicata, Firenze, Italy.

level of nitrogen in the soil. Where rainfall is slightly more abundant (> 400 mm/year), pulse crops replace fallow in crop rotation.

9.2.2. Cultivation

Soil Preparation

In wheat cultivation a precise soil preparation is needed to achieve high yield for two main reasons: first, an even surface can be sown precisely and therefore an optimal plant density can be obtained per unit surface; second, it is useful in controlling the growth of weeds. An optimal soil preparation is obtained not only with a suitable tillage but also with an appropriate protection of the field lay-out to prevent flooding on flat areas and to reduce surface and deep soil erosion on sloping areas.

An efficient system to remove water (ditches, drainage) is essential in flat areas. If the soil is not sufficiently permeable, it is necessary to slope the field to facilitate drainage.

Soil tillage may differ with time, type and depth of tillage, previous crop, crop harvesting time, soil and climate conditions, and other factors.

Ploughing is the main operation: if at the ploughing time the soil humidity is high, a common situation when the previous crop is harvested at the end of summer, the result can be poor. The likelihood of finding the optimal soil moisture condition for ploughing depends on the type of soil: it is higher in sandy permeable soils and lower in heavy clayey soils. When soil moisture conditions are unfavourable, it is better to use tools that open the furrow instead of the plough, even if they require greater efforts to control weed growth, especially when the time for seedbed preparation is short.

For wheat cultivation, deep tillage is often not necessary, especially if the previous crop is a row crop. However, when wheat cultivation is preceded by a hay field, deep tillage becomes necessary.

With clayey soils, the previous crop should be harvested in early summer so that the field can be ploughed in dry soil conditions. This breaks up big clods through the environmental effects of sun exposure and variations of temperature and humidity of day and night. It also favours the germination of weeds after the first summer or autumn rainfalls, weeds that can then be killed in the subsequent seedbed preparation.

After the primary tillage, complementary tillage operations follow, using different types of tools (e.g., grubber, seed belt tiller, rotary plough, disk harrow, tooth harrow) depending on the soil characteristics. This secondary tillage is necessary to prepare a smooth and even seedbed for the seed to find the best conditions to germinate.

The technique of seeding on an untilled soil (sod seeding) has also been tested on wheat by using special seeding machines and killing weeds with herbicides. The minimum tillage technique seems to be more satisfactory, particularly when the previous crop is harvested in late summer or autumn, rainfalls are frequent, and soils are poorly drained.

Seeding

The seed should be healthy, treated with fungicides and characterized by high purity and germination capacity. In Italy, seeding is often done in early autumn, earlier in the north and later in the south. Early seeding is also important at higher altitudes. In colder countries, spring seeding is more frequent than autumn seeding, or it is the only option.

The amount of seed per unit area depends on the desired crop density, the average kernel weight, and those factors that affect the capacity to germinate. Where soil fertility and sufficient water availability are high, a high crop density favours high yields, whereas in arid areas it could cause water deficiency, especially in the kernel filling stage. A high density can also increase the probability for the crop to lodge. In good climatic and soil conditions, the optimal crop density is about 550–600 plants per square metre and can be obtained with a density of 400–500 kernels per square metre. In good seeding conditions, the amount of seed varies from 160 kg/ha in southern areas to 180 kg/ha in northern areas. It is possible to reach 230–250 kg/ha where seeding conditions are more difficult and when sowing is late.

Nowadays, wheat is sown in rows using a wheat-seeding machine. It can be done in simple rows (row spacing 15–20 cm) or paired rows (rows spacing 12–15 cm and paired-row spacing 25–30 cm). The seed is planted at a depth of 3–4 cm. In soft and dry soils, the soil is then lightly rolled so that the kernel will better adhere to the soil and so as to favour the absorption of water necessary for germination.

Fertilization

In wheat cultivation, nutrient uptake depends on many climatic and soil factors and on crop yield. For a seed production of 2.7 t/ha, nutrient uptake is 48 kg/ha nitrogen, 16 kg/ha phosphorus and 29 kg/ha potassium; for a seed production of 6.0 t/ha, it is 132 kg/ha nitrogen, 28 kg/ha phosphorus and 83 kg/ha potassium.

In principle, wheat crop should be supplied with phosphorus and potassium, and also a certain amount of slow-release nitrogen as accumulated fertility. Phosphorus and potassium should come from forage or row crops prior to wheat in the crop rotation, while the slow-release nitrogen should come either from organic residues of the previous forage crop or from manuring of the previous row crop. Phosphorus and sometimes potassium fertilization at wheat sowing time is necessary in case of re-sowing, especially when repeated: the quantity of fertilizer depends on the amount of assimilable phosphorus and exchangeable potassium of the soil.

Nitrogen fertilization is important to achieve a high yield. The requirement is greatest at the raising time. Nitrogen content in leaves and culm is maximum at blooming time. The quantity of nitrogen fertilizer required depends on many factors: cultivar, soil nitrogen content, available nitrogen fraction, organic matter mineralization rate, climatic conditions that influence the soil microbial activity, nitrogen loss due to runoff, and water availability in the soil as the kernels form and ripen.

In northern Italy, the total amount of nitrogen generally used is 150–180 kg/ha, up to 200–250 kg/ha for short lodging-resistant cultivars. In warm, arid southern areas, nitrogen does not exceed 100–120 kg/ha, and it is often conveniently held below 70–100 kg/ha.

A fundamental timing for nitrogen supply is at the greening up (5th–6th leaf); that means before the critical time of the raising. A second critical time is at the beginning of the boot time, in order to satisfy the demand of nitrogen needed to increase the protein content of the seed. Two additional supplies can be considered on a case-by-case basis. When the organic matter content is low, or the wheat crop is re-sowed, 20–30 kg/ha of low-release nitrogen can be supplied at sowing time. In case of cold and wet winters, when plants show broad yellow areas, it is important to apply about 20 kg/ha of nitrogen as nitrates, which can also supply the roots with oxygen.

Weed Control

Since wheat is sown in narrow rows, weeding was done manually in the past, a practice no longer possible because of the high cost of labour. Presently, the only efficient and dependable means is chemical weed control, which leads to pollution, especially if the herbicides are not correctly applied. However, it is important to adopt every direct or indirect method to reduce the consequence of weed infestation, such as crop rotation, correct tillage, false sowing, and use of seed characterized by high quality and purity.

Weeds can be controlled at different stages: pre-sowing, when the technique of sod seeding is adopted, using total-action desiccant herbicides; pre-emergence, when both monocotyledonous and dicotyledonous weed infestation risk is high, using herbicides characterized by broad-spectrum radical and sprouting inhibitor action; post-emergence, more practical and rational, because done when the type of weeds to be controlled is known and it is possible to evaluate whether the level of infestation is worth the expense of weeding. Today, any agronomic operation capable of reducing the use of herbicides, such as even partial mechanical weeding, as enabled by a row seeding, is greatly advisable.

Irrigation

Wheat crop usually does not need irrigation because of the season of its growing cycle. Nevertheless, it is useful to effect a few rescue irrigations during years characterized by drought at the time of ear development and grain filling, as may happen in southern Italy, where rainfall scarcity at the end of winter or beginning of spring can cause grain shrivelling.

Harvest and Storage

Wheat grain is harvested after physiological ripeness and consists of two main operations: cutting of the plant (mowing) and separation of seed from chaff and straw (threshing). These operations are now carried out simultaneously by the combine harvester when the seed is completely ripe (less than 13% of water content, 2–3 weeks after physiological ripeness). The standard moisture of the commercial product is 14.5%, but seeds can safely be stored at water content not higher than 13%.

9.3. Crop Requirements

9.3.1. Climate

Wheat is a viable crop in a wide range of climatic conditions, though its yield is higher in temperate regions: spring wheat varieties are sown in cold areas and autumn varieties in warm-arid and warm-temperate areas. Wheat is not suitable in warm-humid climate because of the high frequency of pest attacks. In Mediterranean regions, wheat is subject to water shortage in the period in which it most needs water, that is, at the stage of seed filling. Durum wheat is less resistant to low temperatures and more resistant to drought than bread wheat. Day-neutral cultivars are used in areas with mild winters and long-day cultivars are used in higher latitudes.

Wheat is resistant to low temperatures especially in the period from the appearance of the first leaves to the beginning of the raising: the most resistant varieties can withstand up to −20°C.

There are four critical stages for water requirements:

- emergence
- beginning of raising
- end of earing—blooming
- grain filling

Wheat is a rainfed crop in temperate regions, whereas in dry and semi-arid regions irrigation is very useful. The average water consumption can vary between 450 and 650 mm

Table 9.1. Climatic requirements for wheat.

Climatic requirements	Class and rating scale			
	Highly suitable	Moderately suitable	Marginally suitable	Not suitable
Mean temperature of growing cycle (°C)	12–23	10–12, 23–25	8–10, 25–30	< 8, > 30
Mean annual precipitation (mm)	350–1,250	250–350, 1,250–1,500	200–250, 1,500–1,750	< 200, > 1,750

(Baldoni and Giardini, 1981). Excessive rainfalls have harmful effects during growth, when they cause lodging, and at the time of harvest (Table 9.1).

9.3.2. Soil

Wheat is not particularly demanding with regard to soil characteristics, provided it has enough water and nutrients and waterlogging is avoided. It can tolerate a wide range of pH, as well as alkaline soils and moderately saline soils, even though a neutral pH (between 6.5 and 7.8) is optimal. In particular, durum wheat is less tolerant of salinity than current new varieties of bread wheat. In Italy, the most suitable soils for wheat are well-drained clay soils.

9.3.3. Nutrients

Wheat needs easily and readily assimilable nutrients, because of its type of rooting system (embryonic roots in the primary stage of development and adventitious roots at bunching time). Nutrient requirement varies according to the cultivar and the yield level. In the past, nitrogen was typically the only fertilizer applied, whereas nowadays phosphorus and potassium are also considered essential to reach an acceptable yield level and to enable a regular bunching. Phosphorus availability also helps root system development and therefore increases drought tolerance.

9.3.4. Water

Wheat usually does not need irrigation, as it is a winter crop. However, during years characterized by insufficient rainfalls, it is advisable to supply water as needed, where an irrigation system is available. The most critical stage for water requirement is the boot stage.

9.4. Land Suitability

Bread wheat is especially viable in fertile soils (residual fertility) having a texture from loam to clay, while it provides inadequate yields in sandy, barren or acid soils. Winter cultivars are well tolerant of cold weather and need increasingly higher temperatures starting from the raising time. At the ripening stage, wheat is favoured by warm and dry weather, while dry hot winds can cause grain shrivelling and lead to low yield. Water requirements are generally moderate and specifically concentrated in the period of time between the raising and the beginning of

Table 9.2. Bread wheat requirements.

Requirements	Value
Climatic	
Air temperature	moderate to high
Air humidity	low
Soil	
Texture	clay to loam
Oxygen availability	high
Water availability	moderate
Soil depth (ground water, soil horizons impenetrable to roots)	–
Nutrient availability	moderate
Soil salinity	moderate
Soil pH	sub-acid to sub-alkaline
Management	
Trafficability	moderate
Workability and mechanization	moderate
Land conservation	
Erosion risk	low
Groundwater pollution risk	moderate to low
Soil compaction risk	moderate to low

ripening. Wheat is only moderately tolerant of waterlogging, especially during cold weather, which causes pest attacks and sparse root growth; it is also poorly tolerant of springtime hard winds and storms, as they can cause lodging. Crop requirements for bread wheat are listed in Table 9.2.

Beyond common crop requirements, various bread wheat varieties can have different production characteristics and different degrees of tolerance to adverse environmental conditions.

9.5. Soil Characteristics and Suitability Assessment Table

In the 1980s, a few experiments on soil assessment for wheat cultivation were carried out in different Italian regions using different approaches. For example, the Emilia Romagna Administration charted soil suitability maps for wheat, covering a few flat areas inside the region (Regione Emilia Romagna, 1984, 1987); in central Italy, Raglione et al. (2002) issued the suitability map for wheat on the Leonessa plateau.

Table 9.3 shows an example of an assessment table based on data from different experiments quoted in literature (Sys et al., 1993; Raglione et al., 2002; Magaldi, 1983).

Table 9.3. Soil requirements for bread wheat.

Soil characteristics	Class and rating scale			
	Highly suitable	Moderately Suitable	Marginally suitable	Unsuitable
Depth (cm)	> 60	40–60	20–40	< 20
Texture[1]	SiC, C, Si, SiL, CL, SC, L, SiCL	SCL	SL, LS	S, $C_{massive}$, $SiC_{massive}$
Salinity (dS/m)	< 4	4–10	10–14	> 14
Gypsum (%)	< 6	6–10	10–20	> 20
pH	6–8.2	5.6–6; 8.2–8.3	5.2–5.6; 8.3–8.5	< 5.2; > 8.5
Sum of bases (Ca, Mg, K, Na) (cmol+/kg soil)	> 5	3.5–5	2–3.5	< 2
Base saturation (%)	> 50	35–50	< 35	
Content of organic carbon (%)	> 1	0.5–1	< 0.5	
Rock fragments (%)	< 15	15–35	35–70	> 70
Exchangeable sodium percentage (%)	< 20	20–35	35–45	> 45
Internal drainage[2]	well drained, moderately well drained (fine and medium soils); moderately well drained, somewhat poorly drained (coarse soils)	somewhat poorly drained (fine and medium soils); well drained (coarse soils)	somewhat excessively drained, excessively drained	poorly drained, very poorly drained
Flooding risk	none	none	rare	occasional or more frequent
Total calcium content: $CaCO_3$ (%)	10–20	0.5–10, 20–40	< 0.5, 40–60	> 60
Available water capacity (mm)	> 125	61–125	< 61	
Cationic exchange capacity (cmol+/kg soil)	> 20	10–20	5–10	< 5

[1] *SiC silty clay, C clay, Si silt, SiL silt loam, CL clay loam, SC sandy clay, L loam, SiCL silty clay loam, SCL sandy clay loam, SL sandy loam, LS loamy sand, S sand, $C_{massive}$ massive clay, $SiC_{massive}$ massive silty clay.*
[2] *From: Sys et al. (1993), modified according to classes quoted in Appendix.*

References

BALDONI, R., GIARDINI, L. (Eds.). 1981. *Coltivazioni erbacee*. Patron Publisher, Bologna, pp. 15–89.

MAGALDI, D. 1983. *Rilevamento del suolo e classificazione del territorio per usi agricoli e non agricoli*. Sviluppo Agricolo no. 8–9 (August–September).

RAGLIONE, M., LORENZONI, P., ANGELINI, R., BONIFAZI, A., FEBELLI, C., SPADONI, M., VENUTI, L., VERZILLI, C. 2002. *Carta dei suoli dell'altopiano di Leonessa (Rieti) e delle loro idoneità per alcune colture tipiche*. Arti Grafiche Nobili Sud Publisher, Rieti, 279 p.

REGIONE EMILIA ROMAGNA. 1984. *Cartografia tematica per la valutazione del territorio del Comprensorio Bassa Pianura Modenese*. Pivetti Publisher, Mirandola, 123 p.

REGIONE EMILIA ROMAGNA. 1987. *I suoli della Bassa Pianura Ravennate*. Gamma, Bologna, 83 p.

SYS, C., VAN RANST, E., DEBAVEYE, I.R.J, BEERNAERT, F. 1993. *Land Evaluation, Part III. Crop Requirements*. Agriculture publications no. 7, General Administration for Development Cooperation, Brussels, Belgium, 166 p.

10. Durum Wheat (*Triticum durum* Desf.)

Giancarlo Chisci[1] and *Letizia Venuti*[2]

10.1. Crop Characteristics

The cultivation of durum wheat (*Triticum durum* Desf.) is continuously expanding because of the constant increase of pasta consumption in the world. This crop is originally from Ethiopia, but now it is mostly grown in the Mediterranean countries, North America, Argentina and Eastern Europe. In Italy, durum wheat is produced in the centre and the south, especially in Puglia and Sicily. Until the 1960s, durum wheat was considered a crop for poor and hot environments, starting in the 1980s it spread into central Italy, initially as a spring crop and later as winter crop. This was made possible by the introduction of new highly productive and winter-hardy cultivars. In the past few years, cultivation has spread to northern areas as well, even though the results achieved are not always qualitatively good.

    The crop is harvested in early June in southern Italy and at the end of June in the north.

10.2. Cropping Techniques

10.2.1. Crop Rotation

Crop rotation has three main objectives. The first is to increase the soil nitrogen accumulation, which can be obtained through the inclusion of leguminous fodder crops (no less than 50% alfalfa or clover) for at least two out of the five-year crop rotation, or of pulse crops (broad bean, pea, lentil and chick pea), for the last three years of the rotation. The second objective is

---

[1] Università di Firenze, Firenze, Italy.
[2] CNR-IRPI—Istituto di Ricerca per la Protezione Idrogeologica, Unità Operativa di Supporto, Pedologia Applicata, Firenze, Italy.

to break the cycle of grain cultivation in order to prevent the propagation of specific pests and diseases. The third is to prevent the spread of weeds while the field is not planted with cereals: weed infestation the year before harvesting can reduce the protein content of the kernel by up to one percentage point.

10.2.2. Cultural Practices

Soil Preparation

Durum wheat cultivation techniques are very similar to those of bread wheat. During the soil preparation, ploughing is typically deeper and the number of tillages is higher, especially in order to increase and preserve the water storage of the soil. In Italy, it is possible to use a shallow ploughing, at a depth of 20–25 cm, and a lower number of tillages, up to two shallow ploughings before seeding. Ploughing deeper is preferable on clay soils in order to help water drainage.

Seeding

Seeding of durum takes place slightly earlier than that of bread wheat, in order to promote bunching and advance blooming and ripening. In the past, the adopted seed density was much lower than for bread wheat, whereas nowadays the amount of seeds has increased, especially in areas that are not dry, although it is still lower than in bread wheat seeding. Usually a density of 350–400 seeds/m^2 is used, equivalent to 180–200 kg/ha. In areas where rainfall is a limiting factor, it is better to adopt a lower seed density (250–300 seeds/m^2) to prevent plant competition for water.

Fertilization

Phosphorus and nitrogen fertilization are recommended in durum wheat cultivation. Potassium supply can also be important, but only where soils are not naturally rich in this nutrient. For phosphorus and nitrogen fertilization, the rules are the same as for bread wheat, that is, the fertilizer supply must be related to the residual content of these nutrients in the soil. In particular, nitrogen should be used in substantial quantity, albeit bearing in mind the risks of lodging and grain shrivelling, especially the latter considering the late ripeness of durum. Short, lodging-resistant cultivars can be supplied with 150–200 kg/ha nitrogen, or smaller amounts in dry areas. Nitrogenous fertilization in durum wheat increases grain production as well as its protein content by decreasing the variation in endosperm protein composition. Late fertilization is also very effective in preventing variation in endosperm protein composition.

Weed Control

Herbicide quantities should be lower than for bread wheat, as durum is more sensitive to herbicide toxicity. Weeding techniques and herbicides are the same as for bread wheat.

Irrigation

Like bread wheat, durum does not usually need irrigation, because of the season of its growing cycle. Nevertheless, drought risks at ear formation and at grain filling times are much higher than for bread wheat especially in southern regions, where durum is widespread and the yield variation depends mostly on rainfall patterns. For this reason, irrigation, whenever possible, is essential for yield preservation particularly in Mediterranean regions, where rainfall scarcity at the end of winter or beginning of spring can cause grain shrivelling.

Harvesting and Storage

Harvesting techniques for durum are the same as for bread wheat and yield levels are today very similar. Therefore, the economic convenience of cultivating either species depends on the local market price, the region of cultivation and the availability of high quality cultivars. From durum seeds comes semolina, the raw material for the production of pasta. Semolina is constituted by various-size, sharp-edged, non-floury endosperm fragments and it is produced via milling techniques different from for those used for bread wheat, which are aimed at flour production.

10.3. Crop Requirements

10.3.1. Climate

Durum wheat has generally a medium-high tolerance to drought. The optimal environment is characterized by precipitations concentrated in winter, without risks of late frost, and by warm springs and summers. Durum is therefore spread out in all the Mediterranean regions, where, however, it is also susceptible to variable and often adverse climatic conditions and therefore to stress risks. In fact, these regions have a scarce average annual rainfall (between 200 and 800 mm) with an irregular distribution; in addition, temperature can at times reach extreme values (hot or cold) during any crop stage. Climatic conditions are the main cause of yield reduction and variation (Nachit and Elouafi, 2004).

Unlike bread wheat, the kernel of durum wheat is characterized by a vitreous fracture. Where the climate is not suitable, presenting low temperature and high humidity, the inner part of the kernel can become floury as a consequence of a variation in the endosperm protein composition.

In difficult areas, the old mid-size cultivars are still widespread. However, new short cultivars have been introduced: they are more productive and more adapted to high levels of nitrogenous fertilizers. In northern Italy, winter-hardy cultivars have also been introduced.

Durum wheat is more drought-resistant than bread wheat and therefore more adaptable to arid areas. The most critical period for durum is the blooming stage, when the climatic and water requirements are most demanding (see Table 10.1).

Table 10.1. Climatic requirements for durum wheat.

Climatic requirements	Class and rating scale			
	Very suitable	Moderately suitable	Marginally suitable	Unsuitable
Mean temperature of growing cycle (°C)	15–25	10–15; 25–27	8–10; 27–30	< 8; > 30
Mean annual precipitation (mm)	300–1250	250–300; 1250–1500	200–250; 1500–1750	< 200; > 1750

10.3.2. Soil

Durum wheat is highly productive in fertile soils, even if here it is prone to fungal diseases (*Fusarium, Septoria, Erisiphe, Puccinia* spp.) and lodging. Soil characteristics are one of the most critical factors for a satisfactory yield. Clay and loamy soils are the most suitable for durum, if they are well drained, more than 50 cm deep, and well structured. When cultivated in sandy soils, durum wheat is less productive and its kernel has lower protein content than bread wheat. It can tolerate soil acidity: pH value should, however, be always higher than 6 (measured in $CaCl_2$) in order to get a satisfactory yield, although similar results have been obtained on soils with a value of pH of 5.5 in the topsoil but 7 or more in the subsoil.

10.3.3. Nutrients

As durum wheat needs a high level of nitrogen content, it is useful to introduce leguminous fodder crops in the rotation, as they leave a high quantity of nitrogen in the soil. Phosphorus deficiency can occur after a long period of fallow and drought. Durum is more sensitive than bread wheat to zinc scarcity.

10.3.4. Water

Being a winter crop, durum wheat does not usually need irrigation in Italy. However, in particularly dry years, it is advisable to provide some irrigation whenever irrigation is available. The most critical stage is at the boot time.

10.4. Land Suitability

Durum wheat has good resistance to drought and gives a high yield on fertile soils. On the negative side, it is often prone to fungi diseases (*Fusarium, Septoria, Erisiphe, Puccinia* spp.) and to lodging. Besides common requirements, different durum cultivars have different characteristics and tolerance to environmentally adverse conditions. Crop requirements for durum wheat are listed in Table 10.2.

Table 10.2. Durum wheat requirements.

Requirements	Value
Climatic	
Air temperature	high
Air humidity	low
Soil requirements	
Texture	clay to loam
Oxygen availability	high
Water availability	moderate
Soil depth (ground water, soil horizons impenetrable to roots)	moderate
Nutrient availability	moderate
Soil salinity	moderate
Soil pH	neutral to sub-alkaline
Management	
Trafficability	moderate
Workability and mechanization	moderate
Land conservation	
Erosion risk	moderate
Groundwater pollution risk	moderate to low
Soil compaction risk	moderate to low

10.5. Soil Characteristics and Suitability Assessment Table

A soil suitability evaluation map of soils on Pliocenic Clay in Valdera, Tuscany, was charted by Lulli and Delogu (1978) and Lulli et al. (1980). Durum quality and yield were related to soil characteristics mainly linked to water and oxygen availability. Raglione et al. charted a suitability map for durum wheat in the Leonessa Plateau (Raglione et al., 2002). An example of suitability assessment table for durum wheat based on data coming from the above quoted studies in Tuscany and on others (Baldoni and Giardini, 1981; Sys et al., 1993; Raglione et al., 2002) is shown below (Table 10.3).

Table 10.3. Soil requirements for durum wheat.

Soil characteristics	Class and rating scale			
	Very suitable	Moderately Suitable	Marginally suitable	Unsuitable
Depth (cm)	> 60	40–60	20–40	< 20
Texture[1]	SiC, C, Si, SiL, CL, SC, L, SiCL	SCL	SL, LS	S, $C_{massive}$, $SiC_{massive}$
Salinity (dS/m)	< 4	4–10	10–14	> 14
Gypsum (%)	< 6	6–10	10–20	> 20
pH	6–8.2	5.6–6; 8.2–8.3	5.2–5.6; 8.3–8.5	< 5.2; > 8.5
Sum of bases (Ca, Mg, K, Na) (cmol+/kg soil)	> 5	3.5–5	2–3.5	< 2
Base saturation (%)	> 50	35–50	< 35	
Content of organic carbon (%)	> 1	0.5–1	< 0.5	
Rock fragments (%)	< 15	15–35	35–70	> 70
Exchangeable sodium percentage (%)	< 20	20–35	35–45	> 45
Internal drainage[2]	well drained, moderately well drained (fine and medium soils); moderately well drained, somewhat poorly drained (coarse soils)	somewhat poorly drained (fine and medium soils); well drained (coarse soils)	somewhat excessively drained; excessively drained	poorly drained; very poorly drained
Flooding risk	none	none	rare	occasional or more frequent
Total calcium content: $CaCO_3$ (%)	10–20	0.5–10; 20–40	< 0.5; 40–60	> 60
Available water capacity (mm)	> 125	61–125	< 61	
Cationic exchange capacity (cmol+/kg soil)	> 20	10–20	5–10	< 5

[1] *SiC silty clay, C clay, Si silt, SiL silt loam, CL clay loam, SC sandy clay, L loam, SiCL silty clay loam, SCL sandy clay loam, SL sandy loam, LS loamy sand, S sand, $C_{massive}$ massive clay, $SiC_{massive}$ massive silty clay.*
[2] *From: Sys et al. (1993), modified according to classes quoted in Appendix.*

References

BALDONI, R., GIARDINI, L. 1981. *Coltivazioni erbacee*. Patron Publisher, Bologna, pp. 15–89.

LULLI, L., DELOGU, F. 1978. Differenze nello sviluppo e nella produzione del frumento duro (var. Appuro) indotte dalle pratiche di livellamento. Annali ISSDS IX, 139–162.

LULLI, L., DELOGU, F., GREGORI, E., MICLAUS, N. 1980. Cartografia di base per la programmazione degli interventi in aree marginali. II. Memorie illustrative della Carta dei Suoli: Carta dei suoli; attitudine dei suoli alla produzione del frumento duro; attitudine dei suoli alle colture legnose. Annali ISSDS XI, 33–96.

NACHIT, M., ELOUAFI, I. 2004. *Durum Wheat Adaptation in the Mediterranean Dryland: Breeding, Stress Physiology, and Molecular Markers—Challenges and Strategies for Dryland Agriculture*. CSSA (Crop Science Society of America and American Society of Agronomy), Special Publication no. 32.

RAGLIONE, M., LORENZONI, P., ANGELINI, R., BONIFAZI, A., FEBELLI, C., SPADONI, M., VENUTI, L., VERZILLI, C. 2002. *Carta dei suoli dell'altopiano di Leonessa (Rieti) e delle loro idoneità per alcune colture tipiche*. Arti Grafiche Nobili Sud Publisher, Rieti, 279 p.

SYS, C., VAN RANST, E., DEBAVEYE, I.J., BEERNAERT, F. 1993. *Land Evaluation, part III. Crop Requirements*. Agriculture publication no. 7, General Administration for Development Cooperation. Brussels, Belgium, 166 p.

11. Maize (*Zea mays* L.)

Letizia Venuti[1] and *Giancarlo Chisci*[2]

11.1. Crop Characteristics

The geographic belt in which maize (*Zea mays* L.) is cultivated is very wide, stretching from the 30th to the 55th parallel, in the southern and northern hemispheres. Most of the maize production, however, is concentrated in the boreal hemisphere, between the 35th and the 45th parallel. In regions further north, the limiting factor is temperature, whereas in southern regions scarce water availability becomes critical and makes irrigation essential.

Maize is a heliophilous crop and has a high photosynthetic efficiency (C4 crop). It is a short-day crop, whereas new varieties are day-neutral. As for rainfall, maize is spread over a wide rainfall belt, stretching from 250 to 5,000 mm/year. Mean water consumption is variable, depending on plant productivity, temperature and air humidity.

11.2. Cropping Techniques

11.2.1. Crop Rotation

Maize is a row crop and it is usually cultivated in rotation. In fact, deep tillage and abundant fertilization leave at harvest plenty of residual nutrients. Because of the type of tillage and the herbicides used, maize also clears fields of weeds. In the rotation, maize is usually followed by a cereal crop (especially wheat) and sometimes by grass. Thanks to new techniques and abundant fertilization, maize is also sown for more years, even if that is not advisable for it has a negative effect especially on fine-textured soils, with a consequent deterioration of their

[1] CNR-IRPI—Istituto di Ricerca per la Protezione Idrogeologica, Unità Operativa di Supporto, Pedologia Applicata, Firenze, Italy.
[2] Università degli Studi di Firenze, Firenze, Italy.

physical properties. Moreover, re-sowing causes maize-specific weed infestation (especially *Sorghum halepense*).

11.2.2. Cultural Practices

Soil Preparation

Maize is the best row crop, and tillage before seeding is often deep (40–50 cm) in order to incorporate manure and to favour root deepening and improve water storage, especially in clay soils of rainfed agriculture in the Mediterranean areas. Deep tillage is carried out by a share plough, but it is advisable to perform a double tillage, both deep and shallow, starting with a grub-breaker ploughing coupled with a skim ploughing. After the late-summer or autumn ploughing, it is necessary to crumble the big clods left by the plough and to control the young weeds that have just come up. This complementary tillage should be performed early enough before sowing so as not to damage the soil structure, which needs months and natural elements to be formed.

Maize does not need a minutely prepared seedbed because the seed is rather large and it is planted deep; however, a well-prepared seedbed favours earlier and regular germination.

When maize is grown as a second crop in the year, the sod seeding technique can be usefully practised in order to save work, speed up operations and get a high yield. This is accomplished by using a special drill provided with small coulters to cut the soil and place the seed at the right depth. After a careful evaluation, it may be useful to use soil-applied insecticides against soil insects, by spreading products in granular formulation on the row, in order to obtain a good product yield.

Fertilization

Being a spring-summer crop, maize benefits from manuring, as the mineralization of organic matter keeps step with nutrient crop requirements. In the past, manuring was the main maize fertilization, whereas now many farmers use only chemical fertilizers and sometimes also non-traditional organic fertilizers such as sewage sludge or compost.

Irrigated maize absorbs on average 300 kg/ha of nitrogen, 144 kg/ha of phosphorus and 240 kg/ha of potassium for a production of 12 t/ha of dry grain. In fertile soils, fertilization should comprise approximately 250 kg/ha of nitrogen (NO_3), 80–120 kg/ha of phosphorus (P_2O_5) and 50–100 kg/ha of potassium (K_2O). When the preceding crop is leguminous hay, nitrogen can be reduced to 150–200 kg/ha. With rainfed cultivations, fertilizer quantity should be halved; otherwise, it could be useless or even harmful. On clay soils where maize is grown as rainfed crop, potassium fertilization can often be omitted.

Manuring and the nitrogen-phosphorus fertilization should be carried out before sowing: manuring at ploughing and the mineral fertilization in part at ploughing and in part at harrowing for the seedbed preparation. Presently, nitrogen fertilization is performed at sowing using slow-release non-leaching nitrogen fertilizers (usually urea).

Seeding

Spring seeding should be done at the earliest time, but only when the temperature is no lower than 12°C. A correct plant density is critical to achieve a high yield: if the density is too high it reduces the LAI (*Leaf Area Index*) and consequently the light capture, cutting down on the ear's nutrient assimilation and fertility, because of excessive shade. Moreover, the low number of ears (modern cultivars have only one ear) reduces the productivity per unit surface.

Maize density depends on different farming systems. For an irrigated maize crop, an adequate density is 6–8 plants/m²; a rainfed crop, 2.5–4 plants/m²; a fodder crop can have an additional plant; a catch grain crop, 7–10 plants/m²; a catch fodder crop for harvesting at the blooming time, 30–50 plants/m².

Inter-row distance should be 0.7–0.8 m in order to allow mechanical farming operations. Seed rate can range from 15 to 24 kg/ha, an approximate rate because maize seeds are sold by number. Sowing depth should be homogenous and properly selected; the recommended depth is 4 cm, or 6 cm if the soil is dry. The seed should be treated with fungicides: hybrid seeds are sold pre-treated by producers.

Weed Control

Weed control ensures high yield in both irrigated and rainfed maize cultivations. In the past, fields were weeded by hoeing and manual weeding.

Mechanical hoeing is not sufficient to control weeds: mechanical tools work only between rows and soil conditions often make it difficult to enter the fields early, before maize covers the ground. Effective herbicides and techniques are currently available to kill maize-specific weeds, such as some monocotyledons (*Sorghum halepense, Echinochloa crus-galli, Setaria* spp., *Digitaria sanguinalis*) and dicotyledons (e.g., *Amaranthus* spp., *Chenopodium* spp., *Convolvus arvensis, Fallopia convolvulus, Polygonum* spp., *Portulaca oleracea, Veronica* spp., *Solanum nigrum*). In the past, the use of atrazine gave the best results in maize-specific weed control, because of its highly effective strength and selectivity. But atrazine was later banned because it contaminated ground water. After the banning of atrazine, other active substances were produced but none of these has the same complete spectrum, and it is necessary to perform multiple treatments with different active substances used individually or blended and at different times.

Herbicides can be applied before or after emergence:

- Pre-emergence application is carried out immediately after sowing or at sowing time if an in-row application is carried out. In-row herbicide application combined with inter-row mechanical weed control can save money and reduce contamination risk.
- Post-emergence application is usually carried out, if necessary, as a complement to the pre-emergence application.

Other Cultural Practices

In the past, farmers used to thin maize plants to adjust plant density. This operation is now superseded because of the advent of the precision seed units. An old practice still common is hoeing, useful to integrate chemical herbicides, which do not ensure full weed control, to reduce evaporation losses, and to aerate soil. Hoeing should be performed when plants are no higher than 0.6–0.7 cm. The addition of ridging, as commonly used in the past to enhance rooting, is very useful, but it interferes with stalk shredding.

Irrigation

Maize water consumption per unit surface is very high. Therefore, in order to obtain high yields, maize needs to be grown in areas where water availability is not limiting. The main critical period is during flowering, i.e., in July in the northern hemisphere, but generally water should be available from tassel emergence to the milk-dough maturity stage.

On clay soils, 500 m³/ha of water is needed to fill the soil profile at 0.7 cm depth; on sandy soils, 250 m³/ha is needed to reach the same result. Assuming that during the summer potential evapotranspiration is 7 mm/d, if rainfall does not occur, the crop should be irrigated every 7 d (clay soils) or every 3 d (sandy soils). Usually sprinkling or furrow irrigation is most suitable for maize.

Harvesting

Maize can be used to produce grain or silage for cattle. Grain maize is harvested by picking of ears or kernels. The first method is traditional: ears are cut manually, the husks are stripped, the ears are dried, and grains are removed manually or mechanically. The second one is much quicker and more common and it makes use of combine harvesters, which carry out harvesting and ginning at the same time. The optimal time for harvesting is when the grain water content is 24–26%; if grains are drier, there is a risk of loss during cutting, whereas if they are wetter, harvesting is difficult and grains are likely to be broken. There are many ways to dry and store grain. For commercial use, grain should have a water content no higher than 13%, even if the standard commercial humidity is 15.5%. To prevent moulding, maize should be artificially dried in hot-air dryers.

Maize grain for cattle feed can be ensiled as follows:

1. wet flour storage in bunker silos;
2. grain storage in metal silos under anaerobic conditions;
3. grain storage in trench silos using preservative treatments.

For farm use, stalks can be harvested at the dough stage, appropriately chopped, and stored in tower or vertical silos, bunker silos or surface stacks.

11.3. Crop Requirements

11.3.1. Climate

The optimal average daily temperature is between 24 and 30°C, depending on the vegetative stage of the plant and the water availability (Baldoni and Giardini, 1981). Low temperatures are critical at the emergence time. The following conditions can damage the crop:

- air temperature > 32–33°C
- high temperature during the night
- soil temperature > 22–23°C

Optimal average daily temperatures at different stages are:

- sowing: 10°C (soil temperature at sowing depth: 12.7°C)
- germination and emergence: 20–25°C

- raising: 22–23°C
- blooming: 24–25°C
- kernel filling: 23–24°C

Frost is always harmful; the minimum temperature for growth is 10°C. Maize can be grown only where the mean annual precipitation is more than 150 mm; in the Mediterranean regions, the mean precipitation requirement for maize ranges from 200–300 to 400–600 mm per year, depending on the soil water-storage capacity. The water requirement increases until the blooming stage, and afterwards it decreases progressively. Abundant precipitation is tolerated only where waterlogging does not occur. Climatic requirements for maize are listed in Table 11.1.

Table 11.1. Climatic requirements for maize.

	Very suitable	Moderately suitable	Marginally suitable	Unsuitable
Mean temperature (°C) during the growing cycle	18–32	16–17; 33–35	14–15; 36–40	< 14; > 40
Mean precipitation (mm) during the growing cycle	500–1,200	400–500; 1,200–1,600	300–400; > 1,600	< 300

11.3.2. Soil

Maize soil requirements are not too demanding. The following types of soil are most suitable:

- loamy
- deep
- soft, not compacted or not subject to summer cracking
- well drained
- having adequate organic matter content
- having neutral or sub-acid pH
- having no salinity, although maize has a moderate salt tolerance

11.3.3. Nutrients

Maize is a highly nutrient-demanding crop. Nitrogen is the most important nutrient. Therefore, maize prefers soils with a high content of organic matter, even if the nitrogen content of soil and the rate of organic matter mineralization are often not enough to meet the nitrogen requirements of maize during its various growing stages. Phosphorus requirement is higher in the first growing stages. Potassium deficiency symptoms are more uncommon and occur only in soils that are coarse-textured or rich in rock fragments, especially in the presence of abundant precipitation. Lime deficiency symptoms are usually rare. Micronutrient shortage is also uncommon; symptoms of zinc and manganese deficiency occur occasionally (manganese deficiency in acid organic soils).

11.3.4. Water

Maize is highly demanding in terms of water and temperature conditions, as it was originally a crop of tropical, humid locations. In the most suitable areas (called "corn belts"), characterized by high yields, summer rains are abundant and regularly distributed and there is no need for irrigation. Water consumption, according to some authors, is 349 g/g dry matter produced but, according to others, only 187 g, which shows that the transpiration coefficient is highly variable depending on climatic conditions. The total water consumption ranges from 6,980 to 2,640 t/ha, considering 20 t/ha dry matter (grains, cobs, stems, leaves, husks, roots).

11.4. Land Suitability

Since maize is a summer crop, the ideal habitat for its cultivation is the subtropical or humid temperate belt, characterized by regularly distributed summer rains (Table 11.2).

Table 11.2. Maize requirements.

Requirements	Value
Climatic	
Air temperature	moderate to high
Air humidity	moderate to high
Soil	
Texture	loam to clay
Oxygen availability	high and constant
Water availability	high
Soil depth (ground water, soil horizons impenetrable to roots)	high
Nutrient availability	moderate to high
Soil salinity	low
Soil pH	neutral or sub-acid
Management	
Trafficability	good
Workability and mechanization	good

Maize is sensitive to cold weather, mostly during the first stages of development. It requires high temperatures during the growing and reproductive periods and a period of growth of at least 120–150 d frost-free. In the Mediterranean region, the most suitable areas have an adequate water availability (soils with good water-holding capacity, or sufficient rainfall, or irrigation during the growing period) and a soil that is deep, well drained, or at least with a sufficiently deep water table, and not prone to summer cracking. Organic matter and nutrient content should be high. Beyond common characteristics, various cultivars may have different production levels and different rate of tolerance of adverse environmental conditions.

11.5. Soil Characteristics and Land Suitability

There is a large body of results on soil evaluation studies for maize crops carried out in various Italian regions, mostly in the 1980s, following different methodologies. Soil suitability maps for maize have been charted in Tuscany by Magaldi et al. (1983) and in some flat areas in Emilia Romagna (Regione Emilia Romagna, 1984, 1987) by the Regional Administration. A suitability map at reconnaissance scale (Ministero dell'Agricoltura e delle Foreste, 1983) was charted following the FAO "agro-ecological" approach (FAO, 1978-81). The most important functional characteristics considered are the ones related to water and nutrient availability.

An example of suitability assessment table compiled from the literature (Sys et al., 1993; Magaldi, 1983; Baldoni and Giardini, 1981) is shown in Table 11.3.

Table 11.3. Soil requirements for maize.

Soil characteristics	Class and rating scale			
	Very suitable	Moderately suitable	Marginally suitable	Unsuitable
Depth (cm)	> 75	50–75	20–49	< 20
Texture[1]	C, SiC, SiCL, CL, Si, SiL, SC, L, SCL	SL, LS	S	$C_{massive}$
Stoniness (%)	0–15	16–35	36–55	> 55
Salinity (dS/m)	0–4	5–6	7–8	> 8
Gypsum (%)	0–4	5–10	11–20	> 20
pH	6.5–7	6–6.4; 7.1–8	< 6; > 8	
Sum of bases (Ca, Mg, K, Na) (cmol+/kg soil)	> 5	3,5–5	2–3,4	< 2
Base saturation (%)	> 50	35–50	20–34	< 20
Content of organic carbon (%)	> 1	0,5–1	< 0,5	
Rock fragments (%)	< 15	15–35	36–55	> 55
Exchangeable sodium percentage (%)	< 15	15–20	21–25	> 25
Internal drainage[2]	well drained, moderately well drained (fine and medium soils); moderately well drained; somewhat poorly drained (coarse soils)	somewhat poorly drained (fine and medium soils); well drained (coarse soils)	somewhat excessively drained; excess-ively drained	poorly drained; very poorly drained
Flooding risk	none		rare	occasional or more frequent
Total calcium content—CaCO$_3$ (%)	< 15	15–25	25–35	> 35

[1] *SiC silty clay, C clay, Si silt, SiL silt loam, CL clay loam, SC sandy clay, L loam, SiCL silty clay loam, SCL sandy clay loam, SL sandy loam, LS loamy sand, S sand, C$_{massive}$ massive clay, SiC$_{massive}$ massive silty clay.*
[2] *From: Sys et al. (1993), modified according to classes quoted in Appendix.*

References

BALDONI, R., GIARDINI, L. 1981. *Coltivazioni erbacee.* Patron Publisher, Bologna, pp. 125–183.

FAO. 1978-81. *Report on the Agro-ecological Zones Project. Vol. 1. Methodology and results for Africa. Vol. 2. Results for southwest Asia; Vol. 3. Methodology and results for South and Central America; Vol. 4. Results for Southeast Asia.* World Soil Resources Report 48/1, FAO, Rome.

GARDIN, L., COSTANTINI, E.A.C., NAPOLI, R. 2002. *Guida alla descrizione dei suoli in campagna e alla definizione delle loro qualità.* Istituto Sperimentale per lo Studio e la Difesa del Suolo, Regione Toscana, 101 p. http://www.issds.it/files/atr-guidasuoliRT02.pdf.

MAGALDI, D. 1983. *Rilevamento del suolo e classificazione del territorio per usi agricoli e non agricoli.* Sviluppo Agricolo no. 8–9 (August–September), 23–27.

MAGALDI, D., BIDINI, D., CALZOLARI, C., RODOLFI, G. 1983. Geomorfologia, suoli e valutazione del territorio tra la piana di Lucca e il Padule di Fucecchio. Annali ISSDS XIV, 21–108.

MINISTERO DELL'AGRICOLTURA E DELLE FORESTE, GRUPPO DI LAVORO PER LA VALUTAZIONE DEL TERRITORIO A FINI AGRICOLI E FORESTALI. 1983. Proposta metodologica di classificazione attitudinale del territorio (esemplificazione per la coltura del mais da granella). Suppl. a Annali Istituto Sperimentale Studio e Difesa del Suolo XIV, 55 p.

REGIONE EMILIA ROMAGNA. 1984. *Cartografia tematica per la valutazione del territorio del Comprensorio Bassa Pianura Modenese.* Pivetti Publisher, Mirandola, 123 p.

REGIONE EMILIA ROMAGNA. 1987. *I suoli della Bassa Pianura Ravennate.* Gamma Publisher, Bologna, 83 p.

SYS, C., VAN RANST, E., DEBAVEYE, I.R.J, BEERNAERT, F. 1993. *Land Evaluation, part III. Crop Requirements.* Agriculture publication no. 7, General Administration for Development Cooperation, Brussels, Belgium, 166 p.

12. Rice (*Oryza sativa L.*)

Giancarlo Chisci[1]

12.1. Crop Characteristics

Rice (*Oryza sativa L.*) is originally from the monsoon area of south-east Asia. It is a marsh species, even though different genotypes have been developed to grow on non-flooded soils. Owing to the presence of many cultivars, rice can adapt to very different environments, but it always needs high temperatures.

Rice is grown both in its originally tropical and subtropical regions and in temperate regions (from the 37th parallel South to the 50th parallel North), but it always needs high temperatures without sharp variations. For this reason, in less suitable environments, it is necessary to partly flood the plant for a period of time, perhaps even the entire growing period. This water layer stores heat during the day and releases it to the soil, the plant and all the surrounding environment during the night, when the temperature drops, acting as a heat reserve, which is very important from the inflorescence differentiation to the blooming stages in order to produce high product yields.

Rice can very well withstand lack of oxygen in the soil, because of the presence of an air-filled parenchyma in the mesophyll and in the cortical part of the stem. Also, the disappearance of root-hairs and the emergence of vertical air-filled channels in roots favour the adjustment to an oxygen-deficient environment. Rice can be grown outside water, like other grain crops, only where the temperature is high enough, the temperature variation between day and night is moderate, and the annual rainfall is at least 1,000 mm, concentrated on the 3–5-month growing cycle.

---

[1]Università degli Studi di Firenze, Firenze, Italy.

12.2. Cropping Techniques

12.2.1. Crop Rotation

Rice does not need crop rotation and can be grown as a monocrop culture. Monocropping is common in Italian rice-growing areas, which have among the highest production rates in the world; rice crop is planted in the same field from 2–3 to 5–6 years, a practice determined by the high cost of soil preparation for flood fields with a homogeneous water layer. Rice may be followed by wheat, mixed hay, white or red clover and Italian ryegrass, summer or winter forage, and maize. Today, monoculture is widespread and rice is an industrial and highly specialized crop in Italian rice-growing areas. Nevertheless, the revival of crop rotation seems to be necessary to find a solution to the consequences of massive use of pesticides and algicides.

12.2.2. Soil Preparation

Because the rice field must be flooded with a layer of water of uniform depth, rice cultivation requires completely flat and horizontal surfaces, bordered by dykes. On rolling landscapes, dykes overlay contour lines and flooded fields have irregular shapes; on flat areas, paddies are regularly shaped. In Italy there are two different types of paddies, according to the landscape characteristics (morphology and altimetry), climate and farm size. The first type of paddy, widespread in the north-western regions, where there are usually mid-size farms, is enclosed by dykes (called "ripe") and divided into smaller fields called "camere" (0.5–5 ha) by transverse smaller dykes; the surface is flat and crossed by furrows that help the flooding-drying sequence. In the north-eastern regions, farms are larger: paddies are flat and large (9–12 ha), regularly shaped and contoured by big dykes on top of which are the headlands; furrows are placed on the top of the field from which other furrows start and enter inside the field. Large longitudinal drains (cross-sectional area: 0.8–1.5 m^2) divide paddies into smaller fields (called "piane" or "pezze"), each subdivided by one or more channels (called "gambini").

12.2.2. Cultural Practices

Soil Preparation

The main soil tillage is a shallow ploughing using a frame plough at the end of winter or beginning of spring. On low support soils, such as peaty soils, spaders and/or rotary hoes are better used. Tillage depth can vary depending on the soil nature (usually 20–25 cm), but it should avoid interfering with any impermeable layer that can reduce water infiltration and consumption. Usually fields are ploughed in a circular fashion and clockwise direction in order not to disturb the soil profile. Before ploughing, dykes are rebuilt or reinforced by adding new soil and pressing it to guarantee efficiency. Harrowing and soil levelling are complementary tillage operations to prepare the seedbed. If sowing occurs on flooded paddies it is usually followed by smoothing and sometimes pore obstruction when the water layer is 2–3 cm thick to reduce water infiltration. On very permeable soils, it is common to pass repetitively over the fields with tractors equipped with cage wheels and sometimes even with a cage roller.

Fertilization

In the past, manuring and green manuring with leguminous crops were adopted, while today chemical fertilization prevails, even if organic fertilization is still traditionally carried out.

Manuring and green manuring are not essential to reach high yields; they can have negative effects especially on silty and clay soils, because organic matter degradation reduces oxygen in an already oxygen-poor environment and consequentially causes irregular and scarce seed germination, and the production and accumulation of toxic substances. Mineral fertilization seems to be more suitable for high yields, when applied in accordance with specific soil, climatic and crop conditions. It should be based on nutrient uptake by the crop, on crop requirements at different stages and on the influence of flooding-drying cycles on soil microflora. Nitrogen and phosphorus absorption is the highest from the bouncing to the raising stages and from the booting to the blooming stages. Flooding causes oxygen consumption by aerobic micro-organisms that mineralize organic matter and are later replaced by anaerobic micro-organisms that reduce mineral and organic oxidized compounds. The redox potential drops and the pH rises to the extent that reduction is high. Flooding also causes denitrification: nitrites are toxic and therefore a high level of soil nitrates is harmful. Other chemical phenomena typical of rice fields are strong ammoniation and increased solubility of phosphorus, silicon and potassium.

Nitrogen fertilization is based mostly on the use of ammoniacal fertilizers such as urea and calcium cyanamide, as during the flooded period rice adsorbs ammoniac nitrogen and only during the drying period can it adsorb nitric nitrogen. If soils are organic or peaty, or the rice crop is preceded by a leguminous crop, fertilization can be avoided. Nitrogen can be applied at a 70–80% level as basal dressing at the sowing time, and the remainder as top dressing at the end of bouncing and beginning of raising time.

Phosphoric fertilizers are usually superphosphate or Thomas phosphate (on acid soils).

Potassic fertilization is not widespread because it does not seem to have direct effect on product yield. Nevertheless, it increases resistance to lodging and pests and favours ripening, balancing the possible delaying effects of an excessive nitrogenous fertilization. Organic, phosphoric and potassic fertilizers must be ploughed in.

Irrigation

In Italy, where the climate is temperate, flooding irrigation is continuous and realized through a slow water flow, interrupted by temporary draining periods of different length and frequency. Paddy irrigation is a rather complex operation; it may vary from year to year and consists in proper and timely water management, which depends on seasonal climate, temperature and amount of available water, soil oxygen scarcity, the need to prevent algal, weed and pest propagation and promote ripening at the right time. Water management in various environments can be very different: this is especially true in the two major Italian rice-growing areas of the Vercelli and Ferrara provinces.

As a common rule, when air temperature drops, flood level is raised and/or water flow is slowed down. Water conditions can influence the effect of flooding rice fields. If the water is too cold, it needs to be heated before being let into the field. The amount of water used depends on soil permeability: water consumption is lower in old paddies because soil pores are filled by the finest particles carried by the water flow. Water consumption ranges from 1 L/sec in less permeable soils to 5 L/sec in light and more permeable soils.

Seeding and Transplanting

In Italy, the time for sowing ranges from the first 10 d in April to the second 10 d in May. In the past, rice was transplanted. Seeds were placed in a nursery having an approximate size of 1/9 of the rice field, over an accurately tilled soil. When the plants reached a height of no less than 15 cm, they were moved from the nursery and planted in the flooded paddy.

Today, seeds are directly sown on flooded paddies is used. There are different types of sowing: (1) broadcast sowing, manually or with a centrifugal seed drill, pre-soaking seeds in order to weight them so that they are distributed regularly and sink quickly; (2) sowing in rows with a seed drill, carried out on non-flooded but water-saturated ground or on dry ground. Flooding immediately follows the sowing or, if seeds have been earthed, fields are flooded after plant emergence (20–25 d after sowing). Seed rate is rather variable: it can range from 170–190 kg/ha to 200–250 kg/ha when difficult germination or poor bouncing is expected.

Weed Control

When other conditions are the same, weed control is 80% responsible for high yield. In fact, there are many different types of weeds that develop in paddies, depending on soil characteristics, water quality, tillage practices and crop rotation. Long flooding periods cause weeds to have peculiar characteristics. Rice-specific weeds are algae, cryptogams, phanerogams, hydrophytic and occasionally hydrophytic plants whose early development is possible only on drained soil.

In the past, paddies were weeded manually by temporary workers (rice weeders), but today weed control is based on the use of herbicides, on a proper water management, especially to kill algae, and on crop rotation to control weedy rice. The widespread use of herbicides in paddies is of concern because of chemical contamination and therefore the use of biodegradable products is strongly advisable.

Harvesting, Seed Drying and Storage

Harvest occurs from the beginning of September (early cultivars) to the end of October (late cultivars). Whereas in the past rice was harvested, today it is performed mechanically. Manual harvesting is still common in some Asian and African rice-growing areas, where only the ear of rice is cut.

Grain water content at harvest is about 22–28%, whereas the optimal humidity level for storage should be around 14%. For this reason, in Italy, artificial drying is necessary. Dried grains are stored in silos or warehouses until they are forwarded to the food-processing industry to transform paddy into rice. During the processing operation, bracts and outer layers are removed by the traditional method or by parboiling. There are a few subsequent steps of cleaning, husking and milling to get the dry dressed rice. This product can be further processed by polishing. In Italy, rice commercial classification is based on processed grain size: common (< 5.2 mm), semi-fine (5.2–6.4 mm), fine (> 6.4 mm), super fine (of high appearance value).

12.3. Crop Requirements

12.3.1. Climate

Rice is an adaptable crop being grown in very different environments, even though it has demanding temperature requirements. It can be grown both in native tropical and subtropical regions and in temperate regions, and at different altitudes, but when the temperature is not high and steady enough, rice needs to be grown in flooded fields and covered with an adequate layer of water for a variable period that may extend to the whole growing cycle. Water accumulates heat during the day and releases it to plants and the surrounding environment during the night. Rice can be grown on dry fields only where the temperature is sufficiently high and without sharp variations between day and night, and precipitation during the growing cycle (3–5 months) is at least 900–1,000 mm. The thermal sum in tropical and subtropical areas ranges from 1,760 to 3,300°C and in temperate areas from 704 to 2,760°C. The minimum temperature for rice growth is 10°C.

Wind can damage the crop especially in the early stages when water waves can uproot the young plants, and also wind can later interfere with insemination and cause lodging. Rainfall has a positive influence on rice growth, favouring water oxygen enrichment, but it may also be harmful, reducing water temperature.

Water requirement is not too different from other grain crops because the transpiration coefficient ranges from less than 500 to 800 L/kg of mineral matter.

12.3.2. Soil

Soil requirements for rice are not very demanding with regard to physical, mechanical, or chemical soil characteristics, nor does rice seem to take advantage of high organic matter content. Rice tolerates acid soils and also, to a lesser extent, alkaline soils (pH tolerance ranges from 3.5 to 8.4). Within limits, rice withstands salinity and shallow soils, but in the latter case it is more prone to lodging because of poor root development.

The ideal soil is sub-acid, silty, or silty-clayey, 30–40 cm deep, and provided with a impermeable layer in the subsoil, which allows good root development but at the same time reduces as much as possible infiltrating water consumption.

12.4. Land Suitability

In Table 12.1, rice requirements are listed in brief.

Table 12.1. Rice requirements.

Requirements	Value
Climate	
Air temperature	high
Range of temperature	moderate
Air humidity	high
Soil	
Texture	silty; silty clayey
Soil depth	30–40 cm
Soil permeability	low
Organic matter content	moderate to low
Soil pH	acid to sub-alkaline
Management	
Trafficability	good
Workability and mechanization	good

12.5. Soil Characteristics and Suitability Assessment Table

Because of its peculiarities, rice is grown in Italy only in a few regions and therefore rice land suitability studies are scarce. An example of suitability assessment table is given in Table 12.2.

Table 12.2. Soil requirements for rice.

Soil characteristics	Class and rating scale			
	Very suitable	Moderately suitable	Marginally suitable	Unsuitable
Depth (cm)	30–60	> 60	20–29	< 20
Texture[1]	SiC, Si, SiL, CL, SC,L, SiCL, SCL	C, peaty	SL, LS	S
Permeability	low, very low	moderately low	moderate	high; very high
Rock fragments (%)	< 5	5–10	> 10	
Salinity (dS/m)	< 5	5–8	9–12	> 12
pH	5–7	4–4.9; 7.1–8	3.5–3.9; 8.1–8.4	< 3.5; > 8.4
Content of organic carbon (%)	1–1.5	1.6–3	> 3	
Exchangeable sodium percentage (%)	< 10	10–15	16–20	> 20
Cationic exchange capacity (cmol+/kg soil)	> 10	5–10	< 5	
Organic matter content (%)	< 2	2–3	> 3	
Redox potential (rH)	high	moderate	low	high oxygen shortage
Internal drainage[2]	poorly drained; very poorly drained	somewhat poorly drained; moderately well drained	well drained	somewhat excessively drained; excessively drained

[1] *SiC silty clay, C clay, Si silt, SiL silt loam, CL clay loam, SC sandy clay, L loam, SiCL silty clay loam, SCL sandy clay loam, SL sandy loam, LS loamy sand, S sand, $C_{massive}$ massive clay, $SiC_{massive}$ massive silty clay.*
[2] *From: Sys et al. (1993), modified according to classes quoted in Appendix.*

References

GARDIN, L., COSTANTINI, E.A.C., NAPOLI, R. 2002. *Guida alla descrizione dei suoli in campagna e alla definizione delle loro qualità.* Istituto Sperimentale per lo Studio e la Difesa del Suolo, Regione Toscana, 101 p. http://www.issds.it/files/atr-guidasuoliRT02.pdf.

SYS, C., VAN RANST, E., DEBAVEYE, I.R.J., BEERNAERT, F. 1993. *Land Evaluation, part III. Crop Requirements.* Agriculture publication no. 7, General Administration for Development Cooperation, Brussels, Belgium, 166 p.

13. Alfalfa (*Medicago sativa* L.)

Letizia Venuti[1] and *Giancarlo Chisci*[2]

13.1. Crop Characteristics

Alfalfa (*Medicago sativa* L.) is originally from Southwest Asia. Today, alfalfa is widely cultivated and adaptable to many different environmental conditions, because cultivars with different levels of cold-resistance and photoperiodism have been selected over centuries. In Italy, alfalfa is grown everywhere, but in acid flat soils and in arid southern areas it is replaced by sulla and sainfoin. It is a perennial plant that may last 4–5 years, although it usually lasts no more than 3 years, because of incorrect management.

13.2. Cropping Techniques

13.2.1. Crop Rotation

Although crop rotation has lost its importance because of new cultivation techniques and mineral fertilization, alfalfa, being a soil-improving crop, usually follows a winter grain crop and precedes a spring crop. Alfalfa has a good weed-suppressive effect and also enriches soil with nutrients, through its symbiotic relationship with nitrogen-fixing bacteria and the plant residues left on the field. Monocropping is not recommended as it causes plant emergence failure and poor plant development.

---

[1] CNR-IRPI—Istituto di Ricerca per la Protezione Idrogeologica, Unità Operativa di Supporto, Pedologia Applicata, Firenze, Italy.
[2] Università degli Studi di Firenze, Firenze, Italy.

13.2.2. Cultural Practices

Soil Preparation

There are two alfalfa sowing techniques: undersowing a winter grain crop, usually wheat, but also rye, barley or oat, or more commonly single cropping on a properly prepared seedbed.

In the case of undersowing, no soil preparation is necessary, but when alfalfa sowing is carried out early within a winter crop at the seedling stage, a light harrowing followed by a light rolling may be advisable. The basal fertilization applied to the previous grain crop is generally sufficient.

When alfalfa is grown as a single crop, deep tillage is necessary to allow the proper development of the tap-root system. The field should be ploughed early during the preceding summer in order to cause the climatic elements to break down seedbed clods into finer size particles, a condition necessary for relatively fine seeds such as alfalfa.

To reduce the seedbed particles, grubbing and harrowing should be carried out in autumn or winter, but suspended before sowing (which is carried out at the end of winter or beginning of spring), so that climatic elements (especially frost) help soil structure formation and favour nitrogen-fixing bacteria symbiosis. There are two sowing methods: broadcast sowing and row sowing. In the first case, sowing should be followed by a light harrowing and then rolling to cover seeds. In the second case, seeds are placed in rows usually spaced 0.15–0.17 m apart and planted at a depth of 0.01–0.02 m by a seed drill equipped with a harrow to cover seed completely. A subsequent rolling enhances the contact between seeds and soil particles.

Seeding

If 15–20 kg/ha of seeds are planted on a carefully prepared seedbed, a yield of 300–450 plants/m^2 can be achieved, which is considered the optimum plant investment. When the seedbed is not sufficiently well prepared, or the soil is clayey and cloddy, or alfalfa is undersown to wheat, the seed amount should be increased to 40 kg/ha.

Fertilization

When alfalfa is grown as a single crop, fertilization consists of an adequate quantity of phosphorus and potassium, applied partly at the ploughing and partly at the secondary tillage; such fertilization will also benefit the crops that follow in a crop rotation. A small amount of nitrogen fertilizer may be useful at the initial stage, when nitrogen-fixing bacteria symbiosis, which later provides alfalfa with nitrogen, is not yet established. Top dressing is also based on the supply of phosphorus and potassium, because of the high demand of these nutrients by alfalfa, especially if highly productive and irrigated. Top dressing is usually applied at the end of winter or after hay mowing.

Weed Control

Usually alfalfa meadows last 3–4 years, until plant density decreases from the initial 300–400 plants/m^2 to less than 80–100 plants/m^2, which makes it more attractive to plough up the meadow: actually the poor yield and the low quality achieved (alfalfa plants are increasingly replaced by other species, even weeds) make unit harvesting and storage costs uneconomic.

Pest diseases, abiotic stresses and competition with various weeds especially after spring mowing, when alfalfa grows slower than dicotyledons and monocotyledons, cause thinning of alfalfa meadows.

The most common weeds starting from the first year are a few annual dicotyledons (e.g., *Stellaria, Capsella, Sinapis, Chenopodium, Amaranthus*) and also annual monocotyledons (*Digitaria, Setaria, Echinochloa*). Later, a few perennial dicotyledons (*Taraxacum, Rumex, Plantago*) or monocotyledons (*Alopecurus, Avena, Lolium, Agropyron*) appear, which have some forage value but less so than alfalfa.

Weeding techniques are influenced by the fact that alfalfa is a perennial crop, so it is necessary to distinguish first-year weeding operations from those of the following years. During the growing season there are two phases: the dormant winter phase and the growing phase, when alfalfa fields are subject to subsequent cutting and regrowth.

During the first year it is necessary to control weeds until alfalfa, characterized by a slow growth, is well established and developed. The suggested weed control treatments are with Benfluralin at pre-sowing, Neburon at pre-emergence, and Imazetapir mixed with 2,4 DB at post-emergence.

Later, when alfalfa is well developed, root-absorption herbicides can be used during the rest period and Imazetapir mixed with 2,4 DB during the growing period, until plants are over 80–100 cm height.

A treatment with Propizamide after every mowing should be applied to suppress cuscuta development, which causes spot thinning in alfalfa meadows starting from the first year.

Harvest

The first-year yield is usually poor, especially if alfalfa has been undersown to wheat. Full productivity is achieved during the second year and, at the third year, yield starts to decrease, because of progressive alfalfa clump thinning. When clump density becomes less than 100–800/m, meadow should be ploughed up as harvesting is no longer profitable. During the growing season the number of cuts may vary from two in dry climate to 4–5 in cool climate or when irrigation is used. Irrigated alfalfa fields in subtropical climate can be cut all year round up to 10–12 times. The best time for cutting is after the beginning of blooming when the relationship between forage quantity and quality is optimal and the carbohydrate accumulation in the roots is maximum, which allows a better regrowth. Alfalfa is usually used as green forage or hay. Silage is not common because alfalfa is rich in proteins and therefore difficult to preserve. Alfalfa should be pastured carefully because it can cause meteorism in ruminants. It is particularly suited to be dehydrated as alfalfa meal. The mean annual yield of an alfalfa field in optimal conditions may reach 13 t/ha, but it is more commonly around 8–10 t/ha (humidity: 15%). Haymaking and picking up should be carried out carefully in order to reduce the loss of dry leaves, for the leaf is the most valuable part of the plant. A good quality alfalfa hay should have a crude protein content of 18–22% of dry matter.

13.3. Crop Requirements

13.3.1. Climate (Table 13.1)

The optimal temperature range for alfalfa cultivation is 15 to 25°C during the day and 10 to 20°C during the night. The minimum temperature for germination is 5°C; physiological activity stops below 5°C and above 35°C. Nevertheless, the cold and heat resistance of alfalfa is very high. During dormancy, alfalfa withstands both low (−10/−15°C) and high temperatures (35–40°C). Cold resistance is variable in different cultivars. Assimilation efficiency, however, decreases remarkably below 5°C and above 30°C. Moreover, plants can be damaged by early winter cold and late spring frost (Baldoni et al., 1974).

Alfalfa tolerates drought much better than waterlogging as it is originally from warm temperate regions. In fact, during an extended dry period, it enters dormancy, whereas waterlogging causes root asphyxia and kills plants especially during the warm season. In winter, when plants are dormant, waterlogging tolerance is much higher.

Alfalfa can be grown in dry areas as long as soil water is available (Baldoni et al., 1974).

Table 13.1. Climatic alfalfa requirements.

Climatic requirements	Class and rating scale			
	Highly suitable	Moderately suitable	Marginally suitable	Unsuitable
Mean temperature of the growing cycle (°C)	20–30	15–20; 30–33	10–15; 33–40	<10; > 40
Mean rainfall of the growing cycle (mm)	800–1,400	600–800; 1,400–1,600	400–600; > 1,600	< 400

13.3.2. Soil (Table 13.2)

As mentioned before, alfalfa has little tolerance of waterlogging, especially during the warm season. Its tolerance of acid soils is also low, whereas its tolerance of alkaline soils, although not excessively alkaline soils, is high. Alfalfa needs deep soils to allow a deepening of its root system and a sufficient but not excessive level of organic matter. Salinity tolerance is quite high, except at a young stage. Alfalfa requires a high level of calcium in order to produce high yields; even a moderate quantity of gypsum has been linked to high production. As for soil texture, alfalfa well tolerates clay soils because of the high penetrating force of its rooting system (Baldoni et al., 1974).

In general, the following soil conditions are preferable:

- deep soils, which allow tap roots to deepen (Baldoni and Giardini, 1981);
- soil texture ranging from loam to clay, provided that the soil contains enough limestone and potassium. Alfalfa well tolerates drought for long periods but in clay soils cracks can damage the root system;
- calcareous and/or gypsiferous soils;
- well-drained soils, to avoid waterlogging;
- soils rich but not excessive in organic matter;

- soils neutral or alkaline but not too alkaline (alfalfa does not tolerate pH values below 6–6.5 because, in this condition, nitrogen-fixing bacteria symbiosis is inhibited and consequently meadow duration is reduced).

Alfalfa cultivation has the following positive effects on soil physical and chemical characteristics:

- increase of organic matter content;
- release of nitrogen fixed by symbiotic bacteria;

Table 13.2. Soil alfalfa requirements.

Soil characteristics	Class and rating scale			
	Very suitable	Moderately Suitable	Marginally suitable	Unsuitable
Depth, cm	> 75	50–75	30–49	< 30
Texture[1]	SL, SCL, SC, L, CL, SiCL, C, SiC, SiL, Si	SL	LS	$SiC_{massive}$, $SiC_{massive}$
Salinity (dS/m)	< 5	5–9	10–12	> 12
Gypsum (%)	< 5	5–10	11–20	> 20
pH	6.5–8	6–6.4; 8.1–8.3	5.5–5.9; 8.4–8.5	< 5.5; > 8.5
Sum of bases (Ca, Mg, K, Na) (cmol+/kg soil)	> 5	3.5–5	2–3.4	< 2
Base saturation (%)	> 35	20–35	< 20	
Content of organic carbon (%)	> 1.2	0.8–1.2	< 0.8	
Rock fragments (%)	< 15	15–35	36–70	> 70
Exchangeable sodium percentage (%)	< 20	20–35	36–50	> 50
Internal drainage[2]	well drained, moderately well drained; somewhat poorly drained	somewhat poorly drained; well drained	somewhat excessively drained; excessively drained	poorly drained; very poorly drained
Flooding risk	none		rare	occasional or more frequent
Total calcium content—$CaCO_3$ (%)	2–5	0–2; 15–25	25–35	> 35
Available water capacity (mm)	> 112	56–112	< 56	
Cationic exchange capacity (cmol+/kg soil)	> 10		5–10	< 5

[1] *SiC silty clay, C clay, Si silt, SiL silt loam, CL clay loam, SC sandy clay, L loam, SiCL silty clay loam, SCL sandy clay loam, SL sandy loam, LS loamy sand, S sand, $C_{massive}$ massive clay, $SiC_{massive}$ massive silty clay.*
[2] *From: Sys et al. (1993), modified according to classes quoted in Appendix.*

- development of soil microflora and microfauna;
- weed-suppressive effect;
- soil structure improvement.

13.3.4. Water

Alfalfa is a water demanding crop because of its long growing cycle. The root depth and the fact that growth slows down or completely ceases during dry periods make alfalfa adaptable to temporary water shortage. The excessive soil moisture is instead more limiting. Poor drainage or excessively shallow ground water damage plants and cause meadow thinning and weed proliferation. Irrigation, when applied, is a rescue irrigation in dry years or in the first year as it promotes quick plant development and optimal plant density. Because of occasional watering and low tolerance to waterlogging, sprinkler irrigation is usually adopted. The highest water need of alfalfa is during flowering (Costantini and Onofrii, 1982). During the first year, fields must be watered carefully to avoid uprooting and damaging young seedlings.

13.3.5. Nutrients

Alfalfa needs fertile soils with a reasonable phosphorus and potassium content. Being a nitrogen-fixing crop, owing to the symbiosis with *Rhizobium meliloti*, alfalfa can meet its nitrogen requirement by taking this element directly from the atmosphere. Alfalfa is therefore considered a self-sufficient crop with regard to nitrogen supply. On the contrary, it has a good response to phosphorus and potassium supply. Because of its high demand for calcium, it needs limestone supplies in calcium-poor soils.

In summary, alfalfa has the following characteristics with regard to nutrient requirement:

- nitrogen: alfalfa demands little nitrogen as it assimilates this element from nitrogen-fixing bacteria symbiosis. An excessive level of nitrogen in the soil instead favours weed infestation;
- phosphorus: uptake is around 50 kg of P_2O_5 per kg of hay;
- potassium: demand is high at the young stage.

13.4. Land Suitability

In brief, the most suitable soils are neutral or alkaline without waterlogging, such as many calcareous soils. Alfalfa has low sensitivity to drought.

References

BALDONI, R., GIARDINI, L. 1981. *Coltivazioni erbacee*. Patron Publisher, Bologna, pp. 823–856.
BALDONI, R., KOEKENY, B., COVATO, A. 1974. *Le piante foraggere*. Reda Publisher, Rome, 290 p.
COSTANTINI, E.A.C., ONOFRII, M. 1982. Ricerche sull'evapotraspirazione effettiva dell'erba medica misurata in cassoni lisimetrici e sue relazioni con quella potenziale. Annali dell'Istituto Sperimentale Agronomico XIII, 89–116.
GARDIN, L., COSTANTINI, E.A.C., NAPOLI, R. 2002. Guida dei suoli in campagna e descrizione delle loro qua*lità*. Istituto Sperimentale per lo Studio e la Difesa del Suolo, Regione Toscana, 101 p. http://www.issds.it/files/atr-guidasuoliRT02.pdf.

MAGALDI, D. 1983. Rilevamento del suolo e classificazione del territorio per usi agricoli e non agricoli. Sviluppo Agricolo no. 8–9 (August–September), 23–27.

MAGALDI, D., BIDINI, D., CALZOLARI, C., RODOLFI, G. 1983. Geomorfologia, suoli e valutazione del territorio tra la piana di Lucca e il Padule di Fucecchio. Annali ISSDS XIV, 21–108.

REGIONE EMILIA ROMAGNA. 1984. *Cartografia tematica per la valutazione del territorio del Comprensorio Bassa Pianura Modenese.* Pivetti Publisher, Mirandola, 123 p.

REGIONE EMILIA ROMAGNA. 1987. *I suoli della Bassa Pianura Ravennate.* Gamma Publisher, Bologna, 83 p.

SYS, C., VAN RANST, E., DEBAVEYE, I.R.J., BEERNAERT, F. 1993. *Land Evaluation, part III. Crop Requirements.* Agriculture publication no. 7, General Administration for Development Cooperation, Brussels, Belgium, 166 p.

14. Potato (*Solanum tuberosum L.*)

Luciano Lulli[1], Enrico Palchetti[2], Giuseppe Vecchio[3] and Adele Maria Caruso[3]

14.1. General Characteristics

The potato (*Solanum tuberosum L.*) is a perennial herbaceous plant for annual cultivation, provided with fleshy underground modified stems, called *stolons*, whose swollen parts, the tubers, constitute an edible product, the potato.

The potato originates in Mexico and South America, where it is diffuse to the southernmost extremities of Chile. The ancient peoples of the Cordillera of the Andes domesticated this plant more than 4,000 years ago, breeding an enormous variety adapted to almost all climates.

Brought to Europe by the Spaniards in 1570, it reached Ireland only in 1586, where it was immediately accepted as an important agricultural product and became a staple for the poor. In Italy, use of the potato spread in the early 1800s with the help of Vincenzo Dandolo of Venice. Today its cultivation is diffuse worldwide and it is one of the foods most important to the human race. Its high content of starch, protein and vitamin C makes it an ideal food. The potato produces more energy per unit of cultivation area than cereals.

In general, potatoes with yellow pulp, compact and semi-compact, are suitable for being boiled, roasted, fried, or baked. White potatoes with mealy flesh are better for preparing mixtures and doughs (e.g., gnocchi, mashed potatoes). However, a lot depends on the dry content of the varieties.

14.2. Cultivation Requirements

The potato requires climatic conditions without sudden temperature variations and without water deficit during all of the phenological phases.

[1] Formerly Istituto Sperimentale per lo Studio e la Difesa del Suolo, Firenze, Italy.

[2] Università di Firenze—Facoltà di Agraria, Dipartimento di Scienze Agronomiche e Gestione del Territorio Agroforestale, Firenze, Italy.

[3] CRA-AB—Istituto Sperimentale per lo Studio e la Difesa del Suolo, Catanzaro, Italy.

Climate

The indications listed in the Rules for Integrated Production (Region Emilia Romagna, 2004: Guide to cultivation techniques—Agronomic techniques—Potato) inform farmers about the conditions needed for correct cultivation of potato (Table 14.1). It must be noted that temperatures near 0°C (1–2°C) cause wilting of the aerial part of the plant, with production delays.

The potato develops within a very limited temperature range (15°C for germination, 20°C for flowering, and 18°C for tuber ripening); therefore, both excessive heat during water deficit and late frosts are to be feared. This is true for all types of potatoes.

Table 14.1. Values of climatic parameters for potato cultivation (Region Emilia Romagna, 2004).

Climatic parameters	Recommended values
Germination temperatures	14–16°C
Minimum biological temperatures	Temperatures lower than 2°C compromise plant survival; avoid zones characterized by late frosts
Optimal ripening temperatures	18–20°C
Maximum temperatures	Prolonged temperatures over 30°C inhibit accumulation of carbohydrates in the tuber with a decrease in specific weight; increased risk of tuberomania
Rainfall	Alternation of periods of high rainfall and drought cause serious malformations in the tubers; it is essential to be equipped for irrigation

Soil

The Region Emilia Romagna also indicates soil parameters (Table 14.2). These parameters must be better defined in order to extend observations on soil suitability to each field of cultivation. The soil is an important factor for the growth of potato because it determines the condition or state that is favourable or not favourable to the respiration of stolons and tubers. An agronomic soil texture classification that is loam or loamy sand is not sufficient to determine good growth conditions. In Tunisia, there are ancient terraced lands that hold very clayey soils

Table 14.2. Optimal pedological parameter values for potato cultivation according to the Rules for Integrated Production, (Region Emilia Romagna, 2004: Guide to cultivation techniques—Agronomic techniques—Potato).

Pedological parameters[1]	Conditions
Texture	loam, sandy loam
Drainage	well drained[2]
Rooting depth[3] (cm)	60–70
pH	> 6–6.5, < 6
Salinity (dS/m)	< 4
Organic matter	good
Total calcium and active lime (%)	< 10

[1] *Refers to layer explored by root system.*
[2] *Well drained: water drains rapidly from soil, and/or there are no periods during growing season characterized by excessive wetness limiting plant development.*
[3] *Depth to restrictive layer.*

that, however, produce a superficial grainy structure (self-mulching) in which it is possible, with the help of irrigation, to cultivate potatoes with excellent organoleptic and marketable qualities. This is one of the many examples that may be drawn of favourable soils with characteristics far from those that are standardized in the suitability evaluation. It is also possible to have good results with pseudo-sandy soils of a volcanic nature that, nonetheless, have a good available water capacity and are certainly aerated.

14.3. Cultivation Techniques

Potato is a renewal crop that requires medium-depth preparation; therefore, to create good cultivation conditions it is necessary to plough to a depth of 30–35cm. After ploughing and harrowing, it is necessary to do basic fertilization operations and then plant seed tubers. If seeds are planted by machine, it is essential to level the land; if they are planted by hand, furrows are made. When the soil is being prepared, it may be treated with specific herbicide products (geo-herbicides). It is a good rule not to plant potato on the same plot before 2 years have gone by, during which other crops have been grown; it is also not advisable to plant potato in succession to other solanaceous plants.

Choice of Variety

The choice of variety must be based on practical considerations, most of all regarding resistance to viroses and length of growth cycle, as a function of the type of environment in which the crop will be planted. The varieties of potato are numerous and some, such as the Spunta, are grown worldwide.

Choice of Seed

There are various classes of seed tubers available, and their price varies with category and size. The seed tuber classes are:

- super elite (SE, viroses 1–3%); Ø 28–35 mm;
- elite (E, viroses 3–5%); Ø 35–45 mm;
- type A certified (CA, viroses 5–9%); Ø 45–55 mm;
- type B certified (CB, viroses 10%); Ø 55–65 mm.

The last type is often cut before use of the seed tuber, because it is large and has many eyes. It is possible to get four portions from one seed tuber with several eyes each. However, this is a risky procedure, because this class allows for a presence of viroses of up to 10%; therefore, even if only one tuber out of ten is diseased, the possibility that the virus spreads is four times as likely.

Cultivation Layout

The cultivation layout has 60–70 cm between rows and 25–35 cm in the row for a planting density of 5–6 tubers per sq m. Usually, seeds are planted with mechanical or semi-mechanical planters (operations of making furrows, planting and covering are mechanical).

Fertilization

Potato needs a high level of fertilization. The uptake per tonne of tubers produced with the corresponding quantity of stems and leaves is 4 kg N, 2 kg P_2O_5, 6 kg K_2O. Potato is a potassium- loving plant, especially the early varieties, and benefits from manuring. It has a low efficiency rate regarding nitrogen (60%), partly because it is almost always grown in permeable lands and, since it does not have a particularly well-developed root system, it tends to lose nitrogen to the soil.

The nitrogen requirements are higher in the first 50–60 d. In order to avoid nitrogen loss, three fertilization operations are carried out: the first for 1/3 of the quantity, placed at planting in ammoniacal form; the second carried out at hoeing; and the third at earthing up (banking).

Eighty percent of the phosphorus is distributed before planting and the rest during planting, to reinforce the mechanical structure of the plant and particularly to favour thickening of the skin (periderm) of the tuber, increasing preservability. Phosphorus is a catalyst for glucide metabolism and therefore particularly important for the production of early varieties.

The quantities per hectare recommended are: 120–150 kg N for early potato and 150–180 kg for the common, preferably ammoniacal, for slow release and non-leaching, 120 kg P_2O_5 and 150 kg K_2O. For early potato, the quantity of nitrogen must not exceed 150 kg/ha distributed in two parts: at planting and at tuberization.

14.4. Cultivation Operations

Tubers are planted in the open field at 8–12 cm depth with seed tubers that were pregerminated in indirect lighting (this allows discarding of anomalous sprouts and shortens the growth cycle by 10–15 d to allow the formation of shapes suitable for consumption). Soon after emerging (15 d), the plants must be earthed up (banked) and hoed to assist stolon development and reduce the weed population.

About 15 d later, a second earthing up and hoeing is done to assist tuberization and prevent greening of tubers, to destroy eventual weeds and to distribute the last amount of nitrogen and phosphorus fertilization.

After these operations, the plants begin to close ranks; therefore, there is no longer the risk of weed competition and the only operations needed are those of irrigation, depending on the different cultivation periods.

Potato and Water

Potato generally does not have a well-developed root system and is almost always grown in loose, rather permeable lands; therefore, it is susceptible to water deficit, which delays the differentiation and growth of the tubers.

In any case, to obtain the *optimum* in both tuberization and tuber development, there must be the necessary conditions for regular growth, connected with a balanced development of the aerial part of the plant. All excesses are negative and have repercussions on both product quantity and quality. Water stress that occurs in the initial phase of stolon development may notably reduce the number of tubers that will form.

Long periods of drought interrupted by irrigation operations lead to tuber malformation, while if soil cooling occurs (irrigation with cold water for leap year potatoes or sudden temperature drop for early potatoes) tuberization of the stolons already initiated may be inhibited. After favourable conditions are restored, the stolons resume growth and begin to form another tuber a little further down from the aborted one, forming a crown of tubers (chain tubers) with no commercial value. Water deficit may occur when the tuber is already formed and is growing; in this case, if the stress is interrupted abruptly, with sudden irrigation or ill-timed fertilization, it may cause excrescences, anomalies, cracks and physiopathologies in the tubers. All of the problems thus described may be avoided by keeping the soil constantly moist to 65–75% of its available water capacity, except for the final phases of maturation. It is a good idea, when the tuber is nearing physiological maturation, to avoid return of vegetation, which would certainly sacrifice product quality. Indicatively, water may be distributed according to some guidelines that vary according to the type of product desired:

- For common potato planted in spring, in the Mediterranean regions at least 450 m³/ha are needed, distributed over the whole period of development.
- For early potato planted in winter, a supplementary addition of a total volume of approximately 150–200 m³/ha may be sufficient since most of the growth cycle takes place during the winter; regarding this type of cultivation, if the temperatures drop much lower than 0°C, there is the risk of tuber freezing and plant loss.
- For potato planted in summer (counter-season), the irrigation should amount to an intermediate quantity between the above-mentioned volumes (200–450 m³) to guarantee survival through the summer.
- For industrial potato, water availability plays a fundamental role in productivity as well as in tuber quality.

Water Control

Different systems may be used for the control of land water reserves to instate an equilibrium that allows development without stress to the crop. One of the most diffuse is the calculation of the potential evapotranspiration. With this system, the data of the daily evaporation of a tank is used, corrected by the crop coefficient, differing with the latitude, the crop and the development phase.

Irrigation operations are commenced when 30 mm of water has evaporated. Experimental research of the Agronomy Institute of the Agricultural Faculty of Florence showed that to allow varieties of a medium-early growth cycle to achieve maximum production capacity, irrigation must be done every 12–15 d in northern Italy for a total of 2–6 irrigations, and approximately every 6 d in southern Italy for a total of 8–9 irrigations.

Pest and Disease Control

Another important aspect in potato cultivation is pest management. Disinfestation of the soil before planting has already been mentioned. Furthermore, it is necessary to keep under control the extremely harmful phyllophagous coleopteran, the Colorado potato beetle (*Leptinotarsa decemlineata*), whose mass attacks are capable of defoliating whole lots in a few days. The attacks of the cryptogams must also be prevented, particularly potato blight (*Phytophthora infestans*), the fungal disease that destroyed all of the potatoes in Ireland in the middle of the 19th century, by specific treatments if and when the conditions favourable to attack occur (high temperature and humidity). Viroses must also be controlled through aphid limitation and selection of resistant varieties.

Harvest

Potatoes are usually harvested when the first layers of leaves start to dry up. However, it is difficult to speak of ripeness in the potato. In fact, even tubers harvested very early (as in the case of early potatoes) are still edible and have a good taste. This phenomenon does not occur in any other organ of herbaceous plants, which must usually reach a minimum level of ripeness before being harvested.

Harvest time for "new" potatoes is springtime, beginning in March until the end of June; the exact time is decided through the evaluation of periderm consistency and tuber size. The common potato is harvested between July and September. According to the Rules of Region Emilia Romagna (2004): "the first parameter to be considered for harvesting potatoes destined to the market for fresh consumption and to industry is that of dry weight. The Region suggests harvesting of the potato when this value is respectively ≥ 18% for the potatoes destined to the market to be used fresh and ≥ 20% for the potatoes to be used in the transformation industry. It is implicit that the periderm must be completely formed and of the right thickness. After harvesting and during transportation it is also necessary to limit the exposition of the tubers to light, because this determines greening and an accumulation of toxic alkaloids." In November-December, the leap year potatoes and counter season potatoes are harvested.

Yields vary as a function of the cultivation environment and the typology of the potato produced. It ranges from 10–15 t/ha in the marginal zones or for early produce to 35–45 t/ha in the intensive cultivation zones of Emilia for the common potato. However, the average is around 24 t/ha.

14.5. Evaluation of Soil/Production Relationships

To find out whether there exists a relationship between production, product quality and land typology, three experimental stations were installed for the duration of five years, comparing three principal soil series—Cecita, Sila, and Croce della Palma—in the potato-producing district of Sila Grande (CS).

Cecita Soil Series: The soils belonging to this series evolved on fluvial-lacustrine deposits made up of gravel, silt and clay. The agronomic possibilities are essentially the potato rotated with pasture, cereals and polyphillous grass covering. From a genetic point of view, the soils of the series belong to the Inceptisols and are classified as Humic Pachic Dystrudept (Soil Taxonomy, 1998).

The soils of the *typical* phase (Fig. 14.1) have a texture between loam and sandy loam, are deep or very deep, are generally well drained, and are very dark brown. Their pH varies between slightly acid and strongly acid. They have an available water capacity between high and very high.

Sila Series: The soils of the Sila series evolved on granite, granite-diorite, quartz-diorite, mica-schist and gneiss. They are found on slopes with a moderate gradient, on sub-plain horizons and in morphologically depressed areas. They have a loamy or sandy loam texture and are moderately deep, well drained and dark brown. Their pH is moderately acid. They have a sufficient available water capacity and therefore a sufficient water supply for plants in the summer. If cultivated they are fit for irrigated crops, while spring-summer cycle renewal crops are represented by the potato rotated with cereals and polyphillous grass covering.

The soils that belong to the *typical* phase (Fig. 14.2) of the Sila series are of a loamy or sandy loam texture and are moderately deep or deep, well drained and moderately acid. They have a sufficient available water capacity and therefore a sufficient water supply for plants in the summer. The soils of the series have a cambic horizon, sometimes an umbric epipedon, fall under the Inceptisols and are classified as Typic Dystrudept (Soil Taxonomy, 1998).

Croce della Palma Series: the soils of the *typical* phase (Fig. 14.3) of the Croce della Palma Series evolved on granite, granite-diorite, quartz-diorite, mica-schist and gneiss, principally

Fig. 14.1. Landscape and typical soil of Cecita series.

Fig. 14.2. Landscape and typical soil of Sila series.

Fig. 14.3. Landscape and typical soil of Croce della Palma series.

along slopes ranging between moderately and strongly inclined. They have a texture ranging from sandy loam to loamy sand and are shallow, well drained and brown. Their pH is between slightly and moderately acid. Their available water capacity is between low and moderate and therefore they offer a scarce water supply for plants in the summer. In some cases they have a xeric type of pedoclimate instead of udic for their low organic matter content and/or limited thickness. From a genetic point of view, soils are not very different from their parent material: for this reason they fall under the Entisols and are classified as Lithic Udorthent (Soil Taxonomy, 1998).

Table 14.3 shows the land suitability evaluation according to the FAO methodology (1976), elaborated for all of the soil phases found in the Sila area, which was modified in its pH limitation value to adapt it to soils destined for potato and the "available oxygen supply" parameter to correct any imprecision deriving from the texture definition.

Regarding texture, it must be noted that the grouping by texture class used for defining the suitability class is indicative. Such groupings are referred to the deviation from a

Table 14.3. Land suitability for potato cultivation.

Soil characteristics	Suitability class			
	S1	S2	S3	N
Texture*	l, sl, cl, sil, scl, sicl, ls	s, sc, sic	c, si	
Available rooting depth (cm)	(> 100)	(100–50)	(50–20)	(< 20)
pH	(5.1–6.5)	(6.6–7.3)(4.5–5)	(< 4.5) (> 8.5)	
Rock fragments	(0–35%) (< 50 mm)	(10–35%) (50–150 mm)	(10–35%) (150–250 mm)	(> 35%) (150–250 mm)
Available water capacity (mm, depth 1 m)	high (> 100)	moderate (100–50)	low (< 50)	
Drainage	well drained	moderately well drained	excessively or poorly drained	
Slope %	(< 20)	(21–35)		(> 35)
Available oxygen supply	good	moderate	poor	imperfect

*c = clayey; l = loamy; si = silty; s = sandy.

granulometric distribution (loam soil) that is kept balanced, but do not consider the real environmental conditions of the rooting zone.

It is easy to guess that rooting depth influences habitability for plant roots and the quantity of water and air available. The pH affects the biological activity and the animal and microbiological populations of the soil, in addition to the availability of nutrients to higher plant species. It should be noted that with a pH less than or close to 4, aluminium tends to become soluble and toxic for the plant, and probably also gives a metallic taste to the pulp. Rock fragmentation may have both negative and positive effects. Fragments in moderate quantities favour the circulation of water and air, but as the fragments increase, the potatoes become less regular in shape. Without irrigation, the available water capacity determines strong fluctuations in potato production. Good drainage is important to avoid water stagnation and the consequent rotting of tubers. Slope becomes a problem most of all if, as soil depth decreases, the tuber is unearthed and turns green, which is associated with the production of toxic substances, mainly solanin. To these characteristics we add the available oxygen supply, which, independently from texture, affects the respiration of the tuber. In clayey soils like Terra Rossa or in sandy soils like a vitric Andosol, the texture characteristics do not count in determining class placement, therefore it is necessary to be accurate in assigning a suitability class to soils destined for potato cultivation.

The fundamentally sandy soils of the Sila tablelands have the peculiarity of having amorphous metals in the parent rock, probably of a pedological origin, which aids the accumulation of organic matter connected with mineral fragmentation and consequently the formation of granular structures that favour soil aeration, soil capacity to hold water available to the plants and better workability. In this case it is inappropriate to define sandy soils as being a limitation for agricultural production.

Potato farm lands were evaluated in the area of Sila Grande. A method developed by Aramini (in G. Vecchio et al., 2000) based on cost-benefit relationships has permitted a comparison of the profitability of the soil series existing on the Sila tablelands, based on previous studies (V. Vecchio et al., 1989, 1991, 1992) as shown in Table 14.4.

Table 14.4. Profitability.

Suitability class*			Marketable gross production			
	t	Price per unit in Italian lire (average value 1997–1999)	Total	Production costs	Net profit	Profitability
S_1	33	60,000	19,800,000	10,970,000	8,830,000	100
S_{2w}	31	60,000	18,600,000	12,238,000	6,362,000	72
S_{2ws}	30	60,000	18,000,000	12,688,000	5,312,000	60
S_{3d}	27	60,000	16,200,000	12,688,000	3,512,000	40

Elaboration by Aramini ARSSA Calabria (Lulli and Vecchio, 2000).

*S = suitability class; 1,2,3 = first, second, third; w,s,d = water, soil and drainage limitations.

Profitability decreases or increases in relation to soil and environment characteristics and, therefore, on the basis of the soil suitability class. However, even soils that belong to the same suitability class may have varying levels of productivity.

Statistical analysis of production data shows that the two soil series that belong to the most profitable class S1 (Sila and Cecita) differ significantly not only in production capacity, but also in the number of irrigations needed to reach a profitable level; therefore, it costs more to produce potato on the Sila series than on the Cecita (Table 14.5).

The very brief analysis of productivity and profitability in Sila Grande in Calabria demonstrates the effect of soil on potato production. The concept may be extended to many other environmental situations or potato-producing contexts.

Table 14.5. Comparison between the average annual production of the Cecita and Sila series in the period 1994–1998. The difference is significant for p > 0.05*.

Land unit	No. of samples	Annual prod., t/ha		Comparison of averages		
		average	standard deviation	Student's *t*	g.l.	Prob.
Cecita	10	36.1	2.4	2.339	6	0.003*
Sila	8	33.0	3.0			

14.6. Evaluation of Product Quality

Even more important than production, in our opinion, are the organoleptic and marketing qualities of the table potato. A series of tests are necessary to define the quality of the potato, whether boiled or fried, and it is necessary to define some descriptors for the gustative/tactile (*flour*) sensations as shown in Table 14.6.

Furthermore, the olfactory sensations of taste and aftertaste (aroma) must be analysed, such as olfactive intensity, starch, chestnut, grass, fruitiness, metal, earth, mould. The following must also be defined: skin colour (chromameter CR 300), washability by evaluation of principal

Table 14.6. Sensorial analysis (evaluation of kinesthetic parameters).

Evaluation	1	2	3	4	5
Texture (pulp firmness)	very tender	slightly tender	slightly firm	firm	very firm
Moisture (presence of liquid on cut surface and on palate)	very dry	slightly dry	slightly moist	moist	very moist
Graininess (presence of grains when chewing pulp)	very coarse	slightly coarse	slightly fine	fine	very fine
Typical taste (presence of the typical taste of the potato responsible for good flavour)	flavourless	→			strong typical flavour
Crunchiness (sensation of crispness perceived at the moment of breaking the product)	not crunchy	→			very crunchy
Typical appearance (refers to typical appearance of french fries)	not typical	→			very typical
Off flavours (refers to the presence of aftertastes with negative connotations on product typicity)	very pronounced	→			none

surface alterations, dry-weight percentage, glucose, browning after frying, and after-cooking blackening. For these characteristics, reference is made to experimental field research carried out in these same localities, on the same lands that were analysed as a function of production.

Three experimental lots were chosen among these pedotypes (the variety used was Agria) that were monitored from the time of seeding to harvest by surveying the following:

Survey and collection of phenotype-biometric data

- Emergence (expressed in number of days to reach 80%)
- Vegetative cover (%) 15 d after emergence
- Vegetative cover (%) at flowering
- Vegetative cover at the end of the phenological cycle (%) and outset of senescence
- Survey of number of principal stems
- Assessment of fresh weight on 10 plants at senescence or at destruction of foliage (defoliation)
- Assessment of colour (aerial part at emergence)
- Assessment of colour (aerial part at flowering)
- Height of vegetation 15 d after emergence
- Height of vegetation at flowering (10% of plants flowering)

Data relating to presence of cryptogams and harmful insects

- Fungus (Perospora—Rizoctonia—Alternaria)
- Insects (Doryphora—Aphids)

From the elaboration of data it is clear that most of the parameters measured do not show significant differences between the three pedotypes compared, while some data show a probable soil effect on plant growth and development. Particularly, tuberization begins first in the Sila series, then in the Cecita series and lastly in the Croce della Palma series. Also, the assessment of fresh weight on 10 plants at defoliation (38 d from seeding) shows significant differences between the three pedotypes. Another substantial difference is in the number of irrigation operations: seven are necessary in Croce della Palma lands, six in the Sila and five in the Cecita.

Fig. 14.4. Differences in potato growth between (a) the Sila and (b) the Croce della Palma series.

The sensorial and organoleptic analyses and related statistical elaboration were carried out by the Mario Neri Interprovincial Agro-environmental Research Center of Imola.

Table 14.7. Comparison of sensorial and organoleptic characteristics of the three series, Cecita, Sila and Croce della Palma.

Soil series	Sensorial and organoleptic characteristics								
	L	a*	b*	C*	h°	Washability*	Dry weight *(%)	VAVI*	Glucose (mg/dl)
Croce della Palma	60.63	−3.12	29.62	29.78	96.04	5.0	24.2	1.5	<20
Sila	61.31	−2.61	32.09	32.20	94.67	4.5	23.9	1.3	<20
Cecita	58.50	−1.31	30.00	30.03	92.50	2.0	23.3	2.0	<20

Comparison significant (P <0.05) for the three soil series examined.
The colour was evaluated using the Minolta CR-300 chromameter. The instrument converts colours into a number code that allows the definition of the three principal physical measurements that characterize it, i.e., luminosity, hue and saturation (represented by the indices L, a*, b* in the C.I.E space). The organoleptic and sensorial characterization is done on one lot of commercial-grade tubers, after a period of conservation lasting approximately 6 months at low temperatures (3–4°C; UR 90%). For the organoleptic characterization the following objective parameters are defined:

- *Washability: a visual evaluation that takes into consideration the incidence of the principal surface alterations due to phenomena of a physiological or a parasitic type. The sample (at least 30 tubers) was examined on a scale of reference that ranges from 4 to 9, formulated by the CNIPT-ITCF.*
- *Dry weight (%): determined through a calculation of specific weight in water. This parameter helps to distinguish the varieties as a function of their final use: marketable fresh (17–20%), pre-fried (19–21%) and chips (21–25%).*
- *Degree of browning after frying: assessed on a 10x10 mm stick extracted from the central part of 20 tubers, washed, dried, and then fried in peanut oil for 5 min at 180°C. The degree of browning was calculated using the Munsell colour system (VAVI) scale of reference;*
- *Degree of blackening after steaming: a defect caused by the formation of melanin compounds due to the complexing of polyphenols (chlorogenic acid) with iron during the cooling of tubers that were steamed or fried and frozen (after-cooking blackening).*

Table 14.8. Comparison of the physical characteristics prevailing among the Cecita, Sila and Croce della Palma soil series (average values).

Soil series	Cecita average value	Sila average value	Croce della Palma average value
texture*	3.21	3.50	2.82
moisture	1.79	2.25	1.91
graininess*	3.00	3.59	3.71
taste	2.96	2.82	2.50
sweetness*	3.21	2.59	2.79

Comparison significant (P < 0.05) for the three soil series examined.

A statistical elaboration of the data coming from the sensorial and organoleptic analyses showed significant differences (P < 0.05) between the tubers coming from the three soil series, regarding the parameters of colour, washability, dry weight, browning, texture, graininess, and sweetness (Tables 14.6 and 14.7). Furthermore, among the prevailing physical characteristics it seems that the variables of texture, graininess and sweetness represent the discriminants between the different soil series (Table 14.8).

14.7. Research Observations on Soil/Potato Quality

On the basis of five years of research, it may be affirmed that potato has different quality-quantity characteristics as a function of the pedotype. In fact, the soils in the Cecita area, characterized by a high content of organic matter and a higher available water capacity than the other soils, need fewer irrigation operations and produce tubers of a medium texture and graininess, with low moisture, a fairly strong typical taste and a chestnut aftertaste, compared to the soils of the Sila series, which have a low organic content and produce tubers that are characterized by a firmer texture and a finer graininess, with a relatively strong typical taste. The differences in plant growth and development between the two soil series are seen in Fig. 14.4.

The Croce della Palma series, because of the variability of its horizon thickness, exhibits also a high variability of organoleptic characteristics and furthermore shows substantial differences in plant growth and development. In fact, the pulp varies from slightly tender to slightly firm, the graininess varies between slightly fine and fine, and typical taste ranges from slight to strong. In the samples examined, no particular aftertaste was found. Finally, it may be hypothesized that lands in Sila produce potatoes with better texture and graininess. Hence, there does exist a soil effect on tuber quality. Another soil effect is on tuberization, which occurs first in the Sila series, then in the Cecita and last in the Croce della Palma series. Furthermore, if we observe the relationships between soils considering two different environments, though similar, one at Camignatello on the tablelands of Sila Grande in the province of Cosenza, and another at Decollatura in the province of Cantanzaro at lower altitudes in the granitic massif, we note that there are significant ($P < 0.05$) differences in those of Camignatello among the three soil series regarding colour, washability, dry weight and browning, for which the Sila soils seem the best. In the same way, the Decollatura seem best for crunchiness and dry weight.

Soils in Sila, both in Camigliatello and Decollatura, enhance typical taste and do not show sensations of aftertaste so strong that they compromise product quality. On the other hand, the samples of potatoes coming from the Cecita soils in both pedo-environments (Camigliatello and Decollatura) show a weak typical taste due to traces of earth and metal. However, in a comparison of the organoleptic profiles it seems that the negative characteristics are accentuated at Decollatura over those on the Sila tablelands. Lastly, a correlation between the chemical-physical characteristics of soils and potato quality may be hypothesized: in fact, soils rich in organic matter and with a soil texture tending towards loamy produce potatoes with an organoleptic profile inferior to those with soils that have a loamy sand texture with a medium-low content of organic matter. All this information leads to the hypothesis that high levels of organic matter content associated with a slightly acid pH and with even very limited contents of soluble aluminium have a negative influence on potato quality.

References

FAO. 1976. A framework for land evaluation. In Soil Bulletin no. 32, Rome.
LULLI, L., VECCHIO, G. 2000. *I suoli della tavoletta "Lago Cecita" nella Sila Grande in Calabria. La geomorfologia e l'idoneità alla produzione di tubero seme.* Monografia ISSDS, sez. Tecnologia. Progetto PANDA, sottoprog. 2, Serie 1, no. 150.
REGIONE EMILIA ROMAGNA. 2004. Norme tecniche di coltura—Tecnica agronomica—Patata.

SOIL SURVEY STAFF. 1998. Keys to Soil Taxonomy (VIII ed.). SMSS Technical Monograph no. 6. Cornell University, Ithaca, New York.

VECCHIO, G., LULLI, L., PAESE, F., BRAGATO, G. 2000. *Attitudine dei suoli alla coltivazione della patata da seme in Sila Grande—Calabria.* In: I suoli della tavoletta Lago di Cecita. Monografia. Progetto PANDA, sottoprog. 2, Serie 1, no. 150.

VECCHIO, V., CASINI, P. 1989. *Patata da seme: effetto della concimazione sui caratteri vegetativi e produttivi.* Verona.

VECCHIO, V., CASINI, P., FERRARO, S.G. 1991. *Impiego di microtuberi di patata nel sistema di produzione del tubero-seme.* Verona.

VECCHIO, V., CASINI, P., PAESE, F. 1992. *Evaluation of NPK application for potato seed-tuber production in Southern Italy.* Abstract of the 11th Triennal Conference of the European Association for Potato Research (EAPR) Edinburgh, UK, 8–13 July 1990, Firenze.

15. Tobacco (*Nicotiana tabacum*)

Fabio Castelli[1] and *Edoardo A.C. Costantini*[2]

15.1. Characteristics

In tobacco (*Nicotiana tabacum*), the commercial product is the leaf, which is the most dynamic part of the plant. It follows that the conditions in which the plant grows and develops have a strong influence on the end product, sometimes with contrasting quantitative and qualitative results (Papenfus and Quin, 1984).

    Tobacco varieties are grouped according to type (flue-cured, light and dark air-cured, fire-cured and sun-cured) and differ in their specific characteristics and pedoclimatic requirements and for the drying or curing technique used on the leaves to obtain a commercially acceptable and storable product (Table 15.1).

    In Italy the flue-cured type is represented by Bright, mainly belonging to the topped type, while the light air-cured varieties include Burley (the neutral type grown in southern Italian regions and aromatic type grown elsewhere) and the much less widely grown Maryland. Aromatic Burley and Maryland varieties are topped, whereas the inflorescence is not removed on neutral Burley variety.

---

[1] CRA-CAT—Unità di Ricerca per le Colture Alternative al Tabacco, Scafati (Salerno), Italy.
[2] CRA-ABP—Centro di Ricerca per l'Agrobiologia e la Pedologia, Firenze, Italy.

Table 15.1. EU classification of tobacco varieties in varietal groups based on curing method and pedoclimatic requirements (Regulation (EEC) no. 2075/92; Davis and Nielsen, 1999).

EU group	Type	Curing method	Variety	Pedoclimatic requirements
I	Flue-cured	Light tobacco dried in ovens with controlled air circulation, temperature and humidity	Virginia, Virgin D and hybrids thereof, Bright	Soils from sandy loam to loamy sand, poor soil; moderately high temperature and air humidity
II	Light air-cured	Light tobacco air-driedunder cover, not left to ferment	Burley, Badischer Burley and hybrids thereof, Maryland	Soils from silty loam (Burley) to fine sandy loam (Maryland), medium-high fertility; moderate temperature and high air humidity
III	Dark air-cured	Dark tobacco air-dried under cover, left to ferment naturally before being marketed	Badischer Geudertheimer, Pereg, Korso, Paraguay and hybrids thereof, Dragon Vert and hybrids thereof, Philippin, Petit Grammont (Flobecq), Semois, Appelterre, Nijkerk, Misionero and hybrids thereof, Rio Grande and hybrids thereof, Forchheimer Havanna IIc, Nostrano del Brenta, Resistente 142, Goyano, Hybrids of Geudertheimer, Beneventano, Brasile Selvaggio and similar varieties, fermented Burley, Havanna	Soils from clay loam to silty loam, medium-high fertility; moderate temperature and high air humidity
IV	Fire-cured	Fire-dried dark tobacco	Kentucky and hybrids, Moro di Cori, Salento	Soils from clay loam to silty loam and sandy loam, medium-high fertility; moderate temperature and high air humidity
V	Sun-cured	Sun-dried tobacco	Xanti-Yaka, Perustitza, Samsun, Erzegovina and similar varieties, Myrodata Smyrnis, Trapezous and Phi 1, Kaba Koulak (non-classic), Tsebelja, Mavra	Sandy clay soil, shallow, poor, without organic matter; high temperature, dry climate

The dark air-cured type includes the varieties Badischer Geudertheimer, Forchheimer Havanna IIc and Paraguay. Although still being grown in considerably large areas, this group is in decline because of the drop in market demand for dark tobaccos.

The only fire-cured type variety present in Italy is Kentucky, a variety that is currently showing a significant revival in Italy due to its increased popularity in Tuscan cigars.

Sun-cured tobacco, also known as Oriental tobacco, is being currently used less in cigarette production because of reduced market demand. The general decline in quality of the varieties Erzegovina, Perustitza, and Xanti-Yaka, once traditionally cultivated in suitable parts of Puglia, and altered EU policies have led to the total disappearance of this type of tobacco.

The most important tobacco varieties currently grown in Italy, in terms of diffusion and economics, are flue-cured Bright (Umbria and Veneto), air-cured Burley (Campania), and fire-cured Kentucky (Campania and Tuscany).

15.2. Cropping Practices

15.2.1. Seedbed

Given their climatic requirements and tiny size, tobacco seeds are sown in a protected environment towards the end of February, in polystyrene trays that contain a substrate that is kept moist (float-system). The seedlings are ready for transplanting when they have reached the 7–8 true leaf stage.

15.2.2. Rotation

The large profits that can be obtained with tobacco, the highly specialized farms and limited availability of suitable soils have led to intense crop exploitation. The most widespread practice is to grow tobacco for three consecutive years, interrupting the monoculture with a winter cereal (wheat or barley) in the fourth year. Such brief interruptions appear to be unable to limit the phenomenon of "soil exhaustion" and, especially in the case of Bright, to control pests such as root-knot nematodes, which are favoured by the highly permeable conditions of the most suitable soils. In the case of Bright and Oriental varieties, tobacco should not be planted after alfalfa or meadow grasses, as they usually release significant amounts of nitrogen, while this is not a problem for the other types of tobacco.

15.2.3. Soil Preparation

Soil preparation consists of a shallow ploughing, in order to incorporate manure as well, followed by secondary tillage (harrowing). Subsoiling or ripping may be useful where waterlogging is expected, as frequently happens in fine sandy loam subsoils, because of the presence of a hard pan. Fumigation is common practice in the soils most suitable for the cultivation of flue-cured tobacco even if expensive and environmentally damaging, because root-knot nematodes are often present because of high sand content and insufficient rotation. The soil is then aerated through shallow ploughing and harrowing to eliminate the residual highly phytotoxic fumigant nematicide, to incorporate the mineral fertilizers, to prepare the soil for pre-transplant weed control and to form a suitable bed for transplanting.

15.2.4. Transplanting and Growth Phase

Transplanting, usually performed by machines, starts at the end of April and is completed by mid-May. Seedlings are planted on flat soil or more often on ridges to prevent or reduce damage from excess water. Plant spacing, and therefore plant population (plants per hectare), depends on the varietal group, variety, date of transplanting, soil fertility and mechanization level. The development of new leaves recommences after a latent period of 2–3 weeks during which, in favourable conditions, new roots grow. Root growth is fast, provided temperature and soil moisture content are adequate. In 5 weeks they reach a depth of approximately 40 cm reaching as far as 90 cm of the soil profile at the end of the cropping cycle. Maximum leaf area is reached within the following 3 weeks. Stalk lengthening terminates with the appearance of the inflorescence at the centre of the apical leaves.

15.2.5. Topping

In some varieties of tobacco (Bright, aromatic Burley, Maryland, and Kentucky), the inflorescences are removed at the start of flowering. Removal of reproductive organs and uppermost leaves leads to an increase in leaf dimension and weight, raises yield, stimulates the growth of new roots and improves the quality of the tobacco as it induces a higher content of nicotine and other aromatic chemical components. Topping breaks apical dominance and induces the rapid emission of suckers from the axillary buds on the stalk, the growth of which must be controlled, usually by chemical applications.

15.2.6. Harvesting and Curing Methods

Harvesting is generally scaled, as the leaves gradually ripen. It normally begins shortly after topping, starting with the lowest leaves, and is completed in three or four steps, usually by September. Another technique, mainly used for Burley and Maryland, is to harvest and cure the plant as a whole. A third alternative involves harvesting the bottom leaves only and then the rest of the plant as a whole (mixed harvest). As soon as they are harvested, the leaves are cured. Throughout the curing and drying stage, the leaf initially undergoes a number of biochemical processes (the breaking down of chlorophyll, proteins and polysaccharides; colour change and partial dry matter loss). The end product is composed of the cured leaves, pressed in bales or packed in cartons. Each batch is evaluated following appraisal of its quality and market value.

Tobacco is mainly used for smoking, with only a minimal amount destined for chewing or snuff. As regards smoking products, cigarettes account for almost 99%, with a net drop in pipe tobaccos and a significant rallying of cigars and cigarillos in recent years.

15.3. Crop Requirements

15.3.1. Climate

Tobacco requires appropriate climatic conditions to sustain fast and uninterrupted growth. In terms of both quantity and quality, it is highly sensitive to even relatively modest fluctuations in environmental conditions and nutrient levels. Steady, moderately high temperatures and high air humidity (> 75%) are generally required especially during leaf ripening (Papenfus and Quin, 1984), except for the sun-cured type, which needs a dry climate.

Tobacco will grow almost everywhere (the cultivation area extends from latitude 55°N to 40°S), provided there are at least 120–140 frost-free days in the season, depending on the type of tobacco (Davis and Nielsen, 1999). In young plants, the temperature strongly influences the transition of the shoot apex from the vegetative (production of leaves) to the reproductive phase (formation of the inflorescence). Sudden temperature decreases lead to early flowering, a serious problem that strongly impairs the yield (Papenfus and Quin, 1984). The optimal air temperature is around 27°C: below 15°C plant growth is very slow, while air temperatures above 35°C are damaging (Jacks, 1936).

In general, insufficient rainfalls cause less damage than excessive rain, as tobacco can survive drought periods, if not too prolonged. Frequent spring rainfalls make soil fumigation and soil preparation for transplanting difficult, while intense and repeated precipitations in late spring, often accompanied by waterlogging and consistent drops in temperature, can have a negative influence on yield. In Italian conditions, summer rainfall is usually low and erratic, i.e., concentrated in a few events, with above 7–8 mm of daily evapotranspiration in the hottest months. Therefore, tobacco should be irrigated frequently, as often as once every 5–6 d in July and August, with the exception of the sun-cured type, which must be dry-cultivated. The seasonal water requirements for Bright tobacco in the Po Valley are estimated at between 400 and 700 mm (Castelli et al., 1991).

Strong wind can cause the breaking of stalks or destruction and removal of leaves from the plant, cold damage or heavy dehydration in young plants. Another serious problem is hailstones, with damage that can cause total crop loss.

15.3.2. Soil

The soil can strongly influence tobacco characteristics such as the size, texture and colour of the leaves. Light-coloured soils, with low nutrient content (especially nitrogen) and reduced organic matter content, are generally suitable for light tobaccos, whereas dark soils are suitable for dark tobaccos (Jacks, 1936). Sandy soils tend to produce relatively large, light-bodied, light-coloured leaves, which develop a delicate aroma during combustion, whereas heavier soil, with higher silt and clay content, induces the formation of small, dark leaves, heavy-bodied and with a high nicotine content, which release a strong aroma when burning (Gisquet and Hitier, 1961). In any case, the soils must be well drained to allow the free movement of water and air. Even if temporary, waterlogging can lead to a considerable yield reduction (Davis and Nielsen, 1999). In fact, the tobacco plant is one of the most sensitive to a lack of oxygen in the soil and requires a minimum air capacity of 15%. A survey carried out on tobacco soils in the Verona area (northern Italy) demonstrated the close relationship between the aptitude for cultivation of Bright tobacco, air capacity and available water content (Costantini et al., 1992).

While the soil surface layers mainly influence the uptake of water by the plant and the maintenance of adequate soil moisture content, good drainage mostly depends on the characteristics of the deeper horizons of the profile. Soil depth is a limiting factor when the water table or a hard pan restrictive to root growth lie at less than about 100 cm. Typical soils for the flue-cured tobaccos are deep, moderately fertile, sandy, loose and porous even if, in the absence of irrigation, they are exposed to the effects of water shortage, with a resulting shorter plant growth and thickening of the leaf lamina. These soils encourage deeper root development, which increases plant stability and the capacity to tolerate brief drought periods. They also provide easy machinery access even after protracted rainfall. Sandy soils behave in practice as an inert substrate and therefore respond well to cropping operations (Morgan et al., 1938). The possibility of obtaining high quality tobacco of the flue-cured type from these soils, despite their low natural fertility, and of supplying nutrients easily and in the required amounts, has led some experts to consider bright tobacco as a form of hydroponic cultivation (Akehurst, 1981). The light air-cured, dark air-cured and fire-cured types adapt better to fertile intermediate loamy-textured soils, with the exception of the Maryland variety, which likes fertile soils, but preferably sandy and deep. The sun-cured type is traditionally grown in poor, rather infertile, shallow soils, containing minimal amounts of organic matter.

15.3.3. Water Requirements

Good water availability provides nutrient transport and ensures full tissue turgidity and good cell expansion, with the consequential formation of large, fine and well-expanded leaves. These leaves generally burn well and have a pale colour, higher filling power, milder aroma (lower resin and gum content), good sugar concentration and relatively low nicotine content. On the other hand, water stress reduces yield; leaves are difficult to cure and are small, thick, tight-textured, stiff, dark, rich in nicotine, but poor in sugar, with a stronger aroma (Davis and Nielsen, 1999; King, 1990).

Good soil moisture content at transplanting creates an ideal environment for the roots; so short irrigation immediately afterwards ensures better survival of the young plants (King, 1990). During the following rooting phase, up to about 4–6 weeks after transplanting, water requirements are low. In fact, during this period, a moderate water deficit induces root growth downwards and the plant thus acquires greater capacity to survive any subsequent periods of water deficit. Limited water stress in this phase also causes small chemical and structural alterations in the leaves that improve their quality, without interfering negatively on yield levels (Papenfus and Quin, 1984).

In the subsequent phase of rapid leaf growth, the crop needs frequent irrigation. Any prolonged water stress reduces the production of leaves, worsens the quality, may greatly delay the senescence processes and prevents the uptake and transport of nitrogen in the leaves. In the last period, during which the leaves are harvested as they gradually ripen, water requirements are low because the transpiring surface reduces with harvesting and the stomatal conductance decreases in the senescent leaves.

When the soil water content exceeds requirements, several problems can arise, such as the leaching of nitrates, which are rapidly transported below the growth limit of the root, resulting in overall plant damage. In the most serious cases, i.e., when waterlogging occurs, the roots are placed in anaerobic conditions, and many of their functions are impaired, including water absorption capacity, while micro-organisms can destroy the roots and plug the xylem of the stem base (Davis and Nielsen, 1999; King, 1990). If this happens at the onset of rapid growth phase, and persists for more than 48 h, the plants appear wilted and lack vigour, with up to 40% in yield losses. When the problem occurs during the phase of leaf ripening and harvesting, the damage is reduced and leaf senescence may be accelerated (King, 1990).

15.3.4. Nutrients

Nitrogen is the key element in the nutrition of the tobacco crop (Papenfus and Quin, 1984). Adequate nitrogen applications provide a remunerative yield and also guarantee high quality. Appropriate nitrogen availability in the early stages of plant growth permits rapid dry matter accumulation and good leaf expansion. An insufficient amount stunts growth and reduces the size of the leaves, which become pale, starting with the lowest leaves (McCants and Woltz, 1967). After curing, the tissue appears substandard, with a low nicotine content and poor aroma. Nitrogen deficiency can occur through insufficient fertilization accompanied by heavy rain or over-abundant irrigation. Especially in flue-cured tobacco, the nitrogen available to the plant must be depleted rapidly at the end of rapid growth phase. Excessive nitrogen can worsen the quality (delayed ripening, thickening of the leaves, excessively strong colour, reduction in burning capacity, pungent smoke) (McCants and Woltz, 1967). It is a fact that excess nitrogen increases production costs (more numerous operations for sucker control, higher fuel consumption for longer curing times) and causes an accumulation of nitrates in the leaves, not to mention pollution of surface and ground waters. In general, a well-developed tobacco crop takes up between 60 and 100 kg/ha of nitrogen from the soil. This amount, for the reasons described above and in flue-cured tobacco in particular, is not always wholly supplied by fertilization; also, the natural organic content of many soils, after mineralization, may take a large amount of nitrate nitrogen from organic matter present in the soil. In the light air-cured, dark air-cured, and fire-cured types, the amount of nitrogen supplied with fertilization can be more than double that of the uptake.

Phosphorus, although taken up in lower amounts than nitrogen (12–20 kg/ha), must be readily available during the whole crop cycle but particularly in the early growth of plants. Any deficiency in the first part of the cropping cycle limits leaf expansion and leads to an unusual dark coloration, formation of necrotic patches on the lower leaves, and stunted and delayed growth and ripening (Papenfus and Quin, 1984). This results in a decline in the yield and quality of the cured leaves. Later in the cycle, the root growth and consequential exploration of a greater soil volume excludes any deficiency phenomena. The supply of phosphorus is generally 80–100 kg/ha, distributed before transplanting, preferably banded not too deeply (McCants and Woltz, 1967) at ridge formation. It is worth considering that the repeated and excessive applications of this element in traditional cultivation areas, with limited uptake and reduced or no losses through leaching, have led to a general phosphorus increase in the soil content, which usually contains more than sufficient amounts (McCants and Woltz, 1967). Nevertheless, phosphorus fertilization shows a positive effect on the crop because phosphorus absorption is strongly limited by low soil temperatures and a non-neutral pH.

The high potassium content in the cured leaves, which varies between 1.5 and 8%, is considered a parameter of good quality and is universally applied to tobacco in all production areas (McCants and Woltz, 1967). It is the element absorbed in the highest quantity by tobacco plants (110–140 kg/ha) (Papenfus and Quin, 1984) and its uptake increases in proportion with the availability in the soil, to levels that can reach 250 kg/ha. Nitrogen in ammonia form and high levels of magnesium, calcium and sulphur can negatively interfere with potassium uptake. The amount of fertilization depends only partly on the soil content and on the amount actually absorbed by the plant because of its recognized beneficial effects on the quality of the leaves such as greater burning capacity, elasticity and intensity of colour (McCants and Woltz, 1967). Up to 200 kg/ha can be applied, mainly in pre-transplanting, and the rest as top dressing.

Calcium is also taken up in large amounts (1.5–2.0% of the leaf weight) and in absolute value it comes immediately after potassium (McCants and Woltz, 1967). Nonetheless, given its wide availability in Italian soils and indirect inputs from phosphate fertilizers, phenomena of calcium deficit are extremely rare if not wholly absent, and may only appear in particularly acid sandy soils.

Magnesium availability is insufficient for the needs of the tobacco plant only in very sandy soils and with particularly heavy rainfall, while the limited uptake is due to antagonism with other ions (very calcium-rich soils, excessive potassium fertilization). It has been demonstrated that when the ratio between potassium and magnesium is higher than 8, symptoms of insufficiency may appear. If there is an excess, on the other hand, it can negatively affect the burn rate (McCants and Woltz, 1967).

Among the microelements, chlorine plays an important role, as it has positive effects if present in small quantities but, on the contrary, it strongly reduces the burning characteristics of the cured tobacco when in excess, perhaps because it causes an increase in the leaf tissue water content (McCants and Woltz, 1967). Excess chlorine may arise in saline soils or if saline water is used for irrigation, when the salinity is due to the presence of sodium chloride. Chlorine deficit is extremely rare (there is a sufficient amount in the soil and it is provided as a secondary element in standard fertilizations in any case), whereas excess can lead to negative effects when fertilizers that are rich in this element are used (for example, potassium chloride).

An insufficient availability of boron is very rare under normal cultivation conditions. It may occasionally occur in coarse and sandy alkaline soils, and as a consequence of heavy rainfall, especially during the phase of rapid plant growth (McCants and Woltz, 1967).

Sulphur is absorbed in low quantities and it is rare to find symptoms of deficit. They may only occur in deep coarse soils, which are poor in organic matter, are unfertilized and have scarce water availability (Reich, 1985).

Relatively large amounts of manganese are absorbed, but situations of deficit are rare in Italian soils, whereas manganese toxicity can occur when tobacco is grown in acid soils (McCants and Woltz, 1967).

15.4. Land Suitability for Tobacco Cultivation

Climate and soil characteristics, hydrological conditions and nutrient requirements all contribute to defining the suitability of a given environment for the cultivation of tobacco. Some of these

Table 15.2. Requirements for the cultivation of tobacco.

Requirements	Conditions
Climatic	
Air temperature	average to high
Air humidity	low to high
Cropping	
Texture	sandy to clay loam
Oxygen availability	high and constant
Water availability	low to average
Soil depth	high
Root growth	rapid
Nutrient availability	low to average high
Peat	absent
Soil salinity	low
Soil pH	neutral to moderately alkaline
Management	
Trafficability	immediate
Workability and mechanization	easy

requirements are common to the different types of tobacco grown in Italy. The essential elements as regards the soil are a deep water table, the absence of plough-pan and waterlogging, a high and steady oxygen availability, as assured by a high air capacity (i.e., high textural or structural macro-porosity) and very scarce to medium organic matter content. Among climatic requirements, a steady and regular air temperature throughout the cropping cycle is considered a necessity for good crop growth. Equally important is water availability, which, not being satisfied by natural rainfall in cultivation areas in Italy, is always ensured by irrigation. For this reason, tobacco, with the exception of the sun-cured type, can only be grown where sufficient and continuous irrigation water is available (Table 15.2). Other characteristics (texture, soil fertility, temperature, air humidity) can differ significantly between the various varietal groups, although always within a relatively restricted range.

15.5. Functional Characteristics of the Soil and Evaluation Tables

The only known evaluation of soil suitability for tobacco growing in Italy was conducted by Costantini et al. (1992) in the flue-cured cultivation area around Verona. In that research, the data obtained from nine years of cured tobacco yields in 49 plots belonging to seven different farms, for a total of 151 observations, were processed and related to the characteristics of nine soils where the crop was grown. The functional characteristics and evaluation tables shown in Tables 15.3, 15.4, 15.5 and 15.6 come from that study.

High soil macroporosity appears to be the main factor for the success of the crop, with the study showing a highly significant proportional relationship between air capacity and crop yield. With flue-cured tobacco, which is very sensitive to a lack of oxygen, the best yield results were achieved with a soil-air capacity of above 30%. Intermediate results were obtained with values between 20 and 30%, while those from soils with an air capacity of less than 20% were poor. Again, with reference to flue-cured crops, the soils with highest available water capacity demonstrated lower production aptitude (Table 15.3). High air capacity satisfies the two main requirements for the cultivation of flue-cured tobacco: a high and steady availability of oxygen (crop requirement) and easily accessible trafficability (management requirement). Surface trafficability is a decisive factor for frequent and timely cropping operations. Soil must

Table 15.3. Air capacity and water availability (% volume). Weighted averages for the root-zone (90 cm) of soils with different suitability levels for flue-cured tobacco growing in the Verona area (northern Italy).

Suitability class	Air capacity	Available water capacity
Very suitable (S1)	34.9	5.0
Suitable (S2)	26.3	8.8
Not very suitable (S3)	26.0	9.1
Unsuitable (N)	16.9	13.0

drain rapidly and have a good load-bearing capacity, so that machinery can gain access after rainfall. Soil should also have limited amounts of gravel and stones (rock fragments), which are a serious obstacle to machinery, especially during transplanting operations.

In terms of physical characteristics, ideal soil will therefore be able to ensure perfect trafficability, optimal oxygenation, good drainage and deep rooting. Root growth is encouraged by a low amount of available water so that, at the start of field cultivation in spring, the roots

Table 15.4. Classes of soil functional characteristics (top 90 cm) for the cultivation of flue-cured tobacco in the Verona area (northern Italy).

Functional characteristics		Classes
Air capacity (% vol.)	a0	high > 30
	a1	moderate 20–30
	a2	low < 20
Rock fragments (% vol.)	p0	absent to frequent (gravel, pebbles and stones < 35%)
	p1	abundant to very abundant (gravel, pebbles and stones > 35%)
Depth*	w0	deep or very deep (> 100 cm from the surface)
	w1	shallow (< 100 cm from the surface)
Peat	t0	absence of peat in the top 100 cm
	t1	presence of peat in the top 100 cm

Water table or root-limiting horizon.

are prompted to explore a large soil volume in order to ensure the necessary water supply for the plant during its later development (Table 15.4). Soil depth and the presence of peat were also considered because of their implications in tobacco cultivation.

The soil nutrient contents in the area under research were not included among the functional characteristics because it was not possible to relate them to either crop yield or soil type. However, it was ascertained that the nutrient requirements of the crop are generally met by the standard fertilizations applied by the farmers.

To evaluate the suitability of land for tobacco growing, the focus was placed on functional characteristics, essentially of a physical nature, which were shown to be particularly important for tobacco cultivation and yield, and which demonstrated a spatial variability that could be described by a pedological map on the scale 1:25,000.

Table 15.5. Average yields of flue-cured tobacco by suitability class.

Suitability class	Yield category	Average yield(t/ha)	P < 0.05 (*)	P < 0.01 (*)	Standard error	Coefficient of variation (%)
S1	Very high	3.2	a	A	140	21.4
	High	3.0	ab	AB	241	24.2
S2	Good	2.6	abc	AB	117	14.1
	Moderate	2.4	bc	AB	179	25.7
S3	Low	2.3	c	B	61	26.6

Averages with different letters differ significantly at the indicated probability levels.

Table 15.6. Guide to the attribution of soil suitability classes for the cultivation of flue-cured tobacco in the Verona area (northern Italy).

Functional characteristics	Classes			
	S1 very suitable	S2 suitable	S3 not very suitable	N unsuitable
Air capacity	a0	a1	a2	–
Rock fragments	p0	p0	p1	–
Depth	w0	w0	w0	w1
Peat	t0	t0	t0	t1

The guide to the attribution of suitability classes for the cultivation of flue-cured tobacco indicates that class S1 (very suitable soils) are soils with no limitations in which the highest yields are obtained (averaging above 3.2 t/ha) (Table 15.5); class S2 (suitable soils) are soils with satisfactory multi-annual average yields of cured tobacco, but significantly lower than S1 (averaging 2.8 t/ha); class S3 (not very suitable soils) are soils with even lower yields (averaging below 2.3 t/ha) that are highly variable according to the farm and year. Soils normally excluded from the cultivation of tobacco are in class N (Table 15.6).

References

AKEHURST, B.C. 1981. *Tobacco.* 2nd ed. Tropical Agriculture Series, Longman, London and New York.

CASTELLI, F., CEOTTO, E., DONATELLI, M. 1991. *Consumi idrici del tabacco di tipo Virginia Bright: un quadriennio di prove nella Bassa Pianura Padana.* L'Inf. Agrario 25, 63–67.

COSTANTINI, E.A.C., CASTELLI, F., CASTALDINI, D., RODOLFI, G., NAPOLI, R., PANINI, T., BRAGATO, G., PELLEGRINI, S., ARCARA, P.G., CHERUBINI, P., SPALLACCI, P., BIDINI, D., SIMONCINI, S. 1992. *Valutazione del territorio per la produzione di tabacco di tipo Virginia Bright: uno studio interdisciplinare nel comprensorio veronese (Italia settentrionale).* Supp. Annali ISSDS XX.

DAVIS, D.L., NIELSEN, M.T. 1999. *Tobacco: Production, Chemistry and Technology.* Blackwell Pub., Oxford, UK.

GISQUET, P., HITIER, H. 1961. *La Production du Tabac.* J.-B. Baillière et Fils, Paris.

JACKS, G.V. 1936. Tobacco. In: Tropical soils in relation to tropical crops. Imperial Bureau of Soil Science, Technical Communication 34, 41–46.

KING, M.J. 1990. Tobacco. In: Irrigation of agricultural crops. Agronomy Monograph 30, ASA-CSSA-SSSA, Madison, Wisconsin (USA), 811–833.

MCCANTS, C.D., WOLTZ, W.G. 1967. Growth and mineral nutrition of tobacco. Advances in Agronomy, vol. 19. Academic Press, New York and London, pp. 211–265.

MORGAN, M.F., GOURLEY, J.H., ABLEITER, J.K. 1938. *The Soil Requirements of Economic Plants.* Soils & Men, Yearbook of Agriculture, USDA, pp. 753–776.

PAPENFUS, H.D., QUIN, F.M. 1984. Tobacco. In: The Physiology of Tropical Food Crops. John Wiley & Sons, New York, pp. 607–636.

REICH, R.C. 1985. *Flue-cured Tobacco Field Manual.* Reynolds Tob. Company, Winston-Salem, North Carolina (USA).

16. Soybean (*Glycine max* L. Merr.)

Fabiano Miceli[1] and Stefano Barbieri[1]

16.1. Characteristics

Soybean (*Glycine max* L. Merr.) is an annual legume that has been cultivated for dozens of centuries in the Far East. Wild *Glycine* species are still common in southern Siberia and Manchuria (Harlan, 1992). Soybean was introduced into Europe in the late 18th century, where it was grown in botanical gardens. The first shipment reached America in 1804 and was used as ballast to balance the vessel. In the beginning, soybean was grown for forage, but after a chemist, Carver, in 1904 succeeded in defining the oil-protein composition of the seed, cultivation turned to seed production. Soybean is now considered the second most important crop in the USA, and the production in the North American continent is the principal source of soybean world-wide. The composition of the seed is illustrated in Table 16.1.

Table 16.1. Average composition of soybeans and seed parts (Cheftel et al., 1985, cited in Berk, 1992).

Seed part	weight % of the whole seed	% dry weight			
		Proteins N x 6.25	Lipids	Carbohydrates (including fibre)	Ash
Cotyledons	90	43	23	43	5.0
Teguments	8	9	1	86	4.3
Hypocotyl	2	41	11	43	4.4
Whole seed	100	40	20	35	4.9

---

[1] Università di Udine—Dipartimento di Scienze Agrarie e Ambientali, Udine, Italy.

The crops grown in Asia, America and Europe are mostly destined for the milling industry for oil extraction, but they can also be used for many other purposes (Snyder and Kwon, 1987; Berk, 1992). Soybean seeds are processed to produce meal and flour for bakery products, but traditional uses of soybean are many: coffee surrogates, sprouted beans, soy milk, shoyu (soy sauce), tofu (soy cheese), tempeh and natto (boiled soybeans that are then fermented using different types of yeast), miso (derived from soy milk) and other typical Oriental products. The oil is used to produce glycerol, fatty acids, sterols, rendered oils, lecithin, antifoams, paints, inks, rubber and wetting agents. Typically, soybean vegetable protein for food, feed and other purposes is obtained from the defatted soy meal.

As regards the strategic importance of soybean and its derivatives on the international market, soybean meal and soybean oil cake are the key products of the oil extraction industry, since they are fundamental for the diet of intensively farmed livestock. Oil, on the other hand, is less important and can be easily replaced with that extracted from other plant species.

The varieties grown in America and Europe derive from a limited number of accessions of Chinese origin that were introduced into the USA in the early 20th century. The varieties are classified into 15 Maturity Groups (from 000 to 12), on the grounds of the length of the growing season, determined at various latitudes ranging from the northern to the southern borders of the USA. The differences in cycle length depend in general on the marked sensitivity to photoperiod of soybean (Cregan and Hartwig, 1984; Danuso et al., 1987; Grimm et al., 1993).

In the 1980s, soybean cultivation became widespread in Italy: after the EEC decided to subsidize oilseed production, soybean increased from 317 ha in 1981 to 513,789 ha in 1991 and then declined to 172,000 ha in 1995. Yield per hectare was unchanged from 1990 to 1995. With 553,002 t, Italy ranks tenth among the 10 most important soybean producing countries (FAOSTAT, 2005), but at an average yield (3.63 t/ha) that is the highest in the world. Sporadic reports of economically important damages do not change the satisfactory health condition of the crop in general. The popularity of crop rotation, the differentiation of germplasm, and high quality of seed most often imported from North America account for the rarity of epidemics (Wrather et al., 1997).

Some 25 years ago, in order to improve the adaptation and the diffusion of a previously underutilized oilseed crop, the Italian Ministry for Agriculture and Forestry funded a specific "Oleaginose" Programme. A series of research and extension activities had been in place for a decade, within a network stretching from Friuli Venezia Giulia to Sicily. Appropriate management techniques for soybean were defined, with a large impact in the agricultural systems of northern Italy (Mosca et al., 1991). Maturity Group I varieties are usually grown in the Po Valley, where most of the soybean area is located. Maturity Group II varieties, planted from April to early May, may give similar yields but the delayed harvests might put the crop at stake, given the heavy rainfalls in autumn. In double-cropping systems, quite frequent in the humid climate of the Eastern Po Valley, soybean is planted after winter cereals with minimum tillage (late June to early July); varieties of Maturity Group 0 (0+) are used. For central and southern Italy, the varietal choice would be larger as Maturity Groups II and III are appropriate. Unfortunately, soybean is almost absent in southern Italy: provided irrigation is available, the high water requirements under Mediterranean climate would imply excessive irrigation, compared to the limited crop returns. On the other hand, at higher latitudes (on the northern side of the Alps, as in Austria or in Switzerland), thousands of hectares are planted with early-maturing soybean varieties (Group 00 and 0).

16.2. Cultivation

16.2.1 Rotation

As previously mentioned, soybean has been grown for over 20 years in the Po Valley (northern Italy). The crop is useful to interrupt the monoculture of summer cereals (maize in Veneto and Friuli, rice or maize in Piedmont and Lombardy) and to extend the timing of sugarbeet presence in Emilia Romagna, south of the River Po. The popularity of soybean among farmers is largely due to the flexibility of sowing, usually performed from early April to late June. Delaying sowing up to late June, as a second crop after barley or wheat, normally causes moderate yield reductions.

This flexibility is much appreciated in the larger farms. Second crops after winter cereals are popular in irrigated lands or areas in which rainfalls are frequent in late summer. A recent problem in northern Italy agriculture is the spreading of Western corn rootworm (*Diabrotica virgifera virgifera*), a beetle of the Chrysomelidae family that causes severe damage to maize and requires interruption in the maize monoculture. Native to America, this beetle appeared in Europe in 1992 and six years later in Italy (Furlan et al., 1998); despite programmes of compulsory pest control carried out in Lombardy, Veneto and Friuli, it has not yet been eradicated. The interruption of maize monoculture may favour a more widespread adoption of soybean in agricultural systems of northern Italy.

16.2.2. Soil Tilling and Preparation

Because of its rooting system, soybean requires medium-deep ploughing. The soil is usually tilled with appropriate machinery using the same techniques adopted for other summer crops. Soil preparation includes medium-depth ploughing for main crops, while minimum tillage or direct drilling may be appropriate for second crops. Clayey soils are generally ploughed in early autumn, whereas sandy or medium-textured ones can be ploughed in early spring. On heavy soils, ploughing is followed by harrowing and levelling of the surface; attention is usually given to avoid non-uniform sowing, as pods from the lower nodes are difficult to harvest from plants growing in soil depressions.

When preparing the soil for second-crop soybean following barley or wheat, time is all important so as to avoid losing soil moisture. Appropriate minimum tillage techniques are thus used. Gaining a few days at sowing time is vital, since the season is shorter and the radiation of warm, sunny days in July should be fully exploited. In general, minimum tillage, carried out with the machinery present on the farm (e.g., disc harrow, rotary harrow), does not cause a reduction in productivity (Ciriciofolo et al., 1992). It is also possible to sow on unploughed land, but no-till techniques require special planters that are not usually available on the farm and must therefore be rented. Winter cereals usually leave the ground in good structural conditions but it is often necessary to use chemical means to destroy the weeds. In these cases, and often even in full-season crops, only limited amounts of phosphorus and potassium should be applied, if not entirely omitted. Given the crop's self-sufficiency for nitrogen and the reduced margin of profit that the crop currently yields, farmers nowadays tend to save on fertilizers and, relying on the residual fertility of the preceding crop, apply only small amounts of phosphorus when sowing. This is particularly common in the case of second crops.

16.2.3. Inoculation

In fields in which soybean has never been cultivated, or else grown more than 4–5 years before, it is advisable to carry out artificial bacterization by inoculating the seeds with a specific Rhizobium (*Bradyrhizobium japonicum*) strain. There are many products available on the market, consisting of a single strain or various combinations: the ideal inoculum must have adequate ability to infect, competitiveness, efficiency, resistance and persistence. The ability of *Bradyrhizobium* to infect roots (and induce nodulation) is obviously a key point, competitiveness refers to the strain's ability to successfully compete against the native microflora, and efficiency is the capacity to fix the largest amount of atmospheric nitrogen under different ecological conditions. Resistance and persistence refer respectively to the strain's ability to tolerate biotic or abiotic stress factors and preserve its viability, infectiveness and efficiency in the soil through the years (Venturi and Amaducci, 1984). Powdered products are available on the market to inoculate soybean: in these, the micro-symbiont is adsorbed on to finely ground peat, at a concentration of at least 10^8 cells per gram. The compound is added to water and then applied to the seeds. Inoculated seeds must be sown as soon as possible; if they are exposed to light or high temperatures or are not sown within 3–4 h, they must be re-inoculated. Despite all precautions, usually the viability of the bacteria decreases by one or two orders of magnitude throughout these operations, and this is why it is necessary to use products having a high concentration of live cells.

16.2.4. Sowing

In the Po Valley, main crops are sown from April to late May. A soil temperature of at least 12°C is advisable; therefore, soybean is usually planted after maize. Second-crop soybean is sown in late June to early July, after the harvest of winter cereals. Group I varieties are used in northern Italy for both main and second crops, although in the latter case early-maturing varieties (0 to 0+) can also be used. An optimal growth cycle lasts approximately 120 d (north and central Italy, crop) or 100 d (north and central Italy, second crop, and southern Italy) (Mosca et al., 1991).

Sowing is commonly performed at about 40 germinating seeds per square metre, so as to have at least 30 plants/m² at harvest, but the literature often indicates a density of 30–60 seeds/m². It must be noted, however, that it is often possible to obtain satisfactory yields with lower densities, i.e., 20–22 plants/m². Indeed, a few options (flower abortion, early pod drop, seed abortion) are available for the plant to precisely and progressively adjust its reproductive potential to the environmental conditions, modulating the capacity of the sink organs to accumulate photosynthesis products (Egli and Crafts-Brandner, 1996). Under low-density conditions, for example, the crop responds to the low inter-plant competition by allowing some intra-plant competition (branching), thus the appropriate number of seeds per square metre is produced at harvest.

The inter-row distance depends on, among other things, the machinery available in the farm, especially that used for mechanical weed control. In the farms south of the river Po, where crops are rotated with sugarbeet and wheat, inter-row distance is approximately 0.40–0.45 m. Smaller distances (0.15–0.20 m) are generally not advisable: mechanical weeding is precluded, fungal diseases may increase and grain yields are hardly increased. Twin-row planting has been proposed (Laureti et al., 1995), to maintain efficient use of available farm machinery and accelerate canopy development and interception of solar radiation. In the central states of the USA, where varieties of groups III and IV are cultivated, the distance between rows can reach 76 cm, although there is a tendency to reduce it to 19 cm to avoid within-row competition. However, the later the sowing, the smaller the distance between rows, so as to ensure that the canopy can rapidly cover the inter-row surface and thus intercept the intense radiation of July.

16.2.5. Weed Control

Whether for genetically modified (GM) or non-GM varieties, appropriate weed control is a key factor for crop profitability. The limited plant height and growth rate at early growth stages favour competition from weeds for water, nutrients and solar radiation. Massive yield reductions are expected without weed control. It has been observed that one *Xantium* plant per 1.5 m row length can reduce productivity by 26% over the entire growth cycle; a weed density of one plant per 0.3 m of row causes a 60% drop in the yield (Scott and Aldrich, 1983). Weeds can be controlled by mechanical and chemical interventions, or by chemical applications only. In various countries, since 1996 it has been possible to use GM varieties that are tolerant to non-selective herbicides such as Glyphosate. Unlike other crop species, transgenic soybeans have met with great success, and in 2007 GM varieties were planted in 91 million ha, i.e., 64 % of the world soybean cropped area (James, 2007). GM seeds are more expensive and not always conductive to higher yields, but those GM varieties are increasingly popular worldwide.

In many countries where GM crops are not allowed, conventional weed control is carried out. On loam soils, pre-emergence applications are popular, since they reduce scalarity of weed emergence and kill off certain weed species; therefore, low amounts of specific molecules can be used in post-emergence application, if needed. In the Po Valley, the diversity of weeds is usually managed by a graminicide (S-Metolachlor) in the pre-emergence stages, in combination with a dicot killer (Pendimetalin, Metribuzin or Oxadiazon), according to the composition of the infesting flora. Mechanical weed control, a common practice in organic farming, is carried out by hoeing and harrowing.

16.2.6. Water Requirements

Soybean is a common crop in the rainfed areas of Veneto and Friuli Venezia Giulia, close to the Adriatic Sea. The heavy and deep soils in this area have a high water retention capacity and can therefore store the water from the frequent rainfalls that occur throughout spring and summer. Indeed, the water requirements for this species are rather high for the biomass produced. Being a summer crop, its water requirements are satisfied only in part without irrigation. With respect to the seed-biomass produced by the crop, water use efficiency is approximately

0.4–0.7 g/kg (Smit, 2000). Given an average efficiency of 0.55 g/kg with a productivity of about 0.5 Mg seeds/ha, seasonal water consumption is around 640 mm. Part of this requirement is satisfied by rainfall and variations in the soil reservoirs. According to the depth and physical characteristics of the soil, the maximum water reserve usually ranges between 100 and 200 mm per metre of depth.

In acidic soils (pH < 5.5), soybean is more tolerant than maize to aluminium and can therefore reach deeper water resources, even when present in acidic sub-surface horizons: unlike the cereal, taproots of this legume can successfully penetrate them (Smit, 2000). Irrigation must be managed during initial growth stages so that it is not excessive. Abundance of water in early stages causes excessive vegetative growth and risks of lodging in later stages and is infrequently linked to high yields. Soybean tolerates short periods of submersion in soils having low infiltration capacity, but scarce drainage causes shallow rooting, stunted growth and the spread of diseases. Conversely, when water applications are scarce, farmers start to irrigate after anthesis. When only one irrigation is possible, water must be supplied at seed filling stages. Finally, when automated irrigation systems are available on a leafy crop such as soybeans, moderate water intensities are preferred in order to avoid lodging. Of course, the crop should be safeguarded from severe water stresses during vegetative stages, but a moderate water shortage at preanthesis stages should not compromise the productive potential and can actually stimulate the root system to grow deeper into the ground. As well as preparing the field for a second crop, irrigation practices can stabilize the yield by improving nitrogen nutrition. Symbiotic nitrogen fixation activity drops and almost disappears in the presence of even modest water stress (Bouniols et al., 1986; Sinclair et al., 1987).

16.3. Crop Requirements

16.3.1. Climate

Early sowing depends on soil temperature and risks of late frosts, whereas late sowing of suitable varieties must take into account the possibility of late autumn frosts. In a species so sensitive to photoperiod, the length of the season depends on interactions between this factor and temperature. Genetic diversity and the array of varieties make it possible to grow the crop from the tropical regions in Brazil to the sub-arctic climates in Sweden and Siberia. Temperature is a key factor in soybean growth. On the whole, air temperatures between 22 and 25°C are just optimal. Minimum temperatures required at various growth stages are 7–10°C for germination and emergence, 16–18°C for flowering and pod formation, 13–14°C for seed formation and development, and 8–9°C for maturation (Raper and Kramer, 1987). Crop performance may be limited by temperatures that are too low or too high. At temperatures lower than 13°C, anomalies may be observed during the reproductive stages; at 40°C, in the absence of water stress, drop of immature pods may easily occur. In soybeans grown for seed production, temperature during maturation can modify the quality of the seed for the following year: optimal temperatures are 25/15°C (day/night).

Development of soybean plants under non-limiting water conditions leads to dense foliage: during full flowering (R2) stage (Fehr and Caviness, 1977), leaf area index values between 5.5 and 7 can be measured. As previously mentioned, this does not necessarily imply record yields, since reproductive structures (pods and seeds per square metre) may not reach top values. Abundant foliage acts as a reservoir in the event of leaf damage caused by biotic or abiotic sources. For instance, a brief hailstorm causes modest damage, as compared to other crop species. On the other hand, soybean is sensitive to wet autumn seasons, with frequent rainfalls close to seed maturation. Under these conditions, the seeds rehydrate and swell up a number of times, the tegument tears and the cotyledons alter and rot. Soils that can withstand the passage of agricultural machinery without being compacted allow earlier harvest. To date, genetic improvement research aimed at lowering seed tegument permeability has had no success, because the seed's ability to germinate must be retained.

16.3.2. Soil

In general, soybean can be grown in various types of soil, since it is not particularly selective. In sandy soils, the limits are set by water availability. Well-structured clay soils ensure greater water retention and therefore regular growth even in the absence of irrigation. A high degree of base saturation in this type of soil also ensures optimal nutrient availability. Silty, loosely structured soils are less suitable, because the reduced aeration limits symbiontic activity and there is a risk of crust formation during seedling emergence. Conversely, the often adequate water retention capacity of these soils might compensate for the above-mentioned disadvantages. With respect to early crop establishment, soybean is better off than maize: seedlings are generally more vigorous, and subsequent growth and branching can cover any bare patches.

Soybean can be grown at different soil pH levels (ranging from 5.2 to 8), but optimal yield is obtained in sub-acidic or neutral soils (pH 6–7.2). In acidic soils (pH < 5.6), although problems may also arise at pH levels between 5.6 and 6, low molybdenum availability can limit nitrogen fixation. On the other hand, soils containing a percentage of active lime above 15% are also scarcely suitable.

In soils with reduced water capacity and in years having maximum temperatures above 33°C for more than a few days, a syndrome is often observed whose aetiology is still unknown. The so-called Green Stem Syndrome or Green Stem Disorder can be recognized by plants in which stems and leaves remain green at advanced growth stages, i.e., close to seed maturation. The plants have numerous pods, but they are often empty or contain undeveloped, mouldy seeds. The damage can be serious, up to crop failure (Miceli and Tassan Mazzocco, 2000). This situation hinders soybean cultivation in the shallow fields at higher altitudes, limiting it only to the deeper, fresher soils of the lower plains in north-eastern Italy (Veneto and Friuli).

Soybean is moderately tolerant to salinity: crop productivity is reduced at ECe values higher than 5 mmhos/cm, and 100% yield reduction at values of about 10 mmhos/cm. The response of the species to waterlogging is complex, given the role and functions of the *Bradyrhizobium* microsymbiont, which is an aerobic micro-organism. Conditions favouring water and air circulation, together with adequate moisture, are considered optimal for the first stages of plant growth and encourage nodulation, since *Bradyrhizobium* requires a liquid film for mobility. Hypoxia and waterlogging during these early stages of growth (VE–V2) can thin out crop density and delay nodulation. A period of submersion during the reproductive stages (R4–R6) does not cause plant death: soybean may adapt to these conditions by differentiating new roots and nodules, even if yields are often compromised (Miceli et al., 1993).

However, conditions close to water saturation are not negative for the species, which indeed prefers absence of water deficits throughout the season. If so treated, grain yields are higher than the well-irrigated controls, and a new growing technique (saturated soil culture) has therefore been proposed for areas subject to these conditions (Troedson et al., 1989).

16.3.3. Nutrients

Nitrogen nutrition plays a pivotal role in soybean productivity. In a well-nodulated crop, no applications of mineral nitrogen are required. However, when large amounts of mineral nitrogen are available at sowing, soybean produces few or no nodules and behaves like a non-nitrogen-fixing crop. The threshold level for NO_3^- is approximately 250 mg/kg soil, a concentration that more or less corresponds to that in the soil solution of unfertilized fields. If, 30–35 d after emergence, the leaves are pale green, the plants may have failed to develop nodules: it is possible to ascertain this by simply uprooting some plants to examine them. In this case, it is necessary to apply at least 150 kg nitrogen per hectare before bloom, in order to obtain an acceptable crop. The amounts applied are equal to those taken up by the plants (seeds and vegetative organs); the yield, however, will be 15–20% lower than in normally nodulated soybean.

In order to fix nitrogen, certain factors are necessary: micro-macro symbiont recognition, completion of the infection process, development of the nodules, synthesis of leg-haemoglobin. On the other hand, throughout its growth cycle, soybean plants exploit different sources of nitrogen, the first in time being allocated into protein reserves in the cotyledons (a concentration of 380–400 g raw protein per kg seed, i.e., over 60 g N per kg, and 20–24 mg N per seed). The catabolism of the storage proteins contained in the seed supplies the necessary material to complete germination and the very first stages of seedling growth. Meanwhile, nitrate reductase activity starts in the root and leaf tissues, which ensures assimilation of nitrate from the soil solution; nitrogen fixation then follows, a process that becomes increasingly important in the later stages of development, peaking at pod formation (R3–R4). Agronomic practices must favour a large nodulation, which is a prerequisite for nitrogen fixation. A good practice is to apply manure to other crops (cereals and forages) grown in rotation, but not directly to the soybean. The ammonia present in animal manure is a major limiting factor: if it comes directly in contact with the micro-symbiont, the latter dies (Ciafardini, 1992).

Symbiotic nitrogen fixation activities are sensitive to plant water status, more than any other physiological activity: moderate conditions of water stress cause a near complete drop of nitrogenase activity (Obaton et al., 1982; Sinclair et al., 1987).

As the seeds fill out (R5), photosynthates are largely diverted to these major sinks. Consequently, root nodules lose their functions and enter senescence processes. Under conditions of monocarpic senescence, an efficient redistribution of N in the plant is the key to ensuring adequate protein supply to the seeds; thus, the enzymes involved in photosynthesis are also degraded. This explains the rapid senescence and abscission of the leaves during the stage of physiological maturation (R7).

In light of the positive effect of its rotations ("soybean effect") often observed in the field, whether soybean leaves a heritage of mineral nitrogen that can be exploited by the following crop is a subject of much debate. Despite the ample variability in the estimates of nitrogen fixed in different agro-ecosystems (Atkins, 1984), in temperate climates values are often lower than the amount of N taken up (3,500 kg dry seed for an average concentration of 62 g N per kg are equivalent to about 220 kg N removed per hectare). The portion of symbiotically fixed nitrogen of that recovered in the plant biomass is variable. However, a well-nodulated crop grown without nutritive stresses leaves a certain amount of organic (and easily mineralizing) nitrogen in the leaves, roots, nodules and exudates. Estimates of 60–70 kg/ha residual N have been calculated as the contribution of soybean to the double-cropped cereal. Conversely, post-anthesis N applications have been proposed also for well-nodulated crops, to cover nitrogen shortage occurring in advanced growth stages, when nitrogen fixation drops rapidly. However, experiments carried out under different ecological conditions have given varying results.

Like most food legumes, soybean responds positively to phosphorus and continues to take up this nutrient until the seed fills out completely. The species has average requirements in terms of phosphate supply. Phosphorus-deficient plants have limited leaf area, are a darker shade of green and smaller because of an unbalanced ratio between biomass above and below ground level (Fredeen et al., 1989). An adequate amount of phosphorus ranges between 25 and 30 mg P_2O_5/kg in sandy and clayey calcareous soils, respectively. The response to phosphorus applications in terms of crop productivity is significant when the soil lacks this nutrient, i.e., at concentrations lower than 18 mg P_2O_5/kg (Olsen method) in medium-textured soils, and 23 mg/kg in clayey ones (Region Emilia Romagna, 2004). Although most Italian farming land is well endowed with phosphorus because of over-fertilization in the past decades, chemical analyses, to be carried out at least every 5–6 years, should indicate how to proceed. Requirements and removal of phosphorus are limited (about 60 kg P_2O_5/ha for a yield of 4 t seeds/ha), however, if the soil does not contain enough assimilable phosphorus, an amount greater than that taken up in calcareous soils balances the back-degradation of phosphates. Moreover, phosphorus seems to increase the number and weight of the nodules and, consequently, the nitrogen-fixing activity is greater. According to the standards for integrated production (Region Emilia Romagna, 2004), high or extremely high levels of assimilable phosphorus (greater than 40 and 50 mg/kg, respectively) do not call for phosphorus applications. If the phosphorus content is normal, the amounts applied are equal to those removed by the crop. On the other hand, if the phosphorus content is scarce or insufficient, the soil is fertilized for a certain number of years, with applications never greater than 250 kg P_2O_5/ha, according to the degree of depletion and the fixation power of the soil. Within the framework of the technical regulations for agro-environments (EC Regulation 1257/1999) in Friuli Venezia Giulia, a region that accounts for about one fifth of the soybean-cultivated land in Italy, under low-input management it is forbidden to apply more than 80 kg P_2O_5/ha (Regione Autonoma Friuli Venezia Giulia, 2004). In the areas in which soybean is grown, it is a common practice to apply about 80–100 kg P_2O_5/ha, most of which is broadcast when the soil is being tilled, whereas only part of it (20–30 kg) is banded near the root zone. Banded applications of phosphorus in soils lacking this element are important during the first stages of seedling development, when uptake is limited by low soil temperatures, insufficient development of the root system and a soil pH varying from neutrality.

Potassium requirements are relatively high (about 140 kg/ha for a production of 4 t seeds/ha). The amounts removed, although not negligible, are limited since about 38% of the element is returned to the soil as crop residues. The potassium content is higher in soybean seeds than in maize: 1 kg soybean contains 5 times the potassium as the same amount of maize seeds (Smit, 2000). In sandy and sandy-loamy soils (clay content lower than 15%) potassium leaching can be significant, i.e., from 30 to 60 kg K_2O/ha per year, whereas in soils with a higher clay content the leaching is very limited. Potassium applications, like phosphorus, are carried out during soil preparation. In-season applications of phosphorus and potassium, as single or multi-nutritive fertilizers, are inappropriate because those amounts are largely unavailable for the roots during the season.

With regard to mesoelements such as calcium, magnesium and sulphur, information on soybean requirements is limited. Calcium and S deficiencies are not described in the literature, whereas Mg deficiency is known to cause interveinal chlorosis of the older leaves of the plant. These symptoms are frequently observed in acidic soils outside Europe but are practically unknown in the plains of Friuli and Veneto, lying at the foot of the Dolomite mountains, typically rich in Ca-Mg minerals. Only in certain cases (very sandy soils), Mg could be insufficient, whereas in calcareous soils a limited Mg uptake can be due to antagonism with other ions. Sulphur deficiency is also uncommon: symptoms include yellowing of young leaves. New food-grade varieties, designed to be processed into tofu and "soy milk", are already marketed in the USA and Canada; for those materials, greater attention may be given to S requirements. For example, the rheological features of tofu require a precise ratio between two classes of storage proteins (glycinine and conglycinine), which in turn depends on the genotype and adequate sulphate availability.

Microelements are as important for soybean as for all other crops. In general, however, the most common symptoms observed in the field are those due to zinc, molybdenum and manganese deficiencies. Zinc requirements of soybean are lower than those of maize, considered a very sensitive species: Zn deficiency, sometimes observed at high soil pH levels, causes interveinal yellowing and a bronze tint in the older leaves. Abundant applications of phosphorus on soils containing a modest amount of available Zn can induce symptoms of deficiency for this microelement (Marschner, 1995). Molybdenum deficiency can occur in soils having low pH levels. The symptoms often coincide with those of nitrogen deficiency: a scarce symbiontic fixation activity is the cause, as Mo (a co-factor of the nitrogenase enzyme) is clearly needed for the process. It is, however, difficult to ascribe the stunted growth to a mere lack of Mo.

Finally, soybean is very sensitive to Mn deficiency (Marschner, 1995); the pathology can be observed in alkaline soils, particularly during the dry season. The use of commercial seeds instead of self-produced ones usually ensures adequate productivity. In general, microelements should be applied as a curative intervention carried out under anomalous conditions and not as a standard fertilization practice (Venturi and Amaducci, 1984).

16.4. Land Suitability for Soybean Cultivation

In an evaluation that includes an assessment of economic as well as biological factors, the suitability of a district for soybean cultivation must take into account climatic and pedological factors, as well as water and nutrient availability (Table 16.2).

Soybean development is optimal at temperatures ranging between 15 and 25°C, with different requirements according to the growth stage (see Section 16.3.1); lower or higher temperatures cause stress and consequently lower yields. Climates characterized by scarce rainfall also limit the range of usable soils or may render irrigation necessary.

Table 16.2. Prerequisites for soybean cultivation.

Prerequisites	Conditions
Climatic	
Air temperature	medium to high
Air humidity	low to high
Cultural	
Soil texture	sandy loam to clay loam
Oxygen availability	medium to high
Water availability	medium to high
Soil depth (presence of water table or horizons that cannot be penetrated by roots)	medium to deep
Rootability	easy
Nutrient availability	low to medium
Peat	absent in the top 50 cm
Soil salinity	low to medium
Soil pH	sub-acidic to moderately alkaline
Managerial	
Trafficability	good
Tillage and mechanization	easy
For preservation of soil and environment	
Risk of water table pollution	low
Risk of degradation by means of soil compaction	low

Among the pedological characters, large macroporosity, common to all well-structured soils with a medium-fine texture, is indispensable for the development of the crop, since adequate aeration is needed for symbiont activity. A good water supply, especially during the stages following flowering, is required in order to reduce water stress, safeguard the process of nitrogen fixation and ensure optimal production. Shallow soils with a coarser texture (i.e., sandy soils or loamy sands) or with a lot of gravel are therefore not suited for this type of crop. Where water reservoirs are limited, and rainfall is scarce, especially during sowing, it is indispensable to irrigate the fields in order to ensure emergence.

Other important factors are soil reaction (excessive acidity, although uncommon in the Po Valley, can hinder the process of nitrogen fixation), and the presence of active lime, which can limit the availability of nutritional elements. Finally, saline or very saline soils can cause significant yield reduction.

16.5. Functional Characters of the Soil and Evaluation Tables

Recently, the Joint Research Centre of the European Commission evaluated the suitability of soils for soybean cultivation (Bagli et al., 2003). The assessment takes into account various climatic indexes, mainly temperature-dependent, but also physical factors, including soil depth and texture. The amount of water available during the different stages completes the picture of the critical elements. Given the scale of the research, other potentially limiting factors are not taken into account. If the soil water reserve seems to be the key factor for the full development of the crop, there are also other chemical and physical characters that can modify the suitability of a land for soybean cultivation. Excessively acidic or saline soils and those having a high active lime content are not suitable. Those with insufficient drainage are only marginally suitable because the movement of machinery is made difficult by water stagnating at the surface: it might be necessary to harvest the crop immediately after rainfall in order to avoid rotting. An excessively rapid drainage, on the other hand, negatively affects water availability. Most suitable soils are, therefore, those having a medium to high water content, a good structure that favours aeration and drainage, and ability to withstand the movement of agricultural machinery. On the grounds of these observations, a range of soil characters, relative to the top 100 cm, has been defined to assess the aptitude of lands to soybean cultivation (Table 16.3). Each character has two or more classes that are used to evaluate suitability.

Table 16.3. Classes of the characters relative to the top 100 cm of soil that are functional for soybean.

Functional characters		Classes
Available water capacity (mm)	d0	high > 150
	d1	moderate 50–150
	d2	low < 50
Stoniness (% vol.)	p0	absent to common (gravel, pebbles and stones < 35%)
	p1	frequent (gravel, pebbles and stones > 35%)
Depth*	w0	deep or very deep (> 100 cm from soil surface)
	w1	moderately deep (50–100 cm from soil surface)
	w2	scarcely deep (25–50 cm from soil surface)
	w3	shallow (< 25 cm from soil surface)
Acidity	a0	pH > 6
	a1	5.2 < pH < 6.0
	a2	pH < 5.2
Active lime	c0	active lime content < 15%
	c1	active lime content > 15%
Salinity (dS/m)	s0	not saline (ECe < 4)
	s1	saline (4 < ECe < 6)
	s2	saline (ECe > 6)
Drainage	i0	moderately well, well, somewhat excessively drained
	i1	somewhat poorly or excessively drained
	i2	poorly or very poorly drained

*Water table or horizons that cannot be penetrated by roots.

Table 16.4. Guideline of the classes of soils suitable for soybean cultivation.

Functional features	Classes			
	S1 very suitable	S2 suitable soils	S3 marginally suitable	N not suitable
Available water capacity	d0	d1	d2	–
Stoniness	p0	p0	p1	–
Depth	w0	w1	w2	w3
Acidity	a0	a1	a1	a2
Active lime	c0	c0	c0	c1
Salinity	s0	s0	s1	s2
Drainage	i0	i1	i1	i2

The suitability classes were obtained by comparing the parameters thus identified (Table 16.4): S1 (very suitable soils) is the class with optimal features for soybean cultivation; S2 (suitable soils) identifies those areas in which one or more characters may limit yield as compared to S1; S3 (marginally suitable soils) are those in which limitations are more severe, making cultivation difficult and compromising the chance of obtaining good yields; class N (not suitable) includes soils that have limitations that would normally make it impossible to grow this crop.

References

ATKINS, C.A. 1984. Efficiencies and inefficiencies in the legume/Rhizobium symbiosis. A review. Plant and Soil 83, 273–284.

BAGLI, S., TERRES, J.M., GALLEGO, J., ANNONI, A., DALLEMAND, J.F. 2003. *Agro-Pedo-Climatological Zoning of Italy—Application to grain maize, durum wheat, soft wheat, spring barley, sugar beet, rapeseed, sunflower, soybean, tomato.* Monograph 20550 EN—© European Communities. Italy.

BERK, Z. 1992. Technology of production of edible flours and protein products from soybeans. FAO Agric. Services Bull. No. 97.

BOUNIOLS, A., CHALAMET, A., LAGACHERIE, B., MERRIEN, A., OBATON, M. 1986. Nutrition azotee du soja: limites et ameliorations de la fixation symbiotique. Informations techniques CETIOM 94, supplement, 157–165.

CIAFARDINI, G. 1992. Contributo della ricerca microbiologica all'ottimizzazione del rapporto simbiotico della soia coltivata in Italia. Agricoltura e Ricerca 135, 35–40.

CIRICIOFOLO, E., MOSCA, G., BENATI, R., BONARI, E., LAURETI, D., LOMBARDO, M.G., POMA, I. 1992. Preparazione del terreno per semine di secondo raccolto in soia. Agric. Ricerca 135, 49–58.

CREGAN, P.B., HARTWIG, E.E. 1984. Characterization of flowering response to photoperiod in diverse soybean genotypes. Crop Science 24, 659–662.

DANUSO, F., BENATI , R., AMADUCCI, M.T., VENTURI, G. 1987. Fenologia della soia (*Glycine max* (L.) Merril): II. Modelli di risposta ai fattori ambientali. Rivista di Agronomia, 1, 37–44.

EGLI, D.B., CRAFTS-BRANDNER, S.J. 1996. Soybean. In: Zamski, E., Schaffer, A.A. (Eds.), Photoassimilate Distribution in Plants and Crops. Marcel Dekker Inc., New York, pp. 595–623.

FAO. 1990. Guidelines for Soil Profile Description. 3rd ed. (revised). FAO, Rome.

FEHR, W.R., CAVINESS, C.E. 1977. Stages of soybean development. Spec. Rep. 80. Iowa State Univ. Coop. Ext. Serv., Agric. and Home Econom. Ext. Stn. Ames, Iowa.

FREDEEN, A.L., RAO, I.M., TERRY, N. 1989. Influence of phosphorus nutrition on growth and carbon partitioning in *Glycine max*. Plant Physiology 89, 225–230.

FURLAN, L., VETTORAZZO, M., ORTEZ, A., FRAUSIN, C. 1998. *Diabrotica virgifera virgifera* è già arrivata in Italia. Informatore Fitopatologico 12, 43–44.

GRIMM, S.S., JONES, J.W., BOOTHE, K.J., HESKETH, J.D. 1993. Parameter estimation for predicting flowering date of soybean cultivars. Crop Science 33, 137–144.

HARLAN, J.R. 1992. Crops and Man. 2nd ed. ASA-CSSA, Madison, Wisconsin (USA), p. 206.

JAMES, C. 2007. Global status of commercialized biotech/GM crops: 2007. ISAAA Briefs No. 37. ISAAA Ithaca, New York.

LAURETI, D., ANGELINI, L., AMADUCCI, M.T., BENATI, R., DANUSO, F., MOSCA, G., PECCETTI, G., VENTURI, G.P. 1995. File binate e controllo delle infestanti nella coltivazione della soia (*Glycine max* (L.) Merr.). Rivista di Agronomia 29, 544–549.

MARSCHNER, H. 1995. Mineral Nutrition of Higher Plants. 2nd ed. Academic Press, London, pp. 324–364.

MICELI, F., MARCHIOL, L., ZERBI, G. 1993. Accrescimento, nodulazione ed azoto fissato in soia (*Glycine max* (L.) Merr.) soggetta a sommersione o a limitato rifornimento idrico post-antesi. Rivista di Agronomia 27, 558–564.

MICELI, F., TASSAN MAZZOCCO, G. 2000. La sindrome del fusto verde della soia. L'Informatore Agrario 10, 47–50.

MOSCA, G., DANUSO, F., BENATI, R., VENTURI, G., AMADUCCI, M.T., TONIOLO, L., MAZZONCINI, M., CIRICIOFOLO, E., LAURETI, D., CAMPIGLIA, E., MARZI, V., PRUNEDDU, G., LOMBARDO, G.M., POMA, I. 1991. Risultati della rete nazionale di valutazione varietale soia (1981–1988): II. Risposta produttiva e stabilità della resa in granella. Agric. Ricerca 117, 59–68.

OBATON, M., MIQUEL, M., ROBIN, P., CONEJERO, G., DOMENACH, A.M., BARDIN, R. 1982. Influence du déficit hydrique sur l'activité nitrate réductase et nitrogenase chez le soja (*Glycine max* Merr. cv Hodgson). Comptes Rendues Academie de Science 294, 1007–1012.

RAPER, C.D., KRAMER, P.J. 1987. Stress physiology. In: Wilcox, J.R. (Ed.), Soybeans: Improvement, Production, and Uses. 2nd ed. ASA-CSSA-SSSA, Madison, Wisconsin (USA), pp. 589–598.

REGIONE EMILIA ROMAGNA. 2004. Disciplinari di Produzione Integrata 2004. www.ermesagricoltura.it/documenti/p_vegetali/disc_p_int/norm_tecniche/erbacee/soia.pdf

REGIONE AUTONOMA FRIULI VENEZIA GIULIA. 2004. Piano di Sviluppo Rurale—Misure agroambientali—Prescrizioni tecnico produttive http://www.regione.fvg.it/progcom/allegati/erbacee/soia04.pdf

SCOTT, W.O., ALDRICH, S.R. 1983. Modern Soybean Production. 2nd ed. S&A Publication Inc., Champaign, Illinois (USA), pp. 137–139.

SINCLAIR, T.R., MUCHOW, R.C., BENNET, J.M., HAMMOND, L.C. 1987. Relative sensitivity of nitrogen and biomass accumulation to drought in field-grown soybean. Agronomy Journal 79, 986–991.

SMIT, M.A. 2000. Soybeans—the Green Gold. ARC-Grain Crops Institute, Potchefstroom (South Africa).

SNYDER H.E., KWON, T.W. 1987. Soybean Utilization. Van Nostrand Reinhold, New York.

TROEDSON, R.J., LAWN, R.J., BYTH, D.E., WILSON, G.L. 1989. Response of field-grown soybean to saturated soil culture. I. Patterns of biomass and nitrogen accumulation. Field Crops Research 21, 171–187.

VENTURI, G., AMADUCCI, M.T. 1984. La Soia. Edagricole, Bologna, pp. 104–108.

WRATHER, J.A., ANDERSON, T.R., ARSYAD, D.M., GAI, J., PLOPER, L.D., PORTA-PUGLIA, A., RAM, H.H., YORINORI, Y.T. 1997. Soybean disease loss estimates for the top 10 soybean producing countries in 1994. Plant Disease 81, 107–111.

17. Sugar Beet (*Beta vulgaris* L. ssp. *vulgaris*)

Piergiorgio Stevanato,[1] *J. Mitchell McGrath*[2] *and Enrico Biancardi*[3]

17.1. Characteristics

The beet (*Beta vulgaris* L. ssp. *vulgaris*) is cultivated for several uses: as vegetable (leaves and/or roots), for fodder, and for sucrose production. For each use, genotypes have been selected for specific morphological and physiological features that define their respective idiotypes and distinguish themselves from their common wild parent, the sea beet (*Beta vulgaris* L. ssp. *maritima* (L.) Arcang). Use of beet goes back to prehistory when the leaves (wild at first and later cultivated) were used as food. Varieties cultivated for roots appeared during the Roman Empire, and other varieties with large roots suitable as winter fodder were used in central Europe from the 1500s onward (Biancardi, 1999a).

By the end of the 1700s, genotypes intended for sugar production had been selected. This industrial use not only required beets with higher sucrose content, but also led to selection for other traits that facilitated sugar extraction. The most salient features of the sugar beet are: (1) high sucrose concentration within the root (sugar content or polarimetric degree); (2) a root morphology that facilitates harvesting with a minimum of soil adhered to the root; and (3) reduced concentration of nitrogen, sodium, potassium, amino acids, reducing sugars, and other compounds that interfere with sugar crystallization in the factory and, more broadly,

[1] Università di Padova—Dipartimento di Biotecnologie Agrarie, Legnaro (Pd), Italy.

[2] USDA-ARS Sugarbeet and Bean Research, 494 PSSB Michigan State University East Lansing, MI, USA.

[3] CRA-CIN—Centro di Ricerca per le Colture Industriali, Sede distaccata di Rovigo, Rovigo, Italy.

that reduce the efficiency of sucrose extraction. The leaves and tops of sugar beets are removed prior to harvest, and this greatly reduces the concentration of substances that interfere with sucrose recovery at the factory. The topping operation is now mechanically performed before lifting the roots, which is also mechanized. All aspects of beet harvest are geared to eliminating foreign materials harvested with the roots (e.g., soil, stones, weeds) while preserving the structural integrity of the root prior to processing (Winner, 1981).

Originally, the natural seed ball containing two or more fruits was planted for sugar beet production. Each seed ball could produce multiple plants competing for limited resources. Manual singling was necessary in order to obtain 8–12 regularly spaced plants per square metre. This expensive operation became unnecessary with the development of the genetically monogerm varieties, characterized by seed balls containing a single fruit (Biancardi, 1999b). With precision seed drills, these varieties no longer require manual intervention to obtain the optimal stand. Because of their naturally irregular shape, monogerm seeds undergo a pelleting process that adds layers of various natural or artificial thickening agents (e.g., clay, starch, polyacrylamide) to give them a spherical shape compatible with precision planting. Since 1960, monogerm varieties have rapidly supplanted multigerm varieties in countries with advanced cultivation systems. Additional benefits from improvements such as hybrid varieties and the introduction of genetic resistance against many severe diseases complete the genetic package required of modern sugar beet varieties (Biancardi et al., 2005; Wilson et al., 2001).

The sugar beet supplies about one-quarter of the world demand for sucrose (145 Mt/y), with the remaining three-quarters being supplied by sugar cane. While cane is best produced in tropical regions with warm, humid climates (e.g., Brazil, India, Australia, Cuba), the sugar beet provides better yields in areas with mild temperatures and evenly distributed rains, such as in northern France, Germany, Denmark, and the north-central United States. The European Union, historically the largest sugar beet producer in the world, has been progressively limiting beet cultivation and favouring sugar cane import. Among the reasons behind this policy are that sugar cane factories are autonomous in energy use, while sugar beet factories use large amounts of fossil fuels to produce sucrose. A modern sugar beet factory needs about 20 Mw of power, and the production costs of beet are consequently higher and very sensitive to energy prices.

By-products of sugar beet factories are important to offset their higher production costs and include (in order of economic importance): (1) molasses, used to produce ethyl alcohol, glutamate and yeast, as well as an animal feed additive; (2) moist and dehydrated pulps for animal feed; and (3) carbonated lime, the residue from the beet juice clarification process, which is often used as an amendment for acid soils. Use of sugar beet leaves and tops that are removed during harvest as an animal feed is currently limited, and these materials are generally left in the field to provide a modicum of organic matter to be returned to the soil.

17.2. Cultivation Techniques

In countries with the highest sugar yields, sowing of beets begins in March, or shortly after any chance of a hard frost that could reduce plant stand after emergence. Harvest usually starts during the first days of September and ends in late October or early November. Where the soil is well drained, the harvest can be extended into December. In Mediterranean areas (Italy, Spain, Greece), sowing starts at the beginning of February, and harvest begins in

August and ends late in October. In Southern Italy, Spain, Morocco, California, and other warm areas, mild winter temperatures enable autumn sowing, and with the benefit of winter rains and the reduced need for irrigation, excellent yields are often achieved in harvesting during the early summer and avoiding the intense summer heat and drought that characterize these growing areas. Winter or spring sowing is also possible in these southern climates, but sugar yields are more likely to be reduced (Casarini et al., 1999).

In developed countries, sugar beet harvest is completely mechanized and exclusively uses genetically monogerm varieties, each with production characteristics tailored to their specific growing region. Root yield and sugar content, the two most important production traits, are negatively correlated. Usually, early harvests use varieties that have greater biomass production of roots (Type E), while the varieties with relatively higher sugar contents (Type Z) are more suitable for later harvests. Varieties with intermediate traits (Type N) are suitable for many situations. This sequence is not regularly followed as harvest time can be affected by other factors, such as the presence of biotic and abiotic stresses, the need for specific resistant varieties independent of physiological type, and, above all, efficient management of factory operations. Type Z varieties are usually inferior for total sugar yield but retain a higher processing quality due to their higher root sugar content and their lower content of impurities that interfere with sucrose crystallization at the factory.

17.2.1. Crop Rotation

The sugar beet requires careful soil management in order to facilitate emergence, development of uniform crop stands, and growth of regularly shaped roots. Consequently, the crop generally follows wheat in a crop rotation to allow adequate time for the necessary tilling and seedbed preparations prior to the onset of frost in the late fall. Rotation with wheat is often advantageous in that it reduces the development of several weed species and allows efficient mechanical and chemical treatments to be applied against them prior to planting sugar beets (Scott and Jaggard, 1993).

Crop rotations of less than three years are to be avoided as they can lead to the spread of severe diseases, such as the sugar beet cyst nematode (*Heterodera schachtii* Schmidt), rhizoctonia, and other root rots. Longer crop rotations can reduce weed pressures and lessen the inoculum load of many sugar beet diseases. However, longer rotations have no effect on decreasing incidence of rhizomania or sclerotium root rot diseases. Because sugar beet is sensitive to over-application of nitrogen (leading to higher impurity levels), it is advisable not to grow sugar beets following legume crops, or crops requiring large amount of nitrogen, such as corn and vegetables.

17.2.2. Soil Tillage

Operations involving soil tillage for sugar beet production are variable and are primarily specific to local conditions. Soil is usually ploughed to 30–45 cm depth (or deeper) with the aim of loosening the deeper soil horizons. This operation is preceded by fertilization with phosphorus and potassium to be mixed with the topsoil. Techniques aimed at reducing ploughing (e.g., low- or no-till) have not generally provided satisfactory results on sugar beets. During tillage operations, small tractors or tractors with low-pressure flotation tires need to be used to prevent soil compaction. The final tillage operation prior to planting involves a light hoeing to prepare the seedbed and incorporate the nitrogen fertilizer.

17.2.3. Seeding

Precision seed drills are employed to form six or more rows. They are usually provided with the following devices: (1) microgranulators that apply insecticides in the seed furrow to control various types of insects during early plant development; (2) fertilizer distributors that apply nutrients in or beside the seed furrow. The use of the monogerm seed, as previously described, requires special attention in order to distribute the correct number of seeds for obtaining a final stand of 8–12 plants/m². If we consider that row spacing generally has a width of 45–50 cm, the distance between seeds in the row should be 14–16 cm, corresponding to 16 and 14 plants/m². The required stand is achieved if the field emergence percentage is average (60–80%). The most significant loss occurs with low germination (Fig. 17.1). Consequently, factors that reduce emergence (e.g., insects, fungi, soil surface crust) need to be carefully controlled. To this end, high seed and seedling vigour is essential, as it ensures rapid emergence for the majority of plants (Wilson et al., 2001). Significant advances have been achieved in this area.

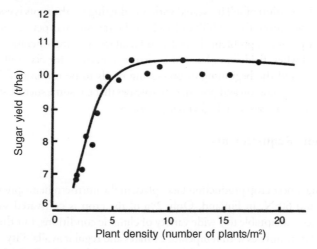

Fig. 17.1. Relationship between plant density (number of plants per square metre) and sugar yield (tonnes per hectare) (Scott and Jaggard, 1993).

In most environments, a stand of 8–12 plants/m² ensures high sugar yield per hectare and a regular shape and dimension of the roots. If plant numbers are less than 6–8/m², among other things, weed control becomes difficult and the root size is not uniform. Higher plant stands, as shown in Fig. 17.1, tend to stabilize sugar production and improve root size uniformity, though they are smaller due to stronger competition. An increase of harvest loss and tare soil also occurs with smaller beets. The timeframe between seeding and emergence depends on soil moisture, temperature, and planting depth. The latter usually varies from 1.5 to 3 cm. Cultivation between rows, sometimes performed before the leaf canopy grows to cover the rows, can be quite useful for aerating the soil and removing a soil crust and weeds.

17.2.4. Disease Control

The pellet coating of the monogerm seed is also a useful substrate to apply fungicides and insecticides, as many species of fungi and insects can damage the plantlets during its early growth stages. Many of the fungi are responsible for damping-off, a hypocotyl rot that kills the seedling. Flea beetles, beet beetles, wireworms, and other insects attack seedlings. Parasites and diseases that damage leaves of older beets (e.g., cercospora leaf spot, powdery mildew, beet armyworms, aphids) can be controlled with chemicals, as necessary, later in the season. Root diseases are more difficult to control. Nematode damage can be reduced with a crop rotation of 3–4 years or by seeding trap crops (e.g., radish, mustard). Nematode-resistant varieties are available, though they still have a limited distribution. For rhizomania, a disease caused by a soil-borne virus (BNYVV) that is present in beet-growing regions worldwide, the only efficient method of control is use of resistant varieties.

17.2.5. Harvest

Beets are harvested with machines that perform the following operations: defoliation, scalping, lifting, and cleaning of the beets. The harvested beets can be piled in windrows in the field or harvested directly into trucks for transport to the factory (Wilson et al., 2001). Throughout these operations, elimination of soil harvested with, or adhering to, the roots is essential. Soil tare is significant since more than 2,000 t of soil can be transported per day to the sugar factories and its disposal is a problem. In order to facilitate cleaning operations, the crop should be harvested when fields have less soil moisture, especially for clay soils. The time between the harvest and the beginning of processing needs to be minimized in order to prevent sugar losses to respiration and post-harvest diseases that consume sucrose stored in the root. These effects are more severe at higher temperatures.

17.3. Cultivation Requirements

17.3.1. Climate

The majority of sugar beet crop production takes place in the northern hemisphere, between 30°N, in Egypt, and 60°N, in Finland. Only 2% of the crop is cultivated south of the equator. The species is adaptable to a wide variety of climatic conditions, but the best sugar yields are achieved in countries with temperate climates and regular availability of moisture during the growing season. Further north, the cold weather limits the length of the production cycle, while in the south the temperatures often exceed the crop's physiological limits and rainfall may not be sufficient.

Temperatures below −1°C can damage the plantlets at the cotyledon stage (Scott and Jaggard, 1993). After the development of the first true leaves, resistance to cold increases and frost damage is less frequent. Low temperatures can lead to bolting (flowering), typical of annual cycle of wild beets. Bolting is detrimental since sucrose reserves in the root are expended for flower and seed production. The sugar beet has been selected as a biennial in which flowering takes place during the second year after a period of cold induction (vernalization). In genetically biennial plants (the annual trait of wild beet is mainly controlled by a single dominant gene), some beets can, under permissive conditions, develop seed stalks with successive production of flowers, pollen and seeds. Premature bolting, which has been greatly reduced under selection for resistant genotypes, can cause production losses and problems with mechanical harvesting, and can spread pollen and seeds in regions where the seed crop and wild species are found in proximity. Wild or bolted beet pollen transmits the tendency to annuality and can seriously damage the commercial seed crops. Moreover, the bolted plants release their seed in the field and result in so-called "weed beets", annual plants that are difficult to eliminate in subsequent crop rotations (Smit, 1983).

Seed germination occurs at a minimum of 5°C. At this temperature, seedlings emerge in 18–20 d. The optimal temperature for sugar production is about 25°C during day and 20°C at night. For sugar concentration within the root (sugar content), the ideal temperatures are 20–23°C during the day and about 15°C at night. At 30–32°C, respiration exceeds photosynthesis with no net gain in sugar storage (Johnson et al., 1971; Elliott and Weston, 1993). The crop is generally not affected by atmospheric humidity. But low humidity is an advantage since it reduces the development of most fungal leaf diseases.

17.3.2. Soil

Sugar beet yields well in sandy as well as clay soils, as long as they are deep enough. Flat fields with uniform soil structure are usually preferred for mechanization purposes. Standing water is particularly damaging at all the stages of the crop, and can be managed by leveling or draining fields. Soils with high water tables need to be avoided. Because of high residual nitrogen and poor mineral nutrition, soils with high organic matter (e.g., peat) or with a pH lower than 5.5 are considered unfit for sugar beet cropping (Märländer, 1991).

17.3.3. Water Requirements

The crop has a water uptake of about 6 mm per day during the summer months. It is necessary to irrigate when the root system can no longer access the required water. Water relations are affected by several factors: climate; soil characteristics; crop growth phase; depth of ground water; irrigation system; and presence of diseases or weeds. The amount of water needed from irrigation is calculated by evaluating evapotranspiration and adjusting for the specific situation using adapted coefficients (Martin, 1983).

In northern areas, the available water is usually sufficient and irrigation becomes necessary only during abnormal periods of drought. Further south, where timely spring and summer rains are often lacking, irrigation is mandatory. Irrigation is sometimes necessary especially where there is a lack of soil water during germination. This is frequent throughout fall seeding climates. If water deficiency happens after germination and before emergence, development of a soil crust can occur and trap the newly emerging crop below the crust. In this case, a few millimetres of sprinkler irrigation can soften the soil surface enough to allow the seedlings to emerge through the crust.

Sugar beets show their highest evapotranspiration rates at the beginning of summer, when the canopy reaches its maximum size. Afterwards, in light of the reduction of solar radiation as well as leaf area, and when the roots have fully developed, the frequency and intensity of irrigations can be reduced. The irrigation system and methods depend especially on economics. In general, crops are irrigated from above with large mobile sprinklers.

Throughout Europe, drip irrigation and sub-irrigation are not commonly used. Elsewhere, furrow or surge irrigation systems are prevalent. Sprinkler irrigation has some negative effects as it can lead to the increase of leaf fungal diseases if leaves remain too wet.

17.3.4. Nutrients

Sugar beet develops a 2.8–3.0 m deep root system (Märländer and Windt, 1996). This deep-rooted feature tends to minimize nutrient deficiencies in the soil, so much so that deficiencies are relatively scarce in the crop.

Nitrogen is a crucial element for sugar beet, which uses about 200–250 kg/ha during the growing season (Draycott, 1972). A deficiency reduces proteins and chlorophyll production, growth is weak, and sugar synthesis is strongly curtailed. The leaves appear light green and are smaller, narrower, and longer than usual. Older leaves tend to yellow and senesce earlier than normal. Excess nitrogen favours luxuriant development of leaves and roots, and the sugar content tends to be lower. A predictable worsening of quality traits such as alpha-amino nitrogen and sodium content occurs, and affects calculation of the purity coefficient (Fig. 17.2).

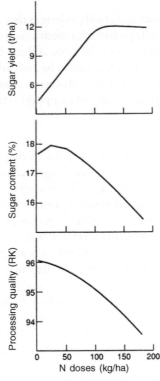

Fig. 17.2. Effects of different levels of nitrogen fertilization on productive and technological traits of sugar beet (Draycott and Christenson, 2003).

The difficulties in determining the right amount of nitrogen to be applied to the crop arise from the complexity of the nitrogen cycle and the difficulty in measuring the precise residual soil nitrogen. Systematic use of soil analyses has narrowed the need to apply nitrogen by considering the residual nitrogen as well as the nitrogen that is really necessary for the crop. Moreover, where the practice of pre-seeding analysis represents a common custom, a progressive reduction of applied nitrogen results in an improvement of purity and an increase of sucrose recovery at the factory (Fig. 17.3). An accurate calculation of residual nitrogen fertility is also difficult and requires the increase of soil sampling depth. In order to reconcile productivity and quality demands, it is necessary to supply sufficient nitrogen for the initial root and leaf system growth and progressively reduce nitrogen supplies through to the harvest period, when the processing quality is often more important than root growth (Johnson et al., 1971). This is only possible within sandy soils with low organic matter and low cationic exchange capacity. Organic fertilizers are not commonly advisable because they continue to release nitrogen close to harvest time.

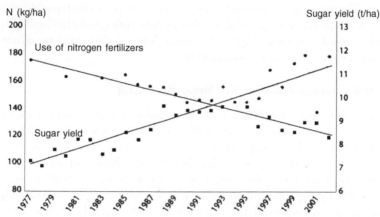

Fig. 17.3. Use of nitrogen fertilizers and sugar yield observed during the 1977–2001 period in France (from Betteraviere Française).

Phosphorus-deficiency symptoms on young plants appear as leaves that are darker than usual. In severe cases, purple-red patches appear on the leaves. The roots grow weakly and tend to develop lateral rootlets. Soils with low phosphorus content are often alkaline, with low organic matter or high percentage of sand. Phosphorus deficiency has diminished in recent years because of the increased use of phosphorus on other rotation crops. Phosphorus must be incorporated during the main tillage operations owing to its relative lack of mobility in the soil or, because of its low toxicity toward the seed, it can be applied directly in the seed furrow. In this case, it can reduce the time to emergence. Excess phosphorus is not detrimental to beet growth or quality.

Potassium has a number of functions in the physiology of the plants. Potassium-deficiency symptoms on sugar beet are rare in clay soils because they usually contain potassium levels in excess of what is sufficient: about 100 ppm of exchangeable element (Draycott, 1972; Draycott and Christenson, 2003). The deficiency, more frequent in sandy soils, initially appears with olive green patches and leathering on the edges of leaves.

Sulphur, sodium, calcium and magnesium deficiencies are quite rare. Only magnesium deficiency can cause leaf symptoms (lemon-yellow patches at the leaf lamina borders) and these symptoms cannot be mistaken for other diseases. Damage due to sodium excess is well known, particularly when available moisture is less than evapotranspiration values, or when saline water is used for irrigation. In these cases, sodium, which moves easy within the soil profile, is likely to concentrate on the soil surface. Though the beet is considered among the most resistant crops to salinity excess, severe physiological problems can occur.

According to Draycott and Christenson (2003) symptoms of deficiency of microelements such as manganese, iron, zinc, copper, chlorine, or molybdenum are often similar to symptoms of other biotic and abiotic stresses. Therefore, leaf or soil available microelements analyses are needed to determine the nature of the deficiency and the necessity of treatment, usually with a solution containing the lacking element (Draycott and Christenson, 2003). Micronutrient deficiency is common in several countries. This is partly due to a high demand for microelements, particularly boron. Boron-deficiency symptoms cannot be confused with those of other diseases; it is also known as "heart rot", which appears as the necrosis and collapse of the apical meristem. The tissue death can extend, in the most severe cases, to the top of the plant and the root. Severe symptoms can be preceded by a characteristic curling of the leaf lamina. Application of solutions containing about 2 kg/ha of boron in June usually reduces the effects of boron deficiency (Draycott and Christenson, 2003).

17.4. Soil Characteristics for the Sugar Beet Crop

Sugar beet cultivation, linked closely to the sugar industry, makes extensive use of technical means for disease and weed control and highly specialized equipment for seeding and harvest (Van Lanen et al., 1992). Consequently, farmers need an adequate level of expertise in order to grow a successful crop. The crop's wide adaptability to local climate and soil conditions needs to be integrated with other social-economic aspects, such as the availability of labour, farm assets, and the availability of specific equipment. In order to contain energy costs, it is highly desirable to minimize, as much as possible, the distance between the beet fields and the sugar factory.

While the sugar beet can adapt to a variety of soils, when demands of mechanization, proximity to sugar factories, and the ability to obtain satisfactory sugar yield with good quality traits are taken into consideration, areas suitable for economic cultivation are considerably diminished. Moisture-retaining clay and loamy soils, in times of wet weather, can limit field access for tractors and other equipment and delay the seeding process and timely applications of herbicides. Prolonged rainy periods can interrupt the harvest and consequently the sugar factory's processing schedule, causing several technical and economic drawbacks. In order to avoid late weather problems, it is often necessary to harvest very early, when sugar production is 60–70% of what could be accomplished at the end of the crop cycle. On more permeable soils, the harvest can be postponed to November or December without problems. In this case, productivity and economic benefits are often maximized.

The crop generally tolerates salinity and active lime. Tables 17.1 to 17.3 describe suitability of a number of soil characteristics that need to be considered. Soils containing even a minimum amount of gravel or rock need to be excluded from cultivation because of the precision needed for seeding, the need to allow smooth functioning of the machine parts lifting beet during harvest, and the adverse effects of rocks and stones on factory operations during root cleaning and slicing.

Table 17.1. Conditions for sugar beet cultivation.

Requirements	Optimal conditions
Climatic conditions	
air temperature	medium
air humidity	low to medium
Soil conditions	
texture	coarse to moderately fine
oxygen availability	good to moderate along the profile
water availability	moderate
depth	> 250 cm
summer groundwater level	> 250 cm
root barriers	absent
nutrient availability	low to medium
peat	absent or very low
salinity	marginal to moderate
reaction (pH)	moderately acid (pH 5.5) to moderately alkaline
surface stoniness	absent
uncontrollable diseases (e.g., sclerotium root rot)	absent
Soil management	
trafficability	well-timed
workability and mechanization	easy
Soil and environment preservation	
groundwater pollution risk	low to medium
risk of degradation due to soil compaction	low

Maximum productivity is achieved when about 600 mm of rain is available from seeding to harvest. If this much rain does not fall, then irrigation becomes necessary. In that case, the availability of water and the timing of irrigation are important considerations for successful production.

Several diseases, such as rhizomania, cercospora leaf spot, cyst nematodes, sclerotium root rot, and rhizoctonia, are likely to reduce sugar production to below profitability. Specific genetic resistances or integrated plant protection practices can reduce, but not eliminate, the effects of these and other diseases. In case of sclerotium root rot infection, where no effective control methods exist, sugar beet cultivation must be avoided.

Soil characteristics need to be carefully considered in order to successfully cultivate sugar beet. The levels of macro- and micro-fertility of the soil do not generally limit cultivation options. In addition to climatic or soil trait variables, other socio-economic factors contribute to a comprehensive evaluation, such as the distance of the fields from the sugar factories, as mentioned above, and/or the location of fields within traditional beet-growing areas.

In the past few years, profitability of sugar beet in the European Union has steadily declined, consequently discouraging sugar beet cultivation in marginal areas or where the sugar yield is less than the European mean. Thus, the economy has been more important in limiting sugar beet production than the availability of suitable agronomic areas for cultivation.

Table 17.2. Soil functional characters and classes (0–250 cm) for sugar beet cultivation.

Functional characters		Classes
Slope	z0	level to slightly inclined surface (0–5%)
	z1	moderately inclined (6–13%)
	z2	highly inclined (14–20%)
	z3	strongly inclined (> 21%)
Stoniness (% vol.)	p0	absent
	p1	very scarce (0.1–0.3%)
	p2	scarce (0.4–1%)
	p3	common (1.1–3%)
Internal soil drainage	d0	good, moderately good
	d1	somewhat poor
	d2	poor
	d3	very poor
Depth	w0	deep (> 250 cm)
	w1	shallow (< 250 cm)
Active lime	c0	low (> 5%)
	c1	medium (5–10%)
	c2	high (> 10%)
Salinity (dS/m)	s0	non-saline (< 2)
	s1	very slightly saline (2.1–4)
	s2	slightly saline (4.1–8)
	s3	moderately and strongly saline (> 8)
Exchangeable sodium percentage (ESP%)	n0	normal (< 5%)
	n1	slightly high (5–10%)
	n2	high (> 10%)
Organic material (peat)	t0	absence (0–250 cm depth)
	t1	presence (0–250 cm depth)
Reaction (pH)	r0	slightly acid, neutral or slightly alkaline (pH > 5.5)
	r1	acid (pH < 5.5)
Uncontrollable diseases	m0	absence
	m1	presence

Table 17.3. Guidelines for soil quality assessment for sugar beet cultivation.

Functional characters	Soil classes			
	S1 (very suitable)	S2 (suitable)	S3 (marginally suitable)	N (unsuitable)
Slope	z0	z0	z1	z2–z3
Stoniness	p0	p0	p0	p1–p3
Drainage	d0	d0	d1	d2–d3
Depth	w0	w0	w0	w1
Active lime	c0	c1	c2	c2
Salinity	s0	s1	s2	s3
ESP	n0	n1	n2	n2
Organic material (peat)	t0	t0	t0	t1
Reaction (pH)	r0	r0	r0	r1
Uncontrollable diseases	m0	m0	m1	m1

17.5. Soil Functional Features and Evaluation Tables

Tables 17.1, 17.2 and 17.3 have been compiled from information found in the literature. The soil and environment data necessary for these evaluations are generally older and out of date. Newer information is less available since the current trend has been to reduce the area under cultivation. This trend does not appear likely to be reversed in the near future.

Table 17.1 presents qualitative soil features needed for suitable beet cultivation and adequate sugar yields. In Table 17.2, each feature is classified according to the intensity of its limitation on sugar production, and these often vary according to soil depth. Features such as rockiness, for instance, are more important within the tilled soil layer and less important in the layers below. Further parameters correlated with production involve the whole layer explored by the roots. Even in this case, their importance decreases as soil depth increases and other traits, such as permeability, need to be evaluated in light of the whole soil profile occupied by the roots. In Table 17.3, intensity levels are presented according to four groups of cultivability:

Group S1: Suitable. The soil does not present limitations and can support more than 80% of maximum production, i.e., the highest sugar yield obtained with commonly employed production means.

Group S2: Moderately suitable. The soils present some limitations and 60–80% of maximum sugar production is possible within these areas.

Group S3: Marginally suitable. The soils present severe limitations that reduce the production to 40–60% of optimal yield.

Group N : Not suitable. The limitations impede any type of beet cultivation.

References

BIANCARDI, E. 1999a. Storia e diffusione della bieticoltura e dell'industria saccarifera (History and development of the sugar beet crop and industry). In: Casarini, B., Biancardi, E., Ranalli, B. (Eds.), *La barbabietola da zucchero in ambiente mediterraneo* (The Sugar Beet in the Mediterranean Environment). Edagricole, Bologna (Italy), pp. 1–38.

BIANCARDI, E. 1999b. *Miglioramento genetico* (Genetic Improvement). In: Casarini, B., Biancardi, E., Ranalli, B. (Eds.), *La barbabietola da zucchero in ambiente mediterraneo* (The Sugar Beet in the Mediterranean Environment). Edagricole, Bologna (Italy), pp. 57–145.

BIANCARDI, E., CAMPBELL, L.G., SKARACIS, G.N., DE BIAGGI, M. 2005. Genetics and Breeding of Sugar Beet. Science Publishers, Enfield, New Hampshire (USA).

CASARINI, B., BIANCARDI, E., RANALLI, B. 1999. *La barbabietola da zucchero in ambiente mediterraneo* (The Sugar Beet in the Mediterranean Environment). Edagricole, Bologna (Italy).

COSTANTINI, E.A.C., SPALLACCI, P. 1987. *Variazioni pedologiche e risposta agronomica di alcune colture irrigue e asciutte nella Bassa Valle Padana* (Soil variations and agronomic response of irrigated and dry crops in the Lower Padana Valley). Annali Istituto Sperimentale Studio e Difesa del Suolo, Firenze, 18, 75–88.

DRAYCOTT, A.P. 1972. Sugar Beet Nutrition. Applied Science Publishers Ltd, London (U.K.).

DRAYCOTT, A.P., CHRISTENSON, D.R. 2003. Nutrients for Sugar Beet Production. CABI Publishing, Cambridge, Massachusetts (USA).

ELLIOTT, M.C., WESTON, G.D. 1993. Biology and physiology of the sugar beet plant. In: Cooke, D.A., Scott, R.K. (Eds.), The Sugar Beet Crop. Chapmann & Hall, London (U.K.), pp. 37–66.

JOHNSON, R.T., ALEXANDER, J.T., BUSH, G.E., HAWKES, G.R. 1971. Advances in Sugar Beet Production. Iowa State University Press, Ames, Iowa (USA).

MARTIN, S.S. 1983. Sugar beet. In: Teare, I.D., Peet, M.M. (Eds.), Crop Water Relations. Wiley, New York (USA), pp. 423–443.

MÄRLÄNDER, B. 1991. *Zuckerrüben* (Sugar Beets). Bernhardt-Patzold, Stadthagen (Germany).

SCOTT, R.K., JAGGARD, K.W. 1993. Crop Physiology and Agronomy. In: Cooke, D.A., Scott, R.K. (Eds.), The Sugar Beet Crop. Chapman & Hall, London (U.K.), pp. 179–237.

SMIT, A.L. 1983. Influence of External Factors on Growth and Development of Sugar Beet (*Beta vulgaris* L.). Pudoc, Wageningen (NL).

VAN LANEN, H.A.J., VAN DIEPEN, C.A., REINDS, G.J., DE KONING, G.H.J. 1992. A comparison of qualitative and quantitative physical land evaluations, using an assessment of the potential for sugar beet growth in the European community. Soil Use and Management, 8, 80–89.

WILSON, R.G., SMITH, J.A., MILLER, S.D. 2001. Sugarbeet Production Guide. University of Nebraska, Lincoln, Nebraska (USA).

WINNER, C. 1981. *Zuckerrübenbau* (Sugar Beet Cultivation) DLG Verlag, Munich (Germany).

PART III

LAND SUITABILITY AND
LAND ZONING

*Land suitability for
small-scale niche cultivations*

PART III

LAND SUITABILITY AND
LAND ZONING

*Land suitability for
small-scale niche cultivations*

18. Emmer Wheat (*Triticum dicoccum,* Schrank)

Giancarlo Chisci[1]

18.1. Crop Characteristics

Emmer wheat (*Triticum dicoccum,* Schrank) is a hulled wheat that originated in the Middle East, like bread and durum wheat. It was the first domesticated crop and for many centuries has been the most important source of food, being the main cereal crop, in the Mediterranean civilizations (Egyptian, Greek and Roman). Today, emmer wheat is reduced to a relict crop because of the spread of free-threshing forms of wheat, which are easier to grow, are more adaptable to technological processing and do not need pounding to obtain grains.

In Italy, emmer is still cultivated in a few marginal areas, where it is part of local cuisine traditions. Recently it is being again appreciated at a national level.

18.2. Cropping Techniques

18.2.1. Crop Rotation

Emmer is usually cultivated in mountainous or hilly areas; the traditional technique of cultivation is rather simplified and extensive and based on use of local cultivars. In the crop rotation, emmer usually follows a meadow (usually a 3-year alfalfa meadow) or sometimes it is cultivated for 2 years. The previous meadow leaves accumulated fertility (field residues) and improved soil structure, which have a low-release beneficial effect on emmer and makes it possible to avoid fertilization. Moreover, meadows have a good weed control power, as they are mowed or grazed.

18.2.2. Cultural Practices

Soil Preparation

Tillage is carried out through ploughing or shallow rotary hoeing, during the summer, to plough up the previous meadow or to incorporate crop residues. After the first autumn rainfalls, it is advisable to till the field again to remove weeds that have grown in the meantime and to prepare the seedbed. All these operations should be carried out at the right time, because plant establishment and density depend mostly on proper and timely sowing.

---

[1] Università degli Studi di Firenze, Firenze, Italy.

Seeding

Hulled seeds are generally sown using a seed broadcaster, at a seed rate of about 150 kg/ha. Sowing is followed by harrowing, using a drag harrow to earth up seeds, followed possibly by rolling, using a ring roller to place seeds at a depth of 2–3 cm and to facilitate the contact between seeds and soil particles. Sowing should be carried out timely, possibly before the end of October, to achieve a satisfactory plant density.

Fertilization

Usually emmer does not need direct organic and mineral fertilizers, because it uses the residual fertility of previous crops. In the case of a repeated emmer cultivation, only organic fertilization can be recommended (20–30 t/ha of manure). Nitrogen fertilizer top dressing is not recommended, because emmer plants are tall and therefore particularly prone to lodging.

Weed and Pest Control

Chemical weed control is not usually adopted for emmer cultivation, as mechanical and cultural weed controls can contain weeds. Use of pesticides is not common in emmer cultivation, with the exception of a storage treatment to preserve seeds for the next sowing season.

18.3. Land Suitability

The cultivation of emmer is localized in a few niche areas of central and southern Italy and employs typically local cultivars. These areas are usually mountainous or hilly, with soils of low fertility, often acid, shallow and rocky. The water requirement is modest and usually satisfied

Table 18.1. Emmer requirements.

Requirements	Value
Climatic	
Air temperature	moderate
Altitude	400–1,000 m
Air humidity	moderate
Precipitation	high
Soil requirements	
Morphology	level to moderately inclined surfaces
Texture	sand to loam
Soil structure	moderate to strong
Stoniness	up to abundant
Soil depth (ground water, soil horizons impenetrable to roots)	moderate
Oxygen availability	high
Nutrient availability	moderate
Soil pH	acid to neutral
Salinity	low
Management	
Trafficability	high
Workability	high
Mechanization	moderate
Land conservation	
Erosion risk	high
Groundwater pollution risk	low
Soil compaction risk	low

by the soil available water during the time the emmer is growing. Emmer is highly sensitive to waterlogging, Emmer is prone to lodging caused by strong winds and spring storms, because of its height.

Tables 18.1 and 18.2 show crop requirements and land suitability for emmer.

Table 18.2. Soil suitability for emmer.

Soil characteristics	Class and rating scale			
	Very suitable	Moderately suitable	Marginally suitable	Unsuitable
Depth, cm	40–60	30–40	20–30	< 20
Texture[1]	L, SiL, SL, SiCL, LS, Si	SiC, CL, SCL	peaty	C, S
Salinity (dS/m)	< 4	4	4–10	> 10
pH	6–7	5–6; 7–8	4.5–5	< 4.5; > 8
Content of organic carbon (%)	1–2	> 2; 0.5–1	< 0.5	
Rock fragments (%)	< 20	20–35	35–70	> 70
Internal drainage[2]	well drained	moderately well drained	somewhat poorly drained; somewhat excessively drained; excessively drained	poorly drained; very poorly drained
Total calcium content —CaCO$_3$ (%)	0.5–5	5–10; 10–20	20–40	> 40
Cationic exchange capacity (cmol+/kg soil)	5–10	0.5–5; 10–15	< 0.5	

[1] *SiC silty clay, C clay, Si silt, SiL silt loam, CL clay loam, SC sandy clay, L loam, SiCL silty clay loam, SCL sandy clay loam, SL sandy loam, LS loamy sand, S sand, C$_{massive}$ massive clay, SiC$_{massive}$ massive silty clay.*
[2] *From Sys et al. (1993), modified according to classes quoted in Appendix.*

19. Edible Truffles (*Tuber* spp.)

Gilberto Bragato,[1] Lorenzo Gardin,[2] Luciano Lulli[3] and Marcello Raglione[4]

19.1. Introduction[5]

Fungi are a kingdom of eukaryotic, heterotrophic organisms in which the vegetative part—the mycelium—is formed by filamentous cells—hyphaecharacterized by perforated septa that allow the intercellular movement of cell structures and substances. Among the many thousands of fungal species, the ascomycetes of the *Tuber* genus spend all their life cycle underground, living in symbiotic association with the roots of several tree species. Their sexual reproduction ends with the release of spores—asci—that are borne by a hypogeous ascocarp, popularly known as truffle.

The little economical and social attention given to fungi not involved in human health or fermentation processes has significantly limited the interest of researchers in truffle species, resulting in an inadequate knowledge of their ecological requirements. This notwithstanding, our report is aimed at providing useful information about the habitats of the five main edible truffle species that produce ascocarps in quite well-defined environments.

In the case of truffle species, soil suitability evaluation is aimed not at characterizing the relationship between soil characteristics and the quality of a crop, but at recognizing the places in which truffle species reproduce by ascocarps. We think two essential conditions must be fulfilled for truffle production: the presence of a host plant and a suitable soil environment. *Tuber* spp. individuals propagate whenever they meet a host, but they produce truffles if and only if the right soil conditions are present.

---

[1]CRA—Centro di Ricera per lo Studio delle Relazioni tra Pianta e Suolo, Gorizia, Italy.

[2]Freelance Forester, Firenze, Italy.

[3]Formerly, Istituto Sperimentale per lo Studio e la Difesa del Suolo, Firenze, Italy.

[4]CRA-Unità per i Sistemi Agropastorali dell' Appennino Centrale, Rieti, Italy.

[5]Prof. Mattia Bencivenga, Dr. Gabriella Di Massimo and Dr. Domizia Donnini of the University of Perugia substantially healped us to deal with the bio-ecological aspects of edible truffle habitats. These notes were jointly written by all authors.

In this report we considered only those truffle species that were the topic of researches aimed at finding the pedogeographic markers of suitable soil conditions for truffle growth.

19.2. *Tuber magnatum* Pico. White Truffle

Also termed white Alba truffle, it is collected in most Italian regions, from Piemonte to Basilicata, as well as in the Balkan countries facing the Adriatic Sea and in the mid-Danube basin. *Tuber magnatum* ascocarps range in size from that of a pea to that of an orange, even though unusually large truffles weighing more than 2 kg have been collected. They are found at depths ranging from 2 cm to more than 50 cm, with more frequent findings between 5 and 20 cm below the soil surface. Truffles have a smooth surface coloured from pale ochre to pale yellow, to soft green. Red shades sometimes occur. The flesh is pale ochre to light brown, displaying many fine white veins that disappear after cooking. Globose, sub-pedunculated asci are 60–70 × 40–65 μm in size. They contain 1–4 reticulated, globose or slightly ellipsoidal spores. The size of spores is 21–30 μm, and 3–4 alveoli of 10–20 μm are visible under microscope.

Tuber magnatum is harvested from October to December under oaks (*Quercus pubescens, Q. robur, Q. cerris*), poplars (*Populus* sp. pl.), willows (*Salix* sp. pl.), hornbeams (*Carpinus betulus, Ostrya carpinifolia*), hazel (*Corylus avellana*), lindens (*Tilia* sp. pl.), stone pine (*Pinus pinea*), and other tree species. Truffles are collected in quite young, well-drained to somewhat excessively drained soils that usually lack gravels. Other important soil characteristics are the slightly to moderately alkaline pH (Fig. 19.1), the large content of calcium carbonate (Fig. 19.2), and the presence of enough water throughout the year.

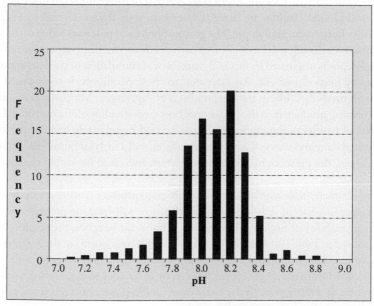

Fig. 19.1. Frequency of pH(H₂O) values in soils suitable for *T. magnatum* (543 sampling locations). The most frequent pH values of soils housing *T. magnatum* are around 8.1, requiring a moderately alkaline reaction for truffle production.

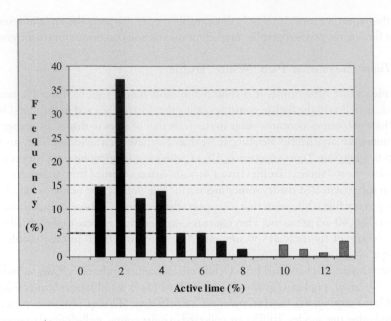

Fig. 19.2. Relative frequency of active lime in soils suitable for *T. magnatum* (124 sampling locations). Carbonates are almost always present in soil and active lime can reach quite high values in marly soils.

Furthermore, truffle-housing soil layers must have a very soft consistency, indicated by bulk density values of 1.0–1.1 kg/L, and a large content of interconnected macropores that makes soil well aerated. Truffles are harvested in moist soils always covered by vegetation, located in valley bottoms and hill slopes. The geomorphological processes acting on these areas are responsible for a disordered distribution of particles due to frequent depositions of alluvial and colluvial deposits originated by fluvial dynamics or accumulation of particles transported along slopes by mass movement. An exception to these conditions is represented by the outcrop of soft marls that, like in the surroundings of Aqualagna, Marche region, originate soft soils becoming productive in the presence of host trees standing alone in crop fields. The microclimate of truffle-producing areas is characterized by relatively low temperatures in summer, limited daily and seasonal temperature ranges, and a high air humidity. Unlike with *T. melanosporum*, the presence of *T. magnatum* is not indicated by visible changes of the herbaceous cover.

The soil characteristics required by *T. magnatum* to produce truffles are quite specific. Limited changes in adjacent soils influenced by the same soil-forming processes drastically decrease the probability of finding white truffles. This is, for instance, the result of the case study summarized in Table 19.1 and Fig. 19.3. Less suitable soils located in the neighbourhood of productive areas can be involved in *T. magnatum* production only when seasonal climate fluctuations depart from the average conditions recorded in that area, as during a prolonged summer dryness or a rainy year. In the latter case, truffle production often migrates from usually productive valley bottoms to adjacent slopes.

The ascocarp of *T. magnatum* could be confused with that of *T. borchii*. The latter, however, differs for the harvesting season, the smaller size of truffles, and the colour of

Table 19.1. Morphological soil properties of the most representative soils of the mean River Mirna basin, Istria, Croatia (from Bragato et al., 2004).

	Horizon	Sector 1 Productive valley bottom	Sector 5 Occasionally productive valley bottom	Sector 3 Unproductive valley bottom	Sector 6 Unproductive slope
OF	Depth, cm			1.5–0	
	Munsell colour			2.5Y 2/1	
A	Depth, cm	0–15	0–8	0–15	0–8
	Munsell colour	2.5Y 4/2	2.5Y 4/2	2.5Y 4/1	10YR 3/3
	Mottles			2 mm, 10 %	
	Structure grade	weak	weak	weak	moderate
AC	Depth, cm		8–30		
	Munsell colour		2.5Y 4/3		
	Structure grade		weak		
Bw	Depth, cm				8–39
	Munsell colour				10YR 4/4
	Structure grade				strong
C	Depth, cm	15–100+	30–100+	15–100+	39–65
	Munsell colour	2.5Y 5/4	2.5Y 5/4	2.5Y 5/5	10YR 4/6
	Mottles	3 mm, 15 %		4 mm, 30 %	
	Structure grade			weak	weak

peridium, ranging from grey to tawny, to brown. Furthermore, the ascocarp of *T. borchii* is never as yellow as that of *T. magnatum*, the latter displaying a homogeneous colour and only an occasional occurrence of red shades. Other diagnostic features are the veins of the flesh, fine and narrow in *T. magnatum*, large and spaced in *T. borchii*, and the scent, which is penetrating and pleasant in *T. magnatum*, garlicky and unpleasant in *T. borchii*. Under the microscope, spores of *T. magnatum* show very large and irregular alveoli, whereas *T. borchii* has a very large number of small, honeycomb-like alveoli.

Only a few truffle orchards planted in the last 20 years have begun to produce *T. magnatum* truffles. They are located in places where the environmental conditions suitable for *T. magnatum* are met: e.g., high shadowing, soft and drained soil. Many truffle orchards instead produced *T. borchii* truffles, and biomolecular assays found roots mycorrhized with this truffle species only. For this reason, no new *T. magnatum* truffle orchard has been planted in the last 10 years and research institutions do not release certificates for seedlings inoculated with *T. magnatum*. Nowadays, an increase in white truffle production may be achieved only in the wild, adopting forestry practices that can preserve the most suitable environmental conditions for *T. magnatum* growth. The presence of mycorrhizae of *T. borchii* or similar species both in truffle orchards and in young host trees of natural *T. magnatum* areas also suggests the need for a sort of biological pretreatment of soil with other mycorrhizal fungi before root colonization by *T. magnatum*.

19.3. *Tuber melanosporum* Vittad. Black Truffle

It is also called black Norcia or Périgord truffle. In Italy, it is harvested from December to March all over the central-northern Apennines under downy oak (*Quercus pubescens*), holm

Fig. 19.3. Soil monoliths from the River Mirna valley (Bragato et al., 2005). Unproductive soils are characterized by a recurrent periods of water stagnation. In some horizons, they otherwise display a well-developed structure where pore size distribution has been modified, decreasing the degree of oxygenation of the soil system. Truffle-producing areas of the Mirna valley are located in recurrently overflooded soils characterized by chaotic sedimentation of solid particles.

oak (*Q. ilex*), hazel, hornbeams, stone pine, rock rose (*Cistus incanus*), and other tree species. The most productive Italian regions are Umbria, Marche, Abruzzo and Lazio. *Tuber melanosporum* is more widespread than *T. magnatum*, being harvested in most of the northern Mediterranean countries: Spain, Portugal, France, Italy, and Balkan countries. It is widely cultivated and harvested in France, which exports its production all over the world.

The truffle is subglobose or lobed, ranging from 1 to 7–10 cm in size. It is usually smaller than *T. magnatum*, rarely exceeding 1 kg, and gives off a delicate and very pleasant scent. Its peridium has small, polygonal warts that are often depressed at the top, giving the ascocarp a less scabrous appearance than *T. aestivum*. Peridium is sooty black, with red shades visible at the base of warts. The flesh is black to violet at maturity, with more or less anastomosed fine white veins that turn red on contact with air. Unripe ascocarps have a red peridium and a white flesh. Asci display a major axis of 25–35 μm and a minor axis of 20–30 μm. They contain 1 to 4 echinulate spores that are dark brown and ellipsoidal.

The soil environment of *T. melanosporum* is in some way very similar to that of *T. magnatum*. It must be carbonated, well drained and aerated but, contrary to white truffle, it cannot be wet for long time intervals. Also, the macroclimate is comparable, as proven by the presence of the two species in nearby areas. However, they require remarkably different microclimates. In summer, *T. melanosporum* locations have air and soil temperature values 4–5°C higher than *T. magnatum* locations, whereas the situation is reversed in winter. There are also differences in water availability in summer, lower in the areas of *T. melanosporum* than in those of *T. magnatum*.

The presence of *T. melanosporum* is indicated by a small, roughly circular surface, 2–5 m, on average, with little herbaceous ground cover around the base of the host tree. It is called *pianello* in Italy and *brûlis* in France. The origin of the *pianello* is still little known. Some laboratory trials showed that *T. melanosporum* extracts limit the germination of grass seeds. The same result was obtained with seeds grown in rooms full of black truffle scent.

Tuber melanosporum prefers a stable soil environment that guarantees the persistence of favourable conditions of growth. For this reason, *T. melanosporum* and *T. magnatum*, even though they have similar requirements, are rarely found in the same areas. In carbonated soils, these conditions are determined by an increased stability of aggregates produced by Al, Fe and organic bridges and, sometimes, by inherited volcanic glass materials. A second essential condition is landform stability. On slopes, it is achieved in fractured limestones, where water circulation is mainly subterranean. In flat areas, it occurs in terraces no longer affected by accumulation of solid particles. A further, non-essential condition is the presence of recurrent freezing-thawing cycles, which increases macroporosity and makes stable both the soil environment and the *pianello*. These conditions are typical of many parts of the Apennines, where limestone outcrops deeply fractured and fragmented by tectonics and water erosion frequently occur.

A further factor of stability is the very large presence of rock fragments both at depth and at the soil surface. It is effective when coupled to stable landforms, when erosion prevails over accumulation, as in inactive fans and debris layers, or on slopes no longer affected by soil creep. Rock fragments are often stratified at the soil surface, originating a sort of mulch that reflects sunlight, decreases evaporation, preserves soil from erosion, and protects aggregates from heavy rains.

Table 19.2 reports the main chemical and physical characteristics of black truffle soils recorded in the Apennine areas of central Italy. In general, a soil suitable for *T. melanosporum* should have a loamy to clay loam texture, a certain amount of carbonates, and a slightly to moderately alkaline pH.

In summary, *T. melanosporum* produces truffles only when the following conditions occur: (1) a landform stable enough to allow the formation of the *pianello*; (2) an aerated, scarcely dynamic and carbonated soil saturated with Ca; and (3) a climate characterized by winter frost, which increases soil softness inside the *pianello* thanks to the size decrease of aggregates, and summer dryness inducing the drought stress probably required by ascocarps before their ripening. In Italy, these conditions are more frequently fulfilled in limestone reliefs of the Apennines.

Tuber melanospourm could be also harvested by hoeing of the *pianello*. In Italy, this deplorable practice is forbidden by both national and local legislation because it irreversibly degrades the suitability of soil for truffle production. A long-term factor of production decrease

Table 19.2. Main chemical and physical characteristics of soils housing *T. melanosporum* in the Apennine (from Raglione and Owczarek, 2005).

Variable	Mean	SD	Min.	Max	CV (%)
Clay (%)	20	13	1	46	62
Silt (%)	51	17	5	83	32
Sand (%)	28	12	11	62	43
Structure index	80.8	13.5	46.5	97.5	16.8
pH-KCl	7.3	0.1	7.1	7.7	1.8
pH-H$_2$0	7.9	0.2	7.5	8.4	2.4
Total CaCO$_3$ (%)	35	22	1	76	63
Organic C (%)	3.2	1.9	0.6	9.8	59.2
Fe (mg/kg)	148.5	90.3	31.7	323.3	60.8
Mn (mg/kg)	237.7	132.5	53.3	534.1	55.7
Zn (mg/kg)	9.0	7.7	1.1	38.7	84.7
Cu (mg/kg)	17.9	17.8	1.1	88.9	99.4

is instead represented by the thickening of woods and the reduction of livestock breeds in the wild that followed the drift of rural populations from the Apennines in the last 50 years.

Tuber melanosporum can be confused with other species of black truffle—*T. brumale, T. brumale* forma *moschatum, T. aestivum* var. *uncinatum*, and *T. mesentericum*—that ripen in the same season. It can be distinguished from them for the shape of warts, slightly pronounced and depressed at the top, the opaque black peridium with red shades at the base of warts, and the black-purple flesh. *Tuber melanosporum* can be further discriminated for its elliptical, brown and echinulate spores.

Tuber melanosporum has been successfully cultivated since the 19th century and the availability of artificially mycorrhized seedlings has significantly increased incomes from truffle-producing orchards. Today, the availability of certified, well-mycorrhized seedlings, along with the knowledge acquired about the soil environment suitable for *T. melanosporum* production, has largely increased the probability of successful cultivation. Nevertheless, some practices are still debated, such as pruning, lopping, irrigation, mulching, and soil tillage. Furthermore, the biological factors driving the formation and maturation of ascocarps remain unknown.

19.4. *Tuber aestivum* Vittad

Popularly known as black summer truffle or *scorzone, T. aestivum* is the most widespread truffle species in Italy. Despite its good taste and smell, it is quite undervalued compared to *T. magnatum* and *T. melanosporum*, fetching 20–25% the price of the latter. Truffles are harvested from May to September under *Cistaceae, Pinaceae* and all other host trees of the other two species. The ascocarp differs from that of *T. melanosporum* for the angular coarse warts displaying horizontally streaked sides, and the bright-black colour shown when washed. The marbled to light brown flesh shows many more or less branched light veins. Asci contain 1–3 irregularly reticulated spores characterized by an ellipsoid or subglobose shape.

Thanks to several aplotypes, i.e., populations, *T. aestivum* grows in various environmental conditions at heights ranging from 100 m to 1,500 m above sea level. It likes carbonated soils but can be found also in decarbonated ones when coarse limestone fragments are present. *Tuber aestivum* originates more or less marked *pianelli* that can be also located in thick deciduous

Table 19.3. Main chemical and physical characteristics of central Italy soils housing *T. aestivum* (from Raglione and Owczarek, 2005).

Variable	Mean	SD	Min.	Max.	CV (%)
Clay (%)	33	14	2	78	42
Silt (%)	42	12	7	78	29
Sand (%)	24	10	12	65	43
Structure index	80.9	14.8	42.8	97.9	18.3
pH-KCl	7.0	0.5	4.9	7.4	7.1
pH-H$_2$0	7.8	0.4	6.3	8.5	5.1
Total CaCO$_3$ (%)	36	23	0	75	66
Organic C (%)	3.4	3.6	0.3	23.2	105.9
Fe (mg/kg)	253.9	191.7	41.5	677.3	75.5
Mn (mg/kg)	267	495.7	72.9	2183.8	185.7
Zn (mg/kg)	12.1	6.1	4.8	22.2	50.4
Cu (mg/kg)	61.1	41.2	7	163	67.4

woods and pine woods. Contrary to *T. melanosporum*, it sometimes does not originate the *pianello* in grasslands or thick tree stands. The continuous decrease of production occurring in no more grazed areas, in particular, indicates the importance of grazing in checking grasses and shrubs.

Even if collected within a depth of 30 cm, truffles are usually harvested within 5–10 cm from soil surface and can partially emerge from it. This behaviour is related to a thicker peridium and may explain the wider diffusion of *T. aestivum* compared to *T. melanosporum*. When the surface horizon is on average more compact than required, the ascocarp of *T. melanosporum* does not reach the size required for maturation, and fructification does not move to the softer and very aerated soil layers in direct contact with the atmosphere. *Tuber aestivum*, on the contrary, is more resistant to dryness, taking advantage of the greater softness at the soil surface.

Tuber aestivum is frequently found in slightly to moderately alkaline soils but, unlike the more valuable truffle species, it grows also in soils with a neutral pH (Table 19.3). Thanks to this flexibility, it can be cultivated in almost all neutral to moderately alkaline soils showing small to medium compactness. Large quantities of truffle have been produced in some reforestations carried out in central Italy with non-mycorrhizated seedlings of deciduous and coniferous trees, in both pure and mixed stands. In many cases, black summer truffles have been harvested in truffle orchards mycorrhizated with other edible truffle species.

These observations can be helpful to plan the cultivation of *T. aestivum* because if production has been obtained in limiting conditions like those described before, there is all the more reason why *T. aestivum* would be cultivated according to modern techniques. Since the intraspecific variability of *T. aestivum* is high, we suggest the use of seedlings inoculated with propagation materials coming from nearby producing areas.

19.5. *Tuber brumale* Vittad

Tuber brumale, its forma *moschatum* in particular, is the least demanded black truffle. It tastes hot and gives off a strong and musky smell. It ripens from January to March under the same host trees and in the same areas as *T. magnatum* and *T. melanosporum*, in particular those in which soil drainage is slower and soil moisture content higher than that required by the other two species. Truffles are almost always smaller than a chicken egg. Their outer skin is blackish-brown, with small warts resembling those of *T. melanosporum*. The grey to brown flesh displays large white veins. Asci contain a maximum of 5 echinulate spores, the colour of which ranges from yellow to pale brown.

Tuber brumale does not originate in the *pianello* and its presence in *T. magnatum* and *T. melanosporum* sites indicates a worsening of their habitats, and a decline of production. In France, *T. brumale* is currently replacing *T. melanosporum* in many truffle orchards, decreasing their economical value. This decline may be related to over-irrigation and increase in soil compactness. French producers have adopted irrigation as a normal practice rather than a supplementary source of water in dry periods, thus raising soil moisture to levels disliked by *T. melanosporum* but not affecting *T. brumale*, particularly when soil drainage is limited. Soil compactness is affected by excess water, wheel traffic in field operations, and the tendency of soil to recover its original porosity that was artificially increased with deep ploughing before seedling plantation. As far as areas of natural production of *T. brumale* are concerned, they are always located in soils of low-lying areas or on foot slopes where soil is wetter, redoximorphic features are almost always present in depth, gravels are rare or absent, and soil aggregates are medium to coarse and firm to very firm when tested for rupture resistance.

19.6. *Tuber borchii* Vittad

Tuber borchii has a remarkable environmental flexibility that allows it to grow from the sea level up to a height of 1,000 m on soil types ranging from sandy soils of coastal dunes to clay soils on hill slopes, and also in slightly acid soils. Truffles of 1–5 cm ripen from mid-January to mid-April, giving off a slightly garlicky smell. They are characterized by a smooth, whitish to tawny peridium. The flesh is light-coloured, tending to tawny or purple-brown, with many branched veins. Asci contain up to 4 slightly ellipsoidal, reticulated spores.

Tuber borchii likes high conifer forests, mainly composed of *Pinus domestica*, *P. nigra* and *P. maritima*: coastal pine woods, afforested stands of inland areas and coppiced pine woods. Truffles are also harvested in woods of coppiced downy oaks and holm oaks, in particular at the boundary with crop fields, in clearings, along tracks, and in areas of wood recolonization. In all these situations, *T. borchii* tends to grow where the wood cover is limited. Very often, when wood cover becomes thicker, truffle production gradually decreases while moving to the wood edge. Production is also related to coppice felling: the number and size of ascocarps greatly increases in the first 2–3 years after felling, then returns to lower values.

Soil surveys carried out in Tuscany have helped to define some specific features of *T. borchii* habitats. It grows and produces ascocarps in almost all climatic types apart from the perhumid mountain climate, living in hot and dry as well as harsh and rainy conditions, with a preference for the former. From the microclimatic point of view, *T. borchii* prefers sunny locations on summits, and soils in which texture or gravel content provides rapid drainage. It dislikes areas characterized by temperature inversion or foggy periods. Its harvesting period ranges from January-February in areas with a Mediterranean climate to March-April in the mountains.

Tuber borchii is more frequently found in coastal dune bars, where it mainly colonizes convex summits and gently sloping shoulders. It is also present in inland sloping areas characterized by the outcrop of marls, turbidites with strongly chipped strata, and ultrabasic

Table 19.4. Main soil chemical and physical characteristics for *T. borchii* cultivation (modified from Gardin, 2005).

Property	Suitable	Unsuitable
Texture	S to SICL	SI to C
pH	6.5 to 8.5	< 6.5, > 8.5
Total carbonates	0 to 10	0

rock formations inhospitable for all other edible truffle species. It is not found in valley bottoms still affected by fluvial activity and only occasionally occurs in terraces with a sand to gravely sand texture. Inland, *T. borchii* is usually found in medium to upper portions of relief not exceeding a 15% slope and is frequently located on convex to linear-shaped crests and summits. The topographical location of natural areas of production suggests the need of *T. borchii* for habitats in which loss of water from soil, insolation, and wind exposure are considerable.

Tuber borchii is probably the least demanding species among edible truffles. Table 19.4 reports some soil characteristics that may affect its cultivation. The pH is generally slightly alkaline and carbonates have some influence on production. *Tuber borchii* is sometimes found in non-carbonated soils, but the presence of carbonates is related to an increase in size and production of ascocarps. Soil texture is another functional property. *Tuber borchii* truffles are more frequently collected in sandy loam, loamy sand and sand soils. When texture is clay loam or silt loam, gravels or well-developed aggregates are present, making soil drainage and aeration easier. Bulk density values further support this observation, suggesting the preference of *T. borchii* for soils with soft, slightly compact surface layers. Bulk density looks also inversely related to the size of ascocarps: the larger its value, the smaller their size. We frequently observed large ascocarps in soils that had been just occasionally tilled and, in general, whenever soil creep occurred. In the end, we can say that *T. borchii* does not grow in fine-textured, slowly draining soils that lack carbonates and have either a slightly acid or a strongly alkaline reaction.

19.7. Comments

Even if the ecological requirements of edible truffle species are well known only for *T. melanosporum*, some soil characteristics common to all of them can be used to infer the environmental requirements of *Tuber* species in general and individual species in particular.

A soil suitable for edible truffles is slightly to moderately alkaline and contains a certain amount of active lime. The combination of bulk density values around 1.0 kg/L, corresponding to more than 20% of soil volume occupied by macropores, and active lime leads to good soil aeration and removal of CO_2 as Ca oxalate that increases the relative content of O_2 in the soil air. These conditions occurred in almost all our observations, apart from a negligible number of locations of *T. magnatum*, *T. aestivum*, *T. brumale* and *T. borchii* that showed a neutral pH and the lack of carbonates in the fine earth fraction. The conditions of soil softness, good aeration, and presence of active lime required by the edible truffle species described in the present chapter can be found in a large range of soil environments, but it is the way soils combine these conditions that plays a decisive role, making a given environment suitable for truffle production.

Tuber magnatum is present whenever a strong aeration is originated in soil by a remarkable slope or fluvial activity—with accumulation of solid particles moved by water erosion, landslides or stream overflow—or pedoturbation due to earthworms, mesofauna, human activity, or other factors. *Tuber magnatum* also requires soil moistening, but does not tolerate water stagnation, so that even when the distance from perennial streams is very short, soils suitable for the fungus are always well drained.

The time required by *T. melanosporum* ascocarp to ripen is probably the longest among edible truffles. The ascocarp starts to develop in late spring, lives out the summer as a saprophytic fruiting body, and ripens in winter. Since *T. melanosporum* needs a somewhat moist soil, summer is a critical season for truffle growth. The loss of water is often limited by the mulching effect of whitish gravels on the soil surface. Mulching is further improved by *T. melanosporum* itself. In the *pianello*, it rules out almost all herbaceous species capable of competing for water. Gravels also prevent erosion and compaction, making soil more soft, drained and aerated. The sites suitable for *T. melanosporum* are located in quite stable landforms, where soils often underwent a partial decarbonation followed by a re-carbonation.

The different processes originating the soil environment of the two most valuable truffle species produce well-distinguished, easily discriminated habitats, whereas the other *Tuber* species live in both environments. Only *T. borchii* is capable of colonizing areas inhospitable for all other edible truffles, i.e., sand bars and coastal plains. When other species are present, *T. borchii* tends instead to colonize sunny crests and summits, upper slopes, and convex or linear surfaces. In these locations, *T. borchii* is found in thin soils affected by slope dynamics or in thicker soils with high gravel content, coarse texture, or strong aggregation. Its production is greatly reduced by wood thickening and increased by grazing, situations that also affect *T. aestivum* production.

Tuber brumale colonizes soil environments completely different from those of *T. borchii*. It is frequently found in the same areas as *T. magnatum* and *T. melanosporum*, living in low-lying areas where soil is wetter and water stagnation occasional. Its soil environment is characterized by a strong, medium to coarse structure, some shrinking-swelling activity of clay particles, and limited aeration.

Tuber aestivum is quite variable from the genetic point of view. Depending on the genotype, its truffles ripen at the soil surface or at a depth of 20–30 cm, and it can live in drier or wetter climates compared to *T. melanosporum* and *T. magnatum*. Irrespective of the climatic conditions, soil must be suitably aerated and well enough drained to prevent water stagnation. *Tuber aestivum* replaces *T. melanosporum* when erosion strongly reduces the soil thickness. It can originate more or less visible *pianelli* depending on wood thickening, and truffles have a very long harvesting season, ranging from June to December.

In conclusion, each truffle species has its own strategy of life, but all of them require moist and draining sub-alkaline soils characterized by good aeration, a full base saturation, and the presence of carbonates, even if some of their aplotypes tolerate neutral, non-carbonated soils.

References

BACIARELLI FALINI, L., BENGIVENGA, M., 2002. Valutazione e tecniche di miglioramento di alcune tartufaie coltivate nello spoletino. Micologia Italiana 31(3), 29–43.

BAGLIONI, F., GARDIN, L. 1998. Carta della vocazione naturale del territorio toscano alla produzione di tartufo scorzone. In: Mazzei, T. (Ed.), *I tartufi in Toscana*. Compagnia delle Foreste, Arezzo, p. 184.

BENCIVENGA, M. 2001. La tartuficoltura in Italia: problematiche e prospettive. Vème *Congres International Science et Culture de la Truffe*, Aix-en-Provence, pp. 27–29.

BENCIVENGA, M., VENTURELLA, G. 2001. Contribution to the knowledge of the genus Tuber in Sicily. Bocconea 13, 301–304.

BENCIVENGA, M. 2005. Stato attuale della tartuficoltura italiana. In: Bencivenga, M., Donnini, D., Gobbini, A. (Eds.), *Seminario sullo stato attuale della tartuficoltura italiana*, Spoleto-Norcia (PG), pp. 8–12.

BONI, C., PUXEDDU, M., TOCCI, A. 1993. Ecologia e possibilità di coltivazione di Tuber borchii Vitt. (Tuber albidum Pico) in Sardegna. Linea Ecologica-Economia Montana 25(3), 44–46.

BRAGATO, G., SLADONJA, B., PERŠURIĆ, Đ. 2004. The soil environment for *Tuber magnatum* growth in Motovun forest, Istria. Natura Croatica 13, 171–186.

BRAGATO, G., PANINI, T., PAGLIAI, M. 1992. Soil porosity and structural conditions in soils involved in white truffle production in the "Crete Senesi" area (Tuscany). Agricoltura Mediterranea 122, 180–188.

BRAGATO, G., GARDIN, L., LULLI, L., PANINI, T., PRIMAVERA, F. 1992. I suoli delle tartufaie naturali della zona di San Miniato (Pisa). Monti e Boschi 43(2),17–24.

CASTRIGNANÒ, A., GOOVAERTS, P., LULLI, L., BRAGATO, G. 2000. A geostatistical approach to estimate the probability of occurence of T. melanosporum in relation to some soil properties. Geoderma, 98, 95–113.

DE SIMONE, C., LORENZONI, P., RAGLIONE, M. 1993. Il manganese nei suoli di produzione di Tuber melanosporum e Tuber aestivum. Ann. Fac. Agr. Univ. Sassari 35, 415–418.

DI MASSIMO, G. 2005. Coltivazione delle piante tartufigene e risultati conseguiti. In: Bencivenga, M., Donnini, D., Gobbini, A. (Eds.), *Seminario sullo stato attuale della tartuficoltura italiana*, Spoleto-Norcia.

DONNINI, D., BACIARELLI FALINI, L., BENCIVENGA, M. 2003. Indagine preliminare sulla affinità di Quercus pubescens Willd. e Ostrya carpinifolia Scop. nei confronti di *Tuber melanosporum* Vittad. *Tuber aestivum* Vittad. e *Tuber brumale* Vittad. Micologia Italiana 23(1), 29–35.

DONNINI, D., BACIARELLI FALINI, L., BENCIVENGA, M. 2001. Analisi della micorrizazione in tartufaie coltivate di Tuber melanosporum Vittad. impiantate da oltre 12 anni in ambienti pedoclimatici diversi. Vème *Congres International Science et Culture de la Truffe*, Aix-en-Provence, pp. 437–440.

DONNINI, D. 2005. Controllo morfologico e certificazione delle piante micorrizate. In: Bencivenga, M., Donnini, D., Gobbini, A. (Eds.) *Seminario sullo stato attuale della tartuficoltura italiana*, Spoleto-Norcia.

DONNINI, D., BACIARELLI FALINI, L. 2006. Micorrizazione naturale di *Tuber aestivum* Vittad. su *Castanea sativa* Miller in Umbria. *Micologia Italiana*, 35(2), 32–35.

DONNINI, D., BACIARELLI FALINI, L., DI MASSIMO, G., BENCIVENGA, M. 2004. Éssais expérimentales de cultivation des truffières en Umbria (Italie du Centre). Ière Symposium sur les Champignons Hypogés du Bassin Méditerranéen, Rabat.

GARDIN, L., BAGLIONI, F., LULLI, L., RISI, B. 1997. Indagine ecologica sul tartufo scorzone (*Tuber Aestivum* Vitt.) in Toscana. Monti e Boschi 48(5), 13–18.

GREGORI, E., LULLI, L., MELI, C., SANI, A. 1997. TIMAGNUM, un prototipo di sistema esperto per la valutazione dell'attitudine della stazione all'impianto di tartufaie artificiali di tartufo bianco (*Tuber magnatum* Pico). Genio Rurale 40(12), 49–60.

GREGORI, E., MELI, C., SANI, L. 1998. Attitudine della stazione all'impianto di tartufaie di Tuber magnatum Pico: un esempio di valutazione basato sulle tecnologie dell'Intelligenza Artificiale. In: Mazzei, T. (Ed.), *I tartufi in Toscana*. Compagnia delle Foreste, Arezzo, 184 p.

GREGORI, E., MELI, C., SANI, L. 1998. La valutazione delle condizioni favorevoli alla fruttificazione del tartufo bianco pregiato (*Tuber magnatum* Pico). Sherwood 8, 35–41.

GREGORI, G., TOCCI, A., BONI, C., PUXEDDU, M. 1995. Il *Tuber aestivum* Vitt in Sardegna: ecologia e prospettive colturali. Linea Ecologica-Economia Montana 27(4), 41–47.

LANZA, B., OWCZAREK, M., DE MARCO, A., RAGLIONE, M. 2004. Evaluation of phytotoxicity and genotoxicity of substances produced by *Tuber aestivum* and distributed in the soil using Vicia faba root micronucleus test. Feb. 12, 1410–1414.

LULLI, L. 1998. Gli aspetti pedoambientali relativi al tartufo bianco ed al tartufo neri pregiati. In: Mazzei, T. (Ed.), *I tartufi in Toscana*. Compagnia delle Foreste, Arezzo, 184 p.

LULLI, L., BRAGATO, G. 1992. Quali caratteri devono avere i terreni naturali per produrre il tartufo bianco pregiato. L'Informatore Agrario 48(47), 69–72.

LULLI, L., BRAGATO, G., GARDIN, L. 1999. Occurence of *Tuber melanosporum* in relation to soil surface layer properties and soil differentiation. Plant and Soil 214, 85–92.

LULLI, L., BRAGATO, G., GARDIN, L., PANINI, T., PRIMAVERA, F. 1992. I suoli delle tartufaie naturali della bassa valle del Santerno (Mugello-Toscana). L'Italia Forestale e Montana 47, 251–267.

LULLI, L., PAGLIAI, M., BRAGATO, G., PRIMAVERA, F. 1993. La combinazione dei caratteri che determinano il pedoambiente favorevole alla crescita di *Tuber magnatum* Pico nei suoli dei depositi marnosi dello Shlier in Acqualagna (Marche). Quaderni di Scienza del Suolo, 5, 143–159.

LULLI, L., PANINI, T., BRAGATO, G., GARDIN, L., PRIMAVERA, F. 1991. I suoli delle tartufaie naturali delle Crete Senesi. Monti e Boschi 42(5), 31–39.

MIRABELLA, A. 1983. Indagine preliminare sui suoli di alcune tartufaie di *Tuber Magnatum* Pico (tartufo bianco pregiato) nella valli del Metauro, Foglia, Marecchia e Savio nelle Marche. Annali Istituto Sperimentale Selvicoltura 14, 389–407.

MIRABELLA, A., PRIMAVERA, F., GARDIN, L. 1992. Formation dynamics and characterization of clay minerals in a natural truffle bed of *Tuber magnatum* Pico on Pliocene sediments in Tuscany. Agricoltura Mediterranea 4, 433–444.

PANINI, T., BRAGATO, G., GARDIN, L., LULLI, L., PRIMAVERA, F. 1991. Suoli e siti tartufigeni di un versante tipico della zona di S. Miniato in Toscana. L'Italia Forestale e Montana 46, 373–393.

PANINI, T., LULLI, L., BRAGATO, G., PAGLIAI, M., PRIMAVERA, F. 1993. Suoli e siti del *Tuber melanosporum* Vitt. sulla Scaglia rossa di Volperino (PG). Monti e Boschi 44(2), 28–34.

RAGLIONE, M., LORENZONI, P., DE SIMONE, C., MONACO, R., ANGIUS, A. 1992. Osservazioni sulle caratteristiche pedologiche di alcuni siti di tartufo nero pregiato (*Tuber melanosporum* Vitt) in provincia di Rieti. Micologia e Vegetazione mediterranea 7, 211–224.

RAGLIONE, M. 1993. L'ambiente e i suoli del tartufo nero pregiato. Tuber 1, 9–12.

RAGLIONE, M., OWCZAREK, M. 2005. The soils of natural environments for growth of truffles in Italy. Mycologia Balcanica 2, 209–216.

TANFULLI, M., DONNINI, D., BENCIVENGA, M. 2001. Problematiche relative alla conservazione delle tartufaie naturali. Informatore Botanico Italiano 33(1), 184–187.

TANFULLI, M., GIOVAGNOTTI, E., DONNINI, D., BACIARELLI FALINI, L. 2001. Analisi della micorrizzazione in tartufaie coltivate di *Tuber aestivum* Vittad. e *Tuber borchii* Vittad. impiantate da oltre 12 anni in ambienti pedoclimatici diversi. Vème Congres International Science et Culture de la Truffe, March 4–6, 1999, Aix-en-Provence, pp. 480–486.

TOCCI, A., VERACINI, A. 1992. Sistema di monitoraggio climatico per un'ipotesi di modello previsionale di produttività di arre tartufigene. Linea Ecologica-Economia Montana 24(1), 29–31.

TOCCI, A., VERACINI, A., ZAZZI, A. 1995. Indagini preliminari sulla ecologia del tartufo bianco (*Tuber magnatum*) in Molise. Micologia Italiana 24(2), 64–69.

20. Ash-tree Manna (*Fraxinus* spp.)

Carmelo Dazzi[1] and Giovanni Fatta Del Bosco[1]

20.1. Classification

Some members of the present Family of *Oleaceae*, genus *Fraxinus L.*, are commonly included under the name "ash-tree manna" and are still cultivated only within the administrative territory of Castelbuono and Pollina, in the province of Palermo (Italy). In these areas, ash-tree manna is represented by the species *F. ornus* L. (orniello in italian) and *F. angustifolia* Vahl. (ossifillo in italian), as well as the sub-species *oxycarpa Bieb.* ex Willd (*F. oxyphylla*) of *F. angustifolia*.

---

[1] Università di Palermo—Facoltà di Agraria, Palermo, Italy.

20.2. Main Botanical Features

Fraxinus ornus L. (orniello, *Amolleo* or *Muddio* in Sicilian dialect) is a tree of medium height (8–10 m), with large opposite leaves with stems, downy buds, and pinnate leaves with 5–9 leaflets. Clusters of small, white and fragrant flowers appear after the leaves in April-May; they are initially erect and then pendulous. Fruits are samaras of 2–3 cm length with a single seed, roundish in section. Pollination is entomophilous.

Fraxinus angustifolia Vahl* (ossifillo) can attain more than 20 m height, with opposite pinnate leaves consisting of 5–13 almost or entirely sessile leaflets, with smooth buds. Flowers appear in clusters at the end of winter before the leaves. The fruit is a single samara of 2–4.5 cm length. The subspecies *oxycarpa* is differentiated principally by the youthful aspect of the proximal parts of the underside of the leaflets. Pollination is anemophilous. The leaves take on a characteristic reddish colour before defoliation in autumn.

20.3. Improvement of Varieties

Improvement of varieties has relied entirely on the results of empirical selection undertaken by local growers, which has therefore achieved only modest results. In fact, the varieties cultivated at present are more or less the same as those reported in the monographs of more than a century ago. Only one cultivar of "orniello" and one of "ossifillo" appear to be recently introduced. On the other hand, if one considers that the most productive trees appear to be those that are sexually sterile, to the extent that seeds of wild trees are needed for reproductive purposes, it seems obvious that over time the natural renewal of cultivars has been extremely limited (Fatta Del Bosco, 1973).

20.4. Cultivars

The overwhelming competition provided by sugary treacle "mannite", which has flooded the market and replaced the natural product, has lead to a gradual diminution of ash-tree groves, reduced today to an estimated surface area of 250 ha (only in Sicily, Italy), with a total of about 70,000 productive trees. This has resulted in the loss of a fascinating heritage of varieties, and it now makes little sense to attempt to retrieve the names of now disappeared cultivars, particularly as by now the cultivation of ash-tree manna has almost disappeared due to reduced productivity and the short harvesting season.

Only a few examples of the most recently identified cultivar, "Serra Casale", exist in a semi-wild state, even if the manna of *F. ornus* must objectively rightly be considered to be of higher quality than that of *F. angustifolia*. The figures reported here, relating to surface area and number of trees, refer only to "ossifillo" in the main, represented by the following cultivars: "Verdello", very recently introduced and overall best; "Abbassa il cappello"; "Russu", a very early and productive variety though not producing such high-quality manna; and, somewhat sporadically, "Ziriddu" (or "Inserillo") and "Niuru".

20.5. Propagation

Seeds or suckers are planted out in nurseries in March. Then the plants are planted and grafted, preferably by budding or wedge or crown methods, when they have reached at least 2–3 cm in diameter.

20.6. Climate and Soil Requirements

Fraxinus ornus grows in southern Europe and is frequently found in Mediterranean coastal areas. In its natural state in Sicily, it lives in temperate woods and prospers both in sub-acid and in sub-alkaline soils. *Fraxinus angustifolia*, with its subspecies *oxycarpa*, inhabits Southern Europe and the Middle East, and in its natural state in Sicily, where it is found on the edges of damp valleys in both flat and hilly areas; it suits different kinds of soils and well tolerates long dry periods.

20.7. Cultivation Techniques

In the following paragraphs we refer only to *F. angustifolia* with its subspecies *oxycarpa*.

20.7.1. Planting

Planting out takes place from December to the end of February with one-year-old plants that are then usually grafted three years after being planted. The distance between them varies according to the soil fertility and can be from 2 m, in less suitable soil, to 3–4 m.

20.7.2. Cultivation Needs

Once planted, the trees' requirements are limited to light hoeing and, frequently, combination with leguminous plants for fertilizing purposes. Pruning is fairly important, aimed at eliminating lateral branches to create smooth columnar trunks that will lean slightly to facilitate, after cutting, the dripping and formation of manna into "cannoli", the most sought-after shape. No anti-parasite treatments are applied, even after severe attacks by *Blemnocarpa melanopigia*, which causes defoliation, or by *Cercospora frassini*, an agent causing trunk rot.

20.7.3. Harvesting

Tappings begin in the third week of July when the foliage takes on a light yellowish colour and continues until the first week of September with one cut per day with 2–3 cm between cuts, taking weather conditions into account. To ensure solidification of the manna exuded, there should be no rain, low humidity, raised temperatures and only moderate temperature changes during the day. During years with wet summers, production is notably reduced or even absent. The equipments used to carry out the tappings, collection and drying of the manna vary slightly from area to another and have typical local names: *mannaluoru, rasula, archetto, stinnituri*.

20.8. The Product

20.8.1. Commercial Uses

Commercial descriptions usually distinguish above all the *cannolo*, which looks like a stalactite and derives from the natural solidification of the liquid exuded from the tappings. It is practically free of impurities, with a high manna content, and is definitely of greater value than the two other types, that is, *drogheria*, which is still useful, and *lavorazione*, which has a lower manna content and contains a degree of foreign bodies such as bark and dirt. The normal total production of these three categories are *cannolo* 7%, *drogheria* 10%, *lavorazione* 83%, while in economic terms the *drogheria* and *lavorazione* are priced at 42% and 58% respectively less than the *cannolo*.

20.8.2. Yields

Production begins about four years after the grafting and ceases within a range of 11–25 years. A good adult tree can supply 2–3 kg manna per year, with the largest yields of 4–6 kg.

20.9. Features, Uses, History and Legend

20.9.1. Features

The best-known natural manna is that which flows spontaneously after insect bites or artificially following tappings during the summer, from the trunks of some species of the genus *Fraxinus* and particularly *F. ornus L.* (orniello) and *F. angustifolia Vahl* (ossifillo), as well as the latter's subspecies, *oxycarpa Bieb.,* all cultivated in a few very limited areas of the Madonie mountain range to the east of Palermo (Fatta Del Bosco, 1989). Raw manna contains mannites (37–43%), organic acids (40%), water (12%), glucose (2%) and, to a lesser extent, levulose, mucilage, resin, and significant traces of *cumarin*-like compounds.

20.9.2. Uses

The uses of the product are principally pharmacological as laxatives and diuretics. But it can also be usefully applied for acute oedemas, in conditions of shock and against ascariasis. Its *cumarinic* derivatives have recognized vasodilatory, anticarcinogenic, antibacterial, anti-tubercular, antifungal and anti-coagulant properties. In addition, the leaves, fruits, seeds, bark and roots of ash-tree manna also have their therapeutic uses. In Table 20.1, preparations from each part of the tree and their effects are given.

Table 20.1. Pharmacological uses of different parts of ash-tree manna (Lentini et al., 1983).

Part of the plant	Preparation	Effect
Manna	decoction, fluid extract, powder, tincture	laxative, purgative, infusion, expectorant, cholagogue
Leaves	decoction, infusion, syrup, tincture	anti-rheumatic, anti-gout, diuretic, purgative
Fruit and seeds	infusion, powder	anti-rheumatic, anti-gout, diuretic
Bark from trunk	decoction, powder	antipyretic
Bark from root	decoction	purgative

20.9.3. History

The literature on ash-tree manna is extensive and during some periods manna has been highly esteemed. In the Age of Reason it formed the base of "Tronchin marmalade", famous for its frequent use by Voltaire during the last years of his life. Less well known is the fact that other trees are able to produce exudations also known as "manna". In the early 1800s, "Brianzone manna" was produced from the European larch, "manna de hatta" from *Cistus ladaniferus* in Andalucia and "Chilean manna" from the *Salix chinensis*.

Before this, in the 16th century, "oriental mannas" from *Alhagi camelorum* (syn. *A. marmifera*), tamarisks and herbs "resembling thyme and marjoram" were known (Gelardi, 1989), while in classical times, Virgil and Ovid observed that "an abundant honey-like dew" could be obtained from oak trees. Finally, according to Gelardi (1989), the manna of the ancient Greeks, used medically for ulceration of the eyes, could be none other than incense powder mixed with bark.

20.9.4. Legend (from Fatta Del Bosco, 1990)

There is wide agreement on the etymology of the word "manna", which derives from "*man-hu?*", "what is it?" or also "this is the gift". In Exodus (XVI: 13,14), Moses relates how such a quantity of manna fell daily, except on the Sabbath, that each Israelite had an omer (2.84 L), which lasted only till the following day; given that the Sabbath was the day of rest, a sufficient quantity fell on Fridays to allow a double ration.

As far as the taste of the divine food is concerned, the Book of Wisdom says that "the manna contained within itself every delicious flavour" and Moses in Deuteronomy (II: 7), says to the Hebrews "that they lacked nothing in the desert". Even the rabbis confirmed that "manna was like oil to the young, honey to the old, bread to the strong", that it possessed every flavour (except that of leeks, onions, garlic and cucumber, vegetables much loved by the Hebrews and therefore irreplaceable) and the different perfumes of an earthly paradise which became, according to personal preference, chicken, partridge, etc. For those who ate it, however, some difficulties did understandably arise, so that Giuseppe Flavio (Antiquit., III, 1) describes the manna as having such a taste that not much of it would be desired, while the Israelites (Numbers XI: 6; XXI: 5) complained, "Such light food makes us sick, our souls are fainting, our eyes see only manna." But Moses admonished them, "You must remember throughout this journey that the Lord God has made you spend forty years in the desert to humble you and to test you [....] he afflicts you with poverty and feeds you with manna, unknown both to you and your fathers, so that you may learn that man does not live by bread alone but by whatever God has ordained" (Deuteronomy, VII: 2, 3).

Akiba's convictions on the origins of the biblical manna are interesting, inasmuch as they deal with appearances of holy light and, confirming this divine nature, the rabbis assert that every grain of manna bore the letter "*vau*", equivalent to "six", to note that it was necessary to collect it during the six days before the Sabbath and that it was one of the things created at the end of the sixth day to perfect the world. Finally, we find manna still within the Christian tradition when in the church's early years it was believed that a condensed dew or "St. John's manna" formed miraculously on this saint's tomb.

20.10. Land Suitability for Ash-tree Manna Cultivation

Climate and soil must be considered the main environmental factors affecting the cultivation of ash-tree manna. Indeed, it is not possible to investigate any reliable method of cultivating the manna tree, apart from its economic aspects, without accurate analysis of the soil and climatic conditions within which the trees must grow and produce. This is so even if the characteristic hardiness of the *Fraxinus* genus allows it to achieve adequate production in environments that are not considered ideal by older cultivators of the ash tree. The climate, more than the soil, appears to be the decisive factor in the cultivation of ash-tree manna (Table 20.2), which in unsuitable climates produces little or nothing. Certain climatic parameters must be respected if the liquid manna exuding from the incisions in the trunk is to coagulate successfully.

Table 20.2. Pedo-climatic requirements for ash-tree manna cultivation.

Requirements	Conditions
Climate	
Temperature	Typically Mediterranean: mild winters, relatively hot summers
Rainfall	Typically Mediterranean, but no rain in July or August
Humidity	Low during July and August
Soil	
Soil depth	Increasing with greater productivity
Texture	Prevalently clayey
Amount of organic matter	Decreasing with greater productivity
Amount of total N	Decreasing with greater productivity
Amount of assimilable P_2O_5	Decreasing with greater productivity
Reaction	Neutral tendency

Some conditions must co-exist, particularly in summer, and can be summarized as high temperatures, absence of rainfall and low humidity. The ash-tree manna does not have specific soil requirements so much as climatic; in its areas of cultivation, soils can be stony and infertile, even in clay soils with noticeable cracks. It is, however, averse to lifeless soil with poor drainage.

20.11. Functional Soil Features and Ash-tree Manna Productivity

In the context of tree cultivation worldwide, cultivation of ash-tree manna is something unique and, perhaps, irreplaceable. The only place where the ash tree is cultivated for the production of manna is the coastal foothills of the Madonie mountain range in Sicily, where the manna-bearing ash still continues to be cultivated according to traditional techniques and harvesting methods by now thousands of years old. As previously mentioned, the area of cultivation of ash-tree manna is presently about 250 ha, almost entirely within the administrative territories of Castelbuono and Pollina (Sicily, Italy), with an average density of 280 trees/ha; the total number of trees in production thus being about 70,000. A study has been made in this unique area to evaluate the soil requirements for the cultivation of ash-tree manna (Crescimanno et al., 1993). Bearing in mind the limited spread of ash-tree manna and in order to define the soil-tree relationship, this study undertakes a direct comparison between soil features and manna production, considering that the non-environmental factors affecting it such as farmer ability, age of tree and density of planting are uniform and therefore not influential.

Three levels of manna production have been defined: high (more than 2 kg/tree); average (1–2 kg/tree); low (less than 1 kg/tree). Among the soil parameters that have been investigated to fully identify those soils that support ash-tree manna (mainly Alfisols, Inceptisols and Entisols), only a few bear any relationship to the levels of production as defined above.

In particular, relationships were discovered to be more or less equated with the following soil parameters: soil reaction, total nitrogen, assimilable phosphorus, organic matter, clay content, soil depth. As far as soil reaction was concerned, it was clear that a certain relationship exists between pH values and manna production.

To be precise, the value of pH was always sub-acid but increased from pH 5.7 for a low productivity level to pH 6.6 for high productivity in the three levels of productivity considered. An unusual link emerged relating the total amount of nitrogen in the soil to levels of production; it appears that lower production of manna occurs on soils with higher values of total nitrogen (0.13%), while higher production results on soils with the lowest amount of total nitrogen (0.094%). Similar findings were shown for assimilable phosphorus: in this case as well the greater amount of phosphorus (136 ppm) corresponds to the lowest production while the lesser amount (86 ppm) relates to higher production. Similar findings were encountered in the relationship between manna production and the presence of organic matter in the soil: greater productivity occurs with soils that have a lower content of organic matter (1.14%) and vice versa (organic matter equivalent to 3.04%). The relationship existing between manna production and clay content appeared to grow in a linear manner from 15% to 46% and then 58% clay content respectively for low, medium and high levels of production. A direct relationship also exists with soil depth, in that the less productive soils are more shallow (< 35 cm) and the more productive are deeper (> 90 cm). The manna ash's degree of suitability to the soils investigated is shown in Table 20.3. It was decided to group Classes S1 and S2 together and place Class S3 with N and not to use numerical values to define the relationships between soil parameters and class of suitability. This is justified by the complete absence of other areas of cultivation with which to compare the soils to which present cultivation is confined, as well as by the limited differences that soil parameters investigated have shown in relation to levels of manna production.

Table 20.3. Suitability of soil for cultivation of ash-tree manna according to some functional soil features.

Parameter	Suitability class S1 and S2 (suitable or moderately suitable)	Suitability class S3 and N (marginally or not suitable)
Soil reaction	Sub-acid to sub-alkaline soil	Acid or alkaline soil
N tot.	Low to medium values	High values
P_2O_5 ass.	Low to medium values	High values
O.M.	Low to medium values	High values
Texture	Fine to medium	Loose
Depth	Deep or moderately deep	Shallow

20.12. Functional Climate Features and Ash-tree Manna Productivity

As well as the soil, the climate must be considered an important factor conditioning the production of ash-tree manna. Where the climate is not suitable, the ash tree has been shown to produce little or no manna. It has been particularly observed that there is a close relationship between rainfall occurring in the period from May to September and the production of manna. It must be emphasized that rainfall should not coincide with the harvesting period, which takes place between the third week of July and the first of September because it will reduce production, which is low or absent during rainy summers, and because the tree requires a degree of water shortage to stimulate its production.

20.13. Conclusions

Considering the relationships that occur between the characteristics of the soil, the climate and manna production, the tables of comparison between the suitability of different types of soil and climate with the cultivation of ash-tree manna (Tables 20.2 and 20.3) indicate that the soils showing features favouring the production of manna must be deep and clayey. There do not appear to be significant relationships with the soil reaction, which does, however, tend to show approximately neutral values where production is higher. Relationships between manna production and the amount of total nitrogen, assimilable phosphorus and organic matter in the soil are unusual. It would be reasonable to expect increased production, within certain limits, to be linked to an increase in these fertility factors. Instead, the reverse is indicated: manna production increases with the decrease of the values of these factors. This can clearly be linked to the extreme hardiness of the ash-tree manna, which needs a low input to produce. The production period of manna actually coincides with the tree's evident state of stress and with arrested vegetative activity, while more generous conditions have negative repercussions and limit production. In fact, the trunk is initially tapped for manna when the tree is "mature", meaning in a state of stress demonstrated by a yellowing of the leaves, which are also dry and crisp.

References

CRESCIMANNO, F.G., DAZZI, C., FATTA DEL BOSCO, G., FIEROTTI, G., OCCORSO, G. 1993. Research on woody plants with low energetic inputs: first results for the "Land Suitability System" on Ash-tree manna (*"Fraxinus angustifolia"* Vahl and *"Oxycarpa"*). In: Paoletti, Foissner and Coleman (Eds.), Soil Biota, Nutrient Cycling and Farming System. Lewis Publisher, pp. 183–192.

FATTA DEL BOSCO, G. 1973. Aspetti e problemi della frassinicoltura siciliana. Tecnica Agricola 1.

FATTA DEL BOSCO, G. 1989. Il frassino da manna tra realtà, prospettive e convegni. Frutticoltura 7.

FATTA DEL BOSCO, G. 1990. La manna da frassino: un excursus tra storia e tradizione. Giornale di Agricoltura 8.

GELARDI, G. 1989. Memorie sulle piogge di manna. Az. Agr., (Ed.), "Dimani".

LENTINI, F., MAZZOLA, P., NOT, R. 1983. I frassini da manna. Natura e Montagna 4.

21. Cactus Pear (*Opuntia ficus-indica* L. (Mill.))

Paolo Inglese[1] and Riccardo Scalenghe[1]

21.1. Characteristics

The cactus pear (*Opuntia ficus-indica* L. (Mill.)), also known as prickly pear, Cactaceae, is a species originating on the plateau of central Mexico. Together with maize, gourd, agave and some legumes, for centuries it has constituted the alimentary base of the local populations (Turkon, 2004). In the 16th century, it was brought to Europe by the Spanish conquistadores. At first the cactus pear was grown in the gardens of the nobles and subsequently in cultivation, naturalizing itself quickly and becoming, in the whole Mediterranean basin, an essential element of the natural and agrarian landscape. Widespread in both hemispheres and on all the continents, it is a constant presence in the livestock and human diets of the aboriginal populations of North Africa, the Middle East and southern Europe, in particular Sicily, Greece and southern Spain. *Opuntia* is cultivated, according to the more reliable estimates, on a total area of about 100,000 ha for fruit production and more than a million hectares of land devoted to pasture or to the production of forage for bovine and ovicaprine fodder in Mexico, South America (in Brazil more than 300,000 ha of *Opuntia* are cultivated for forage production), the USA (Texas in particular), North Africa and Western Asia (the so-called WANA Region),

---

[1] Università degli Studi di Palermo—Facoltà di Agraria, Dipartimento di Colture Arboree, Palermo, Italy.

the Horn of Africa and South Africa. In each of these regions, byproducts are created from the processing of both the fruit and the cladodes,[1] and they are destined for human consumption or for non-alimentary uses (Barbera et al., 1995). The Opuntias have a strategically important role in the diet and therefore in subsistence agriculture of the semi-arid regions, where their cultivation is widespread. Only recently, from the 19th century in Italy and the 20th century in the rest of the world, have they been cultivated on a large scale for commercial purposes. These are two very different cultivation and cultural models, both of equal importance.

For fruit production, the *Opuntia ficus-indica* is cultivated, in specialized plantations, on more than 70,000 ha in Mexico, in the northern central region (Zacatecas, Saltillo, Durango, Saint Luis Potosí, Aguascalientes and Guanajuato) and in the southern central region (Puebla, Mexico, Tlaxcala and Hidalgo). In Chile it is found in the Metropolitana region on a little more than 1,000 ha; in Argentina there are specialized plantations in the regions of Santiago del Estero, Tucuman and Catamarca, covering a little less than 1,000 ha; in Israel, there are approximately 250 ha of specialized plantations in the area of the Negev; in South Africa another 1,000 ha is planted with cactus pear in the Western Cape, the Northern Province and the region of the Ciskei. In the United States, *Opuntia ficus-indica* is cultivated in California in the Salinas plain on more than 300 ha of specialized cultivation. Data are lacking on its diffusion in the North African countries, where it is estimated that not less than 50,000 ha are under specialized cultivation (25,000 in Tunisia alone) (Fig. 21.1).

In Italy, cactus pear is cultivated on a little more than 3,000 ha distributed in the areas of Santa Margherita Belice (AG) and San Cono (CT), as well as on the south-western slopes of Mount Etna (Biancavilla, Belpasso and Paternò), where it has received the acknowledgment of *Protected Designation of Origin (PDO)*.

Considering an average production of 10 t/ha world-wide, it can be estimated that there is an annual production of 1 million t from specialized plantations, while that from spontaneous plants and non-intensive cropping, common in many of the regions of subsistence agriculture where the species is present, must be enormously greater (Pimienta Barrios, 1990; Barbera and Inglese, 1993; Nobel, 1988, 2002).

21.1.1. Botanical Features

Opuntia ficus-indica, family *Cactaceae*, is the most widely cultivated cactus for fruit and forage production, though other species or hybrids are grown in Mexico and South America. It is a perennial species of arboreal-shrubby appearance with a very dense fasciculate and superficial root system. The aerial part is composed of articles, the cladodes, that maintain a photosynthesizing function for 2–3 years from their formation. A characteristic element of the cladodes, apart from the typical morphology of the species, is the presence of two tissues: clorenchyma, responsible for photosynthetic activity, and parenchyma, responsible for the storage of water. On the cladodes, arranged according to a geometry typical of the species, there are the areolae, which are totipotent meristems that are able to generate flowers, new cladodes or, if placed in contact with the ground, roots. The flowers generally develop from the areolae disposed along the crown of the cladode, but the production of flowers from the areolae on the planar surface exposed to the sun is also not infrequent. A cladode can produce

---

[1] The cladode, phylloclade or platyclade of a cactus or succulent is the succulent organ that constitutes the foliage of the plant.

Fig. 21.1. Young orchard of cactus pear on sandy terrains in Tunisia.

35–40 flowers; each one can lose, during the abundant flowering in conditions of high potential evapotranspiration, up to 3 g of water per day, equivalent to 15% of its weight at the time of anthesis (De La Barrera and Nobel, 2004).

The leaves are not persistent; they are found only on young cladodes and fall off when these are developed. The thorns, in the varieties in which they are found, also form on the areolae singularly or in a crown pattern and they differ from the glochids[1] in dimension, colour and shape. The glochids, in fact, are tiny, almost invisible barbed hairs on the areoles.

The species shoots in spring and blooms in May. If all the flowers and the young cladodes put out in the spring are removed before and during (but not after) the flowering, the species re-flowers and the resulting fruit will not ripen in summer (July-August), but in autumn (October-November), giving rise to the so-called "scozzolati" or "bastardoni" hybrid crops. This production constitutes 95% of the Sicilian crop. The species tolerates dryness and is very resistant to high temperatures, but its photosynthetic productivity can diminish

---

[1] Sharp bristles with hooks.

enormously if the temperature exceeds 30°C, even when the water supply is sufficient. Crassulacean acid metabolism[1] allows for considerable water saving because of the nightly opening of stomata. The increased efficiency in the use of water equals 3–4 mg of dry substance per gram of transpired water. The growth period for the fruit is 70–110 d, beginning from the flowering, while the commercial lifetime of an orchard is about 25 years. Ninety per cent of the fruit appears on the one-year-old cladodes, whose average number of flowers is 6–9, with peaks of 30–35. Over 90% of one-year-old cladodes are fertile, while cladodes of 2–3 years are responsible for vegetative renewal.

21.1.2. Varieties

In Mexico, in addition to *O. ficus-indica*, various species such as *O. megacantha* are grown. In the case of *O. ficus-indica* the fruits differ depending on the colour of the pulp on ripening, which can be yellow, red or creamy white. Distinctions can be based on the size of the fruit, on the seed content or soluble sugars of the pulp, on the time the fruit ripens or on the vegetative habits and/or habitus. The greater variability is in Mexico, where the more imposing germplasm exists with more than 50 varieties and clones, of which 14–15 are the most widespread (*Reyna, Cristalina, Burrona, Chapeada, Fafayuco, Naranjona, Roja Lisa, Rojo Pelon, Amarilla Montesa, Blanca de San José, Copena, Cascaron, Liria, San Martin, Esmeralda, Pico Chulo*). In Mexico, many of the varieties cultivated have thorns on both the fruit and the cladodes, while in Italy the varieties cultivated are defenceless and the glochids appear only on the fruit. In Italy, the *Yellow, Red* and *White* varieties occur and the so-called "trunzare" with white or yellow pulp, which are distinguished by a greater crispness of the pulp. The *Yellow* variety covers 90% of the invested surface and the *Red* covers 8%. The *Yellow* is widespread thanks to an elevated re-flowering capacity and to the optimal general characteristics of the fruit. The *Red* variety ripens later, while *White* is more sensitive to post-collection damage and to fruit fly. Among the varieties diffused internationally we note *Ofer*, a cultivar with a yellow pulp grown in Israel; the cultivars *Algerian, Blue Motto* and *Director* prevail in South Africa, where there exists a wide variability in terms of the plant's habits and low temperature requirements; *Andy Boy*, a cultivar with a red pulp, is found in California.

---

[1] Crassulacean acid metabolism or CAM, described originally in the Crassulaceae, is similar to C4 metabolism. CO_2 (in truth HCO_3^-) reacts with a compound with 3 carbon atoms (phosphoenolpyruvate, PEP) in order to form oxalacetate (OAA), a compound with 4 carbon atoms. The C4 metabolism allows the plant to continue to photosynthesize even at low concentrations of CO_2, with the difference that fixation of CO_2 and decarboxylation in favour of the rubisco enzyme are separated in time rather than spatially. This metabolism employs the nocturnal fixation of carbon (generally in malic acid that is temporarily stored), followed by the diurnal incorporation of CO_2 (deriving from the decarboxylation of the malic acid) in Calvin-Benson cycle following the classic C3 method (the number of carbon atoms of the first recognizable stable compound in the Calvin-Benson cycle, phosphoglyceric acid, PGA), following the reaction between the CO_2 and a sugar, ribulose-1,5-diphosphate. Also, this metabolic method represents, in effect, a mechanism of concentration of the CO_2, evolved in response to the dryness in land environments (in such a way that the plant avoids opening the stomata by day to obtain CO_2 and therefore avoids the risk of dehydrating itself) and to the deficiency of inorganic carbon in water environments (Keeley, 1998).

21.2. Nursery Technique and Implantation Material

Propagation and implantation take place in the field, exclusively by means of cuttings, even if grafting is theoretically feasible (by simple split) and it is possible to resort to micropropagation for the purpose of genetic improvement or conservation of the germplasm or for the production of plants in containers and their successive transfer to the field. The cuttings are made up of single cladodes of 1–2 years or of multiple cladodes constituted from a 2-year-old cladode that carries two cladodes, "ears", of one year. After being taken from the mother plant, the cuttings are dried in the shade in order to "heal" the surface of the cut. The ideal is a period that allows dehydration of not more than 20% of the fresh weight; in practice that means 10–12 d at 20–25°C. The cuttings are implanted with the base part in the soil (not more than 50% of the cladode in the case of single cladodes, up to 70% in the case of multiple cladodes) and placed in a perfectly vertical position, with the planar surface normally forming the row. In order to obtain the fastest and most intensive root development, it is advisable to take the cuttings in late spring so as to plant them, if there is irrigation, in May–June. Autumnal implantation slows down the root development, increases competition with the spontaneous flora at a time when the plant is weakly competitive, and increases the risk of rotting.

21.2.1. Genetic Selection and Breeding

Beyond describing, cataloguing and conserving the germplasm worldwide, a fundamental objective of the genetic improvement of the cactus pear is to reduce the number of seeds in the fruit, to diminish their size, and to eliminate the glochid and/or thorns from the fruit and the cladodes. In some cactus pear regions, the specific objectives can be increase of the solid soluble content in the pulp (smaller in the Italian cultivar than in the various Mexican cultivars), increase of the pulp yield, differentiation of the colour of the pulp (green, for example, or roseate, common in various cultivar populations aboriginal to Mexico, Peru and Chile), and increase of the plant's productivity and resistance to the cold, its size and its re-flowering capacity.

21.2.2. Preparation of the Soil

Even in circumstances of mechanization with high energy input, cultivation of the cactus pear does not require any arduous operations. However, there are not yet any comparative tests available in the literature.

21.2.3. Nursery Planting and Pruning

Planting distances vary according to the nursery methods used (in rows or in groups) and the habitus of the variety. In Italy, plants are mainly planted in sixes in 4 × 6 m rectangles (416 plants/ha) or 5 × 7 m rectangles (290 plants/ha), or six to a 6 × 6 m square. If the plants are raised in hedges, the distances come down to 2.5–3 m along the row, and 5–6 m between the rows. The height of the plant should not exceed 3 m. Nursery pruning consists in the elimination of those cladodes not opportunely situated or growing on shadier parts of the plant and in reducing to two the number of new cladodes carried on each "mother cladode". These procedures apply equally for ordinary pruning.

21.2.4. Thinning of the Fruits, Scozzolatura and Collection

Thinning the fruit is necessary in order to obtain a uniform and maximum size (above 120 g). Regardless of the initial number of flowers on the cladodes, it is necessary to leave not more than 6 fruits on each cladode, eliminating all those growing on the planar surface. The fruit is thinned manually, not beyond the second week of flowering. The *scozzolatura* is carried out by removing all the flowers and the new cladodes put out in the spring effusion, during anthesis, but not later. The plant sprouts anew 2–3 weeks after this operation. It sprouts more intensely and quickly according to how early the *scozzolatura* is undertaken. The *scozzolatura* allows the production of extra-seasonal fruits (autumnal). By removing the first spring flowering, and then also the shoots that come after the first *scozzolatura,* it is possible to obtain a third flowering in August that, with opportune measures (tunnel protection of the plants to facilitate fruit ripening and prevent damage from cold), produces a crop in mid-winter (from December to February). The *scozzolatura* is normally carried out by hand with the aid of rudimentary tools.

The fruit is collected the moment the colouring appears at the bottom of the epicarp. At this stage, the colour of the *pulp* is already formed and the solid soluble content is close to the maximum, but the fruit is still firm enough to allow for post-collection handling and conservation. The fruit is collected early in the day in order to avoid the excessive annoyance of the glochids that come away from the epicarp. Also, if fruit is collected when the pulp temperature is high, the post-collection life of the fruit is reduced and its sensitivity to damage from cold is increased. Maturation is generally gradual and therefore collection occurs in two or three stages and can be fairly differentiated, depending on the implantation level at the time of the *scozzolatura*. The ripening rate is a function of the thermal conditions, which, if characterized by high temperatures, induce a very rapid and early maturation. If, on the contrary, temperatures are lower than the necessary 18–25°C, maturation will be delayed and substantial modifications to the shape and characteristics of the fruit will be induced (smaller pulp yield, less sugar content and less colouration of the epicarp).

21.2.5. The Product: Collection and Conservation

Yields of the cactus pear vary in relation to the planting pattern and the intensity of the cultivation. A high-density planting (24 plants/m^2), which ensured almost full interception of incident solar radiation, led to an extremely high shoot dry-weight productivity (50 Mg/ha/y) in the second year and maximal fruit productivity (6 Mg/ha/y) in the third year (Nobel, 1992). Cladode dry weight tended to increase with cladode surface area (Nobel, 1992).

Italy has the highest yields in the world, equal to 20–25 t/ha (Mohamed-Yasseen et al., 1996); the average is 10–12 ton/ha. Wastage is high (10–20%) and has little commercial value. The marketable fruit must weigh at least 120 g, with 13° Brix[1] and a firmness equal to 8 kg/cm^2 on a penetrometer (8 mm tip). In regard to some qualitative aspects, the fruit must have a pulp content of at least 55%; the prevalent reducing sugars are glucose (6–8% of the fresh weight) and fructose (5–6% of the fresh weight), with very low acidity (0.03–0.12%, expressed in malic acid). The pH varies from 5.0 to 7.0 and the nutritional value is equal to 15 kJ/kg of fresh digestible fraction.

---

[1] Equivalent to grams of sugar per 100 grams of aqueous pure sugar solution.

The number of seeds in a fruit vary from 120 to 350 with a particularly high percentage of aborted seeds (50–60%) in Italian varieties (Gorini et al., 1993).

Not climacteric, the fruit of the cactus pear has a low respiration rate and is sensitive to damage by cold and suffers from fungi (*Penicillum*, mainly). For the purpose of conservation, the best results are obtained by conditioning the fruit for 24 to 48 h at 37–38°C or by dipping them in water at 50–55°C for 3–5 min. and then conserving them at 6°C at a relative humidity of 95% for 6 wk after gathering.

The cactus pear fruit is sensitive to fruit fly (*Capitata ceratitis*), which is the only problem in Italy. A more serious parasite, one that is increasing greatly, is the *Cactoblastis cactorum*, which very effectively attacks the cladodes to the point that it is employed, in fact, in the biological control of invasive Cactaceae (Hoffmann et al., 1998; Knight, 2001). A specific entomoparasite is the *Dactilopius opuntiae*. Among the cryptogams that cause root decay or *Armillaria* cancers are *Alternaria*, *Cercospora*, and *Phytophtora cactorum*; among the important bacteria are *Erwinia carotovora* (which causes soft rot on the fruit) and *Agrobacterium tumefaciens* (which causes root tumours).

21.2.6. Secondary Products

Cladodes are rich in mucilages, complex carbohydrates with a large capacity to absorb water that can be considered a potential source of industrial hydrocolloid. The mucilages contain L-arabinoside, D-galactose, L-rhamnose and L-xylose and also galacturonic acid in proportions that vary according to the management of the cultivation, temperature and humidity. In some environments, the mucilages are used in wastewater treatment systems on a commercial and domestic level and recently they have been tried as a means of improving the hydraulic permeability of soils (Sáenz et al., 2004).

Opuntia ficus-indica is also used, exclusively in Mexico, for production of cladodes destined for human consumption (nopalitos or nopals), as well as for the breeding of a cochineal that is specific to the *platyopuntia*, the *Dactylopius coccus*. When this insect is dried and ground, it yields carmine, a natural red colourant (grain cochineal or dacty dye) that, for centuries, has constituted one of the main sources of export wealth from the New to the Old World, and that even today has a significant economic and commercial role as a food colourant and dye in the textile industry of Central and South America (Mexico, Peru and Chile) and on the island of Lanzarote (Canary Islands).

21.3. Cultivation Requirements

21.3.1. Climate

Opuntia colonizes environments characterized by:

- mild winters (mean annual temperature > 10°C);
- a prolonged dry season that coincides with short days; and
- summer rainfall.

These are the conditions of the environments in which the species is autochthonous, while in the Mediterranean basin, the dry season coincides with long days and fruit growth.

CAM metabolism enables the plant to reach maximum photosynthetic productivity with day temperatures of 25°C and nocturnal temperatures of 15°C; temperatures higher than 30°C cause reductions of up to 70% of photosynthetic activity. Temperatures lower than 0°C, even for 4 h, produce irreversible damage to the cladode tissue and the fruit. During the development of the fruit, temperatures lower than 18°C slow down growth, delaying maturation, and diminish the reducing sugars content of the pulp.

21.3.2. Soils

The species grows in all types of soil. It frequently adapts even to soils that are limited by continuous hard rock in the first 25 cm, or that rest on materials with a calcium carbonate content greater than 40% or that contain less than 10% by weight of fine soil (Nobel, 2002). In Lampedusa (Italy), Torta et al. (2007) reported on the vesicular-arbuscular symbiosis between cactus pear and the mycorrhizal fungi population of *Glomeromycota*.

Generally, it does not tolerate salt and a concentration of the soil solution of 50 mM NaCl is considered the threshold for commercial production of the species (Nobel, 2002).

21.3.3. Water Requirements

In spite of the *Opuntiaceae* being typically drought-resistant and able to accumulate in their tissue more than 1500 t/ha of water (Han and Felker, 1997; North and Nobel, 1997), and although Flores-Hernández et al. (2004) showed that drip irrigation (~30%) does not increase cactus pear crop production, *Opuntia ficus-indica* cultivated in the Mediterranean basin is irrigated, although the growth of the fruit comes during the dry period. The seasonal volumes vary from 60–80 mm in Italy to 500 mm in Israel using drip irrigation or fine low-level spray.

21.3.4. Nutrients

In Italy, typical fertilizations include 80–100 kg/ha of nitrogen, with the phosphate and potassium fertilization varying widely. The optimal content of N in the one-year-old cladodes is 0.8–0.9%. In winter, 50 kg N per hectare, 80 kg K and 100 kg P are advisable, and the nitrogen is distributed also in spring-summer in at least two applications (60 kg/ha, using urea). Significant correlations have been verified between the calcium available and the quantity of fruit produced as well as between the potassium present and the quality of the fruit (Galizzi et al., 2004).

In Italy, there is no literature on the effect of fertilization on the out-of-season winter crop. In Israel, Nerd et al. (1991) demonstrated that under continuous NPK fertigation the number of floral buds per plant was much lower in the winter than in the summer crop. Fertilization increased production of floral buds in both crops, but to a greater extent in the winter crop. The increase in floral bud production in fertilized plants was associated with an increase in NO_3^--N content in the cladodes. The fruits of the winter crop ripened in early spring, following the pattern of floral bud emergence the previous autumn. Mean fresh weight and peel to pulp ratio (w/w) were higher in fruits that ripened in the spring (winter crop) than in fruits that ripened in the summer.

21.4. Soil Suitability in Cultivation of Cactus Pear

The suitability of a given environment to the cultivation of cactus pear is defined by soil and climate characteristics, water conditions and nutrient requirements (Table 21.1). Some of these requirements are common to the different varieties of cactus pear cultivated in Italy. As far as the soil is concerned, the cactus pear is very versatile but essential elements are the absence of high salinity and the phenomena of water stagnation. It tolerates high air capacities (high textural or structural macroporosity) and a modest or meagre organic content, while among the edaphic requirements, a good availability of calcium and potassium is considered a useful condition for a good crop. Cultivation of cactus pear can involve a medium to high risk of invasive colonization, while the risk of pollution of the strata and degradation of the ground by tamping is negligible. Risks of "biological invasion" by alien cactus pear are realistic and have been documented (e.g., van Sittert, 2002).

Table 21.1. Requirements for cultivation of prickly pear and conservation .of the pedo-environment.

Requirement	Conditions
Climatic	
Average annual air temperature	18–28 °C
Humidity of the air	very low to high
Cultivation	
Soil texture	sandy to mixed
Structure and porosity	not limiting, except compacted and unstructuredsoils
Water availability	insufficient to average
Salinity of the soil	low
Reaction of the soil	subacid to subalkaline

21.5. Functional Characteristics of Soils and Classes of Suitability

The literature does not contain works specifically on land evaluation suitable for the cultivation of the cactus pear. Nonetheless, on the basis of the cultivation requirements synthesized in Table 21.1, classes of soil suitable to the cultivation of the species can reasonably be proposed. Substantially, apart from thermal and textural limits, the species adapts well to the pedoenvironmental conditions of warm and arid environments, although the low tolerance of salinity remains fixed.

The classes of soil that are suitable for cultivation of cactus pear (Table 21.2) are therefore drawn up as follows: the S1 class (most suitable soils) do not introduce limitations and greater yields on them are hypothetical; the S2 class (suitable soils) have yields that are satisfactory but lower than the optimal yields; the S3 class (marginally suitable soils) have hypothetical yields close to the expediency threshold. The class N encompasses soils that are best excluded from cultivation of cactus pear.

Table 21.2. Functional characteristics and classes of soil capacity for prickly pear cultivation in Italy.

Functional characteristics		Classes			
		S1 most suitable	S2 suitable	S3 less suitable	N not suitable
Min. temp.	°C	> 5	> 5	> -5; < 5	< -5
Mean temp.	°C	18–23	15–18	10–15	< 10
Annual rainfall	mm	400–800	> 250, <1000	250–400	< 250, >1000
Texture		from sandy to mixed		clay-silty	clay and silt
Skeleton		indifferent	indifferent	indifferent	[‡]abundant
Depth		indifferent	indifferent	indifferent	indifferent
Carbonates		indifferent	indifferent	indifferent	indifferent
Reaction	pH $_{KCl}$	5–8	5–8	5–8	< 5; > 8
Organic matter	%	1.0–2.0	0.5–1.0	< 0.5	absent
Available Ca		high	high	mean	insufficient
Available K		high	high	mean	insufficient
[†]EC$_e$	dS/m	< 2	2–4	4–7	> 7
Subsurface ground water		absent	absent	present	present

[†] EC$_e$ *electrical conductivity of the saturate extract of the soil.*
[‡] *Potential difficulties for mechanized cultivation of the soil.*

References

BARBERA, G., INGLESE, P. 1993. La coltura del ficodindia. Calderini Edagricole, Bologna, 188 p.

BARBERA, G., INGLESE, P., PIMIENTA BARRIOS, E. 1995. Agroecology, cultivation and uses of cactus pear. FAO Plant Production and Protection Paper 132, Rome, 216 p.

DE LA BARRERA, E., NOBEL, P.S. 2004. Nectar: properties, floral aspects, and speculations on origin. Trends in Plant Science 9, 1360–1385.

FLORES-HERNÁNDEZ, A., ORONA-CASTILLO, I., MURILLO-AMADOR, B., GARCIA-HERNANDEZ, J.L., TROYO-DIEGUEZ, E. 2004. Yield and physiological traits of prickly pear cactus 'nopal' (Opuntia spp.) cultivars under drip irrigation. Agricultural Water Management 70, 97–107.

GALIZZI, F.A., FELKER, P., GONZALEZ, C., GARDINER, D. 2004. Correlations between soil and cladode nutrient concentrations and fruit yield and quality in cactus pears, *Opuntia ficus-indica* in a traditional farm setting in Argentina. Journal of Arid Environments 59, 115–132.

GORINI, F., TESTONI, A., CAZZOLA, R., LOVATI, F., BIANCO, M.G., CHESSA, I., SCHIRRA, M., BUDRONI, M., BARBERA, G. 1993. Aspetti tecnologici: conservazione e qualità di fico d'india e avocado. Informatore Agrario XLIX (1), 89–92.

HAN, H., FELKER, P. 1997. Field validation of water-use efficiency of the CAM plant *Opuntia ellisiana* in south Texas. Journal of Arid Environments 36, 133–148.

HOFFMANN, J.H., MORAN, V.C., ZELLER, D.A. 1998. Evaluation of *Cactoblastis cactorum* (Lepidoptera: Phycitidae) as a biological control agent of *Opuntia stricta* (Cactaceae) in the Kruger National Park, South Africa. Biological Control 12, 20–24.

KEELEY, J.E. 1998. CAM photosynthesis in submerged aquatic plants. The Botanical Review 64, 121–175.

KNIGHT, J. 2001. Alien versus predator. Nature 412, 115–116.

MOHAMED-YASSEEN, Y., BARRINGER, S.A., SPLITTSTOESSER, W.E. 1996. A note on the uses of *Opuntia* spp. in Central/North America. Journal of Arid Environments 32, 347–353.

NERD, A., KARADY, A., MIZRAHI, Y. 1991. Out-of-season prickly pear: fruit characteristics and effect of fertilization and short droughts on productivity. Horticultural Science 26, 454–608.

NOBEL, P.S. 1988. Environmental Biology of Agaves and Cacti. Cambridge University Press, New York, 270 p.

NOBEL, P.S. 1992. Biomass and fruit production for the prickly pear cactus, Opuntia ficus-indica. Journal of the American Society for Horticulture Science 117, 542–685.

NOBEL, P.S. 2002. Cacti Biology and Uses. California University Press, Los Angeles, 280 p.

NORTH, G.B., NOBEL, P.S.,1997. Drought-induced changes in soil contact and hydraulic conductivity for roots of Opuntia ficus-indica with and without rhizosheaths. Plant and Soil 191, 249–258.

PIMIENTA BARRIOS, E. 1990. El Nopal Tunero. Grafica Nueva, Pipila 638, Guadalajara, Jalisco, Mexico, 246 p.

SÁENZ, C., SEPULVEDA, E., MATSUHIRO, B. 2004. *Opuntia* spp. mucilages: a functional component with industrial perspectives. Journal of Arid Environments 57, 275–290.

TORTA, L., MARTORANA, A., BURRUANO, S., SAIANO, F. 2007. Further studies on the vesicular-arbuscular mycorrhizae of prickly pear in Sicilian areas. Micologia Italiana 36, 47–52.

TURKON, P. 2004. Food and status in the prehispanic Malpaso Valley, Zacatecas, Mexico. Journal of Anthropological Archaeology 23, 225–251.

VAN SITTERT, L. 2002. 'Our irrepressible fellow-colonist': the biological invasion of prickly pear (Opuntia ficus-indica) in the Eastern Cape c.1890–c.1910. Journal of Historical Geography 28, 397–419.

WESSELS, A.B. 1988. Spineless prickly pear. First Perskor Publishers, Johannesburg, South Africa, 61 p.

22. Lentil (*Lens culinaria,* Medicus)

Letizia Venuti[1]

22.1. Crop Characteristics

Lentil (*Lens culinaria,* Medicus), possibly the oldest pulse crop ever cultivated, is originally from the Middle East, from where it spread out to Egypt, central and southern Europe, the Mediterranean area, Ethiopia, Afghanistan, India and Pakistan, so that today it is cultivated in many countries all over the world. In the Mediterranean region, lentil is cultivated in all the countries that border the sea. Particularly in Italy, it is mostly grown in the centre, the south, and the islands, where, however, it has decreased significantly during the last decades and today is considered a marginal crop.

Lentil is an annual crop grown also at high altitudes, from 0 to 3,000 m above sea level, and in completely different environments, from temperate cold steppe to humid and dry subtropical regions, and humid forest regions. In Italy, lentil growing cycle ranges from 3–4 to 5–6 months, depending on the sowing time (autumn or spring).

22.2. Cropping Techniques

22.2.1 Crop Rotation

Lentil is considered a soil-improving crop because, like all other pulses, it fixes nitrogen in the soil. In the crop rotation, lentil usually follows a depleting crop, for instance, wheat, barley or rye (at high altitude). Field residues turned into the soil are a good source of organic matter, characterized by a low C/N ratio, which is beneficial to the humification process.

It is not advisable to grow lentil on the same field in consecutive years because of the increasing risk of pest attacks: a break of 3–4 years is considered adequate.

Sometimes lentil is cultivated on small surfaces and in this case it is considered an orchard crop.

[1] CNR-IRPI—Istituto di Ricerca per la Protezione Idrogeologica, Unità Operativa di Supporto, Pedologia Applicata, Firenze, Italy.

22.2.2 Cultural Practices

Soil Preparation

Soil preparation is carried out as for other row crops: a deep ploughing before autumn rainfalls followed by a shallower tillage such as that done with rotary hoeing in order to crumble clods. The seedbed should be even and firm, with previous crop residues well mixed in the soil. The operations of earthing seeds and later cutting and threshing are more difficult on a rough and cloddy land with abundant rocks and field residues.

Seeding

Sowing time is autumn in mild climates or early spring in colder areas, to avoid summer drought (in Italy, January-February). The sowing depth is in the range of 4 cm, depending on the depth of the driest soil layer, bearing in mind that excessively deep planting would impair seedling emergence. Seed rate ranges from 50 to 100 kg/ha, depending on the seed size, keeping rows 20–30 cm apart.

The use of a precision seed drill is recommended, to keep the sowing depth constant and favour contact between seed and soil particles, in order to help lentil establishment despite competition from weeds. Lentil seed size is rather variable, making it necessary to adjust the seed metering system.

When seed is not inoculated, seed needs to be mixed with *Rhizobium leguminosarum* at least 24 h ahead of sowing. There are different methods of inoculation, including the technique of placing bacterium and seed together in the drill.

Other types of seed dressings, such as treatment with insecticides and fungicides, are not recommended as they can interfere with lentil nodulation.

Fertilization

Nitrogen fertilization should not be necessary because the root inoculation with *Rhizobium* supplies sufficient nitrogen to the crop. Nevertheless, if the organic matter content in the soil is low (< 2%), an application of 30–40 kg/ha of nitrogen can be beneficial to help plant growth until the root nodules develop. It is advisable to place the fertilizer close to the seed but to avoid direct contact. If excessive, nitrogen can inhibit root nodule development and favour excessive vegetative growth at the expense of seed production.

Sulphur demand is rather high, as with other pulses. Phosphorus and potassium applications are recommended when soil is not rich enough in these nutrients. An amount of 30–70 kg/ha of phosphorus and 100–135 kg/ha of potassium are considered optimal for lentil growth. Phosphorus in row application is advisable in phosphorus-poor soils or when the pH value is high (> 7.6). A ternary fertilizer NPK (2-3-1) applied at the sowing stage can suffice to meet lentil requirements, to be integrated by ammonium sulphate, when necessary, at the blooming time.

Weed Control

Lentil is poorly competitive with regard to weeds, especially at the emergence stage. For this reason, mechanical weed control is recommended by means of repeated hoeing and a few passes of earthing up during the growing cycle. If weed infestation is serious, it is advisable to carry out harrowing and rotary hoeing after emergence, an operation to be performed carefully in order to avoid seedling damage. Since a certain number of lentil plants will inevitably be killed together with weeds, it is necessary to account for this loss when calculating the seed rate at the sowing time. A rotary hoeing pass should be carried out 7–10 d after sowing.

Chemical weed control consists in glyphosate application, as a pre-sowing treatment in autumn or post-emergence treatment in spring, in order to kill bermuda grass when it is 15–20 cm high and actively growing. Sethoxydim is applied as post-emergence treatment to kill annual weeds when they are 10–15 cm high.

Irrigation

Both waterlogging and drought stress, especially at the critical stage of reproduction and seed filling, are responsible for yield decrease. A proper use of irrigation makes it possible to reach a satisfactory productive level by reducing the risks of drought stress during the above-mentioned critical periods. Lentil is moderately resistant to drought stress, thanks to the ability to modify its internal osmotic pressure. However, in areas characterized by drought during the critical periods, irrigation is recommended to preserve crop productivity. In fact, in these areas, a yield increase from 450–675 kg/ha (without irrigation) to 2,000 kg/ha (with irrigation) has been observed (Muehlbauer and Tullu, 1997).

Harvesting and Storage

The optimal time for harvesting is when the plant starts changing colour from green to yellow and the bottom pods to brown-yellow. Harvesting should then be done promptly to avoid pod opening and seed losses. The plants should be cut when air humidity is sufficiently high to make pods resistant; it should be avoided during the hottest and driest hours of the day. Typically, a flail mower, with the blades at a low setting because of the fragile lentil stem, is used in combination with a windrower. Windrowing should be carried out over a surface as even and free of rocks as possible, to avoid yield losses. Plants placed in windrows can take one week or longer to dry, depending on the meteorological conditions. Inside the windrow, air circulation is poor and when air humidity is high for a long time, seed bleaching and mould growth can occur.

Dry seeds have 14% water content, but harvesting is preferably carried out earlier (at 18–20% water content), followed by artificial drying; otherwise, seed damage and losses may occur.

If the field has a sufficiently even surface, a combine harvester provided with a flail cutter bar can be used, adjusting the beater at lower speed than used for wheat, in order to avoid breakage of seeds.

Yield varies depending on cultivar, crop management and environment, In Italy, the average national yield is about 1 t/ha, but yields up to 2 t/ha can also be obtained.

22.3. Crop Requirements

22.3.1 Climate

Lentil easily tolerates low temperatures during the growing cycle, and young seedlings can withstand late frost, allowing an early spring sowing. Lentil is more winter-hardy than chickpea but it is heavily damaged by hard and prolonged frost. It is moderately resistant to drought stress and heat, although various cultivars have different tolerance due to their different transpiration mechanism.

Lentil is grown as a winter crop in semi-arid and tropical areas and as a spring crop in colder areas. It is widespread in semi-arid regions, where yield is not high but product is of good quality.

High humidity and rainfall during the growing cycle cause excessive vegetative development at the expense of quality; annual precipitation around 250–300 mm/y is a good compromise between satisfactory yield and good quality. Drought stress and high temperatures during the blooming and seed filling stages reduce yield. Usually, lentil tolerates drought better than it does waterlogging.

The optimal mean temperature during the growing cycle should be in the range of 24°C; germination is possible above 0°C but the optimal germination temperature is about 18–21°C; a temperature > 27°C at the time of germination is harmful.

In Table 22.1, climatic requirements for lentil are shown.

Table 22.1. Climatic requirements for lentil.

Requirements	Highly suitable	Moderately suitable	Marginally suitable	Unsuitable
Mean temperature of the growing cycle (°C)	24	15–23; 25–29	5–15; 29–32	< 5; > 32
Mean annual rainfall (mm)	700–800	600–700; 800–1,000	250–600; 1,000–2,500	< 250; > 2500

22.3.2 Soil

Lentil cultivation takes place on many types of soils, from sandy to clayey, provided that the internal drainage is adequate, because flooding or even waterlogging can kill plants. The most suitable soil is deep, loamy or sandy loam, rich in phosphorus and potassium, and having a neutral pH.

Lentil is well tolerant of alkaline and acid soils (pH: 4.5–8.2), although the ideal pH range is between 5.5 and 7.5.

Salinity tolerance is higher at germination and it ranges between 8.4 and 13.1 dS/m. When salinity is higher, yield is rapidly reduced.

The best quality is obtained on moderately fertile soils, having a non-excessive calcium carbonate content, which may cause seeds to take longer to cook.

22.4. Land Suitability for Lentil

Lentil is a hardy crop that can withstand difficult environmental conditions ranging from cold to drought. Thanks to its short growing cycle it can avoid the coldest months of the year and be harvested before water reserves are depleted. Table 22.2 lists lentil requirements.

In Italy, lentil is cultivated in disadvantaged areas with temperate climate, ranging from cold temperate (at high altitudes) to semi-arid (southern Italy and the islands): in cold areas it is grown as a spring crop and in dry areas as a winter crop. Lentil is typical of uplands where the soil-climate combination makes it possible to achieve high quality levels. Examples are Castelluccio di Norcia and Leonessa Uplands in central Italy.

Current rural development policies aim to make the most of this crop, which is typical of marginal areas.

Table 22.2. Lentil requirements.

Requirements	Value
Climate	
Air temperature	moderate-low
Air humidity	low
Soil	
Texture	moderate
Oxygen availability	high
Water availability	moderate-low
Soil depth (ground water, soil horizons impenetrable to roots)	moderate
Nutrient availability	moderate
Soil salinity	moderate-low
Soil pH	sub-acid to sub-alkaline
Management	
Trafficability	moderate
Workability and mechanization	moderate-high
Land conservation	
Erosion risk	low
Groundwater pollution risk	low
Soil compaction risk	low

22.5. Soil Characteristics and Suitability Assessment Table

Unfortunately, few studies are available on the relationship between lentil quantity, quality and the influence of soil. Most international efforts are directed toward lentil genetic improvement. In Italy, a suitability map for lentil of the Leonessa Uplands has been produced (Raglione et al., 2002). Table 22.3 presents the soil requirements for lentil based on the data coming from different experiences quoted in the literature (Baldoni and Giardini, 1981; Sys et al., 1993).

Table 22.3. Soil suitability for lentil.

Soil characteristics	Class and rating scale			
	Very suitable	Moderately suitable	Marginally suitable	Unsuitable
Depth, cm	> 60	40–60	30–40	< 30
Texture[1]	CL, SiL, SiCL, L, SCL	C, SiC, SL, Si	LS, S	$C_{massive}$
Salinity (dS/m)	< 2	2–4	4–10	> 10
Gypsum (%)	< 0.5	0.5–3	3–15	> 15
pH	6–8.2	5.6–6; 8.2–8.3	5.2–5.6; 8.3–8.5	<5.2; >8.5
Sum of bases (Ca, Mg, K, Na) (cmol+/kg soil)	> 5	3.5–5	2–3.5	< 2
Base saturation (%)	> 50	35–50	< 35	
Content of organic carbon (%)	> 1.2	0.8–1.2	< 0.8	
Rock fragments (%)	< 15	15–35	35–70	> 70
Exchangeable sodium percentage (%)	< 8	8–12	12–15	> 15
Internal drainage[2]	well drained; moderately well drained; moderately well drained; somewhat poorly drained	somewhat poorly drained; well drained	somewhat excessively drained; excessively drained	poorly drained; very poorly drained
Flooding risk	none		rare	occasional or more frequent
Total calcium content— CaCO₃ (%)	< 10	10–50	> 50	
Available water capacity (mm)	> 112	57–112	< 57	
Cationic exchange capacity (cmol+/kg soil)	> 20	10–20	5–10	< 5

[1] *SiC silty clay, C clay, Si silt, SiL silt loam, CL clay loam, SC sandy clay, L loam, SiCL silty clay loam, SCL sandy clay loam, SL sandy loam, LS loamy sand, S sand, $C_{massive}$ massive clay, $SiC_{massive}$ massive silty clay.*
[2] *From: Sys et al. (1993), modified according to classes quoted in Appendix.*

References

BALDONI, R., GIARDINI, L. 1981. *Coltivazioni erbacee.* Patron, Bologna, pp. 347–351.

SYS, C., VAN RANST, E., DEBAVEYE, I.R.J., BEERNAERT, F. 1993. *Land Evaluation, part III. Crop Requirements.* Agriculture publication no. 7, General Administration for Development Cooperation, Brussels, Belgium, 166 p.

RAGLIONE, M., LORENZONI, P., ANGELINI, R., BONIFAZI, A., FEBELLI, C., SPADONI, M., VENUTI, L., VERZILLI, C. 2002. *Carta dei suoli dell'altopiano di Leonessa (Rieti) e delle loro idoneità per alcune colture tipiche.* Arti Grafiche Nobili Sud, Rieti, 279 p.

MUEHLBAUER, F.J., TULLU, A. 1997. *Lens Culinaris.* http://www.hort.purdue.edu/newcrop/Interve.

PART III

LAND SUITABILITY AND
LAND ZONING

*Land suitability for
tree cultivations*

PART III

LAND SUITABILITY AND LAND ZONING

Land suitability for tree cultivations

23. Common Ash (*Fraxinus excelsior* L.)

Sergio Pellegrini[1]

23.1. Botanical Classification and Main Botanical Characteristics

The *Fraxinus* genus includes about 70 arboreal and shrub species, mostly present in the temperate regions of the northern hemisphere. The name comes from the Greek term *phraxis*, which means closing or separation, probably referring to the use of these plants as hedges delineating properties. Ashes are deciduous trees, with opposite, compound leaves adorned with stipules. The flowers are grouped in axillary racemes, sometimes without corolla. The indehiscent fruits hang in clusters and are samaras with only one small seed.

Table 23.1. Botanical classification.

Order: Oleales
Family Oleaceae
Genus *Fraxinus*
Species: *Fraxinus excelsior* L. (common or European ash)

Among the species of ash present in Europe, the ones that grow spontaneously in Italy are the common ash (*Fraxinus excelsior* L.), the flowering ash (*Fraxinus ornus* L.) and the southern ash (*Fraxinus angustifolia* Vahl. and the subspecies *oxycarpa*). While the identification of *Fraxinus ornus* is not difficult, the other species are very similar. *Fraxinus ornus* flowers after it puts out leaves, while in the other two species the flowers appear before the leaves. The best characteristic for distinguishing *F. excelsior* from *F. angustifolia* is the colour of the buds, black or blackish brown in the first and greenish brown in the latter.

The species name *Fraxinus excelsior* in Latin means literally "the tallest", in reference to a superior development in height compared to the flowering ash. The common ash is a tree that

[1] CRA-ABP—Centro di Ricerca per l'Agrobiologia e la Pedologia, Firenze, Italy.

can grow to 30–50 m in height, is straight and elegant, and has a rapid growth rate (up to 50–60 cm per year in adult plants) and a long life span; it rarely reaches two centuries, but specimens aged 300 years have been found in environments having particularly favourable conditions.

The canopy is irregular and sparse, especially in the central part. The bole is almost always upright, straight, and tall with a wide base section. In isolated plants the first branches are found at a fairly low position, while in grouped specimens or those found in forests the trunk is without branches farther up. The principal branches are attached to the trunk at a moderately acute angle; young branches are thin, opposite, shooting upward, smooth, shiny, and of a grey-green colour that tends toward black with aging. The buds are large, pyramidal, opposite, black, velvety and protected by two well-developed bracts.

The bark is smooth and light grey; the adult specimens tend to become grey-brown and black and become finely fissured with tortuous, shallow cracks.

The leaves are odd pinnately compound and formed by 9–11 small lanceolate, elliptical, sessile leaves about 1.5–4 × 4–10 cm in size, minutely serrated with an acuminate tip. The upper side of the blade is glabrous, dark green, and opaque, while the underside of the blade is lighter in colour and pubescent.

Leaf emission is slightly late and usually occurs 2–3 weeks later in isolated specimens, compared to those growing in groups. The flowers are short cobs (racemes) often unisexual, without calyx and corolla and with purple anthers; pollination is anemophilous. Flowering takes place before leafing (first half of March) and therefore flowers are subject to damage caused by late frost.

The fruit is a linear lanceolate samara (7–8 × 30–35 mm). It is light green at the beginning of summer and gradually turns to brown, then dark brown, then blackish after the leaves fall (end of autumn). The samaras remain on the plant during the winter months until the arrival of spring, when spontaneous dissemination takes place. The fruits are a desirable source of food for birds, especially finches.

The root system is strong, with large roots developing horizontally that construct a thick network with strong soil stabilizing effects; on the other hand, they may damage walls, roads and buildings in the vicinity. The taproot in specimens grown from seed penetrates deep into the soil; therefore, the root system of the common ash is able to explore a large soil volume.

The common ash is perfect for use as shade trees and in the consolidation of hillsides subject to landslides, where it is especially indicated for reforestation on slopes with northern exposure. It is also a good choice for wind-breaking hedges and for rural use. In areas of the Alpine, Apennine and Po Valley regions that are not intensely urbanized, it is suited for urban parks and streets without heavy traffic, if the trees are positioned in open areas. It is not advisable for use in parking lots or situations in which it may cause damage to roads, walls or buildings or other trees nearby.

The ash is used for its excellent wood, which is compact, resistant, flexible, and not deformable. It is light brown with shiny highlights and is easy to work. Because of its strength and flexibility, it was used in ancient times to make bows and the shafts for spears and arrows. Today it is used to make oars and masts for boats, tool handles, golf clubs, hockey sticks, pool cues and tennis rackets. It is also employed for the construction of floors, furniture, and plywood. It is not suitable for the construction of window fixtures, because it is not very resistant to weathering.

In the old days it was planted near farmhouses for its foliage, used for livestock forage. Since ancient times, the ash has also been used for its medicinal properties; even today, the bark of branches and roots, the leaves, and fruits are used by herbalists.

23.2. Varieties

Nurseries commonly sell the pure species as well as the subspecies and hybrids that are similar but differ in development and bearing. This may lead to non-homogenous and badly shaped planting. Among the varieties used for urban decoration are the following: F.e. "Aurea", with leaves that are yellow in spring and golden in autumn; F.e. "Aurea pendula", a small, delicate tree with yellow hanging buds; F.e. "Diversifolia", vigorous, with simple or tripartite leaves; F.e. "Pendula", with pendulous foliage and drooping branches, in which the vegetative tip must be pruned to accentuate its "weeping" bearing and which is resistant to low temperatures and stagnant moisture; F.e. "Globosa", which takes the shape of a small round bush; F.e. "Westhof's Glorie", with late leaf emission, narrow crown form, which is straight in its first years and later tends to spread out.

23.3. Propagation

The main propagation method is by seed; seed harvesting takes place at the end of October, followed by layering in sand for one year.

23.4. Pedoclimatic Requirements

The common ash is widely distributed in Europe; its habitat ranges from Scotland and the southern regions of the Scandinavian Peninsula in the north, all the way south to Sicily and Greece; to the east it reaches as far as the Caspian Sea. The tree is spontaneous and frequent in northern and central Italy, rarer in the south, occupying mainly the phyto-climatic zones of the Castanetum and, to the north, of the Fagetum. It rarely forms pure stands, but usually lives together with other broadleaved plants (oak, horn-beam, mountain maples, beeches, silver firs, and spruce) in the mixed woods of the Alps (up to 1,700 m) and the central-north Apennine. It is also grown as an ornamental plant in urban parks, away from houses and streets. It is resistant to cold, wind, snow, and fog, even though it is easily damaged by late frosts once it comes out of dormancy, frosts that are one of the main causes of the bifurcation of the trunk. On the other hand, it has low drought tolerance. Drought conditions lower its resistance to atmospheric pollutants (e.g., ozone). The common ash is a light-seeking plant, and therefore it prefers to be placed in full sunlight; only young specimens can tolerate moderately shady sites. Because of its high moisture consumption, the main factor limiting growth is the soil's available water supply, a direct consequence of the difficulty of transpiration control typical of this species (Del Favero, 1998).

23.5. Cultivation Techniques

23.5.1. *Planting*

Seedlings 1 or 2 years old are usually used for planting in their permanent positions, preferably transplanted in autumn before the first winter frost. The method traditionally used for reforestation or for new wood-producing arboreal stands is digging a hole to plant each plant. It must be noted that experimentation conducted in ex-farming lands in the province of Bologna, Italy, demonstrated that tillage 40 cm deep produced significantly better results in terms of survival and height in the common ash, respectively after 2 and 4 years from planting (Minotta, 1988). Generally, experiments carried out in different Italian pedoclimatic environments showed that good preparation of the soil by tilling at least 70–80 cm deep is necessary before planting; this method of soil preparation produces significantly better results in terms of plant survival and height. Surface tillage, on the contrary, strongly limits the growth of the plants (S. Ravagni, personal communication).

In reforestation, to avoid the frequent bifurcation of the trunk, it is advisable to plant small high-density stands (2,000–2,500 plants/ha) in order to facilitate the selection of well-shaped subjects to be favoured by thinning out; to improve the shape of the boles it is advisable to plant in consociation with beeches (Bernetti, 1995).

23.5.2. Management and Cultivation

Successful rooting is influenced by the age and dimensions of the specimens; 7- to 10-year-old plants, over 4 m in height, such as those used for the planting of urban green areas, must be irrigated for the first 2–3 years following planting, fertilized, and tilled with light hoeing to ensure the best development of the specimens.

In timber-producing lands it is very important, at least in the first years, to take care to prune and to fight pests. Generally, the green pruning is done in spring (May–June), by freeing the bole tip from the competition of the immediately underlying lateral branches. Since the common ash is a sun-seeking species, after an initial phase of establishment for the formation of a straight trunk, without branches for at least the first 4–6 m, the seedling rows should be thinned out to favour the growth of a limited number of specimens, characterized by a well-shaped bole, a wide and well-developed crown, and the absence of narrow bifurcations (Pelleri, 2000).

23.6. Diseases

The ash is a fairly resistant plant; it hardly ever suffers damage by cryptogamic, bacterial, or viral plant diseases; however, it may be attacked by animal pathogens.

To reduce sanitary risks to a minimum and consequently avoid the necessity of using chemical defence, it is advisable to adopt prevention methods such as the use of healthy seedlings, use of correct agronomic techniques (soil management, pruning), and creation of mixed stands.

The diseases that the common ash may be subject to are collar-rot and borers. Collar-rot is caused by various fungi of the *Phytophthora* genus, ubiquitous pathogens that may attack many arboreal species. Characteristic alterations in the collar zone will appear following infection, such as the browning of the bark and presence of wide necrotic zones. These symptoms may also involve the base of the trunk (within the first 20 cm from the ground) and the larger roots. Entrance of the pathogen occurs through the lenticels and, most of all, wounds in the bark; it is therefore necessary to be careful not to damage the trunks during mechanical work. Prolonged moisture stagnation is another factor that leaves the plant open to these diseases.

Plants that have been attacked show symptoms of stress such as yellowing of foliage and extensive desiccation; specimens attacked should be quickly eliminated to reduce the risk of disease diffusion.

Borers are particularly dangerous pathogens for the species grown for its wood, because the larvae bore tunnels under the bark, altering the technological characteristics of the wood and reducing its value. The ash may be subject to attacks from the larvae of the leopard moth (*Zeuzera pyrina*), which bore into the heartwood of branches, causing desiccation and breaking. External signs of the presence of plant diseases are larvae entrance holes, from which issues frass shaped like tiny cubes. One line of defence is mass capture; chemical intervention is not efficacious, therefore not advisable, because it would have to be carried out throughout the summer months, which would damage the useful insect fauna. For mass capture, funnel-shaped pheromone traps should be placed by the middle of May, at least 10 per hectare, and kept active by periodic addition of pheromone capsules (every 4–5 weeks) throughout the summer months. The funnel-shaped traps must be placed with special holders about 1 or 1.5 m above the crown. Among the other insects that may damage the common ash, the *Prays fraxinella* is a lepidoptera that bores the leaves and buds and may cause bifurcation of the trunk.

23.7. Land Capability for Cultivation of Ash

In Italy, the common ash adapts well in the temperate mountain regions (Alps) (Righini et al., 2001), in the hot-temperate sub-continental regions (northern Apennine mountains), and in the temperate sub-oceanic regions (Po river valley); its adaptability in the hot-temperate oceanic regions (northern Adriatic coast), Mediterranean oceanic region (Ligurian and Tuscan Tyrrhenian coast) and Mediterranean sub-continental region (southern Adriatic and Tyrrhenian coasts) ranges from good to poor, while it ranges from poor to scarce in the Mediterranean subtropical climatic regions (Apulia, Sicily, and Sardinia).

Therefore, climate is the decisive environmental factor in the success of new cultivations. In Italy, the only experiment in evaluation of land capability regarding the common ash was conducted by the Pedologic Office of the Cartographic and Geologic Service of the Emilia Romagna Region and the Soil Service of the Lombardy Region Agricultural Development Bureau (ERSAL).

Consequently, the experiment will be used for reference in the identification of the land capability classes for this species. However, this evaluation method is not meant to be exhaustive and represents only a first attempt at identifying the pedologic characteristics that most influence the productive response of this species.

23.8. Functional Characteristics of the Soil

Among the pedologic characteristics that most influence the vegeto-productive response of the common ash are the volume of soil available for root exploration, the available oxygen supply and the salinity of soil (Tables 23.2 and 23.3).

Common ash can reach notable girth and have a well-developed root system, requiring an abundant supply of water. Therefore, it needs an available rooting depth of > 100 cm in order not to encounter significant limitations to its development. Thin soils (< 50 cm) are not considered suitable, for the above reasons and because they do not provide secure anchorage to the plant, with the consequent high risk of instability in bad weather conditions for specimens located on steep slopes (Tables 23.2 and 23.3). However, in geographic areas characterized by particularly high rainfall (around 2000 mm during the growing season, for example, the Pre-alpine region of Veneto and Friuli), it may vegetate well even on lands with a more modest thickness (F. Pelleri, personal communication).

The ash is fairly resistant to bad hydraulic drainage and grows even in the depressions between the dunes in the naturally wooded areas of the Po river valley, such as the Mesola

Table 23.2. Land capability evaluation for common ash in Emilia-Romagna, Italy.

Pedological characteristics	Degree of influence[1]	Intensity of limitations		
		Absent or slight	Moderate	Severe
Texture[2]	++	medium; mod. coarse; mod. fine	coarse; fine	
Available rooting depth (cm)	+++	> 100	50–100	< 50
Available oxygen supply	+++	good	moderate; imperfect	poor; very poor
Chemical reaction (pH)	+	5.5–8.5	4.5–5.5	< 4.5 and > 8.5
Risk of water deficit	++	absent; slight or moderate		strong; very severe
Salinity (EC 1:5 dS/m)	+++		< 0.4	> 0.4
Flooding risk				
-Frequency	+++	none; rare; occasional	frequent	
-Duration	+++	< 7 d	7 d to 1 mon	> 1 mon

[1] *The degree of influence indicates, in synthesis, how much each characteristic conditions the growth of the species: + = minor importance; ++ = medium importance; +++ = major importance.*
[2] *See the Appendix for the description of the general terms used for the texture class.*

Table 23.3. Land capability evaluation regarding the common ash in Lombardy, Italy (ERSAL-Soil Service, 2001).

Pedological characteristics	Degree of influence[1]	Intensity of limitations		
		Absent or slight	Moderate	Severe
Available rooting depth (cm)	++	> 100	50–100	< 50
Drainage	++	good, poor	poor, mod. rapid*	rapid, mod. rapid, slow, very slow, impaired
Flooding risk	++	none, rare, occasional	frequent	
Risk of water deficit[2]	+	moderate, slight, absent	very severe,** strong*	
Texture (100 cm)	+	medium, mod. coarse, mod. fine	coarse, fine	
Chemical reaction (pH)	+	5.5–7.5	4.5–5.5 7.5–8.5	< 4.5 and > 8.5

[1] *+ = secondary limiting factors; ++ = primary limiting factors (ERSAL, 2001).*
[2] *This parameter indicates the risk of insufficient water supply for arboreal cultivations. This figure is expressed by a ratio between hydraulic behaviour of site, climatic influence, and the available water capacity, taking into consideration eventual ground water or capillary water near the surface horizon (ERSAL, 2001).*
If drainage is moderately rapid or the risk of water deficit is rare, but the available water capacity (within the first metre) is > 50 mm, the degree of limitation is moderate.
**If the risk of water deficit is occasional, but the available water capacity is between 50 and 100 mm and ground water is absent, the degree of limitation is moderate.*

Woods. It is so resistant that it encounters severe limitations in development only in the areas where conditions of hydraulic saturation last for very long periods (> 1 month), with consequent scarce or very scarce oxygen availability at rooting level (Table 23.2). The ash, being a plant typical of humid climates, finds favourable conditions for cultivation in leached soils presenting salinity values lower than 0.4 dS/m; higher levels of salinity may hinder absorption of water during droughts (Table 23.2).

Other pedologic characteristics, such as texture and chemical reaction, do not show evident relations with the vegeto-productive response of the common ash (Tables 23.2 and 23.3). In fact, moderate limitations in development are encountered only with soils characterized by coarse or fine texture, or with an acidic soil reaction (4.5 < pH < 5.5). Soils considered inhospitable may be those with extreme pH values (very acidic or very alkaline soils), in which the plants encounter severe limitations in growth. Most literature considers the presence of active lime non-influential (e.g., Franc and Ruchaud, 1996).

Acknowledgements

The author is grateful to Dr Serena Ravagni and Dr Francesco Pelleri (CRA—SEL Centro di ricerca per la selvicoltura, Arezzo) for their valuable comments and critical review of the paper.

References

BERNETTI, G. 1995. Selvicoltura speciale. UTET, Torino.

DAZZI, C. 2000. La pedologia forestale in Sicilia: esperienze e prospettive. Monti e Boschi 51(2), 4–11.

DEL FAVERO, R. 1998. La vegetazione forestale e la selvicoltura nella Regione Friuli Venezia Giulia. Regione Autonoma Friuli Venezia Giulia, Servizio della Selvicoltura, Udine 1, 191–218.

ERSAL—Servizio del Suolo. 2001. Carta di Orientamento Pedologico per l'Arboricoltura da legno della Pianura Lombarda. A cura di: S. Brenna, E. Calvo, M. Sciaccaluga.

FRANC, A., RUCHAUD, F. 1996. Autécologie des feuillus précieux: frêne commun, merisier, érable sycomore, érable plane. Gestion des territoires (18). Cemagref, 15–68.

GELLINI, R. 1985. Botanica forestale. Vol. II. CEDAM, Padova.

MINOTTA, G. 1988. Il rimboschimento dei terreni ex-agricoli dell'Appennino: primi risultati sul confronto tra diverse tecniche di lavorazione del suolo. Italia Forestale e Montana 43(6), 474–483.

PELLERI, F. 2000. Indagine auxometrica e prova di diradamento in un popolamento di neoformazione a prevalenza di Fraxinus excelsior L. Ann. Ist. Sper. Selv. XXIX, 17–28.

PIGNATTI, S. 1982. Flora d'Italia. Edagricole.

POLI, F. Il Frassino. Scheda botanica. [online] http://www.comune.fe.it/aeb/schede/frassino/frassino.htm, verified 2008, march, 27th.

POLUNIN, O., WALTERS, M. 1987. Guida alle vegetazioni d'Europa. Zanichelli.

REGIONE EMILIA-ROMAGNA—Ufficio Pedologico del Servizio Cartografico e Geologico Regionale. Catalogo dei tipi di suolo della pianura emiliano-romagnola. [online] http://gias.regione.emilia-romagna.it/suoli/guidaindice.asp, verified 2008, March 27th.

RIGHINI, G., COSTANTINI, E.A.C., SULLI, L. 2001. La banca dati delle regioni pedologiche italiane. Boll. Soc. It. Scienza del Suolo 50, suppl., 261–271.

24. Common Walnut (*Juglans regia* L.) and Sweet/Wild Cherry (*Prunus avium* L.)

Rosario Napoli[1] *and Roberto Mercurio*[2]

[1] CRA-ABP—Centro di Ricerca per l'Agrobiologia e la Pedologia, Firenze, Italy.

[2] Università degli Studi di Reggio Calabria—Facoltà di Agraria, Dipartimento di Gestione dei Sistemi Agrari e Forestali, Località Feo di Vito (RC), Italy.

24.1. The Common Walnut (*Juglans regia L.*)

24.1.1. Botanical Classification and Principal Botanical Characteristics

The *Juglans* genus, according to the analytic key of Hutchinson's *Juglandaceae* (1959), is distinguished by the indehiscent fruit husk, woody and wrinkly nutshell, subdivided marrow, non-branched staminate aments, 7–105 styles, edible seeds, high oil content, and deciduous leaves.

According to Dode (1906–1909), the *Juglans* genus includes 44 different species divided into 3 sections according to bark colour, which seems to be correlated to the quality of the wood: sect. *Dioscaryon,* the white walnuts with smooth bark, opaque silver in colour in the young plants and later in the adult specimens cracked and grey; sect. *Rhyzocaryon,* the black walnuts with a more or less dark brown bark; and sect. *Cardiocaryon,* the grey walnuts with an ash grey bark. The common walnut is included in the first section.

According to Pardè (1952), the *Juglans* genus includes the following species: *J. rupestris* Engelm. (North America); *J. californica* S.Wats. (North America); *J. cinerea* L. (North America); *J. nigra* L. (North America); *J. mandshurica* Maxim. (North America); *J. hindsii* Rehd. (North America); *J. torreyi* Dode (eastern Asia); *J. stenocarpa* Maxim. (eastern Asia); *J. cathayensis* Dode (eastern Asia); *J. sieboldiana* Maxim. (eastern Asia); *J. cordiformis* Maxim. (eastern Asia); and *J. regia* L. (central and western Asia).

Krüssmann (1986) identifies the following: *J. ailantifolia* Carr. (= *J. sieboldiana* Maxim. non Goepp.; *J. cordiformis* var. *ailantifolia* (Carr.) Rehd.); *J. cordiformis* var. *cordiformis* (Maxim.) Rehd. (= *J. cordiformis* Maxim.; *J. sieboldiana* var. *cordiformis* (Maxim.) Mak.); *J. californica* Wats., *J. cathayensis* Dode (= *J. draconis* Dode); *J. cinerea* L.; *J. hindsii* (Jepson) R.E. Smith (= *J.californica* var. *hindsii* Jepson); *J. major* (Torr.) Heller (= *J. rupestris* var *major; J. arizonica* Dode; *J. torreyi* Dode); *J. mandshurica* Maxim.; *J. microcarpa* Berl. (= *J. rupestris* Engelm.; *J. nana* Engelm.); *J. nigra* L.; *J. regia* L.; NS *J. stenocarpa* Maxim.

All of the species of the *Juglans* genus are diploid (2n = 32) and interfertile. Tutin (1964) includes in "Flora Europaea" only three species: *J. regia* indigena in the Balkan Peninsula, Rumania, and Crete and naturalized following cultivation in various other regions; *J. nigra*; and *J. cinerea* cultivated for its wood.

Order : FAGALES
Family : JUGLANDACEE
Genus : JUGLANS
Species : *Juglans regia* L. (Common Walnut)

24.1.2. Botanical Characteristics

The common walnut is a tree that grows to a height of 20–25 (30) m, with a diameter of up to 1 m. A plant 26 m high with a circumference of 3.80 m was found in Cosenza (Alessandrini et al., 1989). Being a species with a long lifespan, it can reach an age of up to 900 years (Barone et al., 1997). The straight, cylindrical trunk is subdivided at varying heights into principal limbs on which a wide rounded crown develops. The bark remains smooth for a long period, and then becomes longitudinally cracked. Its colour is light grey or opaque silver in the young plants and first dark red then silver-grey on the branches. The glabrous buds are covered with two brownish glabrous perules. The terminal bud is large and accompanied by two smaller lateral buds. The glabrous twigs are shiny brown and carry two large heart-shaped leaf cicatrices. The leaves are alternate, deciduous, and composite, up to 35 cm long, odd pinnate with five to nine small elliptical leaflets with smooth margin, light green and aromatic. Flowers are monoecious. The numerous small male ones (100–160) are grouped in aments 6–8 cm long, are greenish, are pendulous and develop laterally on the twigs of the previous year, before leaf emission or simultaneously. They have no petals and have many short stamens (up to 40) with large anthers containing many pollen grains (900). The female flowers, made up of an ovary and sexual bilobate stigma, are solitary or more commonly in couples. They are at the extremity of the twigs of the current year's growth and may also be at the extremity of lateral shoots in some cultivars (e.g., Ashley, Chico, Payne). It should be noted that the high degree (80–90%) of induction to flower in mixed lateral buds in this type of cultivar means a notable drain of apical dominance and therefore a reduction of overall size and vigour (Germain, 1974), which would make it unsuitable for arboriculture for timber production. Flowering occurs in April–May depending on the cultivar and on environmental conditions. The seed ripens in September–October of the same year, therefore 19–20 wk from flowering.

Fruit: ovoid drupe 3–4 cm long and 2.5–3 cm wide with pulpy epicarp (hull) that is green and rich in tannic acid and detaches when ripe; indehiscent, woody endocarp (nut) formed of two wrinkly valves, cerebrum-shaped seed (kernel) divided into two to four very furrowed lobes, made up of two large, fleshy, oily cotyledons, covered with a thin reddish-brown film. Fruits reach full size about 10 wk after flowering. The hardening of the endocarp is complete between the 12th and the 15th week after flowering.

Plantula: the first leaf is usually composed of five sessile, glabrous leaflets, asymmetric at the base, ovate-lanceolate in shape, of varying size (the apical leaflet is the largest, the ones at the base are smaller). The edge of the leaf is undulate and dentate, the stem glabrous. The successive primary leaves, mainly with seven leaflets, are similar to the first, but larger. Definitive leaves are odd-pinnate, alternate with five to nine leaflets, oval with slightly undulate margin. When rubbed, both the primary and the definitive leaves give off a strong, penetrating odour (Magini, 1975).

Root system: strong, at first forms taproot, then tends to take on horizontal growth between 40 and 90 cm wide. Under ideal conditions the roots may achieve a depth of 3 m and occupy an area of between 2.8 and 8.7 times that of the crown (Chernoboi, 1971).

24.1.3. Varieties

Juglans regia includes many ornamental varieties: var. *monophylla* with simple leaves, very similar to the hazelnut, var. *laciniata* with serrated leaves and considerable development, var. *heterophylla* with leaves varying in shape, and var. *pendula* with pendulous branches. Krüssmann (1986) includes also var. *adspersa* with variegated whitish leaves, var. *praepaturiens* with a bushy habit, the slow-growing var. *purpurea*, with opaque red leaves, var. *recemosa* with fruits growing in racemes (10–15), var. *rubra* with red seeds, var. *corcyrensis* with very large leaves (up to 50–60 cm long), var. *bartheriana* with an almond-shaped nut and the subsp. *fallax* (Dode) Popov *(= J. fallax* Dode; *J. duclouxiana* Dode) of southern China and Himalaya.

Up until recent times, propagation by seed was necessary because of the uncertainty of grafting and determined the diffusion in cultivation activities of numerous genetic types that might also be considered more suited to local conditions. The Italian cultivar population references for fruit production are: *Bleggiana* in the area of Trento, *Feltrina* in Piemonte and Lombardia Regions and the area of Belluno, and *Sorrento*, the most widespread in Italy (Campania Region), which covers 70% of national production.

Among the fruit-bearing cultivars introduced there are the French (*Franquette, Mayette, Parisienne, Marbot, Lara, Solèze, Meylannaise, Ronde de Montignac*) and the American (*Hartley, Payne, Serr, Vina, Pedro, Chico, Amigo, Tehama, Adams10, Chase D9, Geisenheim 139*). It is useful to list the types that are known for the production of hardwood: *San Giovanni* and *Malescia* or *Martellina*. Among the varieties with aptitude for forestry, renowned in international spheres (UPOV), Lowe and Gonzalez (2001) mention the following: Fernette, 4/6/1997, INRA; Fernor, 4/6/1997, INRA; Liba, 2/1/1997, Societe des Mais Europeens; Chereba 2/1/1997, Societe des Mais Europeens; Noba, 21/12/1998, CTIFL.

Krüssmann (1986) lists the following hybrids of *J. regia*: with *J. nigra* one obtains *J. x intermedia* Carr., among which may be identified the varieties *vilmoreana* (Carr.) *Schneid* and *pyriformis* Carr.; with *J. cinerea* the *J. x quadrangulata* (Carr.) Rehd; with *J. hindsii* (Jepson) the '*Paradox*'; with *J. ailantifolia J. x notha* Rehd; and with *J. mandshurica J. x sinensis* (DC.) Dode.

On the French market the hybrids of black walnut with common English walnut (first generation hybrid F1) most used in the production of timber are those marked by the initials NG23 x RA and MJ 209 x RA (Bequey, 1990; Crave, 1991). The French interspecific hybrids MJ x RA and NG23 x RA, characterized by good apical domination and high vigour, have shown good adaptability to the Po River Valley area, where the summer months are not characterized by long periods of drought (Ducci and Tani, 1997). Comparison research on 8-year-old trees carried out in the Arno River Valley (Tuscany) has shown that the walnut hybrid MJ 209 x RA has an initially more vigorous development compared to black walnut, but not in comparison to the common walnut (Buresti et al., 1997).

24.1.4. Propagation

In Italy, the selection of superior phenotypes of the common walnut for the production of wood began in the 1980s by institutions attempting to obtain improved specimens for growth, shape, technological qualities and resistance to the principal adversities (Malvolti et al., 1998). To date there is no improved walnut material for high quality hardwood timber production.

Good results for the production of timber were obtained by cultivation in boxes (Ammannati-Lazzara, 1991; Buresti, 1993c; Ducci and Tani, 1997). With this system, the seedlings are well balanced, with a well-shaped root system, rich in secondary roots. The seedlings are generally larger than those grown with the traditional open field system. Cutting of the taproot may be avoided, which is a practice that can cause difficulties in vegetation restart and leave roots open to fungi causing root rot. Regarding size, the seedlings suitable for use in timber production should have a collar diameter of > 5 mm and be 50–60 cm in height, have a hypso-diametric ratio (h/d) between 60 and 80, be healthy, with a well-shaped apical shoot, a woody terminal shoot, a straight trunk with good apical dominance, and a well-developed root system rich in secondary and capillary roots (Calvo and D'Ambrosi, 1995; Falleri and Giannini, 2003). For this reason, it is necessary to carry out a careful selection in the nursery, discarding defective plants and those that are over- or under-developed.

Agamic propagation techniques by somatic embryogenesis are still at an experimental phase and furthermore the results obtained seem extremely variable as a function of the genotype (Scaltosoyannes et al., 1997; Tang et al., 2001; Cantoni et al., 2005).

24.1.5. Climatic Requirements

The common walnut grows in environments with a mean annual temperature ranging between 10°C and 17°C. As a plant destined for timber use it needs at least 5 months with a mean monthly temperature of 10°C. Resistance to low temperatures varies in relation to the different subspecies, the vegetative phase and the age of the plant: young plants are susceptible to late frosts even at −3°C, while adult plants in full dormancy may resist −35°C. The buds of the same year that vegetate until late autumn, generally scarcely lignified, are more exposed to damage by cold in the following spring. The best vegetative conditions are found in the area located between the cold subzone of the *Lauretum* and the *Castanetum* according to the phyto-climatic classification of Pavari. Walnut is an exigent species when it comes to water requirements. According to Bergognoux and Grospierre (1981) and Montero et al. (2003), it takes up at least 700–800 mm yearly, while the summer water requirements are estimated at 100–150 mm.

24.1.6. Cultivation Techniques

Cultural Models

So far the cultural models that have been most successful are pure plantations, mixed plantations and row plantations (Mercurio and Minotta, 2000):

Pure Plantations. This is traditionally the most diffuse cultural model used in Italy. There are two types, distinguishable by the density of initial planting: (1) plantation with definitive density and (2) plantation with variable density. In the former, density is constant during the production cycle, ranging from 120 to 200 plants per hectare, i.e., 10×10 and 9×9 m, arranged in squares or settonces (Barone et al., 1997). In France, walnut plantations are traditionally planted at a definitive density of 70 plants per hectare, i.e., spaced 12×12 m. This is chosen for economical reasons in order to obtain maximum growth in diameter and in volume, but most of all for phyto-sanitary reasons, to avoid having to thin out the rows and thus discourage the Armillaria fungi, which sometimes crops up on stumps. In this case, the French authors suggest enriching the pure walnut plantations with other broadleaves and/or planting agricultural cultivations in consociation in the first 3–5 years. With variable density plantation, the initial density is higher in order to contain the development of lateral growth, encourage height, and obtain intercrop wood products. In conditions of medium fertility and using common nursery stock, the initial density varies between 625 (4×4 m) and 278 (6×6 m) plants per hectare. In situations of high fertility and with high quality nursery stock, the initial density may run between 123 (9×9 m) and 204 (7×7 m) plants per hectare (Barone et al., 1997). It must be observed that with a plantation density superior to 330 per hectare (6×5 m), the plants start competing early, requiring precocious, non-remunerative thinning (Frattegiani and Mercurio, 1991). Pure plantations should not exceed a surface area of 4–5 ha. Research taking into consideration the concurrent actions produced when growing walnut in small plantations and the benefits of the "cultivation effect" suggest cultivation cycles of 40–50 years for the production of high value material. At present, short rotations of 25–30 years for the production of saw logs show little economical return, except in particularly favourable environments. This is due to the low yield connected with the reduced dimensions of the trunks that it is normally possible to achieve in such a short time. Generally speaking, it is a "technical rotation", aimed to achieve the production of a mass of timber technically and financially more interesting.

Mixed Plantations. Plantations may be mixed by tree or by group (mosaic). Different possibilities for consociation with walnut are being studied: Walnut can be planted with other species yielding high value timber (cherry, ash, maple) in order to differentiate timber production or with nitrogen-fixing species. The French authors were forerunners in this technique (Guinier, 1953; Hubert, 1979). In the case of plantations mixed by tree with *Robinia pseudoacacia*, since it is a very invasive and vigorous plant, Bequey (1991) warns of the high expense of reducing competition between this leguminous plant and walnut. In France, plantations mixed with alder (*Alnus* sp.) may favour the growth of walnut (Bequey, 1991, 1992, 1993). In Italy, experimental plantations mixed by tree using walnut, Neapolitan alder (*Alnus cordata*), and oleaster (*Eleagnus angustifolia*) have shown positive results, at least in the first years of growth of walnut (Buresti, 1993a). It must be noted that use of mixed plantations with the actinorhizal species must still be carefully evaluated, because the juglone produced by the common walnut "exerts a negative action on the development, infective capacity, nodulation and *Frankia*_nitrogen-fixing both in soils and in symbiosis with actinorhizal plants, particularly the *Alnus* and the *Eleagnus* that may have unfavorable repercussions on the plantations where Walnut is in consociation" (Bosco et al., 1997). Contu and Mercurio (2005) have shown that in mixed plantations of the common walnut and Neapolitan alder on calcareous soils there is no nitrogen fixing because of the inhibiting effect of juglone in these soils; Walnut can also be planted with fruit trees or honey-yielding species (apple, hazelnut) to integrate timber production with other products (e.g., fruit, honey) or with accompanying bushes (e.g., elder), that keep the walnut trees from developing vigorous low branches, reducing problems connected with pruning. Furthermore, these bushes may reduce work hours spent on cultivation and soil improvement (Buresti, 1993b).

Row Plantations. Single specimens are placed at a distance of 8–10 m from each other in simple rows. Seedlings must be planted after weak and broken roots are removed, in such a way that the collar is level with the ground. In regions prone to early drought phenomena, the best percentage of successful rooting is obtained when planting in autumn-winter. According to Paris et al. (1995), the use of mulching with black plastic sheets along the rows favours the growth of walnut while keeping the area around the plants free of weeds. Herbicides (with glyphosate base products) may be used to control weeds around the plants. In this case, full wall shelters in the lower part and mesh in the upper part may be useful to protect the saplings from the effects of the herbicide. Since these plantations are destined for the production of hardwood, the saplings must be protected from damage from wild or domestic animals and by the passing of machinery. Mesh guards around the plants have proved efficacious for a well-balanced development of the plants. The use of polypropylene tubes has been rated favourably in Italy for its positive effects on the initial growth of the saplings and for protection against competition from weeds. This has been analogously observed in other countries. Superficial tilling of the land between the rows should be carried out two to four times a year during the spring–summer for the first 5–6 years after planting.

The non-cultivation technique, which consists in keeping the soil covered with grass turf, has generally had negative effects on the growth of walnut, especially in the central-south regions, due to stronger competition for water by the herbaceous species (Minotta et al., 1993; Minotta and Boschetti, 1994). The effective nutrient requirements for timber-producing walnut plantations in terms of quality and quantity have not yet been ascertained, so reference is still made to research results regarding fruit plantations. The rationality of such an operation in terms of costs and benefits must still be evaluated. Regarding requirements for chemical elements, nitrogen is foremost. The quantities of nitrogen to be added vary with soil fertility and the age of the plants. In general, nitrogen fertilization can be done through the first 5 years starting a year after planting with increasing doses of 100 g per plant. (Bergognoux and Grospierre, 1981). Fertilizer must be spread under the protection of the canopy, in a divided dose: two-thirds at the beginning of vegetation restart in February-March and then one-third in April-May. After May, it is absolutely not advisable to use nitrogen, because it could cause a lengthening of the vegetative cycle resulting in weak lignification of tissues, which would increase risk of damage by early frost. The use of urea is not advisable, because it favours the excessive growth of lateral branches. If the plants are also intended for fruit production, after the first 5 years 60–80 kg of P_2O_5 and 80–100 kg of KO may be added per hectare, along with nitrogen fertilization amounting to 80–120 kg/ha (Bergognoux and Grospierre, 1981; Minotta, 1989). However, ternaries are useful, especially those with nitrogen prevalence. Walnut is susceptible to calcium deficiency (Hubert, 1979; Bequey, 1997).

Management and Cultivation

To improve timber production, the plants must be pruned so as to achieve a straight trunk free of limbs for a length of at least 2.5 m. Studies on crown architecture (Barthèlemy et al., 1997), annual growth cycles (Minotta and Boschetti, 1994; Cannavò et al., 2004), and different methods of pruning (Minotta, 1997; Cannavò et al., 2004) have yielded new knowledge about this important aspect of cultivation techniques. Formative pruning begins 2 years after planting; it consists in the gradual removal of broken, excessively developed branches or those with a double tip and working toward achieving equilibrium in the crown. Green pruning is done by eliminating all the lateral shoots while they are still at an herbaceous state. It has a negative effect on the diametrical growth of the trunk, referable to a partial inhibition of the cambial activity due to canopy reduction (Minotta et al., 1993; Minotta, 1997). Excessive pruning causes stress in plants. In fact, the drastic reduction in foliar surface slows growth, alters hypso-diametrical ratios of plants to the detriment of stability (the trunk may be slanted if not supported by a stake) and favours the emission of epicormic shoots along the trunk. Well-balanced formative pruning, in addition to reducing the incidence of knots, minimizes the variation of ring size and wood discoloration. When the plant reaches 3 m in height, improvement pruning or crown raising will begin, consisting in the gradual removal of lower branches and those that are too vigorous, being careful never to eliminate more than 30% of the canopy each time. Pruning usually takes place during periods of dormancy, best just before vegetation activities resume, but it has been shown that it may also be carried out in certain summer periods (Cannavò et al., 2004). The cut must be clean and leave the collar at the base of the branch. The larger scars (those 4–5 cm in diameter) must be covered with mastic.

   To avoid having to irrigate, it is important to choose land with a good available water capacity, or at least sites capable of satisfying the requirements of walnut. If not, irrigation is indispensible during the months between June and August, because then the plant requires a large amount of water to produce most of its annual vegetative mass (leaves, fruit, branches) in the same period that water *deficit* occurs due to strong evapo-transpiration and scarce precipitation. Water requirements may vary in the summer months, according to Bergognoux and Grospierre (1981), between 70 and 100 mm per month, as a function of plant-age, evapo-transpiration, pluviometric tendencies, and the available water capacity of the land. Research conducted by Dettori et al. (1997) on walnut plantations in the hillside-mountainous areas of central-north Sardinia demonstrates that water deficit represents the principal limiting factor for development; therefore, it is necessary to irrigate the plantations with an annual contribution of 200–300 mm from May to September during the first 5 years.

   In high density plantations, the plants may enter into competition very early, so it is necessary to reduce density in order to prevent a perceptible decrease in growth. The degree of stand density may be expressed by the crown competition factor (CCF). For each 10 unit increase of CCF superior to 100, the growth rate decreases by 4–5% per year (Schlesinger, 1987). Based upon this criterion, Frattegiani and Mercurio (1991) have supplied the guidelines for primary thinning. More precisely, when the CCF factor reaches 110, it is necessary to remove about 30–35% of the plants. Thinning should be done with selective criteria, first by eliminating the badly shaped, damaged, or diseased plants, then by favoring the best specimens and identifying plants that show good prospects, i.e., those with a straight trunk, no defects and good development, which might constitute the high quality part of the final harvest.

24.1.7. Pests and Diseases

The principal dangers to timber-producing plantations come from the xylophagous lepidopterans, for example, the *Zeuzera pyrina* L. (leopard moth) and the *Cossus cossus* L. (goat moth). The larvae of the leopard moth attack walnut mostly in the first years, boring galleries inside twigs and trunk. Good preventive measures would be to avoid using land with water deficiencies and avoid planting near fruit or poplar plantations. Direct pest control can be efficiently carried out with traditional methods, including those that entail killing the larvae that have already penetrated inside with specific chemical products. In fact, the use of pheromones and techniques of mass capture and of sexual confusion do not always work. The goat moth attacks walnut plants more than 15 years old and the larvae bore deep, twisting galleries in the wood.

   Among the pathogen fungi are those that cause root and collar rot (*Phytophthora cactorum* —Leb.et Cohn- Schroter) and white rot root disease (*Armillaria mellea*—Vahl.: Fr. Kummer) (Luisi and Magnano, 1997). Bacterial blight, anthracnosis, and white leaf-spots (Anselmi, 2001) are specific diseases that may be damaging to timber production.

24.2. The Wild Cherry (*Prunus avium* L.)

24.2.1. Botanical Classification and Principal Botanical Characteristics

The *Prunus* genus is subdivided into subgenera: the sour cherry belongs to the *Cerasus* subgenus, while the late cherries or bird cherries belong to the *Padus* subgenus. The subgenus

Cerasus includes, among the European tree species, in addition to the wild cherry or sweet cherry (*Prunus avium* L.), two other fruit-bearing species: morello, visciolo, or sour cherry (*Prunus cerarus* L.) and the St.Lucie or Mahaleb cherry (*Prunus mahaleb* L.). The wild cherry may reach a height of 20–25 m and a diameter of 70–80 cm.

It is a tree with a long life span, 90–100 years. The trunk, with good apical dominance, is straight and cylindrical. The principal characteristic of the branching of cherry trees is their tendency to group the branches in pseudoverticils. The crown, especially in isolated specimens, tends to be of an expanded conical type. The bark is smooth, thin, more or less shiny, reddish-grey, with many lenticels arranged in small horizontal lines. As the tree grows in diameter, first the bark flakes, forming strips along the circumference that roll up, then with aging longitudinal cracks form. Twigs are smooth and glabrous. Buds are arranged in groups on brachioblasts and in a spiral along the twigs. They are ovoid, pointed, dark reddish-brown and glabrous.

The leaves are alternate, placed at the base of the lateral buds and pendulous. They have a denticulate blade and generally are obovate-lanceolate in shape and acuminate at the tip, with dimensions of 4–8×10–20 cm. The upper surface of the leaf is shiny and dark green, the underside is light green and slightly pubescent. The stem is 2–4 cm long with two reddish glands near the base of the upper surface of the leaf. The foliage develops immediately after flowering, between March and April. The hermaphrodite flowers are made up of five white petals grouped in a calyx, on which are inserted many stamen. They have pedicels 3–5 cm long and are grouped at the base of the brachioblasts in two to five element umbels. Flowering may take place between the end of February and the end of March. The wild cherry is a self-incompatible species, with entomophilous pollination, generally carried out by bees or bumblebees (Ducci, 2005). The fruits are rounded drupes, with a meaty mesocarp and woody endocarp, and are varying degrees of red when mature. The fruits reach maturity between May and August. The seed is made up of a thin tegument that adheres to a unicellular layer of endosperm that, in turn, surrounds the embryo.

Order : ROSALES
Family : *ROSACEAE*
Genus : *PRUNUS*
Species : *Prunus avium* L. (Wild Cherry)

Plantula with convex cotyledons, ovate, sessile, with pedunculate glands. Primary leaves always have paired glands, serrate margin, and two thread-like stipules. The root system in the seedlings and in the saplings is of a taproot type. Adult specimens produce superficial horizontal rooting from which may develop numerous propagative roots that may develop basal shoots.

24.2.2. Varieties

In the sphere of the *Prunus avium*, among the species of major importance for the production of timber there are the following varieties: *sylvestris* Dierb. (= *avium* L.), *juliana* L., *duracina* L. (Pignatti, 1982).

24.2.3. Propagation

Propagation materials for this species were placed under regulation by the European Policy 105/1999/CE and by DL 386/2003. In this regard, areas have been identified in Italy where seeds should be harvested (Ducci et al., 1988; Ducci and Gras, 1998; Siniscalco, 1997; Calvo et al., 2001; Ducci, 2005).

Planting is usually carried out by using 1-year-old seedlings that have been grown bare root in the open field or in boxes. The quality standards for cultivation referring to arboriculture for timber production include those with a diameter at the collar of > 9 mm and a height of > 80 cm, a hypso-diametric ratio (h/d) of between 70 and 90, besides fulfilling the other morphological and sanitary quality requisites (Calvo and D'Ambrosi, 1995; Falleri and Giannini, 2003). The wild cherry may be propagated by micropropagation, which allows the identification of some clones with excellent wood technological and growth characteristics in the selection phase (Ducci and Santi, 1997; Ducci et al., 2005). Logiurato et al. (1997) observed different behaviours regarding vulnerability to xilematic embolism in some of the clones from the Veneto and Marches areas.

24.2.4. Climatic Requirements

The wild cherry is a species with high sunlight requirements. It prefers sites with a mean annual precipitation of 800–1,600 mm, 100–150 mm during the summer season, when it cannot tolerate drought periods longer than 2 months; its drought resistance is superior to that of the sycamore or the common ash and is comparable to that of the sessile oak. It vegetates in sites with a mean annual temperature between 8°C and 15°C and is able to resist low temperatures (down to –20/25°C).

It may be grown in a wide area including the cold subzones of the *Lauretum* and the warm subzone of the *Fagetum* according to the Pavari phyto-climatic classification.

24.2.5. Cultivation Techniques

Cultural Models

So far the cultural models that have been identified are pure plantations, mixed plantations and row plantations (Mercurio and Minotta, 2000).

Pure Plantations. Planting distance varies from 4×4 to 6×6 m (625 and 278 plants per hectare) if the seedlings normally found on sale are used (AA.VV., 1996). Initial density may be lower (e.g., 204 plants per hectare, 77 m) if selected clones are used (Ducci et al., 1990). Pure plantations should not be carried out on a large scale (not over 5 ha), especially if using materials originated by cloning. In the latter case, multi-clone plantations may be created (mosaic plantations with different clones) to increase genetic variation and raise the resilience of the stand (Ducci and Santi, 1997).

Mixed Plantations. Either by single tree or by group (mosaic), consociation may be carried out, for example, with the common walnut, sessile oak, ash (*Fraxinus excelsior* and *Fraxinus angustifolia*) service tree, sycamore and/or the Neapolitan alder (nitrogen-fixing). In plantations mixed by single tree, a density of 150–200 plants per hectare may be planned (main species 8×8 or 7×7 m), with a distance of 3.5–4 m between the main species and the companion species.

Row Plantations. The seedling trees may be planted at a distance of 6–7 m from each other.

Management and Cultivation

Individual protection (shelters, mesh guards) is useful to safeguard the saplings from damage by animals or by machines. Tests by Andiloro et al. (2000) in Calabria showed that, 4 years from planting, 120 cm high shelters caused development of trunks with a strong hypso-diametrical disproportion. For this reason, the authors advise the use of mesh guards even if the plants show a minor development in height. Dupraz and Bergez (1999) demonstrated that in non-ventilated shelters the reduced photosynthesis rates are due to high temperatures and the scarce radiation energy inside.

Mulching, particularly with plastic sheets, should be done only on well-drained land in order to prevent symptoms of root asphyxia. Minotta et al. (1997) demonstrated the unfavourable effects on the growth of the cherry tree by not cutting the turf, especially when the growth is predominantly herbaceous. In the first 2–3 years, irrigation is indispensible during long drought periods. Trees must be irrigated two or three times over the period between July and September. Drip irrigation tests conducted in France by Belvaux and Tron (1996) proved, after the first 3 years, the positive effects on growth by using a drip-rate of 4 L/h for each plant with two 3 h irrigations per week, instead of using a drip-rate of 2 L/h. For the mountainous-hillside areas of Sardinia, Dettori et al. (1997) believe it necessary to enhance the water supply by 60–120 mm, May through September, for the first 5 years.

In the first 3–5 years from planting, it is preferable to till the soil around single plants (or use chemical herbicide) rather than bring machinery close to the trunks, because the root system and the trunks are susceptible to injury; however, work on the land between the rows may be done by machinery.

The plants should be pruned in the 2nd year from planting to remove damaged or excessively developed branches and double tips (Boulet-Gercourt, 1997). Cuts should involve small branches (2–4 cm in diameter). Afterwards, from the 4th year on, improvement pruning or crown raising must be started to progressively raise the height of the crown, up to 4–6 m. This last operation is usually carried out only on plants that will reach completion of the production cycle. Baldini et al. (1997) found that use of a hand saw instead of various pruning shears made for a faster healing time. The height of the pruning should not be more than 40% of the overall height. The best time for pruning is traditionally in the autumn or the end of winter, but research by Fournier (1989) points to the possibility of also operating during periods of vegetative activity, at the end of the second growth phase (end of July), to limit the emission of buds and permit the healing of wounds within the year. Calvo et al. (2001) believe that the best period for northern Italy is between July 15 and August 7. For Montero et al. (2003) summer pruning is a must, because winter pruning favours bacterial blight. The general criteria to be followed for correct pruning are moderation and gradual intervention, in order to maintain a harmonious equilibrium between the epigeous and hypogeous parts (Mercurio and Tocci, 1982). The need to prevent disease, such as gummosis, and the particular conformation of the crown with distribution of ramifications in "false verticils" (branches inserted in the trunk at the same level) make it difficult to obey the general criteria for pruning. "False verticils" must be eliminated immediately in two successive stages, when the branches are still small. There is more work in pruning pure plantations than in pruning plantations that are mixed by single tree where the companion species, placed at close distance, limits the diametrical growth of the branches.

In pure plantations planted at a relatively high density, thinning is an indispensible operation, because cherry grows rapidly in the first years, has high sunlight requirements and hates competition. Thinnings that are early, selective, moderate and gradual (every 5–7 years) are necessary to leave the crowns completely free and encourage sustained and regular diametrical growth. The cultivation cycle should end before 50–60 years, considering the poor longevity of the species. Depending on site fertility and cultivation techniques, cycles varying from 25 to 40 years are to be expected.

24.2.6. Pests and Diseases

The wild cherry is often affected by root rot caused by *Rosellinia necatrix* (Hart.) Berl., *Phytophthora* sp. and *Armillaria mellea* (Vahl. Fries) Kummer, which causes general weakening, stunted vegetation, chlorosis and progressive wilting of leaves, and eventual death of the plant (Santini and Turchetti, 2005). The most important specific diseases are the agents of gummosis (*Sporocadus carpophilus*—Lev.- V. Arx. =*Coryneum beijerinckii* Oud.) and most of all cylindrosporiosis (*Cylindrosporium padi*—Lib.- Karst), while moniliosis and bacteriosis seem less terrible (Motta et al., 1997; Anselmi, 2001).

More sporadically there may be attacks by xylophagous insects that may cause serious alterations in the trunk: *Cossus cossus* L., *Zeuzera pyrina* L. (leopard moth) and *Capnodis tenebrionis* L. Black aphid, *Myzus cerasi* F., is very common and is responsible for damage to leaves and buds (Francardi and Roversi, 1997, 2005). The wild cherry may be subject to intense defoliation by various insects (e.g., Limantria, Euprottide, Infantria).

24.3. Land Capability for Cultivation of Walnut and Wild Cherry

24.3.1. The Predominant Geographical Areas

The diffusion of arboriculture for wood production of these species on national territories must be referred to two different geographical areas and consequentially distinct morphopedological characteristics. There are (1) plains and alluvial valley floors and (2) hillsides.

The set of local internal characteristics (of the soil) and external characteristics (site morpho-geographic and climatic) assumes a different importance for each of these areas as factors that might limit cultivation growth. It is for this reason that often it is preferable to talk of "limitations" instead of "capability", since the studies carried out so far have highlighted the former over a real identification of characteristics that can "favour" growth. The inverse correspondence between "limiting" factors and maximum capability, in fact, is not always confirmed, since the absence of limitations does not identify unequivocal situations regarding water and nutrition availability during the season cycle.

24.3.2. Plains and Alluvial and/or Lacustrine Valley Floors

Such areas are characterized by soils evolved on alluvial/fluvial deposits of different degrees of energy, in level zones and, in the case of internal valley floors and lacustrine basins, with a topographical layout predisposed to strong local climatic interference (fog and/or frost).

The external characteristics of the site connected to such environments are the presence of flooding areas and/or areas with strong laminar erosion. The fundamental internal characteristics of the soils connected with these types of environments are extremely fine or coarse textures (clayey and/or sandy), surface stoniness and rock fragments, the presence of expandable clays (vertic soils), the presence of organic (histic) horizons, and the presence of water tables during the year (aquic soil water regime). The chemical characteristics (particularly pH and active lime) are often strongly related to and influenced by the above-mentioned physic-hydraulic and mineralogical factors.

24.3.3. Hillsides

Such areas are characterized by soils evolved on incoherent deposits of origins that may be marine and/or fluviolacustrine and/or fluvioglacial and/or volcanic of the Miocene, Plio-Pleistocene and Quaternary-Holocene, and furthermore, in a secondary way, on the silicious, calcareous, or intermediate hillside lithoidal formations, such as torbidites (Flysch), or calcareous marl and clayey calcareous, and limestones of various origin, massive or stratified. The site morpho-geographic functional characteristics that distinguish these areas are the presence of erosion phenomena both laminar and channeled between moderate and strong (e.g., calanchive areas), mass phenomena (surface landslides or soil-slips), local climatic interference connected with the quota or with dominant winds near the peaks of the reliefs, and stone outcroppings (only on the lithoidal formations). Regarding internal characteristics there are extremely fine or coarse textures (clayey and/or sandy), surface stoniness and rock fragments, the presence of rooting limitations due to hard and impenetrable horizons, the presence of expandable clays (vertic soils), and a strongly xeric soil water regime (long periods of water deficit during the summer months).

24.3.4. Functional Factors

The Common Walnut

Since walnut has a root system armed with taproot, sometimes the depth that the root reaches is limited by unsuitable land characteristics, such as shallow rooting depth, insufficient available oxygen supply and the presence of limiting horizons. It is not resistant to long periods of water saturation (Mappelli, 1995). It must be noted that a high water table may cause dark coloration of the timber, which devaluates the stocks. The root system does not develop in zones of the land with a high water table because of anaerobic conditions (Ramos, 1985); therefore, even in deep soils the level of the phreatic surface limits the extension possibilities of the root system and therefore the uptake of water and nutritive elements. The level of the water table should be between 1–1.5 m deep. In any case, lands with strong oscillations of the water table are to be avoided.

The lands most suitable to walnut are those usually defined as "loamy" or medium-textured (loam, silt loam, silt, and clay loam). These are soils with a content of clay under 25%, of silt between 30 and 50%, and of sand between 30 and 50%. A certain percentage of clay is necessary for good water retention. Frattegiani et al. (1996) have found a positive correlation between increase of tree-base metric area and a clay content ranging between 5 and 35%. However, very high clay content causes drainage difficulties and limits soil aeration, therefore the higher the rainfall, the less a high clay content can be tolerated by walnut. On the other hand, soils that are predominantly sandy or have high levels of coarse sand, even if not characterized by problems of available oxygen supply, drain too rapidly and have a low content of nutritive elements. If the insufficient capacity to retain water is not compensated by a fairly rainy climate or frequent irrigation, the growth of walnut is greatly compromised.

According to Becquey (1997), optimal conditions are present when the pH is between 6.5 and 7.5. Plantations on soils with pH superior to 8–8.5 should be avoided, since symptoms of ferric chlorosis may appear.

Unfavourable conditions were found in France in soils that were poorly structured, compacted, hydromorphic (pseudogley phenomena, corresponding to stagnic and gleyic properties according to WRB, IUSS, 2006) and soil horizons on calcareous rocks (rendzina, Bequey, 1991), with scarce available water supply (Renzic Leptosols according to WRB, 2006).

The Wild Cherry

Cherry is relatively adaptable. Deleporte (1977) and Nicot (1983) do not exclude the existence of differentiated ecotypes most of all with respect to the climatic characteristics of the substrate.

It is found on sandy, granitic, calcareous, or clayey substrata. Active lime in the soil profile, even if it does not represent an obstacle to the vegetation of the cherry, does constitute a factor limiting growth (Bosshardt, 1985). In any case, it should not exceed 7%. Cherry finds sufficiently suitable conditions in brown calcareous soils, brown calcic soils, or brown soils that are more or less leached, but not exceedingly acid, even if there are periods of water deficit (Lanier, 1986). On the other hand, if soils are shallow, good timber products are not achieved (Masset, 1979). According to Franc et al. (1992), cherry adapts to soils with a high level of clay content. There are many examples of good plantations in the Mediterranean area (Mercurio, 2003) on clayey soils, albeit lands improved by planned intervention. However, Masset (1979) affirms that on compacted soils the timber may take on "green-vein" defects. The best conditions are undoubtedly found in soils with a sandy silt or sandy clay texture that are deep and well structured, have a good available water supply and available oxygen supply, and are neutral or slightly acid (Scohy, 1989; Loewe et al., 2001). According to Crave (1989), good growth can be achieved also on sandy materials with a pH of around 4.5. For Montero et al. (2003), the best conditions are a pH of between 4.5 and 6 (7.5).

24.3.5. Evaluation Methods Based on Geo-topographical and Functional Land Characteristics

Since land and soil capability for the growth of walnut and cherry is strongly related to the local situation regarding both the predominant geographic area and the soil typologies and associated limiting factors, on the basis of research carried out nationwide it is very difficult, if not impossible, to formulate a single evaluation chart valid for the whole national territory. Furthermore, in many cases a joint study of the relationships between cultivation growth and land/soil characteristics was carried out for the two species together, so as to unify the functional parameters of the charts and the relative limiting thresholds, or at least to have little difference in terms of threshold values or threshold classes considered. For this reason, the following paragraphs show the tables elaborated within the principal research experiences carried out in Italy at a local level.

24.3.6. Research in the Region Emilia Romagna: Pedological Limitations to Growth of Walnut and Wild Cherry for Timber[1]

Introduction

In the past few decades in Emilia Romagna, broadleaf plantations were established over an area of 5,000–6,000 ha using public funds (e.g., Reg. CEE n. 2080/92), at least half with valuable timber species (i.e., common walnut and wild cherry). From 1994 onward, research was carried out on the edaphic requirements of forest species (Regione Emilia Romagna et al., 1995, 1996, 1998, 2002a, b), considering that, in addition to the properly management techniques, it is also necessary to use suitable lands to have satisfactory results with timber plantations. The work method used was based on the involvement of experts in arboriculture for timber and on the definition of evaluation tables that correlate the edaphic requirements of the plants with growth potential classes (Tables 24.1 and 24.2).

Evaluation Charts

The evaluation takes into consideration the soil characteristics capable of directly influencing plant growth and does not consider climatic characteristics and those characteristics (generally external to the soil) that may affect management operations and the risks of land and environmental deterioration; furthermore, they do not take into consideration the interaction among these characteristics. They contain a descriptive synthesis of the edaphic requirements of the plants and represent a "transparent" and "shareable" instrument for the creation of application maps deriving from the Land Maps.

In Emilia Romagna, evaluation charts have been created regarding the principal forest species that are suitable for use in the region for both timber production and the constitution of permanent forests. They may be found in www.gias.net in the *Guida al catalogo dei suoli della pianura emiliano romagnola* (Guide to soil classification of the Emilia Romagna Plain).

The evaluation charts were developed by a working group made up of experts and specialists who operate in the region; the evaluations were made by referring to research done in the region in the forestry sector and by using indications deduced from Italian and international literature. The interdisciplinary working group that participated in the creation of the evaluation charts included: expert pedologists (ITER, RER), expert agronomists (RER), experts in forestry (RER) and in arboriculture (Universities of Turin and Bologna, *Experimental Institute for the Poplar Cultivation*—CRA), specialists of Province Services and of Reclamation Unions. It must be said that, in the territories of the Emilia Romagna plain, there are few walnut and wild cherry timber plantations that have completed a production cycle. Furthermore, research concerning the response of different forest species to different soil characteristics, for example, salinity, active lime, pH, and hydraulic behaviour, is very scarce in this region. For this reason, in formulating the evaluation reference was made to national and international literature, which refers to environments relatively different from those of the Po River Valley. Furthermore, plants of the same species may show different levels of adaptation to soil characteristics due to genetic characteristics (e.g., origins, genera, clones) and, for many species, appropriate knowledge is still lacking.

---

[1]Supervised by: Carla Scotti, I.TER soc.coop, Bologna; Gianfranco Minotta, Dep. AGROSELVITER, Università degli Studi di Torino.

Within the above-mentioned projects, for the integration and partial verification of information gathered from literature, some plantations whose soil profile had been described were visited during the course of the years. These plantations were identified using the initial results of the "Survey of non-fruit-bearing tree plantations (excluding poplars)" carried out by the Department of Arboreal Cultivation of the University of Bologna commissioned by the Park and Forest Service of Emilia Romagna, using information from the Land Maps of the regional plains on a scale of 1:50,000 and the regional land maps on a scale of 1:250,000 of the Geologic, Seismic, and Soil Service of the Region of Emilia Romagna, the pedological characteristics of 10 plantations of the Reno Palata Reclamation Consortium (Region Emilia-Romagna et al., 1998), and the characterization of 30 plantations distributed in the plains and hillsides of the region (Region Emilia-Romagna et al., 2002b).

The threshold values used in these evaluations were therefore "fine-tuned" by the working group during meetings in the assembly room and in the field, and should be considered "guide values" to be used to identify possible limitations to the growth of forest species. Therefore, it is emphasized that these evaluations are susceptible to future improvement and deeper investigation on a local level to highlight the pedological characteristics capable of significantly affecting the growth of arboreal species grown for timber production. Finally, it must be specified that the good outcome of plantations, even in the cases where the species are perfectly suitable to the soil type, is still related to the correct application of the opportune agronomic techniques (e.g., site preparation, weed control, pruning).

The pedological characteristics taken into consideration in the evaluation are texture, available rooting depth, available oxygen supply, chemical response (pH), active lime, salinity, and flooding risk. To each of these pedological characteristics is given an indicator of importance that synthetically expresses the degree of influence (indicated by number of asterisks) of these characteristics on the growth of the forest species under examination.

To evaluate the degree of intensity of the limitations of the single pedological characteristics the following depths were examined:

- 100 cm for texture taking into consideration only the texture classes that characterize horizons with a thickness of at least 50 cm, except for the sandy texture classes for which a thickness of at least 30 cm is sufficient;
- 60 cm for chemical response (pH);
- 60 cm for active lime;
- 120 cm for salinity for all of the forest species except for poplars (clones) for which the first 80 cm are considered.

The evaluation charts (Tables 24.1 and 24.2) include three classes of limitation intensity and they refer to lands managed according to sustainable agronomy criteria, through the following interpretation:

- *Absent or slight limitations:* The lands present no limitation or are suitable to host timber species, favouring full achievement of the quality/quantity production potential; lands may be cultivated with ordinary techniques and do not require specific operations to improve the natural potential of soils.
- *Moderate limitations:* The lands present some limiting factors that require agronomic corrective operations to achieve the full quality/quantity production potential of the timber species.

- *Severe limitations:* Lands present severely limiting factors to arboriculture for timber; any agronomic corrective operations may be too expensive or insufficient to recover the full quality/quantity production potential of the timber species.

Table 24.1. Evaluation of the pedological limitations to growth of the wild cherry for timber production.

Pedological characteristics	Degree of importance	Intensity of limitations		
		Absent or slight	Moderate	Severe
Altitude (m asl)	**	–	–	> 800
Slope (%)	***	< 15	15–30	> 30
Texture	**	medium, moderately coarse, moderately fine	–	coarse, fine
Available rooting depth (cm)	***	> 100	50–100	< 50
Available oxygen supply	***	good		moderate, imperfect, poor, very poor
Chemical response (pH)	*	6.5–8.5	5.4–6.4	< 5.4; > 8.5
Salinity (EC 1:5 dS/m)	***	< 0.4	0.4–0.8	> 0.8
Active lime	*	< 7	7–12	> 12
Flooding risk	***	none, rare	occasional	frequent
Flooding risk (duration)	***	< 48 h	2–7 d	> 7 d

Table 24.2. Evaluation of the pedological limitations to growth of walnut for timber production.

Pedological characteristics	Degree of importance	Intensity of limitations		
		Absent or slight	Moderate	Severe
Altitude (m asl)	**	< 400	400–700	> 700
Slope (%)	***	< 15	15–30	> 30
Texture	**	medium, moderately coarse, moderately fine	–	coarse, fine
Available rooting depth (cm)	***	> 100	50–100	< 50
Available oxygen supply	***	good		moderate, imperfect, poor, very poor
Chemical response (pH)	*	5.5–8.2	4.5–5.5	< 4.5; > 8.2
Salinity (EC 1:5 dS/m)	***	< 0.4	0.4–0.8	> 0.8
Active lime	*	< 7	7–12	> 12
Flooding risk	***	none, rare, occasional	–	frequent
Flooding risk (duration)	***	< 48 h	2–7 d	> 7 d

Here follow the principal considerations regarding the single land characteristics and their potential influence on plant growth.

- Altitude, degree of importance. ** These are orientational indications to be evaluated along with other site characteristics such as exposure of the plantation.
- Slope, degree of importance . *** Slope affects planting and difficulty of work as well as possible erosion processes; when the land is sloping it is necessary to use operations and layouts that favour correct water control through drainage ditches.
- Texture, degree of importance. ** Texture certainly influences the growth of walnut and cherry; in some cases correct cultivation techniques, such as the suitable choice of moisture conditions in which the land should be cultivated, can mitigate the limiting effects, for example, of clay soils. It was decided to take into consideration texture to a

depth of 100 cm, because optimal growth of forest species requires a good rooting depth. It was agreed that sharp changes in texture limit rooting depth since the roots tend to remain in the most fertile zones (unless there is a water supply under the sand) and change the evaluation of available rooting depth to 60–70 cm, which determines an evaluation on the whole of a moderate limitation to the growth of walnut and cherry for timber production.

- Available rooting depth, degree of importance. *** The degree of importance is high considering that, for optimal growth, a good rooting depth is fundamental in the forest species and must be balanced with the growth of the aerial part of the plant.
- Available oxygen supply, degree of importance. *** This is one of the most important factors in the evaluation of factors limiting plant growth; any problems connected with the availability of oxygen can be improved with targeted operations such as double layer tilling and aiding correct water management by reshaping the field surface and digging drainage ditches.
- Chemical response (pH), degree of importance. * In Emilia Romagna, agricultural soils typically have a pH varying between 7.8 and 8.2, which in literature are usually considered moderately limiting values. The group took into account the local pH characteristics in the definition of the threshold values used in the evaluation.
- Active lime, degree of importance. * There is no local experimental data about the actual influence of active lime on the growth of walnut and cherry for timber production. The definition of threshold values was based on reference literature.
- Salinity, degree of importance. *** Considering that in lands with salinity problems forest species do not naturally grow, it was given maximum importance.
- Flooding risk, degree of importance *** In the evaluation charts the duration intervals of flooding risks corresponding to different limitation intensities are there as indicators, but it must be stressed that this characteristic is not described for each soil type since it is not at present possible to give an overall definition, due to the local topographical variability and the scarce information available. For this reason, accurate research is necessary in order to estimate this chamracteristic.

In the evaluation the parameter "water deficit" was not specifically considered since the attribution is complex; furthermore, this parameter is in part connected to other land characteristics, such as texture and available rooting depth.

Plains and Mountain Hillsides

There are undoubtedly environmental differences between the plain, the hillside and the mountain that also influence the growth of walnut and cherry for timber production. From the research carried out in Emilia Romagna it emerged that the increase in diameter is usually greater in the plain than in the hillside for both walnut and cherry, more so in the case of walnut. This confirms the plain as the environment most suitable to achieving the full production potential of the broad-leaved species taken into consideration, which notably gain from the deep soils typical of the alluvial plains. Another characteristic of plain lands is the presence (or not) of hypodermic ground water, in other words a perched water table that may lie within 2 or 3 m in depth. Deep soils with medium textures, with no problems of water drainage within 90 cm and a perched water table fluctuating from 100 to 130 cm, are capable of supplying enough water for the growth of walnut and cherry for timber production.

In the hillside and mountainside areas, agricultural activities encounter more difficulties compared to those on plain lands; furthermore, often timber plantations are located in lots far from the farming centre and are therefore hard to reach, making timely cultivation measures difficult with consequent repercussions on plant development. In the hillside and mountainside areas, in addition to the slope and the altitude, the factors that limit the growth of walnut and cherry are the available rooting depth, which may be affected by the presence of rocky layers that, in addition to blocking root expansion, do not assist the soil capacity to retain water, with a consequent water deficit risk for the plants.

24.3.7. Research in Central Italy: a Study of the Region Abruzzo

Research for the Region Abruzzo, particularly in the forestry sector, was carried out by a multi-disciplinary group made up of soil and arboriculture experts coordinated by the Academy of Forestry Science. The goal of this study was to formulate a research method valid for the Abruzzo territory that could then be used by the Regional Administration as a basis for management criteria regarding future funding in the sector. Along these lines the land capability cartography was formulated for the arboriculture of hardwood timber in two pilot areas at a semi-retail level, on the basis of which work methods for research at a parcel level were identified.

Pilot Areas

The testing areas in which research was carried out at a semi-retail level were: (1) Castilenti Township (TE), representing the Periadriatic Zone with hillside reliefs and sandy clay loam Plio-Pleistocene sediments (Fig. 24.1); (2) S. Demetrio né Vestini Township (AQ), representing the intermountain basins with colluvial-alluvial and karstic dissolution sediments (Fig. 24.2). In these areas, a series of walnut and cherry plantations were described at a pedological level, where a quantity of dendometric research data was gathered regarding a pluriannual time span (Figs. 24.3, 24.4, 24.5).

Methodological Results

For the creation of the Land Unit Map the areas suitable and not suitable to the arboriculture of hardwood timber were selected first of all by excluding areas characterized by slope > 35%, areas subject to diffuse rill and gully water erosion, gully areas, areas subject to the normal wanderings of watercourses, unstable areas subject to solifluxion, creeping mass erosion (e.g., landslides), and areas with commonly acknowledged pedological limitations (e.g., humid zones, high water table, areas with surface stoniness) (see Figures 24.3 and 24.4).

At a parcel level, more detailed pedological research is directed to gaining knowledge of the functional characteristics of lands, to show the limiting factors not reported in preceding analyses for reasons of scale and detail. The characteristics of lands and sites are described following the guidelines contained in the Soil Survey Manual (Gardin et al., 2004).

At least two soil horizons were sampled; in any case, the primary considerations were the presence of cultivated horizons, more significant horizons, and horizons affected by drainage or rooting limitations. Chemical-physical analyses were made on these samples, such as texture, pH (H_2O 1:2.5), total calcium, active lime, and organic matter content, by following the guidelines of Methods for Chemical Analysis of Soils (Mi.P.A.F., 2000).

Fig. 24.1. View of walnut plantation in Castilenti Hill area. The presence of strong hydric erosion features (the "calanchi") represent a strong constraint on correct planning of walnut plantation.

Fig. 24.2. Karst depression with walnut stand in the S. Demetrio de' Vestini area.

The topo-orographic reference parameters believed to be most significant, which in this phase seemed to exclude the possibility of creating arboriculture plantations for hardwood timber, were formulated by a calibration based on local pedological characteristics of the pedo-landscapes examined. In these areas the cultivation response of the test species was controlled in relation to the functional and qualitative characteristics of soils, which according to recent knowledge are considered "limiting". Such a proposal represents a delimitation that does not begin to define a detailed range of optimal characteristics for the various species. Table 24.3 should therefore be considered indicative for the characteristics limiting to the species examined. The limitations have a reference value considering the limited autoecological knowledge of the species examined. Direct observation will determine the definitive choice.

Table 24.3. Evaluation of site pedological and morphological limitations to the growth of walnut and cherry in central Italy (Region Abruzzo) for plantations in hillside margins, in colluvial-alluvial valley floors and terraces.

Site pedological and morphological characteristics	Degree of importance	Intensity of limitations		
		Absent or slight	Moderate	Severe
Distance from crest	**	–	< 30 m	–
Pedoclimate (ST-USDA moisture regime)	*	Udic-Ustic	Xeric	Aquic
Available rooting depth (cm)	***	> 100	70–100	< 70
Internal drainage	***	Well drained	Somewhat poorly or excessively drained	Excessively drained, poorly drained, very poorly drained
Permanent water table (in cm from field surface)	***	> 150	70–150	< 70
Texture class (USDA)	***	L, Sl, Scl, Sil, Sicl, Cl	S, Ls, Sc, Sic, C[a]	C[b], Si
Surface stoniness[c] (%)	***	< 35	35–50	> 50
Percentage on surface of stone outcropping or below limit of rockiness definition (< 50 cm)	***	0	1–2	> 2
Chemical response (pH)	***	6.5–7.5	5-6.5; 7.5–8.5	< 5; > 8.5
Active lime	**	< 7	7–10	> 10
Salinity (mS/cm)	**	< 0.15	0.15–0.4	> 0.4
Flooding risk	***	none, rare, occasional	–	frequent
Flooding risk (duration)	***	< 48 h	2–7 d	> 7 d

[a] *With prevalence of mineralogic clays of the illite group (1:1).*

[b] *Vertic characteristics (presence of slickensides and dynamic cracks > 1 cm wide and > 50 cm deep that open and close with the seasons.*

[c] *Represented by rocks > 7.5 cm average diameter (up to 25 cm).*

Fig. 24.3. Common walnut and wild cherry mixed plantation 10 years old. S. Demetrio de Vestini, Abruzzo Region.

24.3.8. Land Capability Cartography in Piemonte Region[1]

Methodology

The research methodology most suitable to the creation of land capability maps on a scale of 1:50,000 must guarantee the possibility of recovering and extending all of the preexisting information. The methodology used makes reference to the land classification used by the Australian authors (CSIRO) that entails land testing of an integrated type.

It is an interdisciplinary methodical approach: in the "land" concept are included, in addition to soils, the principal environmental factors that most influence the territorial fabric, i.e., geomorphology, lithology, climate, vegetation, and anthropic activities.

Collection and Preparation of Basic Material

First of all, careful literary research was carried out to define the pedo-environmental requirements of the spontaneous wild cherry. Among the documents consulted, precious information about the ecological requirements of this species was found by following the French Forestry specifications (Buolet-Gercourt, 1997) that define precisely what characteristics a site must have to achieve high quality cherry timber production. Furthermore, an important

---

[1] Supervised by Mauro Piazzi - IPLA, Instituto per le Piante da Legno e l'Ambiente, Settore Pedologia, Torino, Italy.

Fig. 24.4. Excellent wild cherry single plant 10 years old. S. Demetrio de Vestini, Abruzzo Region.

contribution to this research came from a preceding work done by the IPLA on the wild cherry (IPLA and Region Piemonte, 1998) suitable to the collection of propagation materials to be used in "forest nurseries", which supplied information about the edaphic and morphological characteristics of many sites in which excellent specimens of spontaneous cherry were present.

Successively three significant portions of the Piemonte territory were chosen in relation to which a pedological map is available, including one plain area and two hillside areas where plantations have recently been concentrated (IPLA and Region Piemonte, 1999, 2000a, b). All of the cartographic units in the study were drawn from these documents.

Fig. 24.5. Common walnut and Neapolitan alder mixed plantation 10 years old. S. Demetrio de Vestini, Abruzzo Region.

Drawing Up of the Land Capability Maps and Caption

Each cartographic unit of the land map may contain one or more pedological types (Phase of Series). Where two or more pedological types are actually present, only the predominant type was considered for the capability evaluation. Data-crossing between predominant soil qualities and morphological characteristics of each unit with the capability tables shown in the following pages made it possible to give each cartographic unit a corresponding capability classification for the cultivation of cherry.

The capability classification makes reference to the FAO outline (1976, 1981), which uses five classes. More specifically to the goals of this paper, the following four were used:

- *Suitable (S1):* includes all those lands with no important limitations to the cultivation of cherry.
- *Moderately suitable (S2):* includes those lands that on the whole have some limitations for the quality cultivation of cherry. The limitations reduce the productivity or the benefits of the plantation. For example, cherry may develop on sandy or gravelly lands, but in drought years stress due to water deficit may occur, which can lead to irregular growth and can be responsible for a decrease in timber quality. In the case of soils with compacted horizons near the surface, the root system of cherry does not develop in depth with several consequences: in years with scarce rainfall the more superficial horizons will dry out rapidly, causing water uptake problems for the root system; vice versa, in years with

abundant rainfall surface moisture stagnation may occur, because the compacted horizons do not permit good drainage of the water to the deeper soil layers. Both these situations tend to weaken the cherry considerably, increasing the possibility of pathogen attacks.

- *Scarcely suitable (S3):* includes lands with limitations that when taken together are important for the cultivation of cherry. For example, on lands with a strong slope there may be problems of soil erosion and water deficit, because the rain water rapidly running off the slopes can erode the surface horizon (especially in arboriculture for timber where the soil goes without the protection of canopies for long periods) and has no time to percolate toward deeper soil layers where it would go to constitute a lasting water reserve. Furthermore, operations of planting, cultivation and cutting are not easy to mechanize and the performance of manual labour is notably reduced by the rough, uneven morphology.

- *Currently not suitable (N):* refers to lands for which the cultivation of cherry is strongly limited or is theoretically possible but not economical. For example, soils with a perched water table may be reclaimed with operations of artificial drainage, but for cultivations of this type, where profits are expected over the long term, it is not economically convenient to execute these operations. The three land capability maps for the cultivation of cherry were drawn up on a scale of 1:50,000. The extension of each area mapped is approximately 15,000 ha and indicatively corresponds to a Table of the C.T.R. (Region Technical Maps) 1:25,000.

In the caption of each map are shown four capability classes with the corresponding soil characteristics that determine classification.

Pedo-environmental Requirements of Cherry in Piemonte Region

Even though cherry grows well on a wide range of soils and in different geomorphological situations, only certain edaphic characteristics allow good growth and sufficient longevity.

Available water supply is a priority, because the soil must stay fresh in the zone explored by the roots during the entire vegetative period. The pedological characteristics that most influence this availability are site morphology and soil depth.

The bases of slopes and often plain areas are the zones in which an available water supply is usually more certain because of the supply of water running off other surfaces. It is therefore possible to find excellent specimens even in sandy soils with a poor capability for water retention if they are situated at the base of a deep valley. Regarding soil depth, the deeper a soil is, the greater possibility it has of storing water useful to root systems. Deep loamy soils, for example, are particularly favourable to cherry, but also well-structured clayey soils without compacted horizons and with sufficient drainage, in which root exploration is possible, ensure a good available water supply. On the other hand, shallow soils with a strong accumulation of calcium carbonates are the cause of plant suffering and very modest growth.

Excess water may create serious problems for cherry: in creating a plantation it is therefore necessary to consider the depth of the water table and its fluctuation over the vegetative period. In fact, in the alluvial zones where there are often problems of hydro-morphing due to the water table, cherry grows only where the soil is sufficiently well drained and the water table is always more than 30 cm deep. In soils where an impermeable layer is present, which allows the formation of a water table in autumn and winter, which then progressively dries in spring and summer, cherry can still grow as long as the saturation does not involve the first 40–50 cm of the soil.

In soils with saturation near the surface, the following problems can occur: susceptibility to pathologies, strong susceptibility to winds due to scarce rooting depth, complete asphyxia of the root system during particularly rainy years due to the total saturation of all soil horizons, and water deficit in very dry summers, since the water table moves lower and away from the root system, which is localized only in the more superficial layers of the horizon.

Another factor limiting the growth of cherry is represented by soils compacted by machinery, in which case asphyxia may manifest itself even without excess water. In such a situation the gaseous exchanges are not efficacious and the water, which percolates very slowly, has little oxygen. Consequently, cultivation operations must be limited on soils with a fragile structure that can easily be compacted under the weight of tractors and other machinery, avoiding activities in conditions of saturated soil.

Regarding the chemical response of the soil, this species has no particular requirements; good specimens of cherry may be found in very acidic soils as well as in the strongly calcareous. However, it is very rare that this happens when the pH goes below 4.5. Therefore, soils that are too acidic do not allow an arboriculture plantation for timber to have good growth, even when the water conditions are satisfactory. As for climate, cherry adapts to the plain, hillside, and piedmont zones, up to 1,000 m asl. It does not particularly fear late/early frosts or heavy snowfall. An available water supply less than 150 mm in the June–August period is a possible limit, particularly if associated with soils lacking a water table or with a coarse texture.

Table 24.4 is the synthetic table of the pedo-environmental characteristics considered significant for assigning capability classification.

The evaluation must be used by considering the pedo-environmental characteristics of each cartographic unit; among these are identified those most limiting to the cultivation of cherry. The corresponding capability class is attributed on the basis of the latter. Example: a unit on an aspect with a slightly inclined surface characterized by silt loam soil, with a rooting depth greater than 50 cm, with no rock fragments, pH above 4.5, without compacted horizons, but with hydromorphic soils between 30 and 50 cm deep, because of this last characteristic rates a low capability even if all of the others would lead to a high capability.

24.3.9. Research in the Region Lombardia Region[1]

Territorial Planning of Timber Plantations: Methods and Goals According to Different Scales

The *Ente Regionale per i Servizi all'Agricoltura e alle Foreste* (Regional Body for Forestry and Agriculture Services) or ERSAF, through several collaborations with Italian and international research institutes, has developed a set of activities with the goals of improving and promoting hardwood tree species in the agriculture and forest milieu. Among the activities undertaken, a network of plantations for timber production was created for experimentation and demonstration in the early 1990s in a normative and operative context antecedent to the introduction of the EEC agricultural support measures. The network was then broadened to include the initiatives financed by European funding.

---

[1] Supervised by Vanna Maria Sale, ERSAF—Dipartimento per I Servizi all'Agricoltura, Milano, Italy.

Table 24.4. Land and soil capability maps for the growth of cherry related to the Region Piemonte. Scale 1:50,000.

Characteristics	Land capability			
	Suitable (S1)	Moderately suitable (S2)	Scarcely suitable (S3)	Currently not suitable(N)
Texture	L, Sil, Sicl, Sl, Scl, Cl	Sic, Si, Sc, Ls	S, C	
Rooting depth	> 50 cm	> 50 cm	< 50 cm	
Rock fragments*	absent to common (0–15%)	frequent (15.1–35%) but not on surface	frequent (15.1–35%) also on surface	abundant to very abundant (> 35%)
Hydromorphic soils	none	present between 50 and 100 cm depth	present between 50 and 30 cm depth	present within 30 cm depth
Chemical response (pH)	> 4.5	> 4.5	< 4.5	
Compacted horizons	none	present between 50 and 100 cm depth		
Morphology	plain, valley floor, aspect with slight slope (5–10°)	aspect with medium slope (11–20°)	aspect with high slope (> 20°)	

* Includes particles with a dimension of > 2 mm in the soil profile.

The regional network has evolved and broadened over time. It encompasses first generation plantations, characterized by simple cultivation models and located in marginal sites and in plain areas, second generation plantations with more sophisticated and rich models in highly fertile plain areas; and third generation plantations using even more complex models and introducing innovative species such as wild pear and wild service tree. In the earliest plantations, initial thinning out has already been done and the later plantations are still in the creation phase. In all the plantations with the regional network of arboriculture for timber, ERSAF has continuously monitored and surveyed their activity.

The publication of results as they are obtained, a goal to which the resources of the institute have been oriented in the past few years, ensures a conspicuous flow of information available to those interested in the subject. The publication of the manual (A.R.F. and Region Lombardia, 2001) showing the techniques advised for the planning and creation of plantations demonstrates this commitment. The manual contains precise indications for the site characterizations necessary and, among other things, instructs users on how to read and interpret pedological and capability maps for the synthetic evaluation of the possible types of use (silviculture, arboriculture for timber, or the absence of suitable parameters for the growth of the species) for walnut, cherry and other arboreal species in different pedo-environments of the regions.

The methodology proposed by the manual for evaluating the suitability of soils for timber arboriculture is based on a point system. The points assigned are directly proportional to the limitations present, for the pedological or site parameters potentially influential for arboreal growth, according to the evaluation in Table 24.5. The total score (sum of the points of the single parameters) determines the capability class; lower scores are better.

The model uses five capability classes. The first three classes (score ≤ 70) are suitable for arboriculture with limitations between none and strong; classes I and II are suitable for all of the species, while class III is only suitable for species adaptable to the limitations present. The other two classes (score > 70) are on the whole not suitable for arboriculture; class IV shows limitations so strong that they compromise the adequate growth of all species, unless considerable correctional operations are carried out; class V is permanently not suitable. The evaluation of the water deficit risk (Table 24.6) based on climatic factors (rainfall during the year and during the growth cycle, annual temperature, evapo-transpiration) allows the definition of the climatic region.

The intersection between the soil capability classes and the climatic region, in relation to the ecological requirements of the species under examination, determines the cultivation or

Table 24.5. Points assigned to various soil characteristics for the calculation of limitations.

Code	Class	Interval	Points
Available rooting depth (cm)			
1	Very shallow	≤ 25	90
2	Shallow	> 25; ≤ 50	30
3	Moderately deep	> 50; ≤ 100	15
4	Deep	> 100; ≤ 150	5
5	Very deep	> 150	0
Drainage			
1	Excessively drained		30
2	Poorly drained		30
3	Well drained		0
4	Moderately well drained		20
5	Somewhat poorly drained		40
6	Somewhat excessively drained		90
7	Very poorly drained		90
Chemical response (pH)			
1	Strongly acid	≤ 4.5	90
2	Acid	4.6–5.5	15
3	Slightly acid	5.6–6.5	5
4	Neutral	6.6–7.3	0
5	Slightly alkaline	7.4–7.8	5
6	Alkaline	7.9–8.4	15
7	Strongly alkaline	≥ 8.5	90
Rock fragments			
0	Absent	≤ 1%	0
1	Scarce	1–5%	0
2	Common	5–15%	0
3	Frequent	15–35%	10
4	Abundant	35–70%	20
5	Very abundant	≥ 70%	40
Slope			
1		0–20%	0
2		20–35%	10
3		35–50%	30
4		> 50%	60
Fertility			
A Horizon OM %	Ca (meq/100 g) first 50 cm	TBS %	
> 2	> 1	> 50	0
> 2	> 1	15–50	5
< 2	< 1	< 15	15

capability unit. The methodology described is that used by forestry experts when choosing and creating sites for plantations; where necessary the information of the soil maps were integrated by targeted, mostly expeditious surveys.

Since 2001, the Region Lombardia has had a pedological orientation map for timber arboriculture on the Lombardy plain on a scale of 1:250,000, drawn up by the joint efforts of the ERSAL Soil Service and the ARF as an information tool for use in preliminary planning. The evaluation methodology elaborated for this document introduces flooding risk among the parameters potentially limiting to arboreal growth and proposes an estimate of water deficit potential resulting from the interaction between some pedological characteristics, such as the available water capacity, drainage and eventual presence of a water table, and the general climatic conditions of the area. The remaining pedological parameters examined are available rooting depth, drainage, texture in the first metre of soil, and chemical response.

Table 24.6. Evaluation of the water deficit risk: parameters concurring in the evaluation for the various classes.

Evaluation of water deficit risk					
Risk class	P year (mm)	P seas. (mm)	EPT (mm)	T mean a. (°C)	Description
1	> 1,000	> 700	< 700	> 9	Areas highly suitable for use in specialized arboriculture, all forest species. Risk of water deficit is absent or slight.
2	> 1,000	> 500 and ≤ 700	< 500	> 9	Areas suitable for use in specialized arboriculture. Risk of water deficit is slight or moderate.
3	> 700 and ≤ 1,000	≤ 500	< 500	> 6 and ≤ 9	Areas marginally suitable for use in specialized arboriculture; tolerated by the more xerophytic species. Risk of water deficit is strong.
4	≤ 700	≤ 500	> 500	≤ 6	Areas not suitable for use in specialized arboriculture unless substantial irrigation measures are taken. Risk of water deficit is very severe.

For the arboreal species under examination, individual evaluation models were elaborated for growth capability, based on the degree of limitation of the lands in the parameters used in relation to the ecological requirements of the species. The adaptability of each species to the conditions present was interpreted by identifying (in plain lands in the pedological survey on a scale of 1:250,000) the presence of limiting factors subdivided into "principal" or strongly influential on development and growth potential, and "secondary" or less incisive in determining adaptability to a pedoclimatic environment, since they may be more easily corrected than the former. In this way, five land capability classes were obtained for each species involved in arboriculture for timber production (Table 24.7).

Land Capability for the Cultivation of Walnut in Lombardia Region

Predominant Geographical Areas. Walnut prefers plains or tablelands; it adapts to slopes while it is intolerant of concave or impluvial morphologies. With reference to the pedo-landscape model of Lombardy it adapts, in arboriculture cultivations for timber, to the following environments: mountain reliefs (slope and piedmont areas); morenic hillsides (intramorenic plains and sandbanks if well drained); paleoterraces and tablelands (intermediate terraces); base level)gravelly high plains and sandy low plains); and stable fluvial valleys and alluvial plains, without flooding.

Table 24.7. Evaluation of the land capability classes for timber arboriculture.

Land capability class	Definition	Degree of primary limitations	Degree of secondary limitations
1	Soils suitable to the development of the species	1	
2	Soils suitable to the development of the species, but show probable secondary limitations	1	1
3	Soils with moderate limitations to the development of the species	2	
4	Soils with moderate limitations to the development of the species, but show probable secondary limitations	2	1; 2
5	Soils little or not suitable to the development of the species	3	1; 2; 3

Climatic and Pedoclimatic Requirements. Walnut prefers the phyto-climatic zone of the *Castanetum*, but may range from the cold *Castanetum* to the cold *Lauretum*; it needs a yearly rainfall of 850–1,200 mm, but also accepts a lower yearly rainfall (not less than 700 mm per year) if well distributed. The average temperature required is over 10°C (with minimums > − 20°C) and occasional late frosts are tolerated.

The ideal environment for the growth of walnut has deep, fresh soils of a medium or moderately fine texture, well drained and with no water table in the first 100 cm, but with a good available water supply, a neutral or slightly alkaline chemical response (the pH tolerance range is 5.0–8.0), and no flooding risks or strong prolonged drought risks.

Land Capability for Cultivation of Cherry in Lombardia Region

Predominant Geographic Areas. Cherry prefers slopes or tablelands and adapts to plains, while it is intolerant of concave or impluvial morphologies. With reference to the pedo-landscape model of Lombardy it adapts, both in silviculture and in arboriculture for timber, to the following: mountain reliefs (slope and piedmont areas); morenic hillsides (intramorenic plains and sandbanks if well drained); paleoterraces and tablelands (both ancient and intermediate terraces); base level (gravelly high plains, only for silviculture) and sandy low plains; and fluvial valleys (fluvial terraces).

Climatic and Pedoclimatic Requirements for Cherry. Cherry prefers the phyto-climatic zone of the *Castanetum,* but may range from the cold *Castanetum* to the warm *Fagetum*; it needs a yearly rainfall of 1,000–1,300 mm, but also accepts a lower yearly rainfall (not less than 700 mm per year) if well distributed. The average temperature required is over 3°C (with minimums > −30°C) and occasional late frosts are tolerated.

The ideal environment for the growth of cherry has moderately deep, fresh soils of a medium texture, well drained and with no water table in the first 80 cm, but with a good available water supply, a neutral or slightly acid chemical response (the pH toleration range is 4.0–8.0), and no flooding risks or strong prolonged drought risks.

Functional Factors

The site and pedological characteristics that influence the possible growth of walnut and cherry are the same and influence the two in a substantially identical way, except for some slight or moderate differences of texture class limitations (Tables 24.8 and 24.9).

Table 24.8. Evaluations for walnut based on the geographic-topographic and functional characteristics of lands.

Type of limitation	Potentially limiting functional characterisitcs	Intensity of limitations		
		1- none or slight	2- moderate	3- severe
Principal	Available rooting depth (cm)	≥ 100	< 100 and ≥ 50	< 50
	Drainage	well drained	somewhat excessively drained*, moderately well drained	excessively drained, somewhat excessively drained, somewhat poorly drained, poorly drained, very poorly drained
	Flooding risk	none, rare	occasional	frequent
	Texture in first 100 cm	medium, moderately fine	moderately coarse	coarse, fine
Secondary	Water deficit risk	moderate, slight, absent	strong*, moderate**	very severe, strong
	Chemical response (pH in H$_2$O)	6.0 ≤ pH < 7.5	7.5 ≤ pH < 8.5	pH < 6.0; pH > 8.5

*If the risk of water deficit is strong but the available water supply (AWC) referring to the first metre is 50–100 mm and the water table is absent, the degree of limitation is moderate.
**If the risk of water deficit is moderate but the AWC referring to the first metre is 50–100 mm and the water table is absent, the degree of limitation is moderate.

Table 24.9. Evaluations for cherry based on the geographic-topographic and functional characteristics of lands.

Type of limitation	Potentially limiting functional characterisitcs	Intensity of limitations		
		1- none or slight	2- moderate	3- severe
Principal	Available rooting depth (cm)	≥ 100	< 100 and ≥ 50	< 50
	Drainage	well drained	somewhat excessively drained,* moderately well drained	excessively drained, somewhat excessively drained, somewhat poorly drained, poorly drained, very poorly drained
	Flooding risk	none, rare	occasional	frequent
	Texture in first 100 cm	medium,	moderately coarse moderately fine	coarse, fine
Secondary	Water deficit risk	moderate, slight, absent	strong,* moderate**	very severe, strong
	Chemical response (pH in H$_2$O)	6.0 ≤ pH < 7.5	7.5 ≤ pH < 8.5	pH < 6.0; pH > 8.5

*If the risk of water deficit is strong but the AWC referring to the first metre is 50–100 mm and the water table is absent, the degree of limitation is moderate.
**If the risk of water deficit is moderate but the AWC referring to the first metre is 50–100 mm and the water table is absent, the degree of limitation is moderate.

Regarding available rooting depth, cherry shows greater adaptability to depths (slightly) shallower than 100 cm, which makes it more suitable than walnut to populate sloping morphologies. On the plains the depth factor is subject to interference by the drainage factor. Since both species do not tolerate excess water well, many of the fluvial plains of Lombardy, where soils are also deep and fertile, are not suitable to the development of walnut and cherry since they are characterized by poor drainage; others are unsuitable because they are affected by occasional or frequent flooding risks. An analogous situation is found in the medium hydromorphic plains (resurgence zones), where plantation of the two species is not advisable because of drainage difficulties and the presence of a perched water table.

In the eastern parts of the region (most of all in the plains near Cremona and Mantua), limitations in available rooting depth may influence arboreal development because of horizons with concentrations of secondary carbonates at a varying depth, which limit root exploration, reduce the available water capacity and may hold seasonal perched water tables. In these territories, as well as in the Province of Pavia, there are wide zones in which the most limiting functional factor is the texture of the first metre of soil. The areas north of the Po River near Pavia and Mantua and wide areas of the plains near Cremona and Mantua are full of lands that are fine in texture within the first metre. They do not offer suitable conditions for the growth of the two species evaluated (also because of their association with difficult drainage). In the plains near Pavia north of the Po River, as well as a good share of the western plains of Lombardia (near Milan, Varese, Brianza, Como and Lecco), soils with a coarse texture predominate, not suitable to the cultivation of walnut and cherry because subject to strong water deficit risk, and characterized by a scarce available water capacity and rapid drainage. Table 24.10a, b and c show in synthesis the areas in Lombardy suitable to the plantation of walnut and cherry.

Use of Agronomic Operations to Reduce Limitations

In Lombardy, a plantation is not established if there are strong site or pedological limitations. In all other cases, one proceeds with the standard agronomic operations, i.e., preparation of the land by tillage and manuring in the pre-planting phase, and then by harrowing, disk ploughing, mulching, and irrigating during normal upkeep.

The territory of Lombardy, particularly the plain, usually has a water deficit risk and in some areas it assumes a conspicuous intensity; therefore, irrigation must be accurately carried out, especially in the first years of the plantation, progressively decreasing with plant growth and then abandoned when the plants reach adulthood.

Table 24.10a. Synthetic summary of the areas in Lombardia Region suitable for walnut and cherry plantations for use in silviculture (code1) and/or in timber production (code 2).

Landscape	Mountain reliefs			Morenic hillsides			
	Slope	Piedmont Areas	Valley floor	Sandbanks	Intramorenic plains	Hydromorphic areas	Alluvial valleys
Species							
Walnut	2	2	–	2	2	–	–
Wild cherry	1–2	1–2	–	1–2	1–2	–	–

Table 24.10b. Synthetic summary of the areas in Lombardia Region suitable for walnut and cherry plantations for use in silviculture (code 1) and/or timber production (code 2).

Landscape	Terrace reliefs above plains			Base level of plains		
	Ancient terraces	Intermediate terraces	Gravelly tablelands	Medium hydromorphic plains	Stable areas	Paleo-river bed
Species						
Walnut	–	2	2	–	2	–
Wild cherry	1–2	1–2	1	–	1–2	–

Table 24.10c. Synthetic summary of the areas in Lombardy suitable for walnut and cherry plantations for use in silviculture (code1) and/or in timber production (code 2).

Landscape	Alluvial valleys				
	Terraced surfaces	Stable alluvial plains	Hydromorphic areas	Open high-water bed	Depressions with very fine sediments
Species					
Walnut	2	2	–	–	–
Wild cherry	1–2	–	–	–	–

24.3.10. Other Experimentation in Italy

The Riselvitalia Project

In the sphere of ulterior ongoing experimentation may be cited the initial results of the Riselvitalia Project, "Subproject 2.1.10—Definition of pedological indicators for the correct evaluation of land capabilities for hardwood timber species" (Giordano et al., 2006). This subproject has the goal of identifying site and pedological parameters significant and/or limiting to the growth of walnut and of fine-tuning keys to the interpretation of pedological maps in terms of land capability for the cultivation of walnut. Nine experimental stations, four located in Piemonte and five in the Marche Region, were examined. The experimental areas were chosen for expression of the maximum variability of the parameters examined and for the presence of longstanding walnut plantations on which to base dendometric measurements. The parameters considered were the characteristics of the station, both topographic and climatic (altitude, exposure, slope, rainfall, temperature), pedological characteristics (water regime according to Soil Taxonomy, available rooting depth, texture, hydromorphic signs, pH, calcium content, cationic exchange capacity in the profile, organic matter, total available nitrogen and phosphorus in the A Horizon), and plantation data (plantation layout, cultivation practices, age, height and diameter).

Analysis of the initial results, even considering the limited number of surveys (seven for Piemonte and nine for the Marche), showed that they were significantly positive, such as favourable water regime (positive gradient passing from the xeric to ustic and to udic), good porosity, and a good enrichment of organic matter and nitrogen connected with species consociation. On the other hand, significantly negative factors were observed, such as the presence of compacted and anaerobic clayey horizons of shallow depth and the improper use of phytocells causing root malformation. Important correlations of soil change and calcium carbonate content were not detected, which seems to be an irrelevant factor. On the basis of these results, an initial approximation of the evaluation and interpretation of land capability for walnut was formulated as shown in Table 24.11, in which the criterion for assigning the S1, S2, and S3 classes (FAO,1976), in relation to all of the parameters considered, was that of maximum limitation.

Other Regional Studies

Concerning other regional studies, it is relevant to cite some general instructions formulated by the Region Molise on the occasion of the Region Forest Plan (Region Molise, 2003)

Table 24.11. Evaluation of land capability for walnut in the experimental areas of the Marche and Piemonte Regions.

	S1 suitable	S2 moderately suitable	S3 scarcely suitable
Water regime (according to ST)	udic	ustic	xeric
Available rooting depth (cm)	> 100	60–100	< 60
Anaerobic horizons within 1m	none	< 5% hydrom-orphic signs	>5% hydrom-orphic signs
Texture	l, sl, scl, sil, sicl, cl	s, ls, sc, sic	c, si
Porosity > 500 μm	> 5	3–5	< 3
Organic matter (%)		2–1	< 1
Nitrogen (‰)	> 1	1 – 0.5	< 0.5

regarding both internal soil characteristics and site characteristics: The species to be used for arboriculture plantations for timber, in the different phyto-climatic zones, should be those that have demonstrated good adaptation in preexisting reforestation. Walnut and cherry can be planted in lands of moderate depth that are permeable, with a not too acidic chemical response (pH > 6), excluding those with clayey soils, those with water stagnation and zones characterized by strong winds. Usually the plantations should be located at an altitude between 300 and 800 m asl. Plantations located at higher altitudes must be adequately justified by specific site conditions opportunely documented in the project; at lower altitudes plantations are permissible if water is available for irrigation. The minimum cultivation cycle of walnut and cherry plantations will be 35 years, unless differently authorized in the case of favourable development of the plants, which may be cut when a diameter of at least 35 cm is reached.

Experimental studies were also carried out in Campania (Di Vaio et al., 2005) on cultivation of walnut for timber, in a series of 13 experimental stations distributed throughout the region territory, which have shown some relationships between the pedological parameters examined, such as texture, pH and active lime content in the surface horizons and dendometric measurements and cultivation growth.

The plantations are located at an altitude between 50 and 700 m asl, with a slope that varied little, level or slightly inclined; in fact, 10 plantations are located on level areas and 3 on slopes of 6%, 7% and 8%. Average annual temperatures range between 12.3°C and 17.3°C. During the research years the minimum annual precipitation measured in the plantations was 800 mm and 853 mm respectively, while the maximum measured was 1,391 mm. In particular, the worst productive results in terms of mean diameter increase were found in plantations characterized by clayey or silty clay soils, while the best growth was registered in lands with a clay content not greater than 25–30%.

It is particularly interesting to note that the above correlations appear less significant when plantations older than 15 years were also included. This could be referred to several factors, such as the reduction of the mean diameter increase during aging and deeper root system depth, which could make the adult plants less susceptible to the characteristics of the more superficial soil horizons and less dependent on rainfall for water supply, unless there are easily accessible water tables.

References

AA.VV. 1996. Guida all'arboricoltura da legno in Abruzzo. Cogecstre (Ed.), Penne (PE).

AGENZIA REGIONALE FORESTE E REGIONE LOMBARDIA, AGRICOLTURA, D.G. 2001. Arboricoltura da legno, technical-operative manual.

ALESSANDRINI, A., FAZZUOLI, F., MITCHELL, A., NIEVO, S., RIGONI STERN, M., BORTOLOTTI, L. 1989. Gli alberi monumentali d'Italia. Il Sud. Abete, Roma.

AMMANNATI, R., LAZZARA, L. 1991. Produzione vivaistica qualificata di latifoglie pregiate. Cellulosa e Carta 42(5), 54–60.

ANDILORO, C., CANNAVÒ, S., MERCURIO, R. 2000. Esperienze sull'uso delle protezioni individuali in piantagioni di ciliegio da legno in Calabria. Legno Cellulosa e Carta VI(1), 2–9.

ANSELMI, N. 2001. Problematiche fitopatologiche in Italia negli impianti da legno di latifoglie nobili. Informatore patologico 7/8, 12–19.

BALDINI, S., BRUNETTI, M., FABBRI, P. 1997. Prove di potatura con impiego di strumenti diversi in una piantagione di ciliegio da legno. Annali Istituto Sperimentale per la Selvicoltura XXV–XXVI (1994–1995), 345–361.

BARONE, E., BURESTI, E., CANNATA, F., DI MARCO, L., MERCURIO, R., MINOTTA, G., PARIS, P. 1997. Modelli colturali e tecniche di coltivazione. In: Giannini, R., Mercurio, R. (Eds.), Il noce comune per la produzione legnosa, Avenue Media, Bologna, pp. 115–163.

BARONE, E., DI MARCO, L., MERCURIO, R., ZAPPIA, R. 1997. Sistematica, caratteri botanici, distribuzione geografica. In: Giannini, R, Mercurio, R. (Eds.), Il noce comune per la produzione legnosa, Avenue Media, Bologna, pp. 25–41.

BARTHÈLEMY, D., MERCURIO, R., SABATIER, S. 1997. L'architettura della chioma dl noce comune In: Giannini, R., Mercurio, R. (Eds.), Il noce comune per la produzione legnosa, Avenue Media, Bologna, pp. 57–65.

BELVAUX, E., TRON, G. 1996. Expérimentation d'arrosage de boisements en forêt méditérranénne. Forêt Méditérranéenne XVII(4), 287–302.

BEQUEY, J. 1990. Quelques précisions sur les noyers hybrides. Forêt Entreprise 69, 15–19.

BEQUEY, J. 1991. Planter des noyers en milieu agriole. Forêt Entreprise 80, 40–47.

BEQUEY, J. 1992. Les plantations de feuillus précieux. D'abord éviter les erreurs. Forêt Entreprise 81, 33–38.

BEQUEY, J. 1993. Quelques éléments de réflexion sur les plantations de noyers à bois à partir de l'expérience française. Linea Ecologica 24(1), 19–25.

BEQUEY, J. 1997. Les noyers à bois. IDF, Paris.

BERGOGNOUX, F., GROSPIERRE, P. 1981. Le noyer. Invuflec, Paris.

BOSCO, M., FAVILLI, F., LUMINI, E., TANI, A. 1997. L'ecologia del noce comune. In: Giannini, R, Mercurio, R. (Eds.), Il noce comune per la produzione legnosa, Avenue Media, Bologna, pp. 43–52.

BOSSHARDT, C. 1985. Etude de quelques feuillus précieux dans le Centre de la France: Frene, Merisier, Noyers, Nogent sur Vernisson: ENITEF\CEMAGREF.

BOULET-GERCOURT, B. 1997. Le merisier. IDF, Paris.

BUOLET-GERCOURT, B. 1997. Le merisier. Les guides du sylviculter (Deuxième édition). Institut pour le Dévelopment Forestier, Paris.

BURESTI, E. 1993a. Orientamenti tecnici sull'impianto e la coltivazione del noce da legno. Linea Ecologica 24(1), 12–18.

BURESTI, E. 1993b. Arboricoltura di pregio. Agricoltura Ricerca 147\148, 67–76.

BURESTI, E. 1993c. L'allevamento in cassone. Agricoltura Ricerca 147\148, 107–110.

BURESTI, E., DE MEO, I., FALCIONI, S., FRATTEGIANI, M. 1997. L'utilizzo del noce comune, dl noce nero e del noce ibrido in arboricoltura da legno. Primi risultati di una prova comparativa in impianti misti di 8 anni di età. Annali Istituto Sperimentale per la Selvicoltura XXV– XXVI (1994–1995), 243–260.

CALVO, E., D'AMBROSI, E. 1995. Proposte di standard di idoneità colturale per il postime vivaistico di alcune ltifoglie nobili. Monti e Boschi 47(4), 22–24.

CALVO, E., D'AMBROSI, E., MANTOVANI, F. 2001. Arboricoltura da legno. II edizione, Regione Lombardia, ARF, Milano.

CANNAVÒ, S., GUGLIOTTA, O.I., LAROSA, S., MERCURIO, R. 2004. Indagini sul ritmo di accrescimento e sull'epoca di potatura in piantagioni di noce da legno. Linea Ecologica 36(2), 45–50.

CANTONI, L., ZAMBONI, A., TONON, G. 2005. Fattori che influenzano l'embriogenesi somatica da embrioni immaturi di Juglans regia L. SISEF, Atti 4, 145–149.

CHERNOBOI, G.M. 1971. The root system of Walnut under unirrigated conditions in the crimean steppe. Tr. Gos. Nikitsk. Bot. Sada 52, 119–123.

CONTU, F., MERCURIO, R. 2005. Risultati di piantagioni sperimentali di latifoglie a legname pregiato in Abruzzo. SISEF Atti 4, 141–144.

CRAVE, M.F. 1989. Erables ondés, Alisier remarqubles, Merisiers. Foret Entreprise 58, 1–4.

CRAVE, M.F. 1991. Noyer hybrides. Foret Entreprise 76, 10–11.

DELEPORTE, P. 1977. Essai d'une typologie des stations a Frene et a Merisier en Nord-Picardie. Nogent-sur-Vernisson: ENITEF + annexes (Memoire), 59.

DETTORI, S., FALQUI, A., MAVULI, S., ORRU, A., PODDIGHE, D., TODDE, M. 1997. Prime esperienze di coltivazione di ciliegio e noce da legno in Sardegna. Annali Istituto Sperimentale per la Selvicoltura XXV–XXVI(1994–1995), 227–242.

DI VAIO, C., MINOTTA, G. 2005. Indagine sulla coltivazione del noce da legno in Campania. Forest@, (Ed.) Italian Society of Silviculture and Forest Ecology 2(2), 185–197.

DODE, L.A. 1906–1909. Conribution a l'étude du genre Juglans. Bull. Soc. Dendr. France 1906, 67–97; 1909, 22–50; 165–215.

DUCCI, F. 2005. Monografia sul ciliegio slvatico (*Prunus avium* L.). CRA-Istituto Sperimentale per la Selvicoltura, Arezzo.

DUCCI, F., DAL RE, L., PROIETTI, R., SIGNORINI, G., GERMANI, A. 2005. Cloni di ciliegio selvatico perché no? In: Ducci, F. (Ed.), Monografia sul ciliegio selvatico (*Prunus avium* L.). CRA-Istituto Sperimentale per la Selvicoltura, Arezzo, 37–52.

DUCCI, F., DE ROGATIS, A., PROIETTI, R. 1997. Protezione delle risorse genetiche di Juglans regia. Annali Istituto Sperimentale per la Selvicoltura XXV– XXVI(1994–1995), 35–55.

DUCCI, F., SANTI, F. 1997. The distribution of clones in managed and unmanaged populations of wild cherry (Prunus avium). Canadian Journal of Forest Research 27, 1998–2004.

DUCCI, F., TANI, A. 1997. Propagazione e tecnica vivaistica. In: Giannini, R., Mercurio, R. (Eds.), Il noce comune per la produzione legnosa. Avenue Media, Bologna, pp. 91–107.

DUCCI, F., TOCCI, A., VERACINI, A.1988. Sintesi del registro del materiale di base di Prunus avium L. in Italia cntro-settentrionale, Bsilicata e Clabria. Annali Istituto Sperimentale per la Selvicoltura XIX, 263–302.

DUCCI, F., VERACINI, A., TOCCI, A., CANCIANI, L. 1990. Primi risultati di una sperimentazione pilota di arboricoltura clonale d legno con Prunus avium L. Annali Istituto Sperimentale per la Selvicoltura XXI, 81–108.

DUPRAZ, C., BERGEZ, J.E. 1999. Carbon dioxide limitation of the photosynthesis of Prunus avium L. seedlings inside an unventilated treeshelter. Forest Ecology and Management 119, 89–97.

FALLERI, E., GIANNINI, R. 2003. La produzione vivaistica per l'arboricoltura da legno nel Meridione d'Italia. Inea, Roma, 83 p.

FAO. 1976. A framework for land evaluation. FAO soils bulletin 32, Land and Water Development Division, Rome, 1976, second printing, 1981.

FOURNIER, D. 1989. Modélisation de la croissance et de l'architecture du merisier (*Prunus avium* L.) Diplome d'Agronomie approfondie. ENSA, Montpellier.

FRANC, A., BOLCHERT, C., MARZOLF, G. 1992. Les exigences stationnelles du merisier: revue bibliographique. Rev. For. Franc. XLIV (n° spec. Les Feuillus Précieux), 27–31.

FRANCARDI, V., ROVERSI, P.F. 1997. Osservazioni sull'entomofauna fitofaga di *Prunus avium* L. in piantagioni da legno dell'Italia centro-settentrionale. Annali Istituto Sperimentale per la Selvicoltura XXV–XXVI(1994–1995), 399–412.

FRANCARDI, V., ROVERSI, P.F. 2005. Difesa dagli insetti fitofagi. In: Ducci, F. (Ed.) Monografia sul ciliegio selvatico (*Prunus avium* L.). CRA-Istituto Sperimentale per la Selvicoltura, Arezzo, pp. 100–106.

FRATTEGIANI, M., MERCURIO, R. 1991. Il fattore di competizione delle chiome (CCF) nella gestione delle piantagioni da legno di noce comune (Juglans regia L.). Monti e Boschi 42(5), 59–62.

FRATTEGIANI, M., MERCURIO, R., PRIMAVERA, F. 1996. Relazione tra tesitura del suolo ecoltivazione del noce da legno. Giorn. Bot. Ital. 130(1), 421.

GARDIN, L., COSTANTINI, E.A.C., NAPOLI, R. 2004. Guida alla descrizione di suoli in campgna e alla definizione delle loro qualità. www.soilmaps.it/ita/downloads.html.

GERMAIN, E. 1974. Création par hybridation de variétés de noyer (Juglans regia L.) associant unefloration tardive a une mise à fruit tres rapide et une productivité elevée. Exposé d'une methode. Ann Amelior. Plant. 29, 187–194.

GIANNINI, R., MERCURIO, R. 1997. *Il noce comune per la produzione legnosa*. Avenue Media, Bologna.

GIORDANO, A., MONDINO, G.P., PELLEGRINI, S., GARAFFI, R., GIAI MINETTI, D. 2006. Attitudine delle terre alla coltivazione del noce. Convegni Progetto Riselvitalia, sottoprogetto 2.1. www.arboricoltura.it/pubblicazioni/DCRiselvitalia2.1.htm.

GRAS, M., DUCCI, F. 1998. Punus. In: Scarascia Mugnozza, G.T., Pagnotta, M.A. (Eds.), Italian Contribution to Plant Genetics and Breeding, Viterbo, pp. 876–878.

GUINIER, P. 1953. Le noyer producteur de bois. Rev. For. Franc. 5(3), 157–177.

HUBERT, M. 1979. Le noyer a bois. IDF, Paris.

HUTCHINSON, J. 1959. The Family of Flowering Plants. Oxford Univ. Press. I. Dicotyledons.

KRÜSSMANN, G. 1986. Juglans. L. In: Manual of Cultivated Broad-leaved Trees and Shrubs, Vol. II, Bt. Batsford Ltd, London, pp. 190–194.

IPLA-REGIONE PIEMONTE. 1998. Caratterizzazione e valutazione di stazioni piemontesi di ciliegio selvatico (*Prunus avium* L.) idonee per la raccolta di materiale di propagazione da destinare allavivaistica forestale (unpublished).

IPLA-REGIONE PIEMONTE. 1999. Studio di caratterizzazione delle produzioni vitivinicole dell'area del Barbera d'Asti (unpublished).

IPLA-REGIONE PIEMONTE. 2000a. Carta dei suoli della pianura Cuneese (unpublished).

IPLA-REGIONE PIEMONTE. 2000b. Carta dell'attitudine alla produzione tartuficola della Regione Piemonte (unpublished).

IUSS WORKING GROUP WRB. 2006. World reference base for soil resources 2006, 2nd ed. World Soil Resources Reports no. 103, FAO, Rome.

LANIER, L. 1986. Précis de sylviculture. Engref, Nancy.

LOGIURATO, A., SARACINO, A., BORGHETTI, M. 1997. Analisi dlla vulnerabilità all'embolia xilematica ed effetto di uno stress idrico moderato in Prunus avium. SISEF Atti I, 321–326.

LOWE, V., PINEDA, G., DELARD, C. 2001. Cerezo comun (Prunus avium). INFOR, Santiago.

LOWE, V., GONZALEZ, M. 2001. Nogal comun (Juglans regia). INFOR, Santiago.

LUISI, N., MAGNANO, G. 1997. Malattie. In: Giannini, R., Mercurio, R. (Eds.). Il noce comune per la produzione legnosa, Avenue Media, Bologna, pp. 181–239.

MASSET, P.L. 1979. Etude sur les liasons entre la qualité technologique du bois de merisier (Prunus avium L.) et la station. Rev. For. Franc. XXXI(6), 491–502.

MAGINI, E. 1975. Guida al riconoscimento pratico dei semi e delle piantine forestali. In: Morandini, R., Magini, E, Il materiale di propagazione in Italia, Collana Verde no. 34, MAF, Rome.

MALVOLTI, M.E., CANNATA, F., GRAS, M., TANZELLA, O.A. 1998. Juglans. In: Scarascia Mugnozza, G.T., Pagnotta, M.A. (Eds.), Italian Contribution to Plant Genetics and Breeding, Viterbo, 872–875.

MAPPELLI, S. 1995. Walnut plant resistance to soil hypoxia: comparaison between J.rgia and hybrid plants. In: Panetsos, K.P. (Ed.), Development of Walnut Trees for Wood and Fruit Production. Dept. of Forestry, Aristotelian University, Tessaloniki, pp. 24–33.

MERCURIO R. 2003. La coltivazione del noce e del ciliegio da legno in alcune regioni dell'Italia centrale e meridionale: Toscana, Abruzzo, Calabria. In: Minotta, G. (Ed.), L'arboricoltura da legno, un'attività produttiva al servizio dell'ambiente. Avenue Media, Bologna, 94–95, 105–106, 108–109, 120–122.

MERCURIO, R., TOCCI, A.V. 1982. Prospettive per l'impiego del ciliegio (Prunus avium L.) nelle piantagioni da legno. Agricoltura e Ricerca 6(30), 41–45.

MERCURIO, R., MINOTTA, G. 2000. Arboricoltura da legno. Clueb, Bologna.

MERCURIO, R., NAPOLI, R., PAOLANTI, M. 2003. Impostazione metodologica per la realizzazione della Carta Regionale delle potenzialità dell'arboricoltura da legno. In: Atti del Convegno: Le Foreste in Abruzzo fra tecnica, economia, ambiente, Direzione Agricoltura e Foreste e Svilppo Rurale (Ed.), Regione Abruzzo, Febbraio 2003.

MINISTERO DELLE POLITICHE AGRICOLE E FORESTALI. 2000. "Metodi di Analisi Chimica del suolo". Collana di metodi analitici per l'agricoltura, Franco Angeli editore.

MINOTTA, G. 1989. La coltura del noce da frutto e a duplice attitudine produttiva in Italia. Frutticoltura 7, 23–29.

MINOTTA, G., 1997. Prove di potatura di allevamento su due progenie di noce da legno. Annali Istituto Sperimentale per la Selvicoltura XXV–XXVI(1994–1995), 323–343.

MINOTTA, G., BOSCHETTI, W. 1994. Indagini sul ciclo vegetativo stagionale di piante di noce (J.regia L.) allevate per la produzione legnosa in alcuni ambienti dell'Appennino settentrionale della Pianura Padana. Atti del convegno "Arboricoltura da legno e politiche comunitarie", 22–23 giugno Tempio Pausania (SS), 123–141.

MINOTTA, G., LOEWE, V., FERRI, D. 1993. Indagine sulla coltivazione del noce da legno (Juglans regia L.) in alcuni ambienti dell'Appennino settentrionale e della pianura padana. Monti e Boschi 44(3), 46–56.

MINOTTA, G., ROFFI, F., BAGNARESI, U. 1997. Indagini su modelli colturali agroforestali nell'Appennino settentrionale. SISEF Atti I°: 199–205.

MONTERO, G., CISNEROS, O., CAÑELLAS, I. 2003. Manual de selvicoltura par plantaciones de especies productoras de madera de calidad. Mundi-Prensa, Madrid.

MOTTA, E., SCORTICHINI, M., BIOCCA, M. 1997. Gravi malattie del ciliegio d legno in Italia centrale. Annali Istituto Sperimentale per la Selvicoltura XXV–XXVI(1994–1995), 373–390.

PARDÉ, L. 1952. Les feuilles. La Maison Rustique, Paris.

NICOT, P. 1983. Etude des exigenses stationelles, des performances de croissance, de la sylviculture et de la qualité du bois du Frene et du Meriser dans diverses stations d'Alsace. DDAF du Bas-Rhin, 238 + annexes (Memoire ENITEF).

PARIS, P., OLIMPIERI, G., CANNATA, F. 1995. Influence of alfalfa (Medicago sativa L.) intercropping and polyethylene mulching on early growth of walnut (Juglans regia spp.) in Central Italy. Agroforestry Systems 31, 139–180.

PIGNATTI, S. 1982. Flora d'Italia. Edagricole, Bologna.

RAMOS, D.E. 1985. Walnut Orchard Management. Coop. Ext. University of California, 178.

REGIONE EMILIA ROMAGNA, SERVIZIO SVILUPPO SISTEMA AGRO-ALIMENTARE, I.TER (SCOTTI, C.), CONSORZIO BONIFICA RENO PALATA (ZAMPIGHI, C), DIPARTIMENTO DI COLTURE ARBOREE, UNIVERSITÀ DEGLI STUDI DI BOLOGNA (MINOTTA, G., BAGNARESI, U.), SERVIZIO PARCHI E FORESTE (BARATOZZI, L.), SERVIZIO DIFESA DEL SUOLO (CAVAZZA, C.), SERVIZI PROVINCIALI AGRICOLTURA. 1995. Catalogo regionale dei principali suoli agricoli di collina e montagna, paragrafo: Scelta specie forestali.

REGIONE EMILIA ROMAGNA, SERVIZIO SVILUPPO SISTEMA AGRO-ALIMENTARE, SERVIZIO GEOLOGICO, SISMICO E DEI SUOLI, I.TER. 1996. Catalogo regionale dei principali tipi di suolo agricoli della collina e montagna.

REGIONE EMILIA ROMAGNA, SERVIZIO SVILUPPO SISTEMA AGRO-ALIMENTARE (GIAPPONESI, A.), I.TER (SCOTTI, C., BUSETTO, R.), CONSORZIO BONIFICA RENO PALATA (GASPARINI, C., ZAMPIGHI, C.). 1998. "Caratterizzazione pedologica 10 impianti del Consorzio di Bonifica Reno Palata (BO)".

REGIONE EMILIA ROMAGNA, SERVIZIO SVILUPPO SISTEMA AGRO-ALIMENTARE, I.TER (SCOTTI, C.), CONSORZIO BONIFICA RENO PALATA (ZAMPIGHI, C.), DIPARTIMENTO DI COLTURE ARBOREE, UNIVERSITÀ DEGLI STUDI DI BOLOGNA (MINOTTA, G., BAGNARESI, U.), SERVIZIO PARCHI E FORESTE (BARATOZZI, L.) SERVIZIO DIFESA DEL SUOLO (CAVAZZA, C.), SERVIZI PROVINCIALI AGRICOLTURA, ISTITUTO SPERIMENTALE DI PIOPPICOLTURA DI CASALE MONFERRATO (FACCIOTTO, G.). 1998. Catalogo regionale dei suoli della pianura emiliano-romagnola, paragrafo: Scelta specie forestali.

REGIONE EMILIA ROMAGNA, SERVIZIO GEOLOGICO, SISMICO E DEI SUOLI (FILIPPI, N., GUERMANDI, M.), SERVIZIO SVILUPPO SISTEMA AGRO-ALIMENTARE (SARNO, G., GIAPPONESI, A.), I.TER (SCOTTI, C.), DIPARTIMENTO DI COLTURE ARBOREE, UNIVERSITÀ DEGLI STUDI DI BOLOGNA (MINOTTA, G., BAGNARESI, U.), SERVIZIO PARCHI E FORESTE (BARATOZZI, L.), SERVIZIO DIFESA DEL SUOLO (CAVAZZA, C.), CONSORZIO BONIFICA RENO PALATA (NEGRINI, C., ZAMPIGHI, C.). 2002a. Metodologie divulgative Carta dei suoli in scala 1:250.000: realizzazione di un protitipo di Cartaapplicativa in area campione.

REGIONE EMILIA ROMAGNA, SERVIZIO SVILUPPO SISTEMA AGRO-ALIMENTARE, DIPARTIMENTO DI COLTURE ARBOREE, UNIVERSITÀ DEGLI STUDI DI BOLOGNA (CINTI, S., BAGNARESI, U.), DIPARTIMENTO AGROSELVITER, UNIVERSITÀ DEGLI STUDI DI TORINO (MINOTTA, G.), I.TER (SCOTTI, C., MORELLI, P.). 2002b. L'arboricoltura da legno con latifoglie di pregio in Emilia-Romagna: analisi di aspetti produttivi, economici, ambientali e vivaistici con riferimento agli impianti attuati ai snsi dei rgolamenti comunitari.

REGIONE MOLISE, A.A.V.V. 2003. Piano Forestale Regionale 2002–2006. Sezione 3: Gli impianti e i rimboschimenti. www.regione.molise.it/pianoforestaleregionale/sezione3/index_sezione_3.htm.

SCOHY, J.P. 1989. Le Merisier. Silva Belgica, 96(3), 47–52, (5), 37–42.

SANTINI, A., TURCHETTI, T. 2005. Malattie dell'apparato radicale. In: Ducci, F. (Ed.).Monografia sul ciliegio selvatico (Prunus avium L.), CRA-Istituto Sperimentale per la Selvicoltura, Arezzo, 91–94.

SCALTOSOYANNES, A., TSOULPHA, P., PANETSOS, K.P. 1997. Effect of genotype on micropropagation of walnut trees (Juglans regia). Silvae Genetica 46(6), 326–332.

SCHLESINGER, R.C. 1987. The effects of crowding on Juglans nigra growth. Proceeding IUFRO Int. Conf. on Thinning, Moscow-Riga (URSS), Sept. 9–17, 1985, Part 2, 2–13.

SINISCALCO, C. 1997. Mobilizzazione per innto e conservazione ex-situ di fenotipi di ciliegio da legno individuati nell'Italia centro-meridionale. Annali Istituto Sperimentale per la Selvicoltura XXV–XXVI(1994–1995), 133–148.

TANG, H., REN, Z., REUSTLE, G., KREZAL, G., GERMAIN, E. 2001. Optimizing secondary somatic embryo production in English walnut (Juglans regia L.). Acta Horticulturae 544, 187–194.

TUTIN, T.G., ET AL. 1964. Juglans. In: Heywood, V.H. (Ed.), Juglandaceae. Cambridge at the University Press, Flora Europaea, I, 56–57.

25. Citrus

Francesco Calabrese[1] and *Carmelo Dazzi*[1]

25.1. Species, Products, Uses

The term *citrus* (*citrus, agrumes, citricos, agrios,* etc.) encompasses a group of species whose fruit is distinctive in having its juice contained within small sacs inside segments attached to a central axis (*medula*) (Calabrese, 1991, 2002). The outer layer consists of an inner white spongy pith with an external peel of various colours that contains minute sacs of essential oils. Every part of the fruit of each species has its commercial uses, including consumption when fresh, juice, essential oils, byproducts (after juice extraction) for animal feed, pectins, colourings. When ripe, the fruit of most edible citrus is rich in sugars (10–14%) and acids (0.8–1.5% in oranges, mandarins, clementines; 5.0–6.5% in lemons), with a prevalence of citric acid and a considerable amount of ascorbic acid (vitamin C) (40–70 mg per 100 cc juice). Two oranges

---

[1] Università di Palermo—Facoltà di Agraria, Palermo, Italy.

a day satisfy a person's daily requirement of vitamin C. They also contain a large portion of vitamin A (carotenes) and some fruit, grapefruit and pink pomelo, contain lycopene, while blood oranges have anthocyanins which, together with ascorbic acid, are powerful antioxidants.

Citruses are the most commonly produced fruits in the world (105,677,706 t in 2003) (FAO, 2003). In Italy, 2,927,794 t were produced overall in 2003, of which 62% were oranges, 18% lemons, 16% mandarins and clementines (ISTAT, 2003). Some countries produce citrus almost exclusively for immediate consumption, others (such as Argentina with lemons, Brazil with oranges, Mexico and Puerto Rico with limes) convert a large quantity into juice and several secondary products (lemonade, candied and dried peel, flavours, essences).

In Italy, citrus fruit processing accounts for about 40% of the production; in detail, 38% oranges, 60% lemons, 14% clementines, 14% mandarins and 72% grapefruit are processed (2001). In Calabria, the entire production of bergamot is used to obtain essential oils for perfume, cosmetics and aromatic food products.

> Genus: *Citrus* (according to Swingle's taxonomy)
> Subspecies I: *Eucitrus* (edible fruits)
> *C. medica*: citron; *C. limon*: lemon; *C. aurantifolia*: lime; *C. aurantium*: sour orange; *C. sinensis*: orange; *C. reticulata*: mandarin; *C. grandis*: pomelo; *C. paradisi*: grapefruit; *C. indica*; *C. tachibana*.

25.2. Varieties

There are many varieties of some species (orange, mandarin and similar), while other species have few varieties (lemon, lime, citron, grapefruit) (AA.VV., 1967; Spina, 1990).

25.2.1. Orange

The orange is the most widespread species in the world and its cultivars are divided commercially into four groups (Calabrese, 1991, 2002): navel, no-navel, blood oranges, acidless oranges. The navel has varieties of great renown that are widespread: Navelina, New Hall, Washington Navel, Navelate, Lanelate (in their order of ripening from the end of October to May around the Mediterranean). These varieties are intended mainly for fresh consumption. Among the blond no-navel varieties, the Valencia dominates and, together with Washington Navel, represents the most widely diffused cultivar in the world. Among others are Shamouti in Israel, Ovale Calabrese in Italy, Verna and Salustiana in Spain, and Pera in Brazil. Valencia ripens late (April to June in the Mediterranean) and has a double destination: fresh market and processing, with fruit rich in juice (48–50%). The group of blood varieties lies almost entirely round the Mediterranean, with its centre in Italy, where it accounts for more than 70% of orange production. The preferred cultivar is Tarocco, with various clones, today's favourites being Scirè and Gallo. The fruit is more acid than Navel oranges. Italy commands the market of blood orange juice. The fourth group, with varieties producing an orange with a lower acid content (0.1–0.2% relating to juice), is little known. It is found in several countries in South America as well as in India, Pakistan, Tunisia, and other countreis. There is one variety in Italy, Vaniglia Apireno, with seedless fruit, which keeps well on tree and is cultivated in southern Sicily, between Ribera and Castelvetrano. This type of orange is beneficial for those with gastric problems.

25.2.2. Mandarin and Similar

Mandarins, known internationally as "easy peeling" and widespread in every continent, can be placed in five groups (Calabrese, 1991, 2002): Satsuma, King, True and Mediterranean Mandarin, Tangerine, and small-fruit Mandarin. Satsumas are well known in Eastern Asia and are of particular importance in Japan. In the Mediterranean, the only country undertaking mass cultivation is Spain.

Satsumas are early-ripening (end of September–November), seedless, and marketed with a green peel at the beginning of autumn. They open the citrus fruit season.

Kings are oriental mandarins, of little commercial value, with a rough peel and seeds.

True and Mediterranean mandarins are widespread in the East (cv. Ponkan) and in Mediterranean countries, now almost exclusively in Italy with the cultivars Avana and Tardivo di Ciaculli.

There are many cultivars and hybrids of *Citrus reticulata* in the tangerine group, the fruit with or without seeds, of various ripening times, heterogeneous in morphology and taste. The clementines of the Mediterranean could be included in this group (genetically they are 'tangors', hybrids between the mandarin and the orange, both bitter and sweet). Clementines are the most widespread easy-peeling citrus fruit consumed in Europe today. There are various cultivars with different ripening times, from October to February, including Marisol, Nules and Henandina.

Small-fruit Mandarins are found in the East. They are of no commercial interest because of their modest size and the presence of seeds.

25.2.3. Lemon

Lemon is almost exclusively cultivated round the Mediterranean, in California and in Argentina. There are few varieties. The Femminello group prevails in Italy, the cultivars Fino and Verna predominate in Spain, Lisboa and Eureka in California, Genoa in Argentina. Lemons are sold for consumption when fresh and for processing (mainly juice and essential oils).

25.2.4. Grapefruit

There are few varieties of cultivated grapefruit: Marsh, with pale flesh, Redblush or Ruby with pink-red flesh in hot countries, Star Ruby with pink flesh also in cooler countries. The pink-red pigmentation is provided by the lycopene, which progressively declines at low temperatures. Narangin, which is more concentrated in the fruit in cooler climate, including the Mediterranean, gives the fruit its bitter taste.

25.2.5. Lime

Limes are acid or acidless citrus fruits. Acid limes take the place of lemons in tropical countries. They have special essential oils, used for perfumes and cosmetics. The most widespread cultivar is Messican lime, a small fruit with many seeds. The Persian or Tahiti and the Bearss, however, produce seedless fruit, about the size of a lemon. The most widespread sweet lime (due to low acidity) is the Palestine lime, also used in the past as a rootstock. 'Limettas' are grown throughout the Mediterranean area, have acidless or acid fruit (Morocco), and have little commercial value.

25.2.6. Citron

Citron has been known since ancient times and is still used in religious ceremonies such as the Jewish Feast of Sukkoth (Calabrese, 2004). It is chiefly used for candied peel as it has a very thick 'albedo'. It is cultivated on a small scale in Italy (in Calabria) and in Puerto Rico. The best cultivar is Diamante or Liscia di Diamante.

25.3. Rootstocks

Today the most widespread rootstocks in the world belong to the citrange group (*Poncirus* hybrids) because of their resistance to a serious disease called Tristeza. The sour orange is still in use in some Mediterranean countries although it is heavily damaged by Tristeza virus when grafted with orange, mandarin and grapefruit.

Other less used rootstocks are Citrumelo Swingle, Rough Lemon, Cleopatra and Sunki Mandarins.

Citranges are less tolerant than the bitter orange to lime, water and soil salinity and exocortis disease.

25.4. Plant Requirements and Cultivation Techniques

25.4.1. Climate

Citrus can be cultivated between the latitudes of 40ºN and 40ºS approximately. Single species, however, have different climatic needs. The most cold-tolerant citrus (to −2ºC) are *Poncirus* (deciduous), some cultivars of *Citrus reticulata* and the genus *Fortunella* (kumquat or Chinese mandarin). The least tolerant are limes, liable to damage around 0ºC, and then lemon. Growth ceases between 12ºC and 13ºC for most species. The best conditions for vegetative activity lie between 24ºC and 30ºC; it ceases at about 38ºC. Temperatures higher than 40ºC are best tolerated when the humidity is high. Damage is suffered if high temperatures are accompanied by low atmospheric humidity (< 50%) and wind. The most heat-seeking are pomelos and pink grapefruit, whose pigment (lycopene) is gradually destroyed by cold (below 14–15ºC). The fruit of these species is more acidic and sour (due to the higher naringin content) round the Mediterranean. The best quality is found in the tropics.

On the other hand, oranges with red pigmentation, due to anthocyanins, need lower temperatures (< 13–14ºC) during maturation to acquire the best colour. In cold environments (winters with an average temperature < 10ºC), however, these varieties may have excellent pigmentation but a very acid fruit when completely mature. Their best habitat, therefore, is geographically restricted. As a general rule, it can be said that carotenes, which provide the orange colour, are best synthesized in areas lying between the 32nd and 40th parallels. Nearer to the equator, the pigment in ripe fruit is lighter or absent.

Fruit of the same species grown in humid tropical countries have more juice and fewer sugars and acids than those in temperate conditions. Overall, the best commercial quality of oranges and mandarins come from these latter climatic areas (e.g., the Mediterranean, California, Argentina).

Citrus is irrigated in the event of drought. There are excellent citrus groves in sub-desert zones.

25.4.2. Nutrient Needs and Fertilization

The most essential nutrients in citrus cultivation are nitrogen, calcium and potassium. Small quantities of phosphorus are necessary. Moderate quantities of microelements (including zinc, manganese, magnesium, iron) have a marked effect on growth and productivity of the plants and on the quality of their fruits (AA.VV., 1968).

Table 25.1 shows percentages of the main nutrients in leaves 4–7 months old, gathered from the central and outer area of the canopy from twigs without fruit. In practice, it is sufficient to check the mineral content of the leaves (referred to dry matter) to be aware of the nutritional condition of the citrus grove. To maintain an ideal nutritional state, there should be annual restitution of what has been removed from the plant in the way of fruit and wood when pruned, and during the processes of leaching and gasification. In the Mediterranean areas, the average annual amount of nutrients removed in the production of 30 t of fruit is shown in Table 25.2 (Calabrese, 1988).

Trees are fertilized after harvest in a single application just before growth re-commences or for nitrogen in two applications: at the end of winter and at the beginning of summer. More frequent applications are needed where the soil has a coarse texture, and these can be administered over three or four occasions. For late-maturing cultivars, all fertilization takes place in the summer once growth has ceased—in Italy about the middle of July. In the increasing cases of fertigation, the nutrients are provided at the beginning of summer (after fruit drop, in June) until the end of summer. This must be well managed, particularly for the cultivars of mandarins, clementines and oranges with unstable growth processes in order to avoid a fall of fruit following the induced growth spurt, particularly in early summer. In some areas, micro-deficiencies of zinc, magnesium and manganese are corrected with spraying the foliage.

Table 25.1. Content of nutrients in orange leaves of 4–7 months, spring growth, from twigs with no fruit (Cohen, Israel, in Calabrese, 1988).

Element	Deficiency value	Moderate lack	Optimum	High levels	Excess value
Nitrogen (%)	2.2	2.2–2.4	2.5–2.7	2.8–3.0	3.0
Phosphorus (%)	0.09	0.09–0.11	0.12–0.16	0.17–0.29	0.3
Potassium (%)	0.7	0.7–1.1	1.2–1.7	1.8–2.3	2.4
Calcium (%)	1.5	1.5–2.9	3.0–4.5	4.6–6.0	7.0
Magnesium (%)	0.2	0.20–0.29	0.30–0.49	0.50–0.70	0.8
Sulphur (%)	0.14	0.14–0.19	0.20–0.39	0.40–0.60	0.6
Boron (ppm)	20	20–35	36–100	101–200	260
Iron (ppm)	35	35–49	50–120	130–200	250
Manganese (ppm)	18	18–24	25–49	50–500	1000
Zinc (ppm)	18	18–24	25–49	50–200	200
Copper (ppm)	3.6	3.7–4.9	05–12	13–19	20

Table 25.2. Quantity of nutrients (kg/ha) removed from a Sicilian orange grove with 30 t of production (Calabrese, 1988).

Removal and losses	N	P	(P_2O_5)	K	(K_2O)	Ca	Mg
Fruits	42	7.50	(17.10)	45.00	(54.20)	17	10.5
Pruning	29	1.00	(2.29)	4.5	(5.40)	66.5	2.0
Leaching	70 ?	4.50	(9.50 ?)	25.00 ?	(30.00 ?)	41.5 ?	6.0 ?
Gasification	30 ?						
TOTAL	171	13.00	(28.89)	74.50	(89.60)	125.0	18.5

? = *uncertain or variable value.*

25.4.3. Water and Irrigation Requirements

The cultivated citrus originated in humid Asian areas. Because of this, the trees have very superficial root systems. In environments with periods of drought, as in the Mediterranean areas, irrigation is necessary. This varies widely in volume and frequency, depending on the soil texture and method of irrigation. The preferred irrigation method in some countries is the micro-sprinkler with distribution of 40–80 L/h, pressure of about 1.5 bar, radial range of about 1.5–2 m. The drip method is preferred where water can be distributed to coincide with plant requirements of frequency and quantity.

In some areas where rainfall is scarce, water used for irrigation is frequently too rich in salts (conductivity may be higher than 2.5 dS/m). This must be borne in mind (Table 25.3) when choosing rootstock; in the past the bitter orange was often selected because of its excellent tolerance for saline waters. The submersion method has now been abandoned due to its low efficiency (about 30–35%); low pressure and localized methods (with efficiency higher than 80%) provide considerable saving of water. In loamy-textured soils, as in part of the coastal Mediterranean Terra Rossa, watering must be sufficient to moisten the soil surface to a depth of 60–80 cm, where most of the capillary roots are found. It is not always possible to water regularly, and water shortages are not infrequent in some countries. It is not advisable to restrict water supply during the fruit's cell multiplication and initial expansion, though, if necessary, some watering can be omitted at the end of summer.

Table 25.3. Loss of production due to increased salinity of irrigation water in rootstock with average tolerance (Ayers and Westcot, 1985).

ECw(dS/m)	Loss of production (%)
1.1	0
1.6	10
2.2	25
3.2	50

ECw = electrical conductivity of irrigation water.

Where soil texture tends to be loose, irrigation is carried out at short intervals (7–8 d or less) with low volume (180–200 m³/ha with microsprinkler). When applying fertigation, frequency and amounts ensure the continuity of nutrient feeding. For the lemon, when trees are "forced" to ensure fruit for harvesting the following summer, irrigation either begins in July or is suspended at the end of June until a noticeable drying out of the leaves occurs. Moderate quantities of fertilizer are applied before re-watering.

Tensiometers and other objective methods for measuring the water content are used in some countries. The choice of time to begin the irrigation is often empirical. Where cooperatives exist, they decide on the opening and closing of the irrigation season according to the amount of water available rather than the needs of the trees.

25.4.4. Soil Management

In citrus cultivation, attention given to the soil varies according to a multiplicity of factors. What prevails in some environments does not in others. Spontaneous or planted grasses, for example, may or may not bring advantages according to specific instances. The following types of intervention are applied: mechanical weeding, chemical weeding, non-cultivation (Calabrese, 2002).

Mechanical weeding is widespread (in dry seasons) and least polluting. It uses mechanical equipment, which, working on the surface as it does, has as its main aim the elimination of weeds when they are truly harmful. Usually two or three very superficial soil tillages are enough in summer. Another tillage takes place at the end of winter to dig in the fertilizer. The idea of keeping the ground permanently free of weeds is gradually being abandoned. There is no reason for them to be removed in the rainy period in the Mediterranean, and indeed they perform the function of reducing compression of the soil caused by people and machinery passing by as well as controlling soil erosion on slopes. When weeds are cut at the end of winter they also provide a noticeable vegetal manure, calculated at around 20–30 t ha of fresh organic matter. This in fact is the only use of organic material in normal cultivation where the practice of animal manuring has more or less disappeared. It must be remembered, however, that where frosts may occur it is not appropriate to leave a layer of weeds because it can cause slightly lower temperatures on the surface of the soil.

Chemical weeding is becoming ever more widespread. In this case the spontaneous weeds are controlled with specific products. In the past, those most used worked selectively before emergencies arose. Today chemicals are used whose active ingredients (such as glyphosate) enter the plant and move towards the roots, where they accumulate and kill the cells. These products are preferred to those that dry up the plant on contact and also have the advantage of degrading rapidly in the soil.

Herbicides are necessary in the case of non-cultivation, when machinery is definitely excluded. This allows for maximum automation of irrigation and fertilization procedures. The result is lower costs when they are considered a priority by the citrus growers.

Citrus cultivation in the Mediterranean countries does not usually include planned or natural grass growing in summer, due to the greater need for water. In these countries the lack of water is a serious problem. Generally, however, growth of grass is justified in lemon groves as grass reduces mal secco infections penetrating the roots.

25.4.5. Planting, Raising, Pruning

Planting takes place with plants raised in plant pots. The distance and spacing are determined by the particular rootstock and the vigour of the cultivar that has been grafted (AA.VV., 1973). Rectangular spacing (6×4 m to 7×5 m) is favoured because it allows the establishment of continuous barriers of vegetation along the rows and permits human and mechanical movement between two rows.

Density arises from the cultivation strategy and can be definitive or dynamic. Dynamic planting allows the removal of some trees when they grow too closely to others and superimpose their foliage. There are some dwarf rootstocks, such as Flying Dragon, that permit close spacing (2.5–3 m in a row, 4 m between two rows).

The preferred shape of modern trees is a globe, with growth reaching to the soil. In some cases, lower branches must be removed (skirt shape) to avoid soil contact and prevent pathogens from attaching themselves to fruit and lower leaves.

Any cutting of the tree in its youthful phase slows its growth and, for this reason, the habit of training the trees to a specific shape is being abandoned. The trees must be allowed to grow naturally and take on bushy shapes. The only essential cutting is the shortening of the leader vertical branch, 3–4 years after planting. In adult trees the only cutting must be to thin out branches that occupy spaces filled with other branches, since citrus trees need thinning out rather than shortening. Cutting can be done every so many years, particularly if the trees are young and belong to certain species, such as lemons and grapefruit. It is principally mandarins and mandarin-type that need an annual pruning.

Suckers are not always removed as they could be productive within a few years. The ones that are part of the trunk or large branches or those that grow in the inner part of the canopy are removed.

Pruning is always done after the harvest and for the most part falls at the end of the winter, except for the later cultivars (e.g., Valencia, Lanelate), which are pruned in summer after spring growth ceases. It is preferable to prune lemon trees at the end of March or beginning of April, just before the emergence of the flower buds, in order to escape as far as possible mal secco infections caused by *Phoma*. Summer pruning was recommended previously but is no longer advisable because the new leaves are attacked by the serpentine miner (*Phyllocnistis citrella*).

Mechanical pruning, which can save as much as 90% of time, has never been adopted in Italy and other Mediterranean countries, but it is applied in the USA and generally in countries where human labour is scarce and expensive.

The machinery does not thin but shortens, and this is not the type of cutting indicated for citrus trees. More appropriate would be making use of pneumatic cutting equipment, which can save about 40% of time spent.

25.4.6. Health Problems and Remedies

There are today three great health problems for citrus growers in the Mediterranean countries: mal secco, Tristeza and fruit fly. The rest are manageable with appropriate treatments. Mal secco (*Phoma tracheiphila*) is an endemic disease that attacks lemon trees with infections at the tip of the branches and progresses downwards, or it attacks the roots (mal fulminante). No attempt to overcome it with chemicals is practicable, either for efficiency or costs. The only method available is to select tolerant cultivars, of which there are some today. This, however, pertains to new plantings.

The Tristeza (CTV virus) problem has emerged in recent years in Spain and Italy, chiefly as a result of importing infected material. Today this is the most serious problem in orange, mandarin and grapefruit cultivation, mainly where the trees are grafted on sour orange with combinations of varieties sensitive to the virus. The only effective remedy, therefore, depends on changing the rootstock and can only be undertaken in new orchards.

The Mediterranean fruit fly (*Ceratitis capitata*) attacks the fruit in the autumn and ruins the peel with prematurely dechlorophyllated stains. The fruit may detach and drop. In this case there are effective chemicals available, but this prevents organic cultivation.

25.5. Land Suitability for Citrus Cultivation

The establishment of quality citrus cultivation must be preceded by a careful analysis of the environment in which the plants must grow and produce. Climate, soil and quality of irrigation water are definitely the main factors to be considered. The choice of rootstock deserves particular consideration (Table 25.4).

Table 25.4. Optimal requirements for citrus cultivation with quality aims.

Requirement	Conditions
Climate	
Temperature	> 3°C; < 30°C
Lowest temperature	0–1°C
Atmospheric humidity	> 50%
Rainfall	with irrigation up to 900-1,000 mm/year
Soil	
Depth	> 60 cm
Texture	loam
Permeability	50–150 mm/h
Drainage	Good
Oxygen availability	high and constant
Soil reaction (pH)	6.5–7.5
Soil salinity	with sour orange < 3.5 dS/m; with citrange < 2.5 dS/m
Active carbonate	with sour orange < 10%; with citrange < 7%
Management	
Salinity of irrigation water	with sour orange < 3.0 dS/m; with citrange < 2.0 dS/m

The climate is a factor of great importance in citrus cultivation. In unsuitable climatic areas, citrus will have feeble vegetation and achieve only modest productivity. Even if single species have different climatic needs, it can generally be said that the minimum and maximum temperatures between which citrus will suffer no damage lie between 2 and 38°C. However, the ideal temperatures for vegetative activity, which commences at about 13°C, and for productive capacity, lie between 18°C and 28°C. In some species, lower temperatures, even very low during the ripening stage, improve the quality of the fruit. Trees will begin to suffer initial damage at temperatures lower than 1–2°C.

Temperatures higher than 40°C will cause damage through drying up the leaves, beginning at the edges of the most tender, halting the development of the fruit and lowering production. For both low and high temperatures it must be emphasized that the negative effects depend on intrinsic factors in the tree, such as species, variety, vegetative state, and rootstock, and extrinsic factors such as cultivation methods, type of soil, rainfall, and wind.

Insufficient and unequally distributed rainfall, which characterizes the Mediterranean environment, provides at the most half of the water requirements of the citrus trees. These are estimated at 800–1,000 mm per year, with a wide variation due to soil characteristics, species and rootstock. Autumn rain has a favourable effect on the size and quality of the products, while plentiful winter rain, particularly on heavy soils, is dangerous because it activate asphyxial and parasitic rot. Considering the climate and its relationship with altitude and aspect, the nearer the approach to the northern limit of cultivation, the 40° parallel, the lower the altitude should be. Windbreaks are installed where citrus groves are exposed to hot or cold winds. The floors of valleys can be subject to frosts and rotating poles, supported by pylons, are sometimes installed to prevent heat from escaping from the lowest areas.

As far as the soil is concerned, citrus trees adapt to widely different soil textures except those that have a marked prevalence of fine texture. Best results are obtained with loamy, fertile, deep soils with good amount of organic matter, permeable and well drained, even if some rootstocks cope with somewhat compact or sandy soils, the former being water-retaining and the latter retaining little water and few nutrients. The most suitable pH lies between 6.5 and 7.5, although there are rootstocks that can easily manage a wider range: the sour orange, for example.

Citrus plants have a great need of oxygen. Where air exchange is lacking, as happens in clay and silty soils, root systems will move on the surface of the ground in search of oxygen.

Irrigation is an indispensable factor in the Mediterranean area. The qualities of the water used must accord with the requirements set out in section 25.6.3.

25.6. Environmental Features Relating to Citrus Productivity

The basic norms of land suitability for citrus cultivation pertain to soil, climate and water quality.

25.6.1. Soil Features

As has been shown, texture, pH and permeability of the soil are the most important parameters affecting citrus productivity. As far as depth of soil is concerned, it must be stressed that citrus groves exist and provide excellent produce on soils of 70 cm depth or even 50 cm, lying on a soft and permeable substrate that allows soil drainage. In order to define land suitability for citrus cultivation, an approximate figure of 70 cm has been selected.

With regard to the texture, the best soils are those with a loamy texture. Definitely unsuitable are those with a high content of fine particles (silt and clay), in which the low drainage causes problems of asphyxia and parasites. Soils with an excessive percentage of sand (higher than 75%) retain little water and few nutrients. Calcareous soils (> 60% $CaCO_3$) are definitely not suitable.

The best pH figures for citrus lie between 6.5 and 7.5, although this range has been slightly widened to prevent the exclusion of soils with pH 8.0–8.2, which are required by plants grafted on sour orange that have good productivity results. It must also be emphasized that soils with pH values higher than 8.4 and lower than 5.5 are not suitable. In order to define the suitability of soils in relation to acidity, Classes S2 and S3 have been grouped together, while there is no distinction between classes for soil drainage; the wide range of types from normal to rapid is considered appropriate, only those with very poor drainage being unsuitable.

Reference has been made to active carbonates when considering carbonate content, the value of the total giving little useful information. Finally, the landscape morphology should be considered when estimating the level of slope as a limiting factor; the most suitable are slightly sloping areas to facilitate the run-off of water without creating problems of cultivation. When assessing the degree of slope a number of problems relative to cultivation operations arise such as erosion and the traffic of workers and machinery. Different levels of citrus adaptability in relation to various functional soil features are considered in Table 25.5.

Table 25.5. Citrus suitability in relation to several functional soil features.

Parameter	S1	S2	S3	N
Depth (cm)	> 70	70–40	40–30	< 30
Texture	Loam; Sandy loam; Sandy clay loam	Sand; Loamy sand; Clay loam; Sandy clay	Silt loam; Silty clay loam	Clay; Silt; Silty clay
Permeability(mm/h)	50–150	15–50; > 150	5–15	< 5
Drainage	← from normal to rapid →			very poor
Reation (pH)	6.0–7.5	5.5–6.0; 7.5–8.4		< 5.5; > 8.4
Active carbonates (%)				
sour orange	< 7	8–10	11–12	> 12
citrange	< 5	6–7	8–9	> 9
Slope (%)	< 15	15–25	25–30	> 30

25.6.2. Climatic Features

When considering climatic conditions and altitude, it is necessary to examine the parameters of temperature, though not of rainfall, given that modern citrus cultivation generally requires artificial water support as naturally occurring water supplies are insufficient (see section 25.4.3). The figures encountered in Mediterranean areas between the 31st and 40th parallels, where citrus cultivation occurs, are adequate but it should be noted that though temperatures above 40°C are relatively frequent they generally last for a short period only and are therefore supported by the trees. It is, however, necessary to protect them with windbreaks along the coast to avoid the effects of wind and increasing dehydration. The climate is also influenced by the interaction established between latitude and altitude.

Differences in suitability to altitude between species and varieties must also be noted. In Sicily, lemon is grown along the coast. Mandarins and mandarin-types can be grown at higher altitudes. Altitude must be considered in relation to latitude: approximately it can be seen that every increase in height of 100–120 m corresponds to one parallel. Thus, the local temperatures must decide the choice (Table 25.6).

Table 25.6. Citrus suitability in relation to some functional climate features in Sicily. Proceeding north, some adjustments should be made according to the latitude.

Parameter	S1	S2	S3	N
Temperature	2	0–2	–2 to 0	< –2
(annual range in °C)	30	31–35	36–45	> 45
Altitude (m asl)	0–100	100–250	250–350	> 350

25.6.3. Quality of Irrigation Water

Citrus is variously sensitive to the quality of irrigation water and in particular to its level of salinity: an increase in saline concentration causes damage shown in loss of production (see section 25.4.3). Excessive salinity of water represents one of the greatest problems in many countries. It is obviously important to consider the varying sensibility of different rootstocks to the presence of salts in irrigation water, which is greater for *Poncirus* and its hybrids including the citranges. The quality of irrigation water and its effects on cultivation is referred to in Table 25.7.

Table 25.7. Suitability of some citrus rootstocks to the salinity of irrigation water.

Class	S1	S2	S3	N
Salinity (dS/m)				
sour orange	< 2.0	2.1–2.5	2.6–3.0	> 3.5
citrange	< 1.5	1.6–1.8	1.9–2.2	> 2.3

25.7. Conclusions

In the evaluation of land suitability for citrus, much depends on the rootstock used, but it must be confirmed that the evaluation of soil profiles must depend chiefly on two parameters: texture and depth. What is detrimental to citrus is a poor soil drainage, which can cause asphyxia and assist the development of pathogens in the soil. In many coastal Terra Rossa in Italy, excellent citrus groves exist in soils of 50–60 cm depth on permeable substrata (for example, *Rhodoxeralfs*), but there are also citrus groves on loam-sandy soils (*Haploxeralfs* in the south-eastern coast of Sicily), or soils with a prevalent silty-clay texture (*Haploxerepts* of the Catania plain), or even soils with a notable clay presence (more than 50%) (*Haploxererts* scattered in bottom-valleys with varying slopes) (Fierotti et al., 1988).

Reaction needs to be carefully considered among the soil quality parameters. In Italy, soil pH varies considerably according to the presence of calcium, mostly between pH 6.5 and 8.2. Unfortunately, the rootstocks deriving from *Poncirus*, like the citranges, as they are tolerant of the presence of Tristeza, feel the negative effects of a high content of carbonates and bicarbonates. There are no problems of boron excess in Italy as there are in other parts of the world with hot dry climates. If there are ever deficiencies of some microelements (magnesium, zinc, manganese) in any areas, they can be corrected effectively by spraying the foliage with specific products.

References

AA.,VV. 1967. Citrus Industry, Vol. I. University of California.
AA.,VV. 1968. Citrus Industry, Vol. II. University of California.
AA.,VV. 1973. Citrus Industry, Vol. III. University of California.
AA.,VV. 1985. Trattato di Agrumicoltura. Edagricole, Bologna.
AYERS, R.S., WESTCOT, D.W. 1985. Water quality for agriculture. Irrig. and drain. paper 29 (Rev. 1) FAO, Rome.
CALABRESE, F. 1988. *Gli agrumi: Nutrizione e concimazione*. Italkali, Palermo.
CALABRESE, F. 1991. Gli agrumi: in Frutticoltura speciale. Reda, Roma.
CALABRESE, F. 2002. Origin; history; soil; cultural practices. In: Citrus, Taylor and Francis. London and New York.
CALABRESE, F. 2004. La favolosa storia degli agrumi. L'Epos, Palermo.
FAO. 2003. Annuario dati statistici. Rome.
FIEROTTI, G., DAZZI, C., RAIMONDI, S. 1988. Commento alla Carta dei Suoli della Sicilia (a scala 1:250.000). Regione Sicilia, Ass. Territorio Ambiente, Palermo.
ISTAT. 2003. Annuario dati statistici. Rome.
SPINA, P. 1990. Gli Agrumi. Edagricole, Bologna.

26. Wine Grape and Vine Zoning

Pierluigi Bucelli[1] and *Edoardo A.C. Costantini*[1]

26.1. Introduction

This chapter concerns themes relating to the evaluation of soil and land suitability for the wine grape and for vine zoning. These themes, which are so very complex and articulated, and for which there is such an enormous number of publications available, are very difficult to synthesize into one chapter of a manual. For this reason, a suitability evaluation made with the use of simple tables seemed to be an excessive simplification in the case of many applications. Therefore, discussion was favoured over schematization, making reference to cultivation limitation factors and to species requirements, taking into consideration the fact that these edaphic limitations need not always be considered negatively, since they may be an important factor in the quality and typicality of the wine produced. Discussion is intended to supply elements useful to the evaluation of those sectors of the wine growing and producing system in which soil research may give particularly important and significant results.

    The factors that act together to determine the conditions and the success of a wine growing and producing system are many and their importance varies according to the different production situations. In addition to physical factors, which are discussed later, there are professional, political and administrative, infrastructural and contextual, and immaterial factors that are continuously gaining importance in determining the commercial success of a wine (Fig. 26.1).

---

[1]CRA-ABP—Centro di Ricerca per l'Agrobiologia e la Pedologia, Firenze, Italy.

Among the factors emerging in importance there is certainly that of sustainability in viticulture. Sustainable viticulture is an activity that satisfies producer and consumer needs

```
o   Physical factors
        – climate
        – topography
        – geology and geomorphology
        – soil
        – landscape
o   Professional factors
        – agrotechnique
        – oenology
        – marketing
        – business management
        – unskilled and skilled labour supply
o   Political and administrative factors
        – European Union regulations and directives
        – State laws
        – regional laws
        – provincial directives and authorizations
        – regulations of production and protection consortium
o   Infrastructural and contextual factors
        – means of transportation
        – relations with wine-making cellar
        – social economic context of the wine farm
o   Immaterial factors
        – consumer taste
        – wine culture, public awareness
        – product image
              • tradition, "wine nobleness"
              • quality and notoriety of rural landscape
              • viticulture sustainability
```

Fig. 26.1. Factors of the viticultural system.

without compromising the capability of future generations to satisfy theirs (Vaudour, 2005). Therefore, it protects agricultural revenue and the quality expectations of consumers, but also conserves soil and environment and protects and improves the landscape. In this concept are included "biological viticulture", "integrated management" and, to a certain extent, "precision viticulture".

Besides being a political and administrative factor, by virtue of the recent changes in the common agriculture policies of the European Union and the relative directives on granting national admittance (MiPAF Decree of December 13, 2004 on the so-called conditionality), sustainable viticulture is certainly also a question of image. In fact, the diffusion in public opinion of a certain environmental ethic promotes an increasing perception by a vaster public, not just by the experts, of the environmental "damage" that follows viticulture carried out without logic.

Furthermore, the possibility in the near future of instating public certification of the environmental sustainability of a business that is "marketable" on the label of a bottle of wine, makes sustainable viticulture increasingly interesting for producers.

o **Survey and mapping**
  – soil map, map of terroirs
  – suitability maps
  – other thematic maps (land capability, management, workability, etc.)

o **Advice for plantation**
  – choice of plots, vine varieties, stocks, vine density, row orientation, choice of grass cover, pest control (chlorosis, Black Goo)
  – land and hydrological setting (levelling and fill, slope stability, density of ditches, etc.)
  – pre-plantation fertilization, green manure

o **Advice for vineyard husbandry**
  – annual fertilization
  – soil conservation measures
  – crop management (tilling, pruning, pesticide treatments)

Fig. 26.2. Pedological evaluations for sustainable viticulture.

Therefore, sustainable viticulture must propose the creation of the production paths most respectful of the environment and the landscape, in which case the contribution of the pedologist may be fundamental in many ways, as shown in Fig. 26.2. This chapter supplies some of the technical elements at the basis of the evaluations.

26.2. Genetic Characteristics of the Species

Every species of grapevine is identified by genetic characteristics that influence its vegetative, productive and qualitative behaviour, factors that are more or less stable in various environments. Many parameters are determined mostly by the genetic patrimony, particularly the trends of phenological phases or the basic characteristics of the cluster and the must (e.g., phenolic and aromatic quality and richness, sugar/acid ratios). However, different varieties have different reactivity and stability with respect to the cultivation site. In fact, there are vine varieties that adapt well even in dissimilar environments, always guaranteeing satisfying results (e.g., Cabernet Sauvignon and Chardonnay, grown worldwide), which interact minimally with the surrounding environment, compared to vine species that instead readily respond even to small pedological and climatic variations, such as Sangiovese. The work of environmental experts may be decisive regarding this second genetic type, avoiding the risk of downgrading varieties and/or environs following bad evaluations. In fact, the vine species can fulfil its potential only if placed in a suitable area, or an environ may stand out and gain renown only if a suitable vine species is cultivated. The same considerations may be valid for clones and their behaviour.

In the sphere of varietal supply, it must be recalled that the strong point of the viticultural sector, particularly the Mediterranean countries, is its high number of existing autochthonous vine species and therefore the enormous genetic variability that underlies a diversification and productive originality difficult to equal. Therefore, it becomes a great advantage to have a vast variety of vine species, selected as a function of determined environments, to which producers can refer for typical, unique, and original productions. Genetic wealth and the consequent variability of the viticultural potential allows, if correctly coupled with a careful qualitative evaluation, the differentiation of productions, linking them increasingly to their natural characteristics and to the ancient history of their place of origin.

o **Climate**
  - solar radiation, temperature, rainfall
o **Topography**
  - elevation, aspect, sunshine duration, distance from bodies of water
o **Geology and geomorphology**
  - soil parent material, weathering, geochemical pool and water quality, deep drainage, surface erosion and landslides, water table and resurgences
o **Soil fertility**
  - soil temperature, water supply, internal drainage and oxygen availability, mineral nutrition and manuring, presence of limiting horizons for vine rooting
o **Landscape**
  - ecosystems (agroecotopes)
  - vineyard length and size
  - crop saturation of the landscape

Fig. 26.3. Functional characters of the physical factors of the wine growing and producing system.

26.3. Crop Requirements

26.3.1. Climate

Crop requirements of the grapevine chiefly pertain to thermal and radiation requisites. Latitude determines the variation in hours of sunlight and consequentially cultivation of the grapevine may not reach farther than 50°N (with special protection it is possible to reach 52°N) and 45°S. Between the equator and the 30th parallel (equatorial and tropical regions), it is possible to cultivate the grapevine with more than one annual production cycle, but with low grape quality, destined only to be consumed fresh or used to make fruit juice. At the latitudes of Italian viticulture, the principal climatic parameters that influence wine growing and producing quality are solar radiation, temperature and rainfall. In Table 26.1 the principal relationships between vegetation cycle and climatic tendencies are synthesized.

For some time the relationship between temperature and the characteristics peculiar to wines has been recognized, characteristics deriving from the different degrees of sugar content reached by the grapes, the acidic contents, the aromatic intensity, the quantity and availability of the anthocyanins and the tannin condensations, the enzyme activities, and other factors. The best wines come from well-ripened grapes, balanced in composition, in which temperature did not limit by defect or by excess the synthesis of its constituents. Intensity in colour and in aroma is diminished by an environment with low ripening temperatures that favour a potential for better delicacy. On the other hand, excessively high temperatures reduce the synthesis of colouring substances and accelerate the degradation of the more thermolabile aromatic compounds and of the essential acids for a good overall balance.

Table 26.1. Principal effects of climatic and pedoclimatic factors on the vegetation cycle (Storchi and Tomasi, 2005, modified).

Parameter and reference month	Positive effect on parameter	Negative effect on parameter
Budbreak date (end of March to mid-April)	High temperatures in the month before it takes place, after an initial period of low temperatures	Low soil temperatures, low air temperatures
Shoot growth (April–May)	High average temperatures	Low average soil and air
Flowering date (May to beginning of June)	Temperatures up to an optimum of 31–32°C	Thermal extremes, thermal deficits
Bud fertility	High temperatures and high insolation during flower differentiation time (May–June of the previous year)	Low temperatures, scarce insolation
Mature cane/veraison (end of July–August)	Not too much rain, high average temperatures but without extremes	Excessive water availability
Ripening date (September–October)	Moderate water deficit stress, high average temperatures, but without extremes	Excessive water availability, low temperatures

The effects of temperature must be analysed as a function of the vegetative state of the grapevine. The minimum temperature level necessary to breaking dormancy and resuming the vegetation cycle (zero vegetation) is usually around 10°C. During winter dormancy the vine is capable of resisting extreme cold (as low as −15°C), while spring frosts often cause damage to young buds when the air temperature falls to −2.5°C. Generally, the best years are those in which the spring temperatures are above average (precocious flowering, usually followed by early maturation) and medium/high temperatures occur in the veraison/harvest period.

However, climatic elements must be evaluated together and in interaction, so the temperature must be directly correlated to the hours of sunlight and of insolation. The combination of these two factors has a direct effect on most of the vegetative activities, of accumulation, reproduction, and sustenance of the vine. The hours of sunlight are positively correlated with the quantity of grapes produced and with their quality, but only if the relative air moisture and temperature are within a favourable range. In hot environments, to achieve full-bodied wines, more hours of sunlight are necessary than in environments less well-supplied thermally, because of a higher consumption for respiration (Gladstones, 1992; Storchi and Tomasi, 2005).

Several authors have worked on the relationships between analytic parameters of the grape (and of wines) and meteorological factors, for harvest predictions and with other objectives. The importance of temperatures in determining quality in northern areas was demonstrated, particularly in direct correlation with sugars and inversely with malic acid in comparison with thermal summations. The scarce malic acid content obtained in hot climates is explainable both with the stimulation of the malic degradation enzymes and with the early ending of shoot growth where the synthesis of the acid takes place.

Sugar synthesis is guided by photosynthesis, which occurs in optimal conditions at 23–27°C and is proportional to sun irradiation up to medium illuminations, along the order of 300 W/m^2, but reaches an upper limit with strong insolation, beyond which it remains constant; in fact, the vine seems incapable of utilizing more than half of the light intercepted by its leaves and therefore the natural availability in light radiation hardly ever constitutes a limiting factor for photosynthesis (Huglin and Schneider, 1998).

Pedoclimatic and agro-technical conditions have a great influence on the synthesis of the anthocyanins. Particularly the ratio between vegetation and clusters, resulting also from the methods of training and pruning employed, has a strong role in the quantity of anthocyanins that can be accumulated. Many authors agree in affirming that the best ratio between leaves exposed to sunlight and grapes is 1.2 m² of leaves per kilo of grapes. The natural factors that most influence the quantity of anthocyanins present in the cell vacuoles of the skin are temperature, sunlight, and available water, which govern the intensity and persistence of the biosynthetic pathways that lead to the synthesis of pigmenting substances, in other words, define the maximum potential achievable in quality. The sequence of these enzymatic reactions involves in its first stages phenylalanine (PAL) and then, through a series of stages catalysed by various enzymes, it comes to the common precursor of all flavonoids, chalcone (Heller and Forkmann, 1988). During this metabolic phase, phenylalanine plays a key role, since it establishes a competition between the primary metabolism of amino acids, which leads to the synthesis of proteins, and the secondary metabolism, which determines the accumulation of polyphenolic substances. If sunlight is not optimal, that is, too low or too high, biosynthesis may turn toward the production of proteins instead of polyphenols.

High nitrogen availability also favours the first pathway and indirectly discourages the accumulation of phenolic substances, increasing the vegetative development of the vine and therefore its shading (Roubelakis-Angelakis and Kliewer, 1986; Price et al., 1992). Daytime temperatures higher than 35°C and nighttime temperatures over 30°C stop the anthocyanic synthesis; the best range for their synthesis is between 17°C and 26°C, while there is inhibition of the formation of anthocyanins in clusters exposed to excessively high temperatures, just like those in too much shade. This opens discussion to a series of considerations tied to the combined effect of sunlight and temperature, to a point at which clusters in sunlight may reach temperatures too high for an optimum biosynthesis of pigmenting substances in the middle hours of the day; in other words, it falls upon the utility of green pruning (hedging and stripping of leaves) done to expose the cluster to sunlight in order to achieve better colour.

Sugar accumulation is usually superior in the berries exposed to sunlight. This may be explained by higher transpiration values and therefore by the more intense phloematic flow in the grape berries in sunlight. High temperatures during ripening determine a clear reduction in malic acid, a compound more sensitive to temperatures than tartaric acid, the decreased presence of which makes wines less fresh and pleasant, especially white wines. Aroma synthesis, even if it has a strong genetic basis, is influenced by the activity of various enzymatic systems and increases with the temperatures up to a certain level, and then decreases. In non-limiting conditions of sunlight and water, temperatures of 20–22°C are considered optimal (Champagnol, 1986; Huglin and Schneider, 1998).

In recent years, new mathematical and statistical models have brought on an increasing trend of defining relationships between climate and genotype and of breaking the vegetative cycle into sub-cycles, which are researched singly for climatic effects. In particular, the climatic conditions of spring and of the period that runs between veraison and harvest are decisive for grape quality. In general, early flowering, slight water deficit stress during ripening, and a not too early harvest are important quality factors.

26.3.2. Topography

Elevation directly influences temperatures. On the average there is a decrease in temperature of 0.6°C for every 100 m in altitude and this corresponds to a 2–3 d delay in vegetation (Kriedmann et al., 1971). In the Mediterranean area, cultivation of the grapevine does not exceed 600 m, except for particular microclimates and solar exposures like that of the Alps and the slopes of some volcanoes (e.g., Etna, Vesuvius).

In hillside areas, the orography of the site brings with it numerous implications of a climatic character, to which must be added that of erosion and gullying. Slope, aspect and elevation are the main factors that influence thermal levels, night/day thermal range, hours of sunlight, inclination of incident rays, movement of water masses, and precipitation. Even inside the area of the same vineyard placed on a hillside, there may be daytime air ascendants and nighttime air descendants that create situations of thermal diversity in all of those cases where the slope gradient allows the air to drain away and rise up again over the span of one day. The part of the vineyard at a lower relief of the valley will therefore be invaded by the cooler air that flows along the slope during the night, which the following morning must be warmed again to invert its route. Consequently, within the same vineyard there may be found areas with different average thermal values and with thermal ranges just as diversified.

It is common knowledge that the solar exposure of a vineyard-covered slope may lead to a harvest of grapes with variable compositional characteristics. In fact, the choice of aspect is important in the interaction between vine species and climate. The aspects more thermally favourable are those facing south and west; in the latter case, insolation, accompanied by the maximum thermal accumulation of the first hours of the afternoon, lasts longer than on the other aspects. The aspect of a vineyard influences the quantity of solar radiation available to the plants and in limit conditions of altitude and latitude it becomes decisive; therefore, at the northern limits of the cultivation or at high relief only southern or south-western aspects are used. The choice of vine species, therefore, may be determined by altitude and aspect. Red grape vine species, for example, have more thermal requirements than the white, while the latter need to preserve the composition and level of varietal aromatic precursors, which are extremely susceptible to high temperatures.

The presence of water bodies (sea, lake) or large wooded areas also has an effect on climate, mitigating thermal excesses and influencing the relative humidity of the air: the water vapour emitted during the daytime blocks part of the solar radiation, but during the nighttime it opposes the irradiation of heat. The distance from bodies of water also has an effect on the diurnal and nocturnal breezes, on the seasonal thermal ranges, on the frequency of rainfall events, and other climatic factors.

26.3.3. Geology and Geomorphology

The geology of a territory influences wine in many ways. The geological nature first of all influences the shape of the landscape, conferring morphology, typical spaces and articulations that characterize a production district. In fact, the wine landscape, with all its visual implications, possible historical ties and cultural contexts of placement, has many relationships with the geology of a place.

The geological nature and evolution of a district not only shape the landscape, but also influence the cultivation of the grapevine and the oenological results. These occur directly through geomorphological processes that, by modelling the relief and determining the meso-climate, influence the cultivation of the grapevine through morphology and erosion processes, but most of all in an indirect way through the soil. In fact, the nature of the rock has important effects on the mineralogical, chemical and physical characteristics of the soil.

Another correlation between the lithological nature of the rock and the cultivation of the grapevine regards the influence on water quality. The origin and the mineral content of the irrigation water may be important since some of the characteristics of the wine may be affected more by the composition of the water than by the parent material of the soil, most of all where irrigation practices cover a long time-span. The composition of the water is strongly influenced by the rocks through which it percolates.

Rocks may have a very different chemical composition, depending on the materials from which they are composed and the relationship of those materials. The dominant component of rocks on the earth surface is silica (SiO_2), in the form of quartz. Quartz is composed of a molecule of covalent bound atoms and is among the most resistant minerals of natural origin: it is not easily alterable and is chemically inert; in fact, most sandy soils contain quartz. Other minerals, mixed with silica in different structures, have weaker bonding and are usually a mixture of magnesium (Mg), calcium (Ca), potassium (K), sodium (Na) and other minor elements that may be released into the environment by mechanical or chemical alteration, providing the nutrients essential to the growth of the grapevine ("geo-chemical pool"). The four principal elements (Mg, Ca, K, Na), in particular, present in variable proportions in the soil, substantially determine the chemical characteristics of the circulating solution and influence the qualitative results of the grapevine.

The mineral content of the rocks also determines the soil texture, that is, the dimensions and form of the mineral particles, and affects soil structure, that is, the arrangement, combination and orientation of the mineral and organic particles, therefore influencing rooting ease and depth and the capability of the grapevine to absorb water and nutritive elements.

Water and mineral nutrition is related to soil water capacity, because water is always the vehicle for mineral nutrition. Regarding soil water capacity, besides texture and structural characteristics, an important role is played by the mineralogy of the clays. This also determines the pool of exchangeable nutrients available to the plant in the circulating water solution, to be evaluated by measuring the cation exchange capacity (CEC). The abundance and type of clays determines the CEC; for example, sediments rich in kaolinite and illite have soils with a low CEC, while those with montmorillonite have high CEC values.

Available water capacity, besides being affected by the nature of the rocks and sediments, is also influenced by their disposition in layers. Impermeable layers retain water and limit deep drainage. When placed on planes that intersect the surface of the aspect, these impermeable layers may intercept infiltration water and conduct it to the surface, causing resurgences that may trigger superficial or mass erosion phenomena. The phenomenon is particularly severe in vineyards where landslides have removed most of the topsoil. In fact, the soil cover is able to regiment the water cycle by absorbing and holding large quantities of water, coming both from rainfall and from ground water, and then by regulating its outflow, particularly surface runoff, in favour of hypodermic outflow. A slope that has had its topsoil removed, showing the outcropping of the substratum that is all but incapable of absorbing and holding water, has very slow permeability, favours rapid surface runoff and triggers gullying and gorging. When the substratum is loamy or sandy, with deep impermeable layers, it can be easily saturated, with consequent landslides.

Since rock weathering affects important qualities in viticulture, such as the chemical and physical nature of the soil, surface and deep drainage, it is possible to attempt a generic association between the geological environment and viticultural vocation. Grapevines placed on very thin soils where rock is near the surface may present excessive water deficit stress. On the other hand, very deep soils with a high water table, such as those on alluvial sediments, produce grape clusters that lack the concentration of juices necessary to obtain high quality wines. Viticulture districts on cracked and karstic calcareous rocks, which avoid hydraulic excesses through deep drainage, often have a good viticultural vocation. Where the parent material is limestone the soils are usually clayey and silty and hold a good quantity of water, while alteration of granites and sandstones produces sands and gravels that dry very easily. On the other hand, soils that are rocky on the surface favour the deep absorption of heat and limit water evaporation and erosion of the surface horizon. Furthermore, not all rock fragments are to be considered inert; those with a fine texture or those showing alteration, even incipient, retain a certain quantity of water and may also exchange nutritive elements. It is also for this reason that some of the viticulture districts on rocky or stony soils are considered of a good quality vocation. Soils deriving from calcareous materials also have the advantage of favouring decomposition and humification of organic materials, for the presence of carbonates. Generally, the presence of a certain quantity of active lime is associated with the production of high quality wines. However, soils on acidic rocks may also produce wines of high quality. In France, for example, those produced on the granites of Haut-Beaujolais, Côte Rôtic and Alsace are famous.

26.3.4. Soil Fertility

The grapevine is a frugal plant that adapts well to growth in many soil conditions. This makes its cultivation possible in many areas. However, different areas have different levels of production vocation for grapes from which it is possible to obtain quality wines. It is for this reason that the territories with the best vocation are usually those where vine-growers have for quite some time successfully found, after a long series of attempts, the right balance between edaphic characteristics, vine varieties, and agro-techniques. Therefore, generalizations are difficult, even if it has by now been demonstrated that the balance of mineral nutrients (particularly nitrogen) and water nutrition are the most determining pedological factors for production and quality. It is furthermore known that soil fertility suitable to quality viticulture must be moderate, avoiding soils that are too fertile as well as those that are too poor. The discussion in the following paragraphs synthesizes the state of the art in knowledge of the principal characteristics and qualities of the soil that determine different fertility conditions, indicating how it is possible to intervene to improve the qualitative results of the grapevine.

Soil Temperature

Soil temperature acts upon the plant metabolism with respect to hormonal synthesis and the absorption of minerals. It is common knowledge that the more abundant the macro-pores, due to coarse texture or to good structure, the more easily the land drains excess water and warms in spring and throughout the season. The thermal behaviour of the land depends on humidity and porosity, which affect the calorific capacity of a volume unit and the soil thermic conductivity (Vaudour, 2005). Experience has taught that precocious land warming in spring, connected with macro-porosity and environmental factors such as aspect and external soil drainage, correspond to early budding, which is a precious factor for the rapid formation of a working leaf system. The importance of soil temperature on the precocity of the grapevine and on the level of maturation reached at harvest was demonstrated in some viticultural areas of southern France (Alsace, Loire Valley). In these areas differences of mean soil temperatures were measured of 1.7°C at a depth of 5 cm and 2.4°C at 100 cm between a sandy soil and a soil on Senonian flint loams. No significant gap was registered during the same period (March–July 1988) for the air temperature, close to 14°C (Morlat et al., 2001). In the cases where budding occurs later, for factors connected to soil temperature and moisture, or due to minor air temperatures, in long pruning (Cappuccina, Guyot, Sylvoz, Pergola) it is common that different vegetation gradients along the fruit cane occur, with repercussions on the quality characters of the wine. In the cases where these phenomena are more evident, it seems suitable to use short pruning of a spurred cordon type and moderate row distance, which favour more phenological and vegetative homogeneity in the fruit canes and along the permanent cordon.

Soil Water Balance and Water Nutrition

In the Mediterranean environment, characterized by a xeric pedoclimate, the phenology and production potential of crops are determined most of all by water availability, in terms of water available to the plants, as well as potential, in other words, the force with which soil retains the water. Also, the vegetative and reproductive activity of the grapevine, which renews a good part of its absorption system each year, is deeply influenced by the rate of water available during the year. The phenological reference model for the grapevine is characterized by break of dormancy in the spring with vegetative resumption lasting until flowering, after which plant growth slows and then stops after the veraison. During ripening, growth ceases and there are slight symptoms of stress. These springtime kinetics of the aerial part of the plant correspond to a strong deepening and absorption by the root system, which is followed by a progressive decrease in activity of the superficial part of the root system, the part most responsible for plant water supply and vigour. In fact, water is the chief regulator of the hormonal equilibrium of the grapevine (Champagnol, 1997). This regulation is carried out through competition between synthesis of cytoquinines or of abscisic acid by the root extremities. In spring, when water is abundant, production of cytoquinine is high and with it plant growth. At the beginning of summer, in correlation with the flower-setting period, water becomes less abundant and is withheld by the soil more tenaciously, in such a way that not even at night does the rooting environment become moist. Under these conditions, cytoquinine synthesis decreases until it ceases, while conditions become right for the production of abscisic acid, first by the roots, then by adult leaves. In this way, plant growth is precociously curtailed, in a way that ripening may be compared to senescence, characterized by hormonal equilibrium in the vegetation and in the fruits. These conditions also induce the accumulation of anthocyanins, tannins and phenolic compounds in the grape berries, while the degradation of malic acid is favoured: processes that contribute to determining the quality of red wines, in particular their fineness and typicality (Champagnol, 1997; Lebon et al., 1997). However, if the conditions of stress occur too early and too deeply, particularly between bud-breaking and flowering, they hinder foliage development and completion of floral organs. The plant is not able to adapt to the hormonal change and compounds in the fruits accumulate in an unbalanced way (Champagnol, 1997; Van Leeuwen et al., 1997). Before veraison, a situation of water deficit has direct, immediate, and sometimes irreversible effects on the quantity and overall quality of production. In fact, during this initial phase the grape is involved in an intense activity of cellular multiplication, fundamental for its final size. After veraison, cell expansion takes place; controlled conditions of slight water deficit do not compromise the quality and weight of the grape and may even lead to a concentration of solutes and a maturation directed more toward the elaboration and accumulation of compounds that are valuable for the quality of the grape.

Ripening rate is mostly determined by the water regime of the grapevine (Van Leewen and Seguin, 1994). A water deficit is favourable to rapid maturation of the grapes since it limits berry size and reduces competition between grape berries and apices for carbon substances. However, it is necessary to avoid excessive stress even late in the season in order not to compromise canopy functionality and in final analysis some of the qualitative characteristics such as aromatic fineness (Poni, 2005).

Soil Available Water and Irrigation of the Vineyard

The vine water status estimate may be evaluated by different methods:

1. The water balance of the soil-plant system, calculated as a function of the water supplied by rainfall and of the consumption due to evapo-transpiration potential (ETP) corrected by the cultivation coefficient Kc.
2. Measurement of soil water saturation or of its potential, force with which the soil retains water.
3. Measurement of the water potential of the vine or of other specific organs of the plant, such as leaves and/or stems, indicating the amount of energy necessary to extract the amount of water a plant contains (Van Leewen et al., 2003).

After twenty years of observation of different varieties and in several vineyard-covered areas, Carbonneau (1998) proposed, in order to assess levels of water deficit stress, the measurement of a base leaf water potential Ψ b (predawn leaf water potential), measured before sunrise, when the stomata are still closed and the vine has been able to re-equilibrate its water state during the night, in relation to the soil water content. Stress is defined as absent or slight if the base leaf water potential stays between 0 and −0.2 MPa, slight or medium within −0.4 MPa, medium or strong within −0.6 MPa and strong under −0.6 MPa.

Ojeda et al. (2003) proposed different water management strategies for white grape vineyards and for red grape vineyards. In the latter, management involves the control of vegetation activity and the concentration of solutes deriving from the shrinking in size of the grapes, brought about by maintaining a water potential level corresponding to a controlled slight or medium water deficit stress in the phase between flowering and veraison, and medium or strong after veraison up to harvest. Instead, for the management of white grape vineyards, which must conserve brightness and freshness in aroma and also safeguard the supply of malic and tartaric acid, the most suitable irrigation strategies entail absent or slight water deficit stress between bud-breaking and veraison and a low water potential (between slight or medium values) from veraison to maturation, so as to gradually slow vegetative growth and obtain a slight shrinking in size of the grape berries. In Table 26.2, there are the cultivation procedures advised for Grenache Noir in a Mediterranean environment by Deloire et al. (2002, 2003), who evaluated soil water status by measurement of the base leaf water potential.

A physiological indicator has been used for describing the vineyard water regime during the ripening period, i.e., the ratio between the two stable carbon isotopes $^{13}C/^{12}C$, called $\Delta C13$, measured in the must sugars upon harvesting (Van Leeuwen et al., 2001). The test is based on the carbon isotope discrimination phenomenon that occurs during the course of chlorophyll photosynthesis: in fact, the grapevine prefers the absorption of the isotope ^{12}C, because in addition to being much more abundant in the atmospheric CO_2 (~ 99%), it is also lighter. Therefore, the sugars formed in the grape have carbon atoms with a higher percentage of ^{12}C isotopes than the atmospheric CO_2, and consequently in the grapes the $\Delta C13$ ratio is lower. In the case of water deficit stress, isotopic discrimination decreases in intensity. In fact, the leaf stomata close during the hottest part of the day, slowing the photosynthetic activity; the gaseous exchanges between leaf and atmosphere are reduced and therefore also the possibility of discriminating the absorption of the carbon isotopes. In these conditions the $\Delta C13$ tends to grow close to that of the atmospheric CO_2 (–8‰).

The isotopic ratio $^{13}C/^{12}C$ is measured by Isotope Mass Spectrometry and the $\Delta C13$ value is expressed as a variable by the thousand (‰) in reference to the international standard V-PDB (Vienna—Pee Dee Belemnite). For the vineyard the range of values varies between –21‰ in the case of strong water deficit stress and –26‰ or more in total absence of water deficit stress (Van Leeuwen et al., 2003). The latter authors have shown the high correlation between the predawn leaf water potential at maturation and the $\Delta C13$ measured in the sugars upon harvesting ($R^2 = 0.86$). Interest for this proxy is due to the simplicity of sampling: it is sufficient to reserve a representative sampling of the grapes or clusters upon harvesting or during the course of ripening.

All of the techniques for gauging regulated water stress in the vineyard are based on the principle of only restoring the amount of water lost to evapotranspiration (ET) in an endeavour to control vegetation growth and production, and achieve more efficient use of irrigation water. Experiments conducted in the past few years on Regulated Deficit Irrigation techniques and other irrigation tests demonstrated the fundamental role that a correct irrigation reintegration may have in the improvement of grape quality, defying the still-widespread prejudice in areas with a great viticulture tradition that resorting to irrigation is always associated with quality decline (Poni, 2000).

Mineral Nutrition and Fertilization in the Vineyard

The mineral elements that explicate essential and specific functions in the grapevine are carbon, oxygen and hydrogen available in the air. Among the macroelements available in water there are nitrogen, potassium, phosphorus, calcium, magnesium and sulphur, and among the microelements iron, manganese, boron, molybdenum, copper, zinc, cobalt and chlorines (Table 26.3). Another three microelements, silicon, aluminium and sodium, are always present in the grapevine but do not seem to be indispensable (Gärtel, 1971).

The quantity of mineral elements in the soil available to the grapevine depends on several fundamental factors:

- Concentration of mineral elements in the soil
- Soil pH
- Soil rooting depth
- Quantity, quality, and state of weathering of rock fragments
- Water and oxygen supply
- Vineyard floor management method (e.g., mechanically tilled soil, grass cover partial or total, temporary or permanent)
- Vineyard agronomic management (e.g., organic and inorganic fertilization, artificial drainage)

Nitrogen

Nitrogen is the principal macroelement. It increases the vigour of the vines and their production capacity, but an excess of nitrogen causes deterioration in wine quality and an increase of susceptibility to cryptogamic diseases and coulure. Most soils supply nitrogen to the vine through the mineralization of organic substances, a phenomenon influenced by the characteristics of the soil and pedoclimate and in other words by the *terroir*.

To obtain grapes with high oenological potential, nitrogen supply in the vineyard must be moderate. If excess nitrogen determines an elevated vigour and susceptibility to *Botrytis cinerea*, nitrogen deficiency may negatively influence the aromatic component of the grapes, particularly the white (Rapp et al., 1995). For vine species producing red grapes, low nitrogen addition favours phenolic richness of the grapes (Van Leeuwen et al., 2000). These authors propose the measurement of the total nitrogen and of the assimilable nitrogen of the must as a pertinent and simple indicator for the correct nitrogen alimentation of the grapevine. The interpretation of data in relation to the addition of nitrogen in the vineyard requires identification of threshold values that vary from species to species. Testing carried out in the Bordeaux region (Masneuf et al., 2000) shows that the threshold of 180 mg/L of assimilable nitrogen in must seems to abundantly satisfy the needs of the grapevine for a quality product. When this threshold is reached it is therefore advisable to cease nitrogen addition for the following year. However, in some quality red wine production situations, the levels of assimilable nitrogen in must may be less than 130 mg/L without diminishing the vigour and productivity of the vine.

Potassium

Potassium is an element of fundamental importance to the grapevine, which is considered the potassium-loving plant *par excellence* (Fregoni, 1980; Mpelasoka et al., 2003). In fact, it favours the accumulation of sugars in the berries and regulates the acid balance by salifying the organic acids of the grape. Furthermore, its sufficient presence in the soil regulates water economy and favours resistance to diseases and frost. During the growth of the canes, the quantity of potassium accumulated by the vine increases, reaching an initial maximum between flowering and curtailing of growth, to then successively decrease and increase again during

Table 26.2. Cultivation operations advised for Grenache Noir with varying water status of the grapevine in Mediterranean regions.

Grapevine water status (Ψb, Mpa)	Phenological status	Characterization of water deficit	Expected consequences	Comment	Advice for cultivation operations*
0 to −0.2 Mpa	From bud break to ripening	Absent to slight	Excessive vigour and dilution of metabolites in the grape berry	Unfavourable to quality	Drainage possible; vertical shoot training system; single or double cordon pruning; equilibrate the bud load at pruning; reduce nitrogen fertilization to a minimum; favour green pruning on the plant to reduce vigour; SFEp/production (m²/kg) ≥ 1; preventive stripping of leaves in the zone of the clusters at cluster setting and /or shutting.
0 to −0.2 Mpa	From bud break to flowering	Absent to slight	Normal growth	favourable	Adapt the yield in relation to the SFEp. Possible use of grass cover non-competitive with the vine (based on leguminous plants) if there are problems of erosion or trafficability.
−0.2 to −0.4 Mpa	From flowering to cluster shutting/ veraison	Moderate and progressive	Controlled vigour (vegetative and fruit growth slowed). No biochemical problems.	favourable	Adapt production in relation to the SFEp.
−0.4 to −0.6 Mpa	From flowering to cluster shutting/ veraison	From moderate to strong	Reduction and/or curtailing of vegetative growth and imbalance of the leaf surface: block of grape berry growth; possible problems in biosynthesis of tannins.	unfavourable	Avoid grass cover; suitable soil tilling; SFEp/production ≥1.25 (for example, 1 m² of exposed foliage surface per 0.8 kg of grapes per m² or per plant); cluster hand thinning to reduce load on the vine; possible gauged irrigation to increase leaf water potential −0.4Mpa (Ψb).
−0.4 to ≥ −0.6 Mpa	From veraison to ripening	From moderate to strong and progressive	Plant: slowing and/or curtailing of vegetative growth; reduction of photosynthesis; possible yellowing of basal leaves. Berries: reduction in growth; reduction in sugar deposits but final grade of alcohol favourable; possible stimulation of biosynthesis of anthocyanins, slow ripening without ulterior blocking; balanced concentration of metabolites; increase of skin/pulp ratio.	favourable	Keep under control grass cover (keep turf between the rows withspecies that are dry during this period to favour trafficability); SFEp/production ≥ 1.25 m²/kg

| > –0.6 Mpa | From veraison to ripening | Strong and very severe | Plant: curtailing of vegetative growth; yellowing and dropping of basal leaves; strong reduction in photosynthesis Berries: strong reduction in growth; possible inhibition of maturation; strong reduction in sugar deposits; problems in biosynthesis of anthocyanins; excessive concentration of metabolites; increase of skin/pulp ratio. | unfavourable in *terroirs* where water deficit is very severe during the course of maturation | No grass cover; SFEp/load > 1.25; cluster hand thinning to be considered; suitable soil tilling; possible irrigation before veraison so that the vines stay at –0.6 Mpa (Ψb) at the beginning of ripening. |

* *It is supposed that the rootstock is suitable to the soil type. The ideal situation is written in bold. SFEp: surface foliage exposure potential.*

Table 26.3. Origin of mineral elements present in the grapevine.

	Macroelements	Microelements
From air and water	carbon, oxygen, hydrogen	–
From soil	nitrogen, phosphorus, potassium, calcium, magnesium, sulphur	iron, manganese, boron, molybdenum, copper, zinc, cobalt, chlorine

ripening. Daily requirements are about 1–2 kg/ha/d during the growth phase and 3 kg/ha/d in the case of highly productive vineyards. The annual consumption per hectare of leaves, branches and clusters is about 25–80 kg (Reynier, 2000).

Potassium absorption differs as a function of the vine species and the rootstock: weak for 1103 P, 140 Ru, 41 B, 110 R; strong for SO_4 and 44–53. In the case of potassium deficiency, during the summer the leaves show discoloration and then yellowing for the white-grape varieties (flavescence), or reddening for the red-grape varieties (mineral red) that progressively reach the veins and all or part of the blade and the margins of the leaves that droop down. During ripening, the very productive plants may show black or dark purple splotches on the upper surface of the leaf, while the underside stays green. The leaves dry up and fall, first the basal leaves and then the higher ones. Causes may be found in the nature of the soil (very clayey or very sandy), in excessive vigour and varietal susceptibility. On the other hand, soils excessively rich in potassium increase competition with manganese, up to its saturation, with possible appearance of pathologies such as rachis necrosis (Bavaresco, 1989).

Magnesium

Magnesium is an important element for the grapevine. Constituent of the chlorophyll molecule, it neutralizes organic acids and joins calcium and potassium in the ionic intracellular balancing. It is the most common activator agent of the enzymes involved in energy metabolism and favours the absorption of other nutrients. Fertilization based only on K reduces the level of Mg in plants, while the opposite does not seem to be true. In soils with a pH of 5–5.5 and in sandy soils the absorption of Mg is notably reduced. In fact, not all of the rootstocks have the same behaviour regarding this element. The SO_4, the Kober 5 BB and the Fercal, for example, absorb it sufficiently, but the 140 Ru and the 1103 Paulsen do not (Boselli, 1986). It is therefore important not to combine varieties susceptible to rachis necrosis with rootstocks that have difficulty absorbing magnesium, for example: Pinot Noir on K 5BB and Riesling on SO_4.

Magnesium deficiency may show in light and acidic soils subject to lixiviation or in soils that have received excessive potassium fertilization. The symptoms first show on the leaves of the base and proceed upwards, in a completely different way from potassium deficiency. At flowering, or even sooner, some necroses appear in a circular position at 10–20 mm from the edge of the margin. After flowering or at the beginning of summer the leaves close to the clusters present yellowing in the white-grape varieties and yellow and then red or directly red in the red-grape varieties. These colorations, marginal at the beginning, advance in digitations between the venations.

Besides being important for the metabolism of the grapevine, magnesium behaves like a glue that aids the aggregation of the soil particles into granular structures. However, in high concentrations Mg may cement the soil structure, slowing internal drainage of the land. If the content in the soil is high compared to K (K/Mg < 0.5) it unbalances the water absorption mechanism of the plant, making the plant seem in need of water even when it is not. A low K/Mg ratio may also give the wine an aroma of green grass or vegetable that may be undesirable. Lands on serpentinite, alluvial soils with sediments deriving from these rocks, often present this problem. Furthermore, these soils may have low structural stability, since excess magnesium may act as a disperser.

Even if potassium fertilization may remedy the excess magnesium, in soils on serpentinite there is still the problem of nickel, contained in the form of trace element that may still be toxic to the grapevine. The choice of rootstock may also be of help in soils rich in Mg, because some varieties are more tolerant than others. Another system is that of preventing water washing over these rocks from reaching the vineyard by runoff or hypodermic flow, using guard ditches and drainage ditches above them.

Calcium

Calcium is the element most absorbed and utilized by the grapevine, but in many Mediterranean territories it is often present at sufficient levels. Lands with calcium carbonate levels of at least 6% positively affect the quality of the grapes in terms of higher sugar content and aromatic fineness of the wine produced. When levels are too high it determines iron chlorosis. Some tests have shown that the ratio of 6:1:1 between Ca, Mg and K is the best condition for fertility in achieving great wines.

Phosphorus

Phosphorus carries out very important biologic functions, entering in the composition of vitamins, nucleic acids and energy transporters that guide sugar metabolism. It does not seem to have an effect on productivity except when it intervenes to correct nitrogen deficiency, reducing plant susceptibility to cryptogamic diseases and coulure. It is, however, considered a regulator of plant development. Vine shoots that are well supplied with P_2O_5 are those that show a higher fertility of the buds. Phosphorus also favours development of the roots and is therefore indispensable in the planting phase of the vineyard.

Phosphorus is generally immobile in the soil and is found in forms that are available to the plant to different degrees: in the water solution it constitutes a limited, but immediately available reserve; absorbed by the clay-humic complex it constitutes the preponderant part of the exchangeable reserve that is easily freed to compensate the loss by subtraction from the circulating solution. This form, together with the previous, constitutes the so-called assimilable reserve. Connected with organic matter, it constitutes a momentarily non-available reserve that is progressively releasable through biological decomposition, while it is totally useless in the form of phosphates, which are insoluble or poorly soluble and are derived from bedrock, the progressive precipitation of phosphates dissolved in calcareous soils, or, on the contrary, very acidic soils (SCPA, 1986).

Organic Matter

The organic matter of a soil comprises the set of living organisms and plant and animal remains present in a soil at a given moment, either added or of a natural origin. Its contribution to soil fertility is common knowledge; it is enough to remark that *humus* combines with the clays to form the clay-humic complex that is at the basis of the pedological structure. Furthermore, humified organic matter improves water retention, buffers the pH, regulates biological activity, and increases the cationic exchange capacity. Finally, its decomposition leads to the release of nutritive elements in the circulating solution. Notwithstanding its important functions, it is difficult to establish thresholds of minimum requirements or ideal supply of organic matter in a soil. This is because its role is always indirect (the plants are autotrophic organisms and therefore do not use the organic matter in the soil) and varies considerably according to soil type, in particular, as a function of the texture and structure, playing an increasing role in soils that are clayey and poorly structured. Also, in lands covered with vineyards the role of organic matter may be negative, by being a source of excessive nitrogen or by inducing conditions of excess fertility. It is a fact that often lands of high quality wines are very poor in organic matter. Table 26.4 shows some indicative values for the levels of organic matter in viticulture districts.

Table 26.4. Levels of organic matter in soils (Bavaresco, 2005).

Organic matter %	
Very poor	< 1
Scarce supply	1–1.5
Medium supply	1.6–2.0
Good supply	2.1–2.5
Rich	> 2.5

Iron

The iron absorbed by the root system is indispensable to chlorophyll synthesis and has a role in processes of respiration and photosynthesis. Iron is present in the soil, where it is generally absorbed without problems by the plant root system. However, an excess of bicarbonate anions and the presence of other unfavourable conditions, such as alkaline pH in the soil, may cause a chlorosis problem.

Iron chlorosis is the direct manifestation of an iron deficiency at a foliage level that reduces or curtails the synthesis of chlorophyll. This iron deficiency may be due to either an insufficient absorption of iron by the roots or by an insufficient migration, or by both phenomena simultaneously. Iron is normally present in the soil in numerous mineral forms, whose solubility is very weak. In calcareous soils under the action of ground water loaded with carbonic gas, calcium bicarbonate (HCO_3^-) is formed, which maintains the pH of the soil solution between 7.5 and 8.5 and reduces the concentration of soluble iron available to the plant. The higher the active lime concentration, the greater the risk of chlorosis problems. This concept was translated into the chlorosing power index (CPI) (Juste and Pouget, 1972):

$$CPI = CaCO_3 \times 10^4 / Fe^2 \qquad (1)$$

where $CaCO_3$ is the percentage of active lime, and Fe is the extractable iron in ammonium oxalate expressed in mg/kg. The maximum threshold of CPI tolerated by the most important rootstocks is shown in Table 26.15.

Other factors that can make the symptoms of chlorosis worse are the following: insufficient supply of starch, for example, because of an excessive harvest the previous year, or because of a weak accumulation of starches in young plants, since plants under 2 or 3 years of age are more subject to chlorosis; the high vigour of a vine that increases its iron requirements; the varietal susceptibility of the species or of the rootstocks (there are grapevine varieties particularly susceptible such as Pinot, Cabernet Sauvignon, Picolit, Moscato d'Amburgo, Ribolla gialla, just to mention the most well known, while rootstocks such as Fercal, 140 Ru and 41B are particularly resistant: Bavaresco, 2002); the agronomic conditions of the soil: the agricultural practices that favour the breaking up of calcareous elements (soil tilling); the release of carbonic gas (addition of fresh organic matter); and plentiful nitric manuring.

Symptoms of iron chlorosis start during the phase of spring growth with the yellowing of apical leaves beginning at the extremities. This discoloration involves the margin of the leaves while the venations stay green. Growth slows, branches are discolored in the same way and in the most severe cases desiccation of the bud occurs, followed by the death of the plant (very rare). The disease makes itself manifest often at break of dormancy when growth resumes and then disappears, but not without harming plant metabolism and affecting production and longevity. Production may be reduced by as much as 50%.

Boron

Boron is a microelement that is indispensable to vegetal organisms, because it affects the metabolism of nucleic acids, protein synthesis, meristematic activity and, in particular, the development and functionality of reproduction organs. It also acts on the translocation of carbohydrates and in general on the transportation of many substances through the vegetal membrane (Jackson et al., 1975).

The physiological role of boron in the grapevine is confirmed by some characteristic deficiency symptoms. Boron deficiency may manifest itself in soils poorly supplied in the element by nature, or in soils from which the boron has been leached. Deficiency is also frequent even when boron is present in sufficient quantities, but in forms not easily assimilable. A considerable amount of boron is contained in organic matter or is absorbed on the surface of humic colloids; the boron content usually follows the humus content in the soil profile and it accumulates in humic horizons. It is also present in variable quantities in the solution circulating in soils under the form of borate solutions or of non-ionized boric acid. Under this form it is more easily absorbed by the plant, but it is just as easily leached away (Corino et al., 1990).

The types of soil subject to deficiencies of boron are those strongly calcareous, with very plastic clays, and sandy soils poor in organic matter (Table 26.5). In fact, the correlation between assimilable boron and organic matter is highly significant. Boron deficiency occurs most frequently in conditions of scarce water availability and high air temperatures and during periods of rapid development: in fact, boron is highly mobile in the xylem and poorly mobile in the phloem. Symptoms include the degeneration of the apical and radical meristems,

Table 26.5. Critical levels of boron in different types of soil (Bavaresco and Vercesi, 1996).

Boron levels (mg/kg)	
Clayey soil	< 0.8
Loamy soil	< 0.5
Sandy soil	< 0.3

with the consequent slowing in growth of buds and roots. Deficiency symptoms appear in the flowering period initially in the weaker stocks and in the poorer zones of the vineyard and are more evident in the younger parts of the plant and at resumption of vegetation. The primary leaves remain small and the successive ones, in the distal part of the buds, present characteristic mosaic discoloration with yellow or bright red spots between the veins, first distinct, which then tend to form digitations between the venations. The principal buds often have crowded internodes and necrotic apices because of obstruction in the vascular tissues. In severe cases the flowers fall and inflorescences dry up, the berries involved remain small and take on plumbeous tones (Corino et al., 1990).

Copper

Copper is an essential element for the vegetation life of the common grapevine. It influences almost all of the metabolic activities and acts directly upon lignin synthesis and on reproductive processes. It also appears to be indispensable for the synthesis of chlorophyll, proteins and vitamins A and C (Fregoni and Corallo, 2001). Copper deficiency, rather rare in vineyards, causes pollen sterility and manifests itself by spotted necrosis at the margin of the blades, precocious fall of apical leaves and tip dryness.

Treatments against downy mildew with cupric products may cause an accumulation of assimilable copper in the upper layers of soil, especially in soils with low pH, poor in organic materials, active lime and mineral colloids. Testing carried out in several vineyards demonstrated that copper content may notably vary, from a few units to hundreds of ppm (Costantini, 1999). The normal ratio is between 20 ppm (sandy soils) and 100 ppm (clayey soils). A content greater than 100 ppm is considered abnormal. As for surplus, the grapevine eludes high levels of copper in the soil, because it has a root system distributed between 20–30 and 70–100 cm, in which copper levels are generally low.

Copper is an antagonist of Fe, blocking its absorption and use by the plant. Also, high doses of phosphorus associated with excess copper increase the risk of iron chlorosis. Excess Cu may be dealt with by use of organic matter, because the humic acids are capable of absorbing the Cu, making it insoluble.

Zinc

Zinc is one of the principal oligoelements. The grapevine needs 100–500 g of zinc per hectare each year. Higher concentrations may cause toxicity. Zinc tends to accumulate in soils with a pH above 6 and in the proximity of galvanized vine stakes, where 2,300 ppm concentrations have been registered (Müller and Breiter, 1999). The effects of toxicity may appear in the form of delay in growth or growth restricted to roots. Toxicity may manifest itself in the plant not only due to excess zinc in the tissues, but also indirectly by influencing other essential elements. In fact, high levels of zinc may have a negative influence on the iron content of the plants, it being an antagonist to iron absorption.

Exchange Capacity and pH

The nutrients are present in the circulating solution mostly in the form of cations and are held by the negatively charged sites of the clays and organic materials, which determine the cationic exchange capacity (CEC) of the soil. The minimum level of 10 meq/100 g in fine soil is generally considered sufficient for the grapevine, which has limited requirements. Some very sandy soils with limited organic materials may, however, have such a low CEC that it has a negative influence on production and on wine quality.

The pH of a soil in water suspension may be considered a synthetic indicator of chemical fertility. The reference value for the grapevine is between 5.5 and 8.2. Outside this range, toxicity and deficiency phenomena and nutritive imbalance are probable, which may influence the grapevine directly or indirectly by acting upon the microflora.

All of the metallic microelements are more soluble at low soil pH (Fig. 26.4), while calcium, magnesium, phosphorus, and potassium deficiencies may occur. Nitrogen availability may also suffer, because the bacteria capable of transforming organic substances develop with greater difficulty in very acidic soils (Tables 26.6 and 26.7).

Table 26.6. Sufficient levels of exchangeable K_2O, MgO and CaO as a function of soil cation exchange capacity (CEC) (Cottenie, 1980).

CEC meq/100 g fine soil	exchangeable K_2O meq/kg	exchangeable MgO meq/kg	exchangeable CaO meq/kg
< 10	0.37–0.74	0.84–2.08	9.99–24.97
10–20	1.27–2.23	3.27–6.60	49.93–78.46
> 20	1.91–3.82	4.96–9.92	99.86–149.79

Table 26.7. Optimal supply of some nutritive microelements in soils with different pH values (ppm assimilable-soluble EDTA) (Vercesi, 1987).

pH in KCl	Fe^{+2}	Zn^{+2}	Mn^{+2}	Cu^{+2}
4.5–5.5	25–30	2.5–4.5	25–35	1.0–2.0
5.6–6.5	30–35	2.0–2.5	20–30	1.0–1.5
6.6–7.5	35–50	2.5–3.5	25–40	1.0–1.5
7.6–8.5	50–75	3.5–7.0	40–55	1.5–2.0

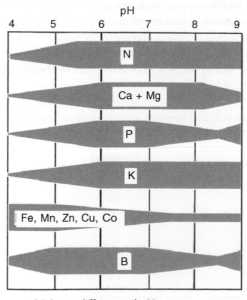

Fig. 26.4. Mineral element availability at different soil pH.

Salinity and Sodicity

It is assumed that the grapevine is a species with moderate salinity tolerance, being able to tolerate a salinity content proportional to an electric conductivity of up to 1 dS/m (Sbaraglia and Lucci, 1994). The depth of the soil profile needed to measure salinity is about 80 cm, which often corresponds to the available rooting depth. To determine the salinity of a soil use an instrumental measurement at 25°C of the electric conductivity using solutions of soil in water at different concentrations.

The presence of sodium in the circulating solution and in the soil exchange complex significantly limits rooting depth of all plants. For the grapevine in particular, the level of sodium in the exchange complex should not exceed 8% (exchangeable sodium percentage, ESP < 8).

Manuring

The grapevine is a plant with limited nutritional requirements (Table 26.8), especially regarding nitrogen and phosphorus, while compared to other agricultural cultivations it requires more potassium and calcium. Fertilization management entails the addition of nutritional substances that are scarcely available in the soil, such as P, K, Mg, and B if their deficiency is suspected, to allow easier access to young vines with underdeveloped root systems. The addition of organic substances, recommended in the past to improve land structure, should in fact usually be avoided, so that problems of excess soil fertility are not created. It must also be considered that it is increasingly difficult to find organic substances that have been well humified and consequently the use of low quality organic materials is increasingly widespread. The spreading of organic mass that is not well humified on soils before planting may disturb the equilibrium of biological activity, or reduce it, particularly in somewhat poorly drained soils. In these soils fermentation may take place, with a reduction in oxygen availability and production of toxic gases that can cause the death of the young plants. In general, it is always preferable to preserve the natural soil horizon for cultivations, instead of attempting to build up fertility of soils or sediments from which the upper stratum has been removed because of stripping. Fertility management must be geared as a function of the nutritional state of the vineyard, evaluated by physical-chemical analysis of the land, by leaf and/or petiole diagnosis and by keeping record of uptake by the production of grapes, wood and leaves.

Table 26.8. Minimum and maximum uptake of micro and macroelements in the production of grapes, wood and leaves (between 70 and 250 q/ha) (Fregoni, 1980).

Nutritive elements	Minimum	Maximum
Nitrogen kg/ha	22	84
Phosphorous kg/ha	2	15
Potassium kg/ha	34	123
Calcium kg/ha	20	146
Magnesium kg/ha	3	15
Boron g/ha	37	228
Iron g/ha	292	1,121
Manganese g/ha	49	787
Copper g/ha	64	910
Zinc g/ha	110	585

26.3.5. Landscape

The need to follow a multifunctional and environmentally sustainable model for development in viticulture brings the farmer and the territorial planner to consider aspects of management and improvement of the rural landscape, in addition to the protection of the environment. Landscape protection is an increasingly affirmed value following the profound changes that in the agricultural world have brought about the disappearance of characteristic "cultural landscapes" (ESDP, 1999), such as the typical terraced and mixed cultivations. Intensive plantation models, which strive to optimize labour organization, have been widely applied, often not taking into consideration soil conditions and in many cases causing a drop in soil quality, particularly in the hillside districts where orchards are widespread, particularly vineyards (Costantini, 1992). Besides the aspects connected with soil quality, landscape should be considered an economic asset to be preserved. As quoted from the "Regional plan for the restructuring and re-plantation of vineyards" of the Region of Tuscany (2001), "Recent studies state that two thirds of the value of a great bottle of wine is to be attributed to immaterial factors. The CENSIS concluded that in orientation and behaviour the imaginary prevails upon the material and wine increasingly involves an emotional experience; wine evokes environments, local landscapes and cultures and wine-cellar atmosphere; wine districts promote themselves as tourist attractions." The image of the hillside landscape evoked by a great wine is characterized by the agricultural structure typical of the area. In Tuscany, for example, there is the simultaneous presence of vineyards, olive groves, wooded and cultivated areas that alternate within the landscape in a rational and balanced way, and that is the context in which the wine cellars, buildings and farming structures are placed.

Vineyard placement in the landscape is not yet evaluated in an integrated manner, so each specialist tends to give greater importance to his or her own field. It is a subject that may be examined from many viewpoints: from the choice of the lot, its dimensions and its aspect, its distribution among other cultivations, the preservation of hedges and trees, hydraulic-agricultural systems, orientation of the rows, and even the choice of the materials for the vine stakes.

Below are presented some simple evaluation methods for the environmental and landscape impact of viticulture in hillside areas that were developed in the province of Siena (Costantini et al., 2004).

Evaluation of Environmental Impact of Agro-ecosystems (Agro-ecotopes)

This method was developed in order to evaluate the changes that occurred in Tuscany following the diffusion of specialized cultivation during the 1970s, financed by the FEOGA (European Agriculture Guidance and Guarantee Funds) of the EEC, but it may also be used for evaluating different possible scenarios for agricultural development. It is based on the creation of soil maps, agro-hydraulic system maps, and land use maps. Between 1965 and 1994, the use of lands in the test hydrographic basins studied in San Gimignano has changed profoundly: specialized tree plantations, in particular, have increased tenfold, while the traditional mixed cultivations have decreased by about 80% (Figs. 26.5 and 26.6). Always in the same time period, the traditional transversal layout of fields *(cavalcapoggio, girapoggio)* has been almost totally replaced by that of *rittochino* (cultivation up and down the slope). Also, in fields sited on plains, ditches were placed further apart and, on the more gentle slopes, ditches often disappeared completely.

Fig. 26.5. South-east slope at San Gimignano: traditional agro-ecotopes, with mixed cultivation.

Fig. 26.6. North-west slope at San Gimignano, the new agro-ecotope, with specialized vineyards saturating the landscape unit.

The most visible change in the agricultural landscape (Table 26.9) is the appearance of the agro-ecotope characterized by specialized arboreal cultivations (mostly vineyards) with a rejuvenated soil profile because of anthropic erosion. In this area, sometimes unstable due to geomorphological causes, the placement of specialized cultivations with the *rittochino* layout and the intensivee remodelling of the horizon by land levelling and deep tilling have caused the outcropping of the pedogenetic substratum, favouring horizon erosion and in some cases increasing the tendency for landslides to occur. The agro-ecotope most characteristic of the "Tuscan landscape", the traditional mixed cultivation on hillsides in areas that are stable, preserved from erosion and on soils with good fertility and hydraulic characteristics, have decreased almost by 79%.

Table 26.9. Variations of the agro-ecotopes in the basins of Vergaia and Borratello in San Gimignano (SI).

Agro-ecotypes	1965		1994	
	ha	%	ha	%
1: mixed cultivations on stable land	393.9	18.5	83.2	3.9
2: traditional fields with trees on plains (Tuscan banks)	83.1	3.9	36.4	1.7
3: bare sowing on plains (wide fields)	63.7	3.0	143.0	6.7
4: specialized cultivation in stable areas	15.9	0.7	217.6	10.2
5: specialized cultivation in unstable areas	31.7	1.5	287.0	13.5
6: bare fields on hillsides in unstable areas	525.7	24.6	496.7	23.3
7: bare fields on hillsides in stable areas	107.2	5.0	71.7	3.4
8: urban areas	45.7	2.1	49.3	2.3
9: water bodies	5.1	0.2	6.1	0.3
10: wooded areas (also poplars)	274.2	12.9	335.9	15.7
11: mixed cultivations on hillsides in unstable areas	516.7	24.2	91.3	4.3
12: fallow and fields on escarpment	62.8	2.9	136.3	6.4
13: mixed areas	7.3	0.3	9.4	0.4
14: specialized tree cultivations in areas presenting anthropic erosion	0.0	0.0	169.2	7.9

Evaluation of Cultivation Length

The length of the vineyard has been chosen as a landscape indicator because of the evident scenic impact of very long plots. To this must be added that of soil conservation, which should, however, be complemented by the indication of the type and spacing of hydro-agricultural layouts.

The dimensions of vineyards were evaluated using the Corine (1990) soil use data bank, on the aerial photo AIMA on a scale of 1:10,000 (1987). To evaluate in particular the length of the vineyard, 3–7 transects were measured for each polygon of vineyard-covered land, traced along the line of maximum gradient of the vineyards, starting from the crest and stating the results as maximum and average gradient values of the vineyards present in the polygon. The values were grouped into the classes 150, 300 and 450 m, corresponding to a limited, medium and strong impact.

The results indicated that in the province of Siena the areas with vineyards 150 to 300 m long predominate, therefore, there is medium landscape impact. However, there are also vineyards with a strong impact, most of all in the township of Montalcino (Fig. 26.7).

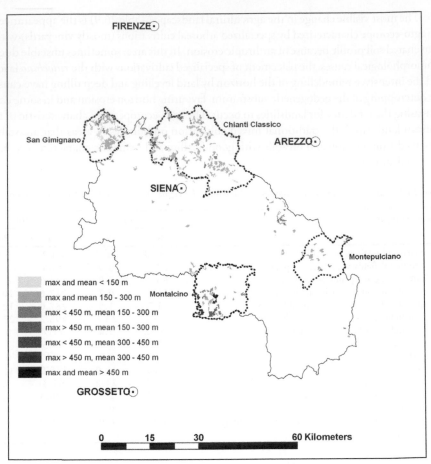

Fig. 26.7. Length classes in the vineyards in the province of Siena.

Viticulture Saturation of the Landscape

The landscape impact of the vineyard may also be seen as a reduction in cultivation diversity. For this reason, an indicator was developed for "viticulture saturation of the landscape".

Viticulture saturation of the landscape around Siena was obtained by crossing the Corine data on specialized vineyard with the data bank of the Territorial Systems (scale of reference 1:100,000) of the Region of Tuscany. The data indicates the percentage of vineyards specializing in a certain landscape system, i.e., in a characteristic combination of factors with reference to scales of lithology, morphology, and land use, to which corresponds a certain association of lands. This indicator showed that the Siena landscape is more saturated by the traditional DOCG areas, over the non-denomination areas or the recently denominated (Fig. 26.8). Among the DOCG areas, those most saturated are San Gimignano, Chianti, and Montalcino, while the best situation is present in Montepulciano.

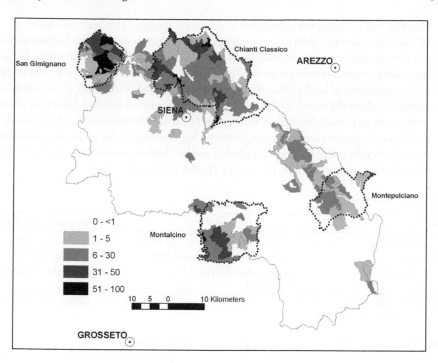

Fig. 26.8. Viticulture saturation in the Siena landscape (%).

26.4. Soil Survey and Viticulture

During a soil survey, it is possible to observe a series of site and soil profile conditions that may be useful to the planning of viticultural operations. This section illustrates the details of some aspects particularly important to the decisions to be made before planting a vineyard, decisions that involve the evaluation of difficulties in the agro-technical management of the land, modalities and depth of ditches and soil and water conservation measures, in addition to the choice of optimal vine density, rootstocks and grass cover.

26.4.1. Evaluation of Difficulties in Agro-technical Management of Land

- Slope is the factor that most affects the potential to use mechanization and the safety of agriculture workers.
- Surface stoniness indicates the percentage of area (around the observation site) occupied by rock fragments of diameter greater than 75 mm.
- Rockiness indicates the percentage of area (around the observation site) occupied by materials with diameter greater than 500 mm, not removable by normal work methods.
- Coarse fragments in surface horizons are expressed as the percentage of volume occupied by rock fragments of diameter greater than 2 mm in the horizon surfaces that are involved in common operations of soil turning by farm machinery.
- Soil depth indicates the depth from the surface to the first continuous and coherent rock-bed underneath, or the depth of contact with the paralytic limit, such as a densely packed soil layer (hardpan).

- Flooding risk indicates the probability of this event during a 100 year time span.
- Workability and trafficability are complex properties that are defined on the basis of not a single property, but a combination of several physical-mechanical characteristics. For an approximate evaluation, the property of texture is used, which is the data most influential in the estimation of land workability. By definition, a soil is workable if the friability conditions allow the creation of the desired structural conditions after the action of the farm machinery. Such conditions are hindered by a high content of clay that increases the degree of cohesion between aggregates, by a high content of clay and loam combined that increases the risk of sinking after rainfall, and by a high content of loam that increases the risk of crust formations on the surface (Tables 26.10 and 26.11).

Table 26.10. Workability classes.

Easy (W0)	Difficult (W1)	Very difficult (W2)
clay + loam ≤ 65% sand total ≥ 35%	clay + loam between 65 and 85% sand total between 35 and 15%	clay + loam > 85% sand total < 15%
texture class (USDA)*: s, ls, sl, scl, sc, l, sicl	texture class (USDA): sicl, cl, sil	texture class (USDA): c, si, sic

*s=sand; ls=loamy sand; sl=sandy loam; sc=sandy clay; l=loam; sicl=silty clay loam; cl=clay loam; sil=silty clay; c=clay; si=silt; sic=silty clay.

Table 26.11. Table of difficulty classes for intensive agricultural management.

Quality and characteristics	Suitability class			
	S1	S2	S3	N
Slope%	≤ 14	15–21	22–35	> 35
Surface stoniness % (rock fragments between 250 and 500 mm)	< 0.3	0.3–15	16–50	> 50
Rockiness % (rock fragments > 500 mm)	< 2	2–10	11–25	> 25
Rock fragments % horizon Ap or A	< 35	35–70	> 70	
Rooting depth (cm)	> 100	50–100	25–49	< 25
Flooding risk	none	rare	occasional	frequent
Workability, trafficability (class)	work without particular problems(W0)	work difficult due to clayey or loamy texture (W1)	work very difficult due to clayey or loamy texture(W2)	–

26.4.2. Soil and Water Conservation Measures

Soil and water conservation measures to be operated before the installation of a new vineyard must be regulated as a function of the horizons and other factors limiting rooting depth. As a standard, a horizon is considered explorable by root systems when at least 30% of its volume is penetrable to roots. Therefore, the presence of a hard rock without cracks is not the only factor that can limit rooting depth. The usable depth may be inferior to the real depth of the soil owing to the following limitations:

- Presence of a shallow water table
- Presence of cemented horizons
- Scarce macro-porosity and oxygenation

- Excessive compaction
- Excessive rock fragment content
- Vertic (shrink-swell) characteristics
- Unfavourable chemical conditions

On soils found on reclaimed lands, where a shallow water table is present (within 1 m of the surface), viticulture is usually discouraged because of the difficulty of obtaining a high quality product. However, even on the hillsides there may be perched water tables, even if temporary, over low permeability horizons or in morphological situations where the ground water circulating tends to pool, keeping the roots from going deeper and making piping or ditching drainage necessary.

In some soils there is a calcareous B horizon characterized by high total calcium and active lime content, as well as by difficult penetration of grapevine roots. In these situations it is opportune to effect suitable operations, such as surface harrowing accompanied with deep tilling by a subsoiler (two-layer cultivation), so as to break the continuity of these horizons. It is important, however, to avoid bringing to the surface the layers with undesirable characteristics. It is also important to use rootstocks resistant to high levels of active lime. The same may be said of soils formed on saline substratum, where the climate is not moist enough for the leaching of salts and where the removal and consequent leveling of the land may cause the outcropping of sediments still rich in salts. This is the case of many clayey sea sediments of the Pliocene and the Miocene geological eras, rather diffuse in central and southern Italy.

Many Italian lands, poor in organic matter and intensively cultivated for many years, subject to the traffic of increasingly heavy farm machinery, present low macro-porosity and are highly compacted, which can make them difficult to be penetrated by roots, especially by those of young grapevines. It is precisely to increase macro-porosity that the soil is broken, by harrow, subsoiler, or ripper, before new vineyards are planted. However, it must be taken into consideration that the soil found under the topsoil may be more compacted and impenetrable than the topsoil itself and, if brought to the surface, can worsen soil aeration conditions.

In paleosoils, that is, in very ancient soils, in particular those having horizons cemented by iron oxides, there are problems related to both physical and chemical fertility. In fact, impenetrability can be accompanied by a chemical poverty, due to low exchange capacity and base saturation. In these cases, besides operating to break the ferric crust, it may also be useful to add lime and manure. On the other hand, other paleosoils present a compacted but not completely impenetrable B horizon, which may limit excess fertility of the soil and contribute to the creation of high quality wines. This is the case of some paleosoils of Montepulciano and of Montalcino in Tuscany (Costantini and Lizio-Bruno, 1996; Costantini and Priori, 2007).

Lands with dynamic clays, like smectites, present swell-shrink phenomena, called "vertic" phenomena, which lead to the formation of deep cracks in the summer and the slipping of soil macro-aggregates in the wet season. These phenomena can lead to the breaking of vine roots, particularly the large, deep ones. Soils are called "vertic" when there are cracks in the summer and pressure faces that tend to give aggregates a wedge shape due to slickensides leaning on each other. These conditions must be evident in a horizon at least 15 cm thick within 125 cm from the surface.

In soils evolved on high-energy alluvial sediments, for example, fluvial terraces and alluvial cones, it is sometimes possible to find under the topsoil just a few centimetres thick a very strong layer of gravel or stones, often calcareous, almost devoid of fine earth, that makes root deepening very difficult. In this case the roots can explore only the mass of fine earth, which should not be further reduced or mixed with the underlying stones. This is the case of the famous "Grave" of Friuli. However, in other environs, like those of marine conglomerates, the presence of rock fragments and surface stoniness, even in abundance but immerged in a fine matrix, should be considered a positive factor for the grapevine, because it limits excess fertility in soils, favouring drainage and protecting the horizon from erosion (mulching effect).

26.4.3. Soil and Plant Density

The choice of plant density, or in other words vine spacing and row spacing, is a function for every area of land capability in terms of capacity to sustain the development and productivity of the grapevines (Silvestroni and Paliotti, 2005). In more fertile soils, where the development possibilities of the plant are better, spacing must be wide in order to optimize quality and quantity of the grapes. In fact, grapevines that are too close together would present excessively dense foliage, resistant to the penetration of sunlight, and would require mass operations of rebalancing such as lopping, leaf-stripping, and disbudding. In fact, achieving high quality levels in grapes seems strictly related to a right equilibrium between the root system and the crown, i.e., between quantity of water, mineral elements and carbohydrates capable of allowing a correct growth of shoots up to the veraison, when it is necessary that the vegetative activity be curtailed to favour ripening of the grape.

Poorer lands require a higher planting density. In fact, with crowding in the rows, the dimensions of the root system decrease, because the space available decreases and competition between plants increases. Consequently the development of the aerial system is limited, reducing on one hand the quantity of metabolites available for the growth of the roots and, on the other, the capacity to absorb mineral substances for the growth of shoots, until it reaches a condition of equilibrium. The vine spacing considered advisable at present on the basis of eco-pedological characteristics of the vineyard is shown in Table 26.12.

However, the choice of planting distance is influenced by the trellis system, type of pruning and vineyard mechanization. The limits imposed by mechanization are strong when considering the distance between rows, which in general is decided on the basis of the width of the machinery operating in the farm. Since this width varies between 1.8 and 2.2 m, for the small and medium farms where the machinery is employed also for other cultivations, the free space between rows becomes 2.5–3.0 m. Today there are tractors available on the market designed for use in the vineyard that can work with a space of about 2 m between the rows. Under this distance it is necessary to use over-the-row machines for all of the vineyard operations, which are very costly and therefore amortizable only for big businesses. This shows that the distance between rows presents, on the same level as all the other factors, a

Table 26.12. Vine spacing according to soil fertility (by Silvestroni and Paliotti, 2005).

Fertility	Distance on the row (m)
Poor soils	0.80–1.00
Soils with medium fertility	1.10–1.40
Very fertile soils	1.50–2.00

fundamental variable for the increase or decrease of productivity per unit of surface without altering grape quality (Intrieri et al., 2004).

In high-density vineyards with a distance of 1–2 m between rows (close and short vineyards), root density is important and the vegetation ensures a foliage cover that is more homogenous if the rows are closer, because loss of illumination between the rows is less. In fact, the height of the vegetation must be proportional to the distance between the rows, to avoid shadow zones from one row to the next and strong water deficit stress: the *optimum* is situated between 0.8 and 0.6 times that distance (the height of the vegetal wall is measured from the highest tip after hedging and the lowest point of vegetation). In vertical shoot positioning the rows are very close, just 1 m apart. This seems suitable to areas with very low fertility, where the grapevine presents a limited development of shoots that stop growing early and form crowns with a height that may easily be contained within 0.5–0.6 m (Silvestroni and Paliotti, 2005).

If the distance between the rows is more than 2 m, a vineyard is called tall and wide. Each plant explores a wider volume of soil, but root density is weaker. The overall vigour is high and there is the risk of leaf crowding with the consequent formation of an unfavourable microclimate both for the leaves and the clusters. On the side with sun exposure the first layer of leaves and, intermittently, the second layer receive energy for a sufficient photosynthetic activity. The other layers of leaves, located inside the vegetation, have a very weak photosynthetic activity. The stomata remain open more frequently favouring intense transpiration and high water consumption, therefore a microclimate suitable to cryptogamic diseases. Furthermore, they overshadow the clusters, slowing their ripening processes. When the distance between rows is sufficient to avoid interference between roots and reciprocal shading between vines of contiguous rows, there is no alteration in the quality of the grape. In lands of medium fertility such interference tends to disappear using a distance between rows slightly over 2 m (Intrieri et al., 2003). Therefore, since in the majority of vineyards the distance between rows is 2.5 to 3.5 m, this variable does not influence the quality of the grape, but changes the total length of the production walls per unit of surface, affecting the quantity of grapes produced in the vineyard (Tables 26.13 and 26.14).

Table 26.13. Comparison between close, short vineyards and tall, wide vineyards with equal yields (Reynier, 2000).

Characteristics	Close, short vineyards	Tall, wide vineyards
Vine variety	Cabernet sauvignon/SO_4	
Density	5050 (1.8 × 1.1 m)	2525 (3.6 × 1.1 m)
Canopy management	Vertical shoot	
Height from soil to leaf	1.50 m	2.10 m
Vine/load (average)	9.5 buds	16.8 buds
Trellis system	Guyot	Arcuate
Results		
Yield hectoliters/ha	40.5	39.4
Alcoholic proof	11°	10°
Wood weight kg/ha	2,414	2,411
Wine type	Quality wines, e.g., premières côtes of Bordeaux	Lighter wine, Bordeaux denomination

Table 26.14. Comparison between different distance on the row in wide vineyards with the same load per hectare, same cultivation methods, Ugni blanc a Cognac (Reynier, 2000).

Distance	3.2×0.8 m	3.2×0.95 m	3.2×1.15 m	3.2×1.50 m
Density	4,000	3,200	2,700	2,000
hL/ha	150	167	164	181
Alcoholic proof	7. 09	6.81	6.70	6.70
hL/ha of alcohol	10.37	10.94	11.23	11.67
Wood weight kg/ha	3,260	3,160	2,950	2,875

26.4.4. Rootstock Choice

The rootstock, employed in viticulture from the end of the 1800s to defend the European grapevine from the fatal effects of the phylloxera invasion beginning in France and rapidly spreading throughout the viticultural areas of the old continent, must be capable of guaranteeing a good varietal adaptation to soil conditions, contributing to a better vegeto-productive balance of the plant. It must no longer be considered a "necessary evil", but an important agronomic instrument to achieve goals of productivity and vineyard system quality (Di Lorenzo and Sottile, 2000).

The criteria to be followed in the choice of the rootstock most suitable to the different pedoclimatic conditions are many: physical-chemical soil characteristics, varietal affinity, vigour conferred and influence upon the vegetative cycle of the variety (e.g., accelerating or delaying ripening), toleration to active lime, drought and high soil-moisture conditions, selectivity in uptake of certain nutritive elements, possibility of inducing a certain tolerance to specific viroses, and choice of sanitarily certified material.

The cultivation surface area of rootstock mother plants in Italy declared by nursery owners up to the year 2002 are shown in Fig. 26.9. The rootstocks most employed in viticulture today are the single hybrids *V. Berlandieri x V. Riparia* (Kober 5BB, SO$_4$, 420A) and *V. Berlandieri x V. Rupestris* (1103P, 140 Ru). The latter were rediscovered in recent years thanks to their diffusion in Sicily due to their resistance to salinity and drought.

The sharp drop of the use of Kober 5 BB appears evident. In 30 years it has gone from 1,100 ha to less than 400 ha of surface area invested. It is very vigorous and grows rapidly and gives the European scion a considerable development, which in deep, fresh lands may be excessive.

Table 26.15 synthetically describes the characteristics of the rootstocks most used in Italy. The vigour, not only of the rootstock, but also of its combination with the scion variety, represents one of the decisive choice parameters and is closely tied to that of vine spacing and trellis system. From a quality viticulture point of view, the rootstock must supply the plant with a medium-low vigour that allows for a higher planting density with a lower production per stock with the same yield per hectare. The knowledge of the physical-chemical conditions of the soil supplies precious information regarding the adaptation capability of a certain rootstock, in terms of a more or less high uptake capacity for certain nutritive elements over others, in terms of resistance to situations of scarce or high water availability and tolerance to more or less high levels of active lime.

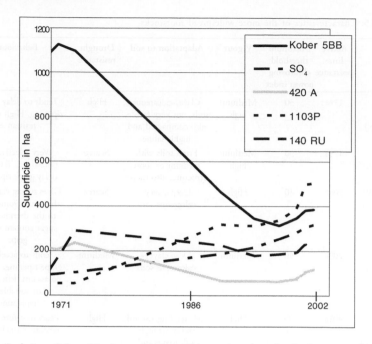

Fig. 26.9. Evolution of the cultivation surface area of the mother plants for the five rootstocks most used in Italy (Nursery control service, 2002).

Mineral nutrition is certainly a very complex phenomenon that varies as a function of the characteristics of the rootstock, the soil, the vine variety and the climate. Rootstocks such as SO_4 and 110 Richter show high selectivity in potassium uptake. In these conditions the plant might develop magnesium deficiency, unless this element is not present in the soil in very high quantities. Magnesium deficiency leaves the plant open to rachis necrosis, while a high potassium uptake, triggering the salification of the tartaric acid contained in the grapes, leads to a decrease in the acidity levels of the must, which may notably affect the organoleptic characteristics and the conservable quality of the wine to be produced.

Rootstocks such as the 140 Ruggeri and the hybrids *V. vinifera x V. Berlandieri*, 41 B (*Chasselas x Berlandieri*) and Fercal [333 EM (*Cabernet x Berlandieri*) *x Berlandieri Colombard* BC1] are characterized by their tolerance to very high levels of active lime (30–40%) that allow them, even in extreme conditions, to absorb iron in an efficient manner, avoiding the manifestation of chlorosis phenomena. The 41 B, however, has some rooting difficulties and is susceptible to the downy mildew, while the Fercal is highly susceptible to magnesium deficiency.

26.4.5. Choice of Grass Cover

The use of covering techniques has evolved greatly from the 1960s to today, going from spontaneous grass cover to artificial cover with types normally used in forage cultivation. The type of cover may be temporary or permanent, depending upon the water availability of the

Table 26.15. Characteristics of the more widespread rootstocks.

Rootstock	Active lime resistance	Maximum threshold chlorosing power index	Vigour	Adaptation to soil	Drought resistance	Behaviour
1103 PAULSEN (Ber. X Rup.)	17%	30	Medium/ high	Clayey-calcareous soils, not very fertile, also compacted and highly saline	High	Tends to delay grape ripening. High salinity resistance
SO₄ (Berl. X Rip.)	17%	30	Medium/ high	Fresh soils with medium or loose structure, also moist	Scarce	Very selective to potassium. Tends to delay grape ripening
KOBER 5BB (Berl. X Rip.)	20%	40	High	Deep clayey-calcareous soils	Scarce	Gives strong vigour to the scion, sometimes to the detriment of sugar content of the grape
420 A (Berl. X Rip.)	20%	40	Medium	Deep clayey-calcareous soils	Medium	Tends to accelerate grape ripening. Suffers potassium deficiencies, not suitable in replanting
140 RUGGERI (Berl. X Rup.)	40%	90	High	Soils tending towards clayey, strongly calcareous and chlorotic, dry	High	Fears moisture. High resistance to chlorosis
110 RICHTER (Berl. X Rup.)	17%	30	High	Clayey-calcareous soils, not very fertile	High	Tends to delay grape ripening. Suitable also to acidic soils. Low yield in nursery

different areas. Table 26.17 shows the strong points and the weak points of the different solutions.

The use of a grass cover, because of its function in environment protection and for the positive effects that it can have on product quality as an efficient agronomic method for the control of vigour and productivity of the grapevine, has come to the attention of researchers and technicians as a substitute for tilling and to avoid its negative effects. Experiments are ongoing on management methods, type of grass cover, choice of species and other aspects. The methods proposed for the cold/humid viticultural areas of the temperate climate and the Mediterranean ones are very different. Taking as reference the climatic subdivision of Europe (Finke et al., 1999), it is possible to choose different species and practices.

Temperate Suboceanic Climate

In the vineyards of temperate suboceanic climate the use of grass cover is long since well established in regulating the relationship between the productive and the vegetative activity of the plants while respecting the environment. The natural grass cover, however, is generally constituted of species that are very aggressive in the summer season, which may have a negative effect on the vegeto-productive equilibrium of the grapevine; it therefore becomes fundamental to choose the species, single or in association, most suitable a artificial grass cover (Lavezzi et al., 2000).

Table 26.16. Adaptability of the principal rootstocks to functional characteristics of the soil (Melotti and Scotti, 2002, modified).

Pedological characteristics	Rootstocks	Excellent	Moderate	Scarce
Soil rooting depth (cm)	K5BB; SO$_4$; 140 Ru; 1103 P	> 100	50–100	< 50
	420A; 110 R	> 50	25–50	< 25
Texture	K5BB; SO$_4$; 420A; 110 R	medium, mod. fine, mod. coarse	coarse, fine	
	140 Ru; 1103 P	coarse, mod. coarse, mod. fine, fine	medium	
Cracking	K5BB; SO$_4$; 420A; 110 R	low	medium	strong
	140 Ru; 1103 P	low	medium, strong	
Salinity (1:5 dS/m)	K5BB; SO$_4$; 420A; 140 Ru 110 R;	< 0.4	0.4–0.8	> 0.8
	1103 P	< 0.8	0.8–1	> 1
Exchangeable sodium percentage (ESP)	K5BB; SO$_4$; 420A; 110 R; 140 Ru;1103 P	< 8	8–10	> 10
Oxygen availability	420A; 110 R; 140 Ru; 1103 P	good, moderate	moderate	imperfect, very poor
	K5BB; SO$_4$	good	moderate	imperfect, very poor
Response (pH)	420A; 110 R	6.5–7.5	5.5–6.5; 7.5–8.5	< 5.5; > 8.5
	K5BB; SO$_4$	6.5–7.5	5.5–6.5; 7.5–8.0	< 5.5; > 8.0
	140 Ru; 1103 P	6.5–8.0	5.4–6.5; 8.0–8.8	< 5.4; > 8.8
Active lime (%)	110 R; 1103 P; SO$_4$	< 12	12–17	> 17
	K5BB; 420A	< 12	12–20	> 20
	140 Ru;	< 15	15–40	> 40

Table 26.17. Comparison of grass cover strategies (Valenti et al., 1999, modified).

	Types		Management	
	Natural	Artificial	Temporary	Permanent
Strong points	Species well adapted to the environment No installation costs	Choice of most suitable plant species Quick cover. Large quantities of organic matter produced Weed control	No competition for water and nutrients	Better defence against erosion
Weak points	Species very competitive for water and nutritive elements Management difficulties of the grass cover Needs frequent mowing	Installation costs Difficulty in choice of best species, most adaptable to environment and soil	Erosion in periods without cover Greater soil compaction	Competition for water and nutrients Management costs

Among the more interesting Graminaceae, *Lolium perenne* in the turf varieties (with low development, suitable for frequent mowing) showed high bedding ease and speed, so as to allow an efficient control of the spontaneous flora. Its durability is limited in environs characterized by low rainfall and high summer temperatures (Benati and Maggiore, 1999). It best shows its potential mixed with *Festuca ovina*, a hardy and trafficable species. The fundamental turf variety *par excellence* is the *Festuca rubra* spp., hardy, small, non-competitive, and suitable for poor and dry soils. Its limitation is its slow bedding speed (Scienza et al., 1988). In fertile environs with well-distributed rainfall the *Poa pratensis* may also guarantee a good cover.

Among the Leguminosae there is *Trifolium repens*, a species that, though not perennial, may last a long time because of its vegetative reproduction systems. It prefers loose soils. Not being particularly aggressive, it is better used mixed with graminaceae. Mixing with *Lolium perenne* and *Festuca arundinacea* showed interesting aspects of cover persistence (Benati and Maggiore, 1999).

Trifolium subterraneum showed good behaviour on well-drained soils, such as those tested in some vineyards of the Colli Orientali in Friuli and on the karstic tableau, where it ensured good soil cover and turf durability (Lavezzi et al., 2000). In zones with more limited water availability, the present trend is to look for annually self-seeding species that develop in the autumn-spring period and disappear in the summer months, allowing the grapevine to make the most of the entire environment.

Mediterranean Oceanic and Subcontinental Climates

The use of grass cover in central Italy can no longer be delayed, because of the presence of vineyards in declivous areas with high slope gradients and *rittochino* rows that have soils containing large amounts of clay and little organic matter and are therefore subject to erosion and insoluble nutrients. The high climatic variability that characterizes central Italy, from the hillsides of Umbria and the Marches to those on the coastline of Tuscany and Latium, serves to demarcate three large zones where certain species are installed and managed: (1) internal areas with scarce Mediterranean influence (Mediterranean subcontinental); (2) sub-humid Mediterranean areas with relatively cold winters; and (3) Mediterranean areas with mild winters and dry summers (Mediterranean oceanic). The species that have given the best results are shown in Table 26.18.

There are a few species that are generally of little interest, but could be useful in particular conditions: *Poa pratensis* (moderately aggressive to weeds and scarcely adaptable to shade); *Bromus catharticus* (large size and low trafficability). Among the species mentioned, *Festuca arundinacea* is strongly competitive with the grapevine for water supply, the opposite of the leguminosae, particularly the annually self-seeding. The latter prove useful also for weed control, soil protection, trafficability and persistence of the grass cover.

Research carried out on Sangiovese in Montalcino (Siena) has confirmed the high rivalry between *Festuca arundinacea* and leguminosae-based cover in terms of bud growth and total foliage area (Storchi et al., 2000). Recently, other species of Festuca have been widespread (dwarf *Festuca longifolia*) that compete less with the grapevine, which would allow the use of grass cover even in environs characterized by very scarce rainfall during vegetative activity of the grapevine (Mattii and Ferrini, 2005). Other research has been done on two vineyards of Sangiovese in the Chianti Classico area of Tuscany that were grown on spurred cordon with grapevines placed at a distance of 2.80 × 1.00 m, situated in the zones of Radda in Chianti and of Castellina in the province of Siena (Pisani et al., 2000). Different types of partial grass cover (strips of 2.00 m in the area between the rows and free under the rows) were compared

Table 26.18. Characteristics of some species indicated for grass cover in vineyards of central Italy. Scale of reference 0–3 (Santilocchi and Talamucci, 1999, modified).

	Lolium perenne	Lolium rigidum	Festuc arubra	Festuca arundinacea	Trifolium subterraneum (early cv.)	Medicago polymorpha	Leguminous mixture self-seeding
Internal areas	2	0	3	3	1	1	2
Sub-humid Mediterranean areas	2	1	1	3	2	2	3
Mediterranean areas	0	3	1	1	3	3	3
Calcareous land	3	3	3	3	1	3	3
Size	2	2	1	3	1	3	2
Resumes vegetation early	2	3	1	3	3	3	2
Weed control	2	1	2	3	2	3	3
Competition with grapevine	2	1	2	3	0	0	0
Soil protection	2	2	3	3	3	3	3
Trafficability	2	2	2	3	2	2	3
Persistence	2	2	3	3	3	2	3

with the traditional methods: (1) spontaneous grass cover; (2) grass cover with 80% *Festuca rubra* (var. *Rapid*) and 20% *Lolium perenne* (var. *Naki*); (3) grass cover with *Trifolium subterraneum* (var. *Clare*); (4) grass cover with *Bromus catharticus*. The differences in soil management considerably influenced the productive activity of the grapevines in the two fields tested. In particular, production was lower in the ones subjected to spontaneous grass cover and to a grass cover with *Festuca rubra* + *Lolium perenne*. The latter guarantees a good turf throughout the year, while lower values were found, especially in summer, in the tests using *Festuca rubra* + *Lolium perenne* due to desiccation of the aerial part when drought occurs. Among the different theses tested, widely different contributions of biomass to the land, and therefore of organic matter, were found: grass cover with *Trifolium subteranneum* ensured the best contribution (9 t/ha), followed by the natural and the *Festuca rubra* + *Lolium perenne*, while the lowest quantity was registered with *Bromus catharticus*.

Mediterranean Sub-tropical Climate

In southern Italy, prolonged summer droughts and high temperatures increase the negative effects of using grass cover in grapevine production in many zones typical of certain species. The choice of species therefore becomes even more important. The use of graminaceae should therefore be excluded, considering their high water requirements, but annual self-seeding leguminosae with an autumn-spring vegetative cycle and ability to establish quickly may be useful (Corleto and Gristina, 1999). Alfalfas such as *Medicago polymorpha*, *Medicago truncatula* and *Trifolium subteranneum* respond well to these requisites. Among the latter, the *Trifolium brachycalcycum* is the species most used in the Mediterranean area. Suitable to land that is neutral or tending toward alkaline, it has given good results when combined with the cv. Clare on pastures in the Murgia and Gargano areas where flowering begins in April and seed ripening is completed in the first ten days of June. This variety adapts well to shading and could be useful in the covers under pergolas.

26.5. Application of Studies on the Plant/Environment Relationship: *Terroir* and Grapevine Zoning

Many Italian wine growing and producing districts are facing the problems of replanting vineyards installed between the 1960s and 1970s, years in which a strong expansion of viticulture occurred, subsidized by measures of the FEOGA and the EEC. It has been well established that this sudden increase in the area covered by vineyards, along with its undeniable merits such as higher income for farmers and renewed interest for territories with an otherwise stagnant economy, has also had numerous disadvantages, such as overproduction and the use of lands not suitable to quality products. Furthermore, the use of more or less identical layout models for the installation of vineyards in all of the different environmental situations has sometimes markedly increased soil erosion and lowered quality levels of the wines produced (Costantini, 1992). The replanting of vineyards can therefore be an opportunity to re-calibrate production, reducing surplus and minimizing the environmental impact of the vineyards. This prospect must entail the selection of areas in which new plantations should be situated and, in order to ensure objectivity and secure the desired results, the selection must be founded on scientific bases.

In the past twenty years, studies to determine the suitability of different zones for viticulture have taken on an integrated and interdisciplinary character, involving subjects regarding environmental factors (soil and climate, foremost), ecology, agriculture methods and viticulture genetics, and the transformation and oenological evaluation of the product (Lulli et al., 1989; Scienza et al., 1990; Costantini et al., 1991; Fregoni et al., 1992). The definition of environmental suitability is derived from integrating the climatic, pedological, and cultivation information with the vegetative, productive and qualitative expression of the vineyards. The concept of "vine suitability" is coupled with that of "zoning", meaning the subdivision of a territory based on its eco-pedological and geographic characteristics, with verification of the adaptation response of the different vine varieties. Zoning is a very complex process and consists of integrated and interdisciplinary study aiming, through different analyses, to subdivide a territory as a function of its vocation for grapevine cultivation. Through viticultural zoning, the constituents of the *terroir* are studied and divulged as well as the relationships that involve these factors and the types of wine to be obtained.

From a juridical point of view, the *terroir* is a well-defined area; it is precisely for the correct delimitation of an area that the denominations of origin may be protected only on the basis of a precise knowledge of the *terroir*. If knowledge of the *terroir* is scientific and supported by precise data, the denomination will be to that extent less vulnerable and more defensible. The laws regarding Origin Denomination (e.g., in Italy, Law no. 164/92) enter in this category and constitute the original, principal instrument for the defence and optimization of the viticultural territory scientifically qualified in the manifold characteristics that combine to create a quality product.

Terroir: tract of land whose natural characteristics, namely soil, subsoil, relief and climate, form a unique assemblage of factors that give to the agricultural product, through plants or animals, specific and high quality characteristics. Farmers have geared agricultural husbandry and processing technologies to the particular natural environmental conditions to enhance the quality of food and confer on it peculiarity and exclusivity.

From an economic point of view, defining a *terroir* makes it possible to group a set of units according to their potential for certain cultivations, define a particular and specific elaboration method for the product, and find a particular market for it. It is, therefore, important that the size of the property and the harvesting methods should be in agreement with possibilities for marketing, workforce and territory structure. In this way, high-specificity products will be created that can take full advantage of original and competitive agricultural economic policies. From an oenological point of view, in producing the best wines the selection of a bunch from a harvest with regard to its origin is as important as the quality and the variety of the grapes. In fact, it is better to produce wine separately from grapes coming from different *terroir* to make available a wide range of wines to be assembled as a function of the wines desired (Laville, 1990).

Taking Italy as an example, the importance of pedological analysis in grapevine zoning was clearly affirmed in the early 1980s, to then be further developed in the past ten years. The studies on vine zoning, especially on a detail and semi-detail scale and in the prestigious grape and wine producing districts, DOC and DOCG areas, have supplied useful elements to the comprehension of the environmental factors functional to the production of excellence for many varieties cultivated (Table 26.19). Today there are tools to better study the pedological environment in vine zoning such as Geographical Information Systems, Digital Elevation Model and satellites (Vaudour, 2005).

Viticultural zoning is comparable to a land suitability classification for viticulture. The evaluation is more precise to the extent that knowledge of the factors involved is extensive and more profound: in our case, principally climate, soil, grapevine physiology, and agronomic techniques. It must be emphasized that the procedure of evaluation must also take into consideration the finalities of the evaluation and the scale on which the surveys are carried out: e.g., farm, consortium, province. For each size of scale used, the finalities change and the research parameters used may also be different.

For farm zoning, the objectives turn to maximizing the interaction between the vine variety and the different environmental conditions prevailing inside the farm. Therefore, the farmer needs information about the correct dislocation of vine varieties in the different farm lots, the choice of rootstocks, suitable choice of planting density and productive loads in relation to site potential, an appropriate harvesting date, a planned oenological differentiation of the grapes that reach the wine-cellar, and other factors. All of these are examples and results of a deep knowledge of one's own *terroir*, of a knowledgeable and modern use of one's resources, for a final result that is planned, not merely hoped for.

In the case of a district-scale evaluation, the finalities are to be found in territorial planning, for example, the study of the possible extension of viticulture in the district jurisdiction, as

Table 26.19. Principal viticultural zones published in Italy in the past 15 years.

Project	Doc/Docg	Scale	Period	Results published
Oltrepo Pavese (PV)	Oltrepo Pavese	1:25,000	1983–1987	Scienza et al., 1990
S.Gimignano (SI)	Vernaccia di San Gimignano	1:10,000	1984–1986	Lulli et al., 1989
Forlì	Sangiovese, Albana	1:10,000	1986	Antoniazzi and Bordini, 1986
Faedo (TN)	Trentino	1:10,000	1988–1990	Falcetti, 1994
Val Tidone (PC)	Colli Piacentini	1:5,000	1989–1991	Fregoni et al., 1992
La Toscanella	Chianti Classico	1:5,000	1989–1990	Costantini et al., 1990
Montepulciano (SI)	Vino Nobile di Montepulciano	1:25,000	1989–1992	Costantini et al., 1996
Isonzo (GO)	Isonzo	1:50,000	1989–1993	Michelutti et al., 1996
Franciacorta (BS)	Franciacorta	1:25,000	1992–1994	Panont et al., 1997
Valli Cembra e Adige (TN, BZ)	Trentino, Alto Adige	1:10,000	1992–1995	Falcetti et al., 1998
Marche (AN, MC)	Verdicchio dei Castelli di Jesi e di Matelica	nd	1993	Intrieri et al., 1993
Castagneto Carducci (LI)	Bolgheri	1:10,000	1993–1995	Bogoni et al., 1997
Val d'Illasi (VR)	Soave, Valpolicella	1:25,000	1993–1995	Failla, Fiorini, 1998
Barolo (CN)	Barolo	1:25,000	1994–1996	Regione Piemonte, 2000
Trentino (TN)	Trentino	1:10,000	1995	Bertamini et al., 1995
Colli Euganei e Piave (PD, TV, VE)	Colli Euganei, Piave	1:25,000	1995–1996	Falcetti, Campostrini, 1996
Grave (PN)	Grave del Friuli	1:50,000	1995–1997	Colugnati et al., 1997
Val Lagarina (TN)	Trentino	1:20,000	1994–1996	Porro et al., 2002
Barbera (AT, AL)	Barbera d'Asti	1:25,000	1997–1998	Regione Piemonte, 2001
Cirò (Kr)	Cirò	1:25,000	1998–1999	ARSSA, 2002
Conegliano (TV)	Prosecco	1:10,000	2000	Tomasi et al., 2004
Arezzo Province	Chianti, Cortona, Valdichiana	1:50,000	1999–2001	Scienza et al., 2003
Siena Province	Various with reference variety Sangiovese	1:100,000	1999–2003	Costantini and Sulli, 2000; Costantini et al., 2001; Costantini et al., 2006

well as the creation of various territorial coordination plans and territorial information systems. In fact, zoning may be an instrument for economic planning that allows an evaluation where best to invest, whether in improving knowledge, encouraging the choice of cultivations coherent with proven aptitudes, or discouraging (or at least submitting to more accurate evaluations) contrasting cultivation choices. However, it has been widely demonstrated that the control of environmental factors that influence product quality, more than quantity, is possible only at detail and large detail, and that the agro-technical capabilities and the ability to transform the business may have an important role in the final result (Costantini and Campostrini, 1996; Lebon et al., 1997; Champagnol, 1997). Therefore, it must be emphasized that from the results of the territorial evaluation on a reconnaissance level it is not possible to give a negative or positive judgment on grape and wine producing businesses.

A potentially "vocated" area is therefore a territorial area where there is a good probability of finding soils that in most years produce in a satisfactory way, with good results in quality, without the need for particularly onerous agrotechnical operations, without geomorphological risks or risks to soil conservation. Viticultural zoning tests have indicated that each single environmental situation must be evaluated in terms of potentiality regarding the agronomic reference model and the environmental characteristics that allow its creation, in other words, the oenological result corresponds to a model of growth and maturation of the plant determined by the agricultural practices, climate, and soil conditions (Van Leeuwen and Seguin, 1997; Costantini, 1998). In fact, environmental factors influence the hormonal equilibrium of each variety, which regulate the expression of the genotype. The evaluation of soils must therefore be effectuated in relation to the distance that runs between the specific conditions and those of reference, in other words, as a function of the limitations that the natural conditions oppose to reaching the agronomic objective. For example, Sangiovese, when grown in more fertile soils, i.e., those lacking permanent limitations, produces unsatisfactory oenological results due to excessive productivity. Better results may be obtained in rather fertile soils with pedological limitations that induce a moderate stress. Soils with little fertility, for example, severely eroded soils, always produce less than those better preserved, but supply widely variable results depending on the climatic trends of the year (Campostrini and Costantini, 1996).

In the case of the grape and wine producing districts, the territorial evaluation process takes the form of an estimate of land evaluation for the production of high quality wine (land suitability for product quality). The evaluation is carried out on the basis of the permanent physical characteristics of the territory itself; a methodological example is shown in Figs. 26.10, 26.11, and 26.12. In any case, alongside its specific finalities, the procedure of evaluation must consider the principle of sustainable use of the territory, i.e., a use that does not determine a severe or permanent deterioration of its quality; in other words, a study of the environmental impact of the new vineyards is needed. In fact, since each agricultural operation brings about an alteration of the environment, whether natural or anthropic, it is important that during the evaluation the consequences of such operations be taken into consideration, in order to declassify lands that may undergo severe deterioration phenomena with the type of use planned. Practically, zoning begins with a preliminary socio-economic survey, followed by a pedological characterization and the acquisition of multi-annual climatic data already available for the sites in the area under study. Last of all, there is the critical phase of the choice of the test vineyards.

The choice of the test vineyard is strategic in that it determines the relevance and representativeness of the results reached. This choice must be made only after the survey of as wide as possible an area of the vineyards present in the district, endeavoring to check the

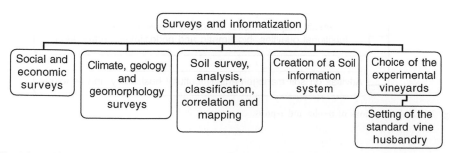

Fig. 26.10. Phases of a vine zoning project. Data collection and plot selection.

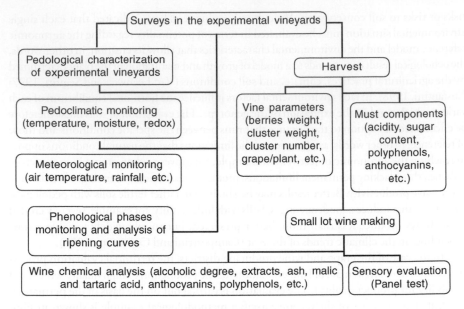

Fig. 26.11. Operations in the experimental vineyards.

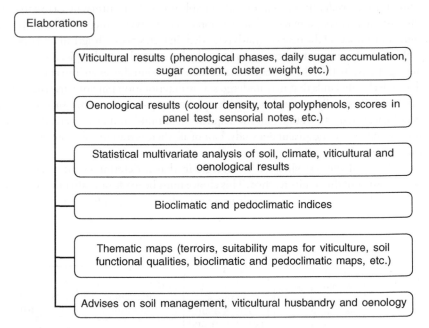

Fig. 26.12. Elaborations of results and reports.

pedologic, climatic and agro-technical variables. It is usually necessary to examine many possible vineyards before choosing the most suitable. These should be placed in different farms, in order to consider the effects of farm organization. It is a good rule, then, to plan a series of annual surveys (at least 3–4 years) to achieve an estimate of interaction between vine variety and environment.

Surveys in the test vineyards and the elaboration of findings will lead to the identification of *terroirs*, spatial units of homogeneous function of the viticultural territory. Inside this area, all the plants of the same vine-rootstock combination will have a comparable behaviour and the viticulturist will be able to carry out optimal common cultivation techniques on all of the *terroirs*. Land suitability classification should always be considered probabilistic. The various names of the areas indicate and findings show that there are soils present with different dominant vocations. Land suitability classifications therefore express a classification of areas by different uses, corresponding to a minor or major probability of finding lands with a determined capability (Fig. 26.13).

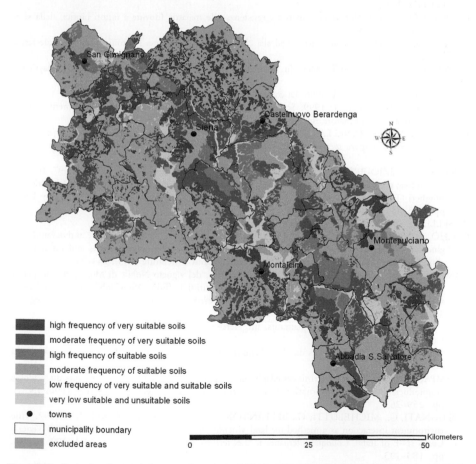

Fig. 26.13. Example of a suitability map for wine production (Sangiovese vine) in the Province of Siena (Italy). Legend accounts for the probability of finding more or less suitable soils.

Besides the land suitability classifications that give a synthetic vision useful mostly to planning needs, a series of classifications useful to management needs may be produced, which illustrate the environmental characteristics that may be of interest in the agronomic management of the vineyard, or identify some important qualities of the lands. In the first case, some examples may be soil stoniness, coarse fragment content, internal drainage, salinity, rooting depth, vertic characters, available water capacity, and pH. In the second case, examples are agrotechnical management difficulties, environmental sustainability maps, viticulture landscape maps, and purifying capability of soils (see, e.g., Costantini et al., 2006; Morlat, 2001; Vaudour, 2005).

References

ANTONIAZZI, A., BORDINI, R. 1986. *Indagine sulle vocazioni viticole della provincia di Forlì*. Camera di Commercio I.A.A., Forlì.

ARSSA - SERVIZIO AGROPEDOLOGIA. 2002. *Carta dei suoli e zonazione viticola del Cirò DOC*. Regione Calabria, 229 p.

BAVARESCO, L. 1989. Nutrizione minerale e resistenza alle malattie (dovute a fattori biotici) della vite. Vignevini 9, 25–35.

BAVARESCO, L. 2002. Sensibilità dei vitigni alla clorosi ferrica da calcare. Phytomagazine, Speciale ferro. Cartacea 1, 7–12.

BAVARESCO, L. 2005. La fertilizzazione. In: Manuale di viticoltura (a cura di Matteo Marenghi), Edagricole, Bologna.

BAVARESCO, L., VERCESI, A. 1996. La concimazione della vite. L'Informatore Agr. 8.

BENATI, R., MAGGIORE, T. 1999. Scelta, impianto e gestione delle specie da inerbimento. Italia settentrionale. L'Informatore Agr. 38, 47–49.

BERTAMINI, M., BAZZANELLA, G., MESCALCHIN, E. 1995. *Viticoltura ed ambiente trentino. Primi riscontri di un progetto sperimentale-dimostrativo in aree marginali*. Provincia autonoma di Trento, 134 p.

BOGONI, M., LIZIO-BRUNO, F., MACCARONE, G., NICOLINI, G., SCIENZA, A. 1997. Interaction among grapevine cultivars (Sangiovese, Cabernet Sauvignon and Merlot) and site of cultivation in Bolgheri (Tuscany). Proc. Colloque international "Les terroirs viticoles", Angers, 17, 18 July 1996, pp. 423–431.

BOSELLI, M. 1986. Il magnesio, elemento primario per elezione. Vignevini 11.

CACHO, J., FERNANDEZ, P., FERRIERA, V., CASTELLS, J.E. 1992. Evolution of five anthocyanidin 3-glucosides in the skin of the Tempranillo, Moristel and Garnacha grape varieties and influence of climatological variables. American Journal of Enology and Viticulture 42(3), 244–248.

CAMPOSTRINI, F., COSTANTINI, E.A.C. 1996. Gestione del vigneto Nobile di Montepulciano per la valorizzazione delle risorse naturali del territorio. In: Costantini, E. (Ed.), "Vino Nobile di Montepulciano: zonazione e valorizzazione delle risorse naturali del territorio" a cura di Campostrini, F. Regione Toscana, Firenze, pp. 110–120.

CARBONNEAU, A. 1998. Aspects qualitatifs. In: Tiercelin, J.R. (Ed.)., Traité d'irrigation, Tec & Doc Lavoisier, pp. 258–276.

CHAMPAGNOL, F. 1986. Eléments de physiologie de la vigne et de viticulture générale. Saint Gely du Fesc.

CHAMPAGNOL, F. 1997. Caractéristiques édaphiques et potentialités qualitatives des terroirs du vignoble languedocien. Colloque international "Les terroirs viticoles" Angers, 17–18 July 1996, INRA, pp. 259–263.

COLUGNATI, G., MICHELUTTI, G., BELLANTONE, P., BULFONI, D., ZANELLI, F. 1997. Vine environment interaction as a method for land viticultural evaluation. An experience in Friuli Venezia Giulia (N-E of Italy). Colloque international "Les terroirs viticoles", Angers, 17–18 July 1996, INRA, pp. 183–193.

CORINO, L., LUZZATI, A., SIRAGUSA, N., NAPPI, P. 1990. Osservazioni sulla carenza borica della vite in alcune zone del Piemonte. Vignevini 4, 39–49.

CORLETO, A., GRISTINA, L., 1999. Scelta, impianto e gestione delle specie da inerbimento. Italia centrale. L'Informatore Agr. 38, 51–52.

COSTANTINI, E.A.C. 1992. Study of the relationships between soil suitability for vine cultivation, wine quality and soil erosion through a territorial approach. Geoökoplus III, 1–14.

COSTANTINI, E.A.C. 1998. Le analisi fisiche nella definizione della qualità dei suoli per la valutazione del territorio. I Georgofili, Quaderni 1998 III, La normalizzazione dei metodi di analisi fisica del suolo, Firenze, pp. 33–57.

COSTANTINI, E.A.C. 1999. Soil survey and chemical parameters evaluation in viticultural zoning. Atti del 2° convegno internazionale "zonazione viticola", Siena, May 1998, pp. 127–137.

COSTANTINI, E.A.C., BARBETTI, R., BUCELLI, P., CIMATO, A., FRANCHINI, E., L'ABATE, G., PELLEGRINI, S., STORCHI, P., VIGNOZZI, N. 2006. Zonazione *viticola ed olivicola della provincia di Siena.* Grafiche Boccacci editore, Colle val d'Elsa (SI), 224 p.

COSTANTINI, E.A.C., BARBETTI, R., IORI, M. 2004. Valutazione di impatto ambientale e paesaggistico della viticoltura ed olivicoltura nella provincia di Siena. Boll. S.I.S.S. 51(1–2), 13–19.

COSTANTINI, E.A.C., BAZZOFFI, P., PELLEGRINI, S., STORCHI, P. 2001. Dove è possibile estendere la coltivazione ecocompatibile del Sangiovese nella provincia di Siena? Atti Simposio Internazionale "Il Sangiovese", Firenze, 15–17 February 2000, ARSIA, Firenze, pp. 185–194.

COSTANTINI, E.A.C., CAMPOSTRINI, F., ARCARA, P.G., CHERUBINI, P., STORCHI, P., PIERUCCI, M. 1996. Soil and climate functional characters for grape ripening and wine quality of "Vino Nobile di Montepulciano". Acta Hort. 427 ISHS, 45–55.

COSTANTINI, E.A.C., LIZIO-BRUNO, F. 1996. I suoli del comprensorio vitivinicolo di Montepulciano. Le loro caratteristiche, gli ambienti, i caratteri funzionali per la produzione di Vino Nobile di Montepulciano. In: Campostrini, F., Costantini E., "Vino Nobile di Montepulciano: zonazione e valorizzazione delle risorse naturali del territorio". Regione Toscana, Firenze, pp. 47–74.

COSTANTINI, E.A.C., LULLI, L., MIRABELLA, A. 1991. First experiences to individuate soils suitable for the production of high quality Vernaccia of San Gimignano. Atti del simposio internazionale: La gestione del territorio viticolo sulla base delle zone pedoclimatiche e del catasto. S. Maria della Versa (PV) 29–30 June 1987, pp. 125–135.

COSTANTINI, E.A.C., LULLI, L., PINZAUTI, S., CHERUBINI, P., SIMONCINI, S. 1990. Indagine sui caratteri funzionali del suolo che agiscono sulla qualità del vino. Atti del X° incontro su: Contributi ed influenza della chimica nella produzione, conservazione e commercializzazione del vino, Ist. Chimica org. Univer. Siena, pp. 27–40.

COSTANTINI, E.A.C., PINZAUTI, S. 1992. L'importanza dell'indagine pedologica in viticoltura. Atti del convegno: La zonazione viticola tra innovazione agronomica, gestione e valorizzazione del territorio, L'esempio del Trentino, San Michele all'Adige, 28 August 1992, Ed. Falcetti, M., Trento, pp. 137–156.

COSTANTINI, E.A.C., PRIORI, S. 2007. Pedogenesis of plinthite during early Pliocene in the Mediterranean environment. Case study of a buried paleosol at Podere Renieri, central Italy. Catena 71, 425–443.

COSTANTINI, E.A.C., SULLI, L. 2000. Land evaluation in areas with high environmental sensitivity and qualitative value of the crops: the viticultural and olive-growing zoning of the Siena province. Boll. S.I.S.S. 49(1–2), 219–234.

COTTENIE, A. 1980. Soil and plant testing as a basis of fertilizer recommendations. FAO Soil Bull. 38, 64–65.

DELOIRE, A., LOPEZ, F., CARBONNEAU, A. 2002. Réponses de la vigne et terroir. Eléments pour une méthode d'étude. Progr. Agric. Vitic. 119(4), 78–86.

DELOIRE, A., SILVA, P., MARTIN-PIERRAT, S. 2003. Terroirs et état hydrique du Grenache Noir. Premiers résultats. Progr. Agric. Vitic. 120(17), 367–373.

DI LORENZO, R., SOTTILE, I. 2000. *I Portinnesti.* In: Contributo della scuola italiana al progresso delle scienze vitivinicole. Acc. It. Vite Vino (Ed.) 1, 111–130.

ESDP. 1999. *European Spatial Development Perspective.* http://ec.europa.eu/regional_policy/sources/docoffic/official/reports/som_en.htm (verified 6.12.2007).

FAILLA, O., FIORINI, P. 1998. *La zonazione viticola della Val d'Illasi.* Manuale d'uso per il viticoltore, Cantina sociale di Illasi, 178 p.

FALCETTI, M. 1994. *Faedo e il suo vigneto. Annotazioni geografiche, storiche ed agronomiche sulla viticoltura e l'enologia del conoide trentino.* Stampalith (Ed.), Trento.

FALCETTI, M., CAMPOSTRINI, F. 1996a. *I suoli dell'area a DOC del Piave. Provincia di Treviso*. Ente di Sviluppo Agricolo del Veneto, Serie pedologica no. 2, 91–108.

FALCETTI, M., CAMPOSTRINI, F. 1996b. *I suoli dell'area a DOC del Piave. Provincia di Venezia*. Ente di Sviluppo Agricolo del Veneto, Serie pedologica no. 3, 85–102.

FALCETTI, M., CAMPOSTRINI, F. 1996c. *I suoli dell'area a DOC dei Colli Euganei*. Ente di Sviluppo Agricolo del Veneto, Serie pedologica no. 4, 103–118.

FALCETTI, M., DE BIASI, C., ALDRIGHETTI, C., COSTANTINI, E.A.C., PINZAUTI, S. 1998. Progetto di zonazione delle Valli di Cembra e dell'Adige. Analisi del comportamento di Pinot nero in ambiente subalpino. Atti Simp. Intern. Territorio e vino, Siena, 19–24 May 1998.

FINKE, P., HARTWICH, R., DUDAL, R., IBÀÑEZ, J., JAMAGNE, M., KING, D., MONTANARELLA, L., YASSOGLOU, N. 1999. *Georeferenced Soil Database of Europe. Manual of procedures. Version 1.0.* Scientific Committee of European Soil Bureau, EUR 18092 EN, 184 p.

FREGONI, M. 1980. *Nutrizione e fertilizzazione della vite*. Edagricole, Bologna.

FREGONI, M., ZAMBONI, M., BOSELLI, M., FRASCHINI, E., SCIENZA, A., VALENTI, L., PANONT, C.A., BRANCADORO, L., BODONI, M., FAILLA, O., LARUCCIA, N., NARDI, I., FILIPPI, N., LEGA, P., LINONI, F., LIBÉ, A. 1992. Ricerca pluridisciplinare per la zonazione viticola della Val Tidone (Piacenza, Italia). Vignevini 11.

GÄRTEL, W. 1971. Données fondamentales sur l'alimentation minérale des vignes et emploi des engrais en viticulture. Bull. O.I.V. 978–1003.

GLADSTONES, J. 1992. *Viticulture and environment*. Winetitles, Adelaide.

HELLER, W., FORKMANN, G. 1988. Biosynthesis. In: Harborne, J.B., *The Flavonoids*. Chapman and Hall Ltd, London.

HUGLIN, P., SCHNEIDER, C. 1998. *Biologie et écologie de la vigne*. 2nd ed. Tec & Doc Lavoisier, Paris.

INTRIERI, C. 2003. Distanze di impianto nel vigneto. L'Informatore Agr. 48, 41.

INTRIERI, C., FILIPPETTI, I., RAMAZZOTTI, S. 2004. Concetti di base sulle densità di impianto in viticoltura. L'Informatore Agr. 16, 5.

INTRIERI, C., FILIPPETTI, I., SILVESTRONI, O., MARCHEGIANI, E., MURRI, A. 1993. Zonazione bioclimatica e primi rilievi fenologici nella viticoltura della Regione Marche. Vignevini 6, 62–68.

JACKSON, J.F., CHAPMAN, K.S.R. 1975. The role of boron in plants. In: Nicholas, D.J.D., Egan, A.R. (Eds.), *Trace Elements in Soil-Plant System*. Academic Press, New York.

JUSTE, C., POUGET, R., 1972. Appréciation du pouvoir clorosante des sols par un novel indice faisant intervenir le calcaire actif et le fer facilement extractible. Application au choix des porte-greffes de la vigne. C.R. Acad. Agric. de France 58, 352–364.

KRIEDMANN, P.E., SMART, R.E. 1971. Effects of irradiance, temperature and leaf water potential on photosynthesis of vine leaves. Phothosynthetica 5, 6–15.

LAVEZZI, A., COLUGNATI, G., ALTISSIMO, A. 2000. Effetti dell'inerbimento sulla vite. Nord-est. L'Informatore Agr. 2, 65–67.

LAVILLE, P. 1990. Le terroir, un concept indispensable à l'élaboration et à la protection des appellations d'origine comme à la gestion des vignobles : le cas de la France. Bull. OIV 709–710, 217–241.

LEBON, E., DUMAS, V., MORLAT, R. 1997. Influence des facteurs naturels du terroir sur la maturation du raisin en Alsace. Colloque international "Les terroirs viticoles" Angers, 17–18 July 1996, INRA, 359–366.

LULLI, L., COSTANTINI, E.A.C., MIRABELLA, A., GIGLIOTTI, A., BUCELLI, P. 1989. Influenza del suolo sulla qualità della Vernaccia di San Gimignano. Vignevini 1/2, 53–62.

MASNEUF, I., MURAT, M.L., CHONE, X., DUBORDIEU, D. 2000. Dosage systématique de l'azote assimilable: détecter les carences au chai. Vitis 249, 19–22.

MATTII, G.B., FERRINI, F. 2005. La gestione del suolo. In: Marenghi, M. (Ed.), Manuale di viticoltura. Edagricole, Bologna, pp. 157–171.

MELOTTI, M., SCOTTI, C. 2002. *Considerazioni sulla scelta del portinnesto in fase di impianto della vite.* Bollettino di produzione integrata della Provincia di Modena, 29.

MICHELUTTI, G., BELLANTONE, P., MION, T., BULFANI, D., MENEGON, S., COLUGNATI, G. 1996. *Comprensorio di produzione dei vini Doc Isonzo. I suoli e la vocazione viticola.*

MORLAT, R. 2001. *Terroirs viticoles: étude et valorisation*. Oenoplurimedia, Chaintré.

MPELASOKA, B.S., SCHACHTMAN, D.P., TREEBY, M.T., THOMAS, M.R. 2003. A review of potassium nutrition in grapevines with special emphasis on berry accumulation. Australian Journal of Grape and Wine Research 9, 154–168.

MÜLLER, D.H., BREITER, H. P. 1999. Zinc in vineyards. Bull. OIV 72, 821–822, 485–497.

OJEDA, H., DELOIRE, A., CARBONNEAU, A. 2003. Determinacion y control del estado hidrico de la vid: efectos morfologicos y fisiologicos de la restriccion hidrica en vides. Proc. Curso Internacional de Vitivinicoltura, Neuquen 26–29 August, section 2.3.

PANONT, C.A., BOGONI, M., MONTOLDI, A., SCIENZA, A. 1997. *Improvement of sparkling wines production by a zoning approach in Franciacorta (Lombardy, Italy).* Colloque international "Les terroirs viticoles", Angers, 17–18 July 1996, INRA, pp. 454–460.

PISANI, P.L., LORETI, F., SCALABRELLI, G., BANDINELLI, R., BOSELLI, M., MANCUSO, S., PORCINAI, S. 2000. Risultati di ricerche sull'inerbimento del vigneto nel Chianti Classico. In: Chianti Classico 2000. Consorzio Vino Chianti Classico—San Casciano V.P. (FI), 4, 7–21.

PONI, S. 2000. Sensibilità della vite allo stress idrico e implicazioni per la tecnica irrigua. L'informatore Agr. 47, 71–80.

PONI, S. 2005. La gestione idrica del vigneto. In: Marenghi, M. (Ed.), Manuale di viticoltura. Edagricole, Bologna.

PORRO, D., STEFANINI, M., MENEGONI, R., PINZAUTI, S., CAMPOSTRINI, F., FALCETTI, M. 2002. *La zonazione del vigneto.* SAV. Società Agricoltori della Vallagarina, Rovereto (TN).

PRICE, S.F., YODER, B., BREEN, P.J., WATSON, B.T. 1992. *Solar radiation effects on anthocyanins and phenolics in skins of "Pinot noir" and "Pinot gris" berries.* Proc. IV Simp. Int. Fisiol. Vite, Torino, pp. 565–570.

RAPP, A., VERSINI, G., ENGEL, L. 1995. Nachweis und Bestimmung von 2-Aminoacetophenon in vergorenen Modellösungen. Vitis 34(3), 193–194.

REGIONE PIEMONTE. 2000. *Barolo. Studio per la caratterizzazione del territorio, delle uve e dei vini dell'area di produzione.* Quaderni Regione Piemonte—Agricoltura.

REGIONE PIEMONTE. 2001. *Barbera. Studio per la caratterizzazione del territorio, delle uve e dei vini dell'area di produzione del barbera d'Asti.* Quaderni Regione Piemonte—Agricoltura.

REGIONE TOSCANA. 2001. *Deliberazione Consiglio Regionale 17 gennaio 2001, n.21 "Piano regionale per la ristrutturazione e riconversione dei vigneti (art.11 del Reg.CE n.1493/99)".* www.rete.toscana.it/sett/agric/vino (verified 6.12.2007).

REYNIER, A. 2000. *Manuel de viticulture.* 8° Ed., TEC & DOC (Ed.).

ROUBELAKIS-ANGELAKIS, K.A., KLIEWER, M. 1986. Effects of exogenous factors on phenylalanine ammonia-lyase activity and accumulation of anthocyanins and total phenolics in grape berries. American Journal of Enology and Viticulture 37(4), 275–280.

SANTILOCCHI, R., TALAMUCCI, P. 1999. Scelta, impianto e gestione delle specie da inerbimento. Italia centrale. L'Informatore Agr. 38, 51.

SBARAGLIA, M., LUCCI, E. 1994. *Guida all'interpretazione delle analisi del terreno ed alla fertilizzazione.* Studio Pedon, Pomezia (Roma), 123 p.

SCIENZA, A., BOGONI, M., VALENTI, L., BRANCADORO, L., ROMANO, F. 1990. La conoscenza dei rapporti tra vitigno ed ambiente quale strumento programmatorio in viticoltura: stima della vocazionalità dell'Oltrepò Pavese. Vignevini 12, 4–62.

SCIENZA, A., MARIANI, L., FAILLA, O., BRANCADORO, L., PRIMAVERA, F., GIULIERINI, P., BERNAVA, M., FASOLI, V., TONINATO, L. 2003. *Arezzo: Terra di vini. Dalla zonazione al manuale d'uso del territorio.* Provincia di Arezzo.

SCIENZA, A., SICHER, L., VENTURELLI, M.B., MAGGIORE, T., PISANI, P.L., CORINO, L. 1988. L'inerbimento in viticoltura. L'Informatore Agr. 21.

SCPA, 1986. *La fertilisation raisonnée de la vigne.* Pamphlet.

SILVESTRONI, O., PALIOTTI, A. 2005. Distanze di impianto. In: Marenghi, M. (Ed.), Manuale di viticoltura, Edagricole, Bologna, pp. 135–155.

STORCHI, P., MATTII, G.B., FERRINI, F. 2000. Effetti dell'inerbimento sulla vite. Centro-ovest. L'Informatore Agr. 2, 70–71.

STORCHI, P., TOMASI, D. 2005. Ecologia viticola e zonazioni. In: Marenghi, M. (Ed.), Manuale di viticoltura, Edagricole, Bologna, pp. 17–34.

TOMASI, D., CETTOLIN, C., CALÒ, A., BINI, C. 2004. I suoli e i climi della fascia collinare del comune di Conegliano e loro attitudine alla coltivazione del vitigno Prosecco (Vitis sp). Comune di Conegliano (Ed.) (TV).

VALENTI, L., MAGGIORE, T., SCIENZA, A. 1999. Tecniche di gestione del suolo in viticoltura. L'Informatore Agr. 38, 35–37.

VAN LEEUWEN, C., FRIANT, P.H., SOYER, J.P., MOLOT, C.H., CHONE, X., DUBOURDIEU, D. 2000. L'intérêt du dosage de l'azote total et de l'azote assimilable dans le moût comme indicateur de la nutrition azotée de la vigne. J. Int. Sci. Vigne Vin 34(2), 75–82.

VAN LEEUWEN, C., GAUDILLERE, J.P., TREGOAT, O. 2001. Evaluation du régime hydrique de la vigne à partir du rapport isotopique $^{12}C/^{13}C$. J. Int. Sci.Vigne Vin 4, 195–205.

VAN LEEUWEN, C., SEGUIN, G. 1994. Incidences de l'alimentation en eau de la vigne, appréciée par l'état hydrique du feuillage sur le développement de l'appareil végétatif et la maturation du raisin. J. Int. Sci. Vigne Vin 2, 81–110.

VAN LEEUWEN, C., SEGUIN, G. 1997. Incidence de la nature du sol et du cépage sur la maturation du raisin, à Saint-Emilion, en 1995. Colloque international "Les terroirs viticoles" Angers, 17–18 July 1996, INRA, pp. 154–157.

VAN LEEUWEN, C., TREGOAT, O., CHONE, X., JAECK, M.E., RABUSSEAU, S., GAUDILLERE, J.P. 2003. Le suivi du régime hydrique de la vigne et son incidence sur la maturation du raisin. Bull. O.I.V. 867, 868, 367–379.

VAUDOUR, E. 2005. *I terroir. Definizioni, caratterizzazione e protezione*. Edagricole, Bologna, p. 295.

VERCESI, A. 1987. Gli assorbimenti radicali della vite: meccanismi e fattori influenti. Vignevini 4.

27. Olive Tree (*Olea europea L.*)

Elena Franchini,[1] Antonio Cimato[1] and Edoardo A.C. Costantini[2]

[1] CNR-IVaLSA—Istituto per la Valorizzazione del Legno e delle Specie Arboree, Sesto Fiorentino (Firenze), Italy.
[2] CRA-ABP—Centro di Ricerca per l'Agrobiologia e la Pedologia, Firenze, Italy.

27.1. The Olive Tree

In its botanical classification, the olive tree (*Olea europaea L.*) belongs to the *Oleaceae* family, to the *Olea* genus, to the *europaea* species and to the *sativa* subspecies.

In its genealogy, the *Oleacee* family includes some spontaneous kinds, ornamental evergreen shrubs (e.g., *Ligustrum, Syringa, Jasminum, Fraxinus, Phillyrea, Forsythia*), as well as some arboreal trees, among which the olive tree (*Olea europaea L.*) is the only one that produces edible fruits (olives).

For the *europaea* species, whose taxonomic boundaries are complex, literature makes reference to two subspecies of the Mediterranean area: the first, *Olea europaea* subsp. *sativa*, definitely classified, and the second, *Olea europaea* subsp. *oleaster*, with a more uncertain classification. The *sativa* subspecies designates the cultivated olive trees (cultivar) and encompasses the results of spontaneous hybridizations, natural or man-induced selections enhancing the productive ability of this tree as well as its adaptation to different pedoclimatic environments. Therefore, it encompasses trees that, over time, have been maintained by cultivators in an agamic way. The word *oleaster*, on the contrary, is indicative of shrubs or spontaneous trees being specific of maquis, with short branches having thorns and small leaves; small drupes, generally round and with a big stone, that show their limited fructification.

The olive tree is useful for farming practices in many countries and, over the centuries, has reached a paramount importance, economically and socially, playing a considerable role in food and cultural tradition.

A fascinating 6,000-year history, documented by legends, traditions, religious texts and archaeological discoveries, sets the origins of the olive tree in the Middle East and describes its establishment in all the countries bordering the Mediterranean Sea and in many other areas that are suitable to its cultivation. Over time, the olive tree spread first through the Fertile Crescent area, Turkey, Palestine, Israel, Lower Egypt, the territories of Greece, the coast of the Balkans, Italy, Spain and Portugal. Spurred by favourable recognition and by the encouraging environmental conditions of the Mediterranean climate, its cultivation radiated out to Morocco, Algeria, Tunisia and the oases of Libya. After the discovery of America, olive growing spread to Peru, Argentina, Chile and Uruguay and to the coastal regions of Mexico.

More recently, extra virgin olive oil has been much appreciated in international markets, ranking first in the Mediterranean diet; moreover, the accreditation of medical research on its nutritional and health properties has led to a remarkable increase of world consumption of this food.

Figure 27.1 shows the shift from an average consumption of 1.6 million t in 1990 to more than 2.8 million t (2005) over the last 15 years.

During the same period, this event was followed both by an expansion of the olive crop areas (Fig. 27.2) and by an increase in the number of olive trees (Fig. 27.3).

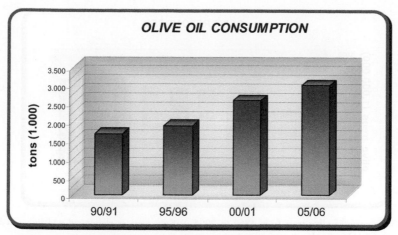

Fig. 27.1. Variations of world consumption of olive oil *(COI Source).*

Figure 27.2 shows that, since the 1980s, the introduction of modern agronomic techniques (e.g., irrigation, automated harvesting), the increase of plantations and the implementation of new development plans in several regions not earlier known for olive cultivation (Chile, southern California, South Africa, Australian coasts, New Zealand and China) have determined a greater spread of the olive crop areas, reaching more than 10 million ha in 2005.

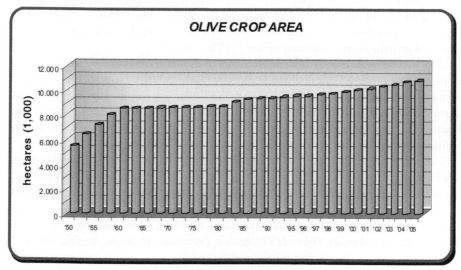

Fig. 27.2. World spread of olive cultivation from 1950 to 2005 (various sources elaborated and integrated with documents of FAO and other organizations).

Obviously, the spread of plantations implied a proportional increase in the number of olive plants internationally. In 2005, 865 million olive trees were registered, compared with 720 million trees registered in the 1960s, with an average increase of 13.3%. This trend has been confirmed and intensified particularly over the last decade (Fig. 27.3).

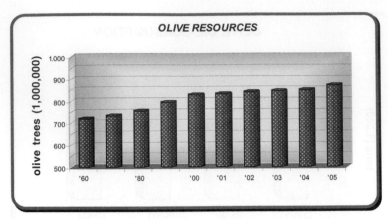

Fig. 27.3. Increase of olive trees from 1960 to 2005 (various sources elaborated and integrated with documents of FAO and other organizations).

The olive tree, which is integrated in the Italian landscape, distinguishes itself from other arboreal fruit trees for its capacity to be cultivated in different environments, to produce even in arid and cold conditions. It has always been a functional tree in the agricultural system, finding an equilibrium between an anthropic environment and a natural one. In Italy, the olive tree is present in almost all the Italian regions, although with different economic and social importance, where it occupies areas with different pedoclimatic characteristics, counting a cultivated area of about 1.167.675 ha and 180 million trees (ISMEA, 2004). Olive cultivation is practised more in the south (724.098 ha) and on the islands (197.759 ha), compared with the central-northern regions (245.818 ha).

Italian olive cultivation is different from that of the other Mediterranean countries for the presence of a wide genetic resource. It includes "varieties" with manifold biological and productive properties (cultivars), "improved trees" that are obtained by hybridization, mutagenesis interventions or clone selection (clones) and age-old olive trees (genotypes) that, although they have no significant impact on olive production quantities, are numerous in the traditional plantations, representing a rich "autochthonous" biodiversity.

According to literature, in Italy there are more than 600 varieties of the olive tree. Some varieties produce fruits for direct consumption (table olives) and, among the most widespread, we point out *Ascolana, Bella di Cerignola, Moresca, Nocellara del Belice,* and *Nocellara Etnea.* The production of many other varietie is preferable for oil extraction (e.g., *Bianchera, Biancolilla, Bosana, Canino, Carolea, Casaliva, Cellina di Nardò, Coratina, Correggiolo, Frantoio, Gentile di Larino, Leccino, Moraiolo, Ogliarola, Ottobratica, Pisciottana, Racioppa, Rosciola, Taggiasca, Tonda Iblea*).

27.1.1. Tree Morphology: Crown, Trunk and Roots

The epigeal portion of the olive tree consists of trunk, limbs and branches. The stem portion, between insertion of the primary limbs and the collar (the area connecting the tree to the root system) identifies the trunk. Limbs and branches, together with the organs they bear (buds, leaves, shoots, blossoms and fruits) represent the crown.

The olive crown, if it develops naturally, can have different forms: "pendant" when branches grow drooping; "spreading" with branches slightly bent in the direction of daylight and space; and "rising" when the main branches grow vertically (this is a characteristic typical of trees with a great apical dominance).

Tree development, which is determined by the length and diameter of limbs and branches as well as by the abundant vegetation, exhibits two characteristics: crown vigour and density. These natural properties may vary over time, because they are determined by pruning, environmental conditions (tree height tends to decrease in areas where climate is not favourable to the olive tree) and also agronomic practices (e.g,. fertilization, irrigation, cultivation).

As in other fruit trees, the olive crown is divided into limbs, branches and shoots. Limbs are particularly developed branches, with a special form, that represent the structure of the tree; primary limbs are inserted into the trunk and are no younger than two years, and secondary ones (sub-limbs shoot laterally from the primary ones. Branches are growing axis with buds. In olive trees there is a sharp distinction between branches that are one year old and the older ones: generally the one-year-olds are less vigorous, have a pendant direction and have several buds displaying flowers, except for the bud placed at the apex (shoot) that holds growing characteristics, ensuring tree growth during the season. The year-old branches are characterized by the presence of fruits, which are in competition with the vegetative growth of the tree. It is easy to identify one-year-old branches, because of their vigour and because they have no fruit-bearing buds: with regard to terminology, there is a distinction between "suckers", which emerge from the foot and/or root area, and "young shoots", deriving from buds hidden in the trunk or placed on the dorsal side of limbs.

The olive tree has simple leaves, with a smooth edge and short stem. They can be identified by their shape and dimensions (length and width). The shape can be lanceolate, elliptic-lanceolate or elliptic. The leaf blade dimension, that is to say the size of its surface (varying from 5–7 cm in length and 1.2–1.5 cm in width through the centre) can be small, average, large or very large. The upper face of the leaf is grayish-green, while the underside is whitish and felt-like, because of the presence of many multicellular hairs.

In the olive tree, the buds (fruit or flower buds) are placed in the leaf axils, while the one-year branch has a terminal bud (flower) whose goal is to ensure annual plant growth. Axillary buds generally evolve for reproduction, becoming, in springtime, fruit buds and therefore inflorescences (florescence) with a central axis and lateral ramifications (rachis). On more vigorous branches there are often "mixed" buds that give both a shoot and small inflorescences.

The typical olive inflorescence consists of a florescence, generally having a flower at the apex of the rachis, while other groups of flowers (from two to five) are located on the secondary axes of lateral ramifications. Each inflorescence may have a different form and, therefore, may have a very variable number of flowers (10–40). This variability depends on soil agronomic fertility, crown light conditions and plant genetic characteristics (cultivar). The flower has a white corolla with a light green flower-cup and is hermaphrodite, i.e., it has both female (pistil) and male (stamen) sexual organs. The former consists of a superior ovary, a stylus and a stigma that catches pollen; the latter consists of two filaments surmounted by yellow and bilobate anthers.

The fruit is a drupe that can be morphologically distinguished by shape (extended, ovoid or spheroid), weight, symmetry, shape of top (rounded or pointed) and shape of base (cut or rounded) as well as by its peduncular cavity (different for dimension, shape and depth). In the olive it is possible to distinguish the epicarp, the mesocarp and the endocarp or stone that protects the seed. The epicarp (peel), consisting of a thin layer of spherical cells cemented with cutin, must protect the fruit, like a film. Initially the epicarp is green; then, as ripening progresses, it first becomes reddish and then totally black. The mesocarp (pulp) is a material whose thickness may vary and represents 70–80% of the weight of the whole drupe. The pulp cells are rich in protoplasm, are spherical in shape and nearing the stone become smaller and smaller, more and more spindle-shaped and with a palisade frame. The mesocarp, during ripening, also undergoes changes in thickness and colour. The endocarp (stone) consists of cells that, after the first development stages of the fruit, harden their membranes, giving a wooden thickness to the envelopment in order to protect the seed. The endocarp has morphological features that identify and distinguish the different cultivars. The endocarp may be distinguished not only by shape (extended, elliptical, ovoid or spherical), but also by weight, symmetry, position with a maximum transverse diameter, shape of top and base, for the fibro-vascular tracks and the top end (rostrum). Inside the endocarp there are the endosperm that plays the role of an energy-gathering tissue and the embryo that, in special conditions, will give rise to a new seedling.

The epigeal part of the olive tree is completed by the trunk, whose bark is characterized by its grayish-green color. The trunk starts from the insertion of the main limbs and ends with the foot (or tree stump), which connects the tree to its radical apparatus.

For olive trees, root apparatus morphology and development initially depend on the propagation method applied; afterwards these features are regulated by soil properties, plant metabolism, genetic factors and the agronomic technique used. The plantlets obtained by seed have a root apparatus that is dominated by a taproot that, in favourable soil conditions, can reach 1 m depth at the end of the first year. If the plant is not transplanted, this central root does not change for 5–6 years (Morettini, 1972); during the following years the taproot is gradually replaced by a new root apparatus consisting of adventitious roots, which start from the trunk in the more superficial layers of the soil, and of lateral roots inserted into the taproot. Trees shooting from a self-rooted cutting, on the contrary, develop a multidirectional system of roots along the main axis where they are inserted. Over time, these differences tend to diminish, till they disappear (Guerriero, 1972), while internal factors such as the genotype energy tend to prevail on the morphology of the root apparatus, as well as external factors such as the soil physical and chemical conditions and the cultivation techniques being applied to the olive tree.

The root system serves several functions: anchorage to the soil, vascular connection to the stem, and absorption of water and of other nutrients in the soil. These tasks are carried out by main roots, leading roots, transition roots and absorbing roots; these roots varry in age, diameter and capacity to suberize. The main roots, as well as the leading roots, ensure anchorage and vascular continuity with the stem; the others together make up the root hairs, which are the part most active in absorbing water and nourishment.

The root system is also able to synthesize hormones (gibberellin and cytokinins) that control the physiological activities of the plant and help to gather reserve substances during the autumn that are absolutely necessary to the vegetative flush as well as to the completion of biological phenomena that ensure the production cycle of the olive tree.

For the olive tree, as for other arboreal trees, root development follows a seasonal periodicity. Activity starts when soil temperatures reach 14–16°C and continues following a bimodal growth model with two different periods: the first "spring-summer" period has intense root development and the second "autumn" period is characterized by a slower and altogether less important development.

27.1.2. Reproductive and Vegetation Cycle

It is crucial to know the processes that cyclically determine plant vegetation growth dynamics and fructification effectiveness in order to decide the right and sustainable agronomic management of a plantation.

For the olive tree, shoot growth ends the same year that the terminal bud opens; for reproduction, on the contrary, two subsequent seasons are necessary. In fact, the buds giving rise to flowers and fruits develop during the first year, while the phenomena ruling autogenous induction, flower biology and plant fructification end during the following season.

Branch, limb and stem vegetation growth depends on the alteration that in turn is directly controlled by bud development. In springtime, flower buds give rise to new shoots. This phenomenon, regardless of the position on the tree of the year-old branch (vertical, horizontal or bent), always starts from the bud placed at the tip of the branch and then recurs for other flower buds inserted in the middle section; the few that occupy the branch base area are often hidden. After budding and growth of the tip, it is possible to see new leaves appear and extend to their complete formation and the lengthening of internodes. Shoot development is at its peak in springtime, slows in summertime and starts again, with some intensity, after the end of August–September rains and almost disappears in October when temperatures start decreasing. Shoot growth and total development are influenced by genetic control, along with several other factors that are not easily distinguished; however, they depend on soil characteristics and properties, agronomic land management, nutritive conditions of the plant, water supply, annual intervals for pruning and various climatic factors (light and temperature, in particular).

Most of the buds located on the sides of the year-old branches evolve in a reproductive sense (florescence). This phenomenon, named bud induction and antogenous differentiation process, starts in autumn during the year that the shoot forms; it becomes more effective during the period before springtime (Cimato et al., 1985, 1986). During this time of the season, the bud tops undergo a physiological guidance that prepares them for the following steps of morphological differentiation. Generally, this phenomenon, identified with flattening of the bud top, is followed by the slow appearance of flower primordia, which afterwards turns into inflorescences (florescences).

For the olive tree, florescence distribution on year-old branches differs. The buds that develop into inflorescences are more numerous in the terminal and middle areas, whereas the branch base area is occupied by buds that are often hidden. Together with the hereditary feature (genetic control), the induction and antogenous differentiation phenomena are controlled by light (total energy flow detected by the crown, photoperiod), environmental conditions (e.g., temperature, CO_2), by the availability of nutritive elements as well as by specific hormone ratios. Therefore, the transformation from a neutral bud to a flower bud results from a complex interaction between factors that are exogenous and endogenous to the plant.

With this phenomenon, the evolution of buds into flowers reaches its end (March–April) and the complex fructification cycle of the plant starts. Fruit bud growth slowly continues until the opening of the inflorescence (florescence) that, during its development, takes on features similar to a cluster (phenological state of the florescence). The following month (April–May, depending on the climatic area) is devoted to the formation of pollen (microsporogenesis) and of the reproductive apparatus (macrosporogenesis) that ensure flowering first and then pollination and fecundation and therefore the productivity of the olive tree.

For the olive tree, depending on climatic conditions, flowering (anthesis) takes place in springtime. This is a crucial step of the fructification cycle, because tissues forming the flower stop growing so that special stimuli intervene, such as pollination and fecundation. From the biological viewpoint, flowering consists of three subsequent steps: corolla opening (stamens and pistils free themselves from the petal protection), anther dehiscence (consequent pollen emission) and pistil susceptibility. In Mediterranean areas, corolla opening lasts on average 7–12 d and generally takes place between mid-April and the first week of June. For the adult olive tree, anthesis is a complex phenomenon, because it does not involve the whole plant simultaneously. The first florescences to flower are those more exposed to sunlight and thus the more external branches of the crown; afterwards, flowering involves the lowest part of the crown and then the florescences of the more internal and shaded branches. This progressive phenomenon also concerns year-old branches (the corolla opens first on terminal florescences and then on those placed on secondary axes) and flowers of the same inflorescence (the flower of the rachis terminal area opens first and then the other groups of flowers inserted on secondary axes). The beginning of the phenomenon is genetically controlled, but it also depends on favourable climatic conditions. If the season is adverse (frequent rains, low temperatures, weak light), flowering may take place some days later and its beginning may continue over time.

It is important to know the seasonal period when plants blossom in order to support pollination, a mechanism through which the pollen, being given out by ripe anthers, reaches receptive stigmas to pave the way for fecundation. In the olive tree, the pollen of a flower, although it is vital and germinable, is not always able to give out pollen granules that are appropriate to fertilize its own gynoecium or one on plants belonging to the same cultivar (self-incompatibility); in fact, this biological process is more likely to succeed with flowers of other cultivars (intercompatibility). The conclusion of fecundation is often hampered by sterility: morphological sterility because of poor or absent development of ovary (flowers with abortive ovary), cytological sterility for anomalies arising during microsporogenesis and macrosporogenesis (poor or lacking pollen germinability, degeneration of the female gametophyte and of the embryo) and factorial sterility or incompatibility, because of the presence of genetic factors of sterility both in the genetic set of pollen granules and in the genetic set of the stylar tissue.

Incompatibility may concern both flowers of the same cultivar (self-compatibility) and flowers of different cultivars (intercompatibility). It is clear that olive flowers belonging to a self-compatible cultivar, even if they have germinable pollen, are not able to fertilize themselves or to ensure, with their pollen, the fecundation of flowers of plants belonging to the same cultivar. From a practical point of view, this knowledge is necessary to ensure the presence of pollinator olive trees at the time of planting. For this biological process as well, the final result is connected with the nutritional state of the plant and with the coincidence of favourable climatic events such as light breezes during flowering time that ensure anemophilous transportation of pollen from one flower to another and ensure the pollen reaches the ovary. Unfavourable climatic conditions are rains during flowering time and spring frosts. Setting ends in about 10 d and then dropping of stamens and petals coincides with the first step of fruit development.

The complete growth of olives lasts about 5 months and there are significant differences between cultivars and different growing environments. Olives are already large enough to be visible 10–15 d from setting. Generally, by careful observation of the fruit-bearing branch one can tell a good setting by the beginning of July. During the following period, appraisal of plant output is put into a different perspective due to plentiful falling of the scarcely grown smaller fruits. It is a phenomenon meeting specific nutritional needs of the plant, needs of the fruits in strong competition with each other and with the growth and development of new vegetation. This phenomenon is influenced by environmental conditions and by the position of the drupe on the crown (exposure), the branch (terminal, middle or basal area) and the inflorescence. In adverse conditions, because of poor water supply or excessive summer temperatures, it is clear that the initial growth of the drupe is hampered and its achievement of final dimensions is not ensured. Fruits in the internal areas of the crown, being less exposed to light (shading), grow more slowly than those placed externally and better able to catch the energy flow that ensures quicker, better growth. Likewise, fruits with a terminal position on the branch grow more, thanks to their better exposure to light and because they are close to the most active metabolic centres (ongoing growth of tips and young leaves). Finally, fruits on the top of the inflorescence always develop more because they receive more direct nourishment from the plant.

The growth model of the olive, like that of other drupaceous species, can be shown as a double sigmoid curve. The first section of the curve shows that at the beginning of fruit development, an interval during the first 48 d after flowering time, cell activity is characterized by strong multiplication (cytokinesis); this serves to increase the number of cells of the ovary tissue as well as starting the inolition process in the drupe (Tombesi, 1994). The second section of the curve shows a moderate growth ending approximately 100 d after flowering time. For the fruit, it is possible to distinguish two moments when endocarp lignifies (mid-August) and to recognize the embryo and the endosperm in the seed (end of August). The last section of the curve, starting from September and lasting about 80 d, represents the most complex step of fruit growth: during this period the last intense development of the drupe takes place, with changes in the morphology itself, together with physiological changes (photosynthetic and respiratory activities), histological transformations (differentiation of the embryo, of the vascular system of the stalk) and changes in the metabolism involving proteins, carbohydrates (e.g., glucose, fructose, saccharose and mannitol), lipids (Cimato et al., 1997, 2001). These changes also involve all the components of the "unsaponifiable" fraction (e.g., hydrocarbons, sterols, minor polar compounds, tocopherols, chlorophylls, carotenoids, phytol, aliphatic acids, terpenic hydrocarbons, esters, aldehydes, ketones, ethers).

During the lignification stage of the endocarp, significant nutritional competition takes place in the plant between growing drupes and developing shoots. Once physiological equilibrium is reached between these two "sinks" (shoots and growing fruits), olive growth is ensured as well as the extension of the shoot on which depends the production of the plant for the following year. Therefore, it is absolutely necessary that water and nutritional shortages do not occur throughout these steps of the plant biological cycle, hampering these processes and giving rise to phenomenon of production alternation.

From May to June, when florescence, flowering, pollination and setting take place, water shortages can cause anomalies in flower formation with consequent ovary abortion, lack of stamens, decrease of flowers in inflorescences and low setting rate. Nevertheless, it is not unusual for the Mediterranean region to have water shortages during this phenological step, and inadequate nitrogen supply may occur. If water and nutritional needs are not met during the season following, which coincides with the appearance of small fruits and ends with olive picking (July–December), that year's fructification will be compromised (limited growth of olives, early falling of fruits) as well as plant output for the following season (poor development of shoots). Metabolism of fruit ripening will be jeopardized.

From September on, drupe cells increase their volume and the conditions for the synthesis and the storing of oil occur (Lavee et al., 1991). The mechanisms underlying triglyceride gathering in the olive are not yet clear, even if some people are persuaded that this process is under genetic control (Patumi et al., 1990). In the beginning the phenomenon manifests itself through very small drops whose volume increase until they occupy most of the intracellular space. Generally, 70–85% of the oil existing in a fruit is inside a vacuolar and membranous structure (available or free oil), while the remaining fraction is lost, by tiny drops, in the cytoplasm. This last fraction (stiff oil) is not extractable by the pressing systems currently available. As for the fruit, morphological changes and biological activities characterizing the ripening phenomenon are shown by changes in the colour of epicarp and mesocarp. These changes, genetically controlled and influenced by environmental and anthropological events, mark the differentiation of four phenological stages for the olive tree.

Olive Herbaceous Stage

The fruit is green, its weight increases over time, its pulp is hard, its endocarp is formed and lignified, and its seed actively grows. At this phenological stage, there is chlorophyll active in the fruit that ensures photosynthesis and the synthesis of soluble sugars (e.g., glucose, fructose, saccharose, mannitol). For the olive, triglyceride collection is functional; it controls the development and activities of cell membranes (energy and structural function). These compounds vary in composition and in their respective ratios (Frias et al., 1975; Sanchez Raya et al., 1992). For instance, while the synthesis of mono-unsaturated fatty acids (palmitoleic and oleic ones) starts, the accumulation of saturated long-chain fatty acids (palmitic, heptadecanoid, stearic, arachic and behenic acid) is very high. The herbaceous stage ends when the epicarp colour becomes a lighter shade of green.

Olive Veraison Stage

The olive turns yellowish. This change is linked with the continuous degradation of chlorophyll into pheophytin and phytin and to the increasing carotene content. Photosynthetic activity is weakened in the fruits; the weight increases slightly and many lipids are stored (this is the time when the synthesis of oleic acid is at its peak and the palmitic and stearic acids decrease) (Williams et al., 1993). This is the most intensive time of metabolism, leading to the formation of hydrocarbons (squalene) and most of the main components of the unsaponifiable fraction such as minor polar compounds (Solinas et al., 1975; Amiot et al., 1989), tocoferols, chlorophylls, and sterols. The formation of anthocyanic pigments triggers the appearance of small dots that are first reddish and then violet and that in time spread over the whole fruit. During this stage, a waxy layer develops (bloom), protecting the peel, which has the function of preventing excessive water loss in fruits due to transpiration.

Olive Ripening Stage

Colour changes are more intense and involve the whole fruit, which turns black, reaching its largest size and highest levels of oil content (expressed as a percentage of the dry substance). Ripening metabolism concludes the oleic acid synthesis and is very active in forming polyunsaturated fatty acids (linoleic and linolenic acids) (Di Matteo et al., 1992; Williams et al., 1993). Other elements, like minor polar compounds, undergo a strong hydrolysis and a set of products are derived from oleuropein, for example, elenolic acid, hydroxyl tyrosol, demetiloleuropein, tyrosol, vanillic acid, protocatechuic acid (Servilli et al., 1995; Romani, 1996; Vinceri, 1997). Sterols, which originate during the early stages of fruit development, at this stage undergo slight changes in content of β-Sitosterol (decreasing) and of Stigmasterol and δ-Avenasterol (increasing). The constituents of the alcoholic aliphatic and triterpenic fraction increase up to the time of maximum inolition (Camera et al., 1976; Tiscornia et al., 1978), like volatile compounds (Bertuccioli et al., 1978; Olias et al., 1980). The chlorophyll and carotenoid concentrations decrease while percentages of xantophile, often esterified, increase (Mosquera et al., 1989; Modi et al., 1992).

Olive Over-ripening Stage

As ripening progresses, pectolytic enzyme activity increases with cell-wall degradation (Montedoro, 1981). Inside the fruit, which is still attached to the branch, triglycerides start degrading, total contents of natural antioxidants (e.g., chlorophyll, anthocyanin, tocoferols, minor polar compounds) decrease, with negative impacts on organoleptic properties and on the hedonistic assessment of oil (Vasquez Roncero et al., 1971). Notable changes do not generally concern fruit dimension, which has already become stable by the end of ripening; on the other hand, respiratory activity and water shortage decrease, with an apparent increase of oil inside the fruit.

Morphological changes and biological activities manifest in fruit ripening are variable, because the specific phenomenon of cultivar (genetic control) is connected with different interactions that control oil development and accumulation in olives (Cimato, 1988). In particular, it is a matter of environmental conditions (e.g., cultivation area, rains, temperatures), root-soil relationship, entrepreneurial choices (olive picking time) and practical choices concerning plantation management (e.g., manuring, watering, pruning, protection).

In summertime, cultivation interventions of leaf manuring delay the synthesis of minor polar compounds, thus delaying fruit ripening (Cimato et al., 1994). Similar changes to the biochemistry of these compounds are indicated for table olives (Patumi et al., 1998) with irrigation during summertime.

27.2. Crop Requirements

27.2.1. Tree and Environment

The olive tree succeeded in perfectly adapting itself to the Mediterranean environment, characterized by notable changes in soil and climate; its production significantly influenced the development of rural civilizations in several and "different" conditions.

As a rule, in Italy the olive tree occupies surfaces located at higher altitudes and on steeper slopes when compared to other agriculture crops as a whole. In fact, olive groves are located at an average altitude of 300–400 m, but in many areas in southern and central Italy this crop reaches 600–800 m on slopes with more than 15% gradient.

Production quality and quantity depend on specific physiological processes such as photosynthesis effectiveness, biological transformations (flowering and fruit development), absorption phenomena, and transportation and assimilation of nutritive elements. These processes in turn are determined by environmental conditions such as thermic regime, light and water supply, and chemical, physical and hydrological features of the soil.

27.2.2. Temperature

Air temperature plays a key role since it is the first parameter that rules the geographical distribution of the olive tree; therefore, currently thermal limits for its development are set between 46°N and 35°S, thus including both areas in northern regions of Italy and areas in South Africa and Australia. Although it is not always easy to distinguish the specific effects of temperature from those of other climatic factors, it is clear that temperature has a significant influence on the regulation of several physiological processes such as transpiration, breathing, photosynthesis, enzymatic activity, the formation and degradation of different compounds existing in the fruit (e.g., fatty acids, polyphenols, chlorophyll), and cell division and expansion. Therefore, temperature acts on vegetation growth, on the morphological differentiation of different plant organs (buds, flowers), on fruit setting and ripening and consequently on the overall productive capacity of the plant. For photosynthetic activity, the thermal optimum is set at 25–28°C. Almost all the metabolic processes are altered with average temperatures lower than 5°C or higher than 35°C. In fact, the olive tree is sensitive to winter and spring cold. The leaves of these trees are damaged whenever the temperature is below 5°C. Limbs and the whole crown may be seriously damaged if the temperature goes below –10°C.

In Italy, the olive tree is often exposed to winter frosts, with minimum temperatures that may damage cultivations. Risk areas are Friuli and Veneto, some areas of Liguria and Romagna, Tuscany, Umbria, upper Latium and Marche. Frosts in 1956 and 1985 showed that minimum temperatures (–6°C or –8°C), although prevailing for a short time while the vegetation lay fallow, were enough to cause defoliation, bark damage, parching of the youngest branches, and even loss of production for the following year.

The same temperature values are more detrimental if they occur during vegetative flush or during full vegetation. Factors that can worsen damage due to low temperatures are air stagnation or cold draughts; within the same cultivar, young or very old trees are more sensitive. Damage from cold is more evident in strongly hydromorphic soils, that is, those that have problems of water stagnation: these conditions may occur both in morphologically lower areas and in the lower parts of slopes, but also wherever rain intensity and volume override soil permeability.

Frost can cause heavy defoliation, more or less intense dewatering of cortical tissues with cracking of the bark and formation of necrotic plates in the youngest cortical areas, as well as a consequent crown parching and death (Ferrini, 1996). In new plantations it is therefore necessary to avoid soils that are very hydromorphic and climatic areas where thermal needs cannot be met, preferring the so-called "thermal area" corresponding to hills between 150 to 450 m asl, and on aspects facing west or south-west, which are less prone to temperature extremes.

The negative effects of high temperature are more common whenever there is not an appropriate supply of water and a low transpiration rate. These cause an increase of the surface temperature of several organs and cause burns on leaves, trunk and limbs. In places where the daily mean summer temperature is higher than 30°C, vegetation development of the plant takes place in spring and autumn, while it slows to a stop in summer. In more temperate areas, vegetation activity slows only in winter.

Climatic conditions are also very important during flowering time. According to some authors, the *optimum* occurs when temperatures vary between 2–4°C minimum and 14–18°C maximum, on a daily basis. Wintertime minimum temperatures are the key limiting factor to cultivating this species in cold climates; however, a period with low temperatures is also absolutely necessary to flower formation. Ten to twelve weeks at temperatures lower than 12°C are needed for the success of the flowering process. This is why the olive tree grows without flowering in tropical climates. At a constant temperature of 13°C, the olive tree blossoms a great deal, but many of the flowers are staminiferous; therefore, fruit development is affected and fructification is insufficient. On the other hand, low temperatures during flowering time delay the development of the pollen granule inside the ovary and often lead to a reduced setting rate.

Vegetation shoots, unlike the flower shoots, have fewer needs in terms of temperature, since their opening and development do not need a cold period. They only need mild temperatures (higher than 18–20°C). Temperature also has a direct impact on photosynthesis and on the division of assimilated elements.

Oil qualitative parameters, such as fatty acids composition and polyphenol content, are also under climatic control (Tura et al., 2005). In particular, the degree of olive ripening is positively linked to the overall temperatures during ripening. Dry regimes, together with high temperatures, reduce the lipoxygenase enzyme activity, thus determining the decrease of volatile oil compounds.

In cold environments, fruits are less rich in oil but provide a product of better quality due to the higher percentage of unsaturated fatty acids (oleic, linoleic and linolenic ones) as well as the content of total polyphenols in the oil. However, early cold winters that coincide with the last stages of fruit ripening may seriously damage olives, causing drawbacks during the extraction stage. The oil takes on the typical "dry" taste and, as a consequence, has poor organoleptic properties. Moreover, it cannot be preserved for a long time.

27.2.3. Light

Light, i.e., the total energy flow captured by the crown, can affect oil production in different ways. Light is essential for biosynthesis of carbohydrates, chlorophyll and anthocyanin formation, stomatic opening, transpiration and, indirectly, also the supply of nutritive elements. At the same time, it has significant morphogenetic effects on the development of different organs, thus regulating plant vegetation growth and production.

For the olive tree, maximum photosynthesis net values vary with genotype and environmental conditions (Gucci, 1998). These values are never high because of the CO_2 resistance spread over the mesophyll, because of the stoma morphology, as well as because of the tomentous character of leaves (Bongi and Palliotti, 1994). Shoots that are always exposed to sunlight, as a result of leaf direction and bending, receive an average light intensity equal to a degree of saturation; those in shadow that are located inside the crown or shaded by nearby trees may have a negative photosynthetic balance for most of the day.

Areas of the crown that are less exposed to sunlight exhibit clear-cut morphogenetic characteristics (Guerriero and Vitagliano, 1973; Tombesi and Standardi, 1977). In particular, decrease has been reported in leaf dimensions, branch extension, percentage of flower induced buds, number of flowers with totally developed ovary, percentage of set flowers, and fruit size.

27.2.4. Water Supply

Approximately 90% of world olive cultivation is located in the Mediterranean area. The olive tree, which is adapted to the long dry summer periods characterizing this climate, is even able to survive with annual rains of only 200 mm; on the other hand, it badly tolerates root asphyxiation. The daily water consumption of plants in good nutritional and health conditions is equal to 1–1.2 L water per m^2 of leaf surface (Gucci, 2003). In addition, the olive tree capacitance (the quantity of water that different tissues may give out from their stocks by transpiration flows) is very high and helps to decrease the water content and the formation of a high potential gradient between leaves and roots. This gradient allows the plant to absorb water even when the soil gradient reaches values equal to –2.5 Mpa. Nevertheless, if in springtime water supply is insufficient, roots definitely absorb less nourishment, causing a decrease in plant vegetation growth (Xiloyannis et al., 2004). The year following there is limited growth of shoots and a decrease in the leaf blade (alternate bearing), with a consequent decrease in production.

Problems in water nutrition also jeopardize the productive process. The frequency of flowers with abortive ovaries increases, pollen production is limited, germinability is partly damaged and therefore the plant has a shorter setting period. Whenever water shortage continues for the whole summer, fruit growth decreases, while physiological fall increases.

If water is lacking in autumn, since the olive tree cannot make its fruits larger, it speeds up the ripening metabolism so that the drupes will have a low pulp/stone ratio at harvest. These phenomena are more serious in young plantations (Doorenbos et al., 1977).

27.2.5. Wind

Wind is also an important "environmental factor" for vegetation growth as well as for the production of the olive tree, given the effects it may have directly or indirectly on plant metabolism. Moderate breezes, in particular during anthesis, aid anemophilous pollination, with positive repercussions also on the phenomenon of crop alternation.

Although we have little information on the effects that wind has on the plant (mechanical damages, such as branch and limb breaking, defoliation and parching of the areas exposed to sea winds in particular) extreme summer and autumn winds cause stigma parching, excess transpiration and drupe shriveling. In windy environments, it is better to create a high density plantation and to employ appropriate windbreak barriers.

27.2.6. Hail and Snow

Hail and snow mainly cause mechanical damage. Hail causes damage to tree organs and defoliation: if it hits the main trunk and the leaves of young plants, it may cause damage that compromises planting success. For adult plants, damage is more easily repaired, but it may facilitate attacks by scab, as well as by other pathogens and parasites. Cuts to the drupes represent irreparable damage to plantations for the production of table olives and imply reduced quality if oil is produced, since the incisions trigger oxidation processes in the fruits. Heavy snowfall is rather uncommon: it may happen occasionally in areas located at higher altitudes or in the northern edges of the olive tree cultivation area. Snowfall may cause branches to stretch downward and break.

27.3. Tree and Soil

The olive tree is more tolerant of nutrient deficiencies or surplus and of stony, arid and scarcely fertile soils than are most of the fruiting trees. Rock fragments ensure better internal drainage conditions and, if calcareous, aid carbonic acid precipitation and soil oxygenation.

27.3.1. Soil Physical Factors

The root activity of the olive tree lasts all year long. High permeability and high water retention ensure the intensive gas exchanges necessary to root development. The main soil physical factors that have an influence on root development and plant growth are texture, structure, porosity, oxygen capacity, and available rooting depth.

Texture, Structure, Porosity and Air Capacity

Texture and structure influence water and nutrient retention, water movement, oxygen availability, workability and trafficability. Generally, for rain-fed olive tree cultivation, soils must have water retention in proportion to the amount of rainfall. As soil available water content decreases, the olive tree needs more rainfall, in order to grow and produce profit (Palese et al., 1996). In the Mediterranean countries, many olive tree soils developed on fluvial or structural terraces, where there may be soil horizons with strong texture contrasts. For instance, it is possible to find a horizon with fine texture (argillic horizon, Bt) under a coarse-textured one (Fig. 27.4).

Although the argillic horizon increases water and nutrient retention, it may cause problems for drainage and root aeration. The olive tree can grow well in soils with a clayey subsurface horizon, only provided that it is well structured, with high macroporosity and a good percentage of long and interconnected pores. For soils with a contrasting texture profile, it is advisable to avoid deep cultivation, which would bring to the surface portions of the clayey horizon.

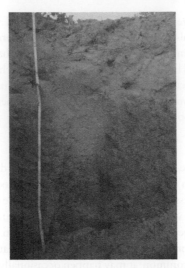

Fig. 27.4. Soil with strong textural contrast in an olive grove of Sicily.

Fig. 27.5. Soil with hard setting problems in an olive grove of Sicily.

There are also other soils that give rise to special cultivation problems; these form very big clods that severely harden after ploughing, even if ploughing is carried out in conditions of low moisture (soils prone to hard setting) (Fig. 27.5).

Air passing through soil pores ensures soil root and microorganism breathing.

The olive tree is sensitive to root asphyxia and this sensitivity differs with the cultivar (Navarro and Parra, 2001): *Frantoio, Leccino* and *Mignolo Cerretano* proved to be the least sensitive (Corti and Cuniglio, 2000), given their capacity to form roots that are close to the soil surface. Plant resistance depends on tree age as well: young trees are characterized by very active growth and may not survive even if their roots stay in water-saturated soil only for a few days.

Fig. 27.6. Winter water stagnation in an olive grove.

The soil may fill with water in winter or in spring due to heavy rains (Fig. 27.6) or when surface and subsurface water flow accumulates in swales. Water saturation may occur in a soil also when there is a poorly permeable subsurface horizon. During very rainy periods, since water does not percolate through the lowest horizons, it hampers soil aeration. The soil may still be saturated even a long time after the rain ends, forming a perched water table. In these cases the soil shows redoximorphic features, i.e., the pedogenetic blackish, yellow-reddish or red mottles associated with grayish, greenish or bluish colours, caused by cycles of iron and manganese oxidation and reduction (Fig. 27.7). Roots do not develop where aeration is poor also because of the possibility that some toxicants are produced and adsorbed. However, whenever water is lacking the olive tree discontinues aerial growth, but not root growth. This modifies the root/crown ratio in favour of the roots, therefore limiting development of the leaf area and allowing the root apparatus to survive (Nuzzo et al., 1995).

In the case of a shallow water table, roots extend superficially because soil saturation may cause asphyxia. In such soils, it is necessary to provide for proper drainage. However, it is better not to use soils that have a water table at less than 1 m from the surface.

Available Rooting Depth

A key soil property is the actual volume of soil that may be explored by roots. Roots grow in the soil following very different models and reaching different levels of specialization and development according to soil nature and availability of nutrients (water and oxygen included).

Fig. 27.7. Subsurface soil hydromorphy in an olive grove of Tuscany.

Fig. 27.8. Whitish calcic subsuperficial horizon in an olive grove.

In fact, the root is very plastic in responding to different nutritive conditions of the soil; as a rule, the plant needs an absorbing mass inversely proportionate to soil fertility.

Whenever placed in poor or insufficiently fertile conditions, the root branches out constantly and intensively, so that the root mass is bigger than the aerial part of the plant. In fertile soils, the root system has an absorbing mass that is almost half the volume of the aerial one, since it is not necessary for the apparatus to develop greatly to satisfy the needs of the plant.

A good supply of nitrogen and phosphorus favours the development of the roots, while potassium has no prominent effects. Calcium shortages in the deepest profiles of the soil restrain primary root extension (Coutts and Philipson, 1980). A good supply of soil organic matter acts positively on root growth, because it reduces soil resistance to root penetration and increases the water retention capacity.

Most olive tree roots, although not in a homogeneous way, explore the soil layer between 15 and 20 cm deep (these are mainly adventitious roots resulting from foot hyperplasia) and between 60 and 100 cm in depth. This is why 1 m of available depth is usually enough for normal root growth. However, it may be insufficient in soils that are rich in rock fragments, or wherever dense horizons limit root penetration. As a rule, a horizon with a successful rooting rate lower than 30% is considered impenetrable.

Horizons rich in salts also limit root development; among these the calcic horizons are the most widespread throughout Italy, mainly in the centre and in the south, on carbonate lithologies and on level morphologies (Fig. 27.8).

The presence of a continuous hard rocky substrate, as in the case of many hilly areas with a strong slope, also influences the available rooting depth (Fig. 27.9).

Together with rock, there also may be pedogenetically hardened horizons that hamper root development. These are the horizons of salt accumulation, particularly petrocalcic and petrogypsic horizons (Fig. 27.10) deriving from a strong accumulation of calcium or gypsum, which are mainly found in southern and insular regions. There are also horizons rich in iron and manganese nodules and petroferric horizons, present in some paleosols.

Fig. 27.9. Limited rooting due to surface rocks in an olive grove.

Fig. 27.10. Limited rooting due to petrocalcic subsuperficial horizon (Btkm) in a soil of Sardinia.

Fig. 27.11. Evident damage to tree vegetation caused by vertic shrink and swell of soil masses of a soil in Sicily.

The presence of vertic horizons, i.e., of dynamic clays that crack during the summer and swell in winter, also represents a serious restriction to olive tree roots. Mediterranean environments are characterized by a strong contrast between the seasons. Vertic shrink and swell of soil masses are able to break the roots of trees that, like the olive tree, may have a wide and deep root system (Fig. 27.11).

27.3.2. Chemical and Biological Factors

The olive tree requires the elements available in the soil that are essential to carry out all the metabolic activities responsible for growth, as well as for the productive cycle; in particular, it needs nitrogen, potassium, phosphorus, calcium, magnesium, iron and boron. Inside the plant, each of these elements has a well-defined function that is never independent, but always connected in a synergic or in an antagonistic way to one of the other nutrients.

An assessment of the soil must take into consideration fertility elements as well as the following soil chemical factors: pH, salinity, excess sodium and possible toxicity due to boron and chlorides.

Nitrogen

Nitrogen is the inorganic nutrient to which the olive tree reacts most quickly. It is also the most important, since it improves the plant vegetative and reproductive activities, limiting the undesirable phenomenon of production alternation. Nitrogen stimulates growth by supporting the production of new shoots, controls apical dominance together with phosphorus, and aids the flower setting and fruit development processes. Moreover, it is a component of chlorophyll and increases the quantity of chlorophyll in leaves, thus aiding the absorption of other elements. Nitrogen absorption is closely linked to the available water supply of the soil. In spring, early nitrogen flow toward the buds is essential for their opening and for the first stage in forming new vegetation. Lack of nitrogen reduces the growth of leaves and particularly fruits and is also associated with a reduced transfer of nutritive substances to these organs, as well as an increased transport towards roots. However, very rich nitrogen supply may favour the attack of pathogenic agents, such as those responsible for sooty mould.

Potassium

Potassium is the element of fruit output and ripening. It is present in the centres of most intensive biological activity and is essential in the phenomena involved in plant water metabolism, heightening resistance to drought and to fungal diseases. Potassium favours the synthesis of sugars, their accumulation as starch and the formation of fat, increasing olive oil yield. During ripening, this nutrient neutralizes uronic acids resulting from protopectin degradation and controls enzyme activities regulating the synthesis of amino acids and phenolic acids. Potassium is very mobile in the olive tree and, in fact, it is hardly present at constant levels in the plant. A low potassium content in leaves, if depending on low phenol levels, reduces the flowering process (Gonzales et al., 1976). Potassium content in leaves is generally inversely proportionate to that of calcium and magnesium. Excessive availability of potassium in the soil usually results in greater difficulty in absorbing magnesium.

Phosphorus

Phosphorus is the key element of enzymes and proteins, playing a role of paramount importance in the cell division process and in the development of meristematic tissues. As to the influence of phosphorus on the output of the olive tree, apparently this nutrient favours phenomena linked with flowering, setting, and metabolic processes that speed up fruit ripening (Hartmann et al., 1966). Phosphorus consumption is not high and its content in leaves does not undergo notable changes during the year. Phosphorus is seldom lacking in soils where olive trees are grown; however, this element may be blocked and be made unavailable by calcium, iron and aluminum oxides, and the amorphous and poorly crystalline minerals that are widespread in many volcanic soils.

Calcium

Among the arboreal fruit trees, the olive tree is the most sensitive to lack of calcium. Its main function is connected to the mechanical resistance of tissues. In this plant, the concentration of calcium is inversely related to that of potassium: when percentages of calcium are high, low percentages of potassium are found in the leaves of cultivars *Frantoio*, *Ascolana* and *Coratina* (Crescimanno et al., 1975). Above all, assessment of calcium is important when using acid or neutral soils that do not contain lime. A direct correlation has been assessed between calcium and potassium.

An unusual Ca/Mg ratio (lower than 0.9–1.0) may cause a special magnesium toxicity; this phenomenon never occurs if this ratio reaches approximately 2. It is possible to find unusual values of the Ca/Mg ratio in soils developed on green rocks (ophiolite) as well as in colluvial and alluvial deposits originating from these rocks. In some extremely leached paleosols we can also find residual accumulations of magnesium and lack of calcium, which is preferably absorbed over magnesium.

Magnesium

Magnesium is one of the constituents of chlorophyll, takes part in the RNA synthesis processes and is an activator of many enzymes. Magnesium is seldom lacking in the soils where most Italian olive groves are located. However, magnesium deficiency in leaves may occur due to leaching in very permeable soils, due to sub-acid pH, or in soils with high potassium concentrations that hamper magnesium absorption.

Boron

Boron is important to the olive tree and takes part in carbohydrate metabolism, activates enzyme systems and hormone functions, and aids the synthesis of flavonoids and pyrimidinic bases (DNA and RNA) and the transport of sugar through the phloem. Boron may be lacking in calcareous soils rich in clay, with high pH, and in conditions of water stress. Whenever there is a lack of boron, between May and June, the plant shows a huge decrease of flowering and setting, followed by an intensive summer fall of small fruits, diminishing production (Cimato et al., 2002).

Iron

Iron contributes to a correct function of photosynthesis and thus has an influence on the formation of the photosynthesizing surface and, consequently, on the fructification cycle. In calcareous soils, a nutritional physiopathy called ferric chlorosis may occur due to the non-absorption of iron in its reduced form (Fe^{2+}), which is easily absorbed by the plant. This phenomenon happens in olive groves in Andalusia, where it seriously jeopardizes successful production.

Below we report the main soil chemical properties found in excellent areas for olive tree cultivation (Table 27.1).

Table 27.1. Values of macro and microelements in soils having good productive qualities.

Element	Values
Total nitrogen	1.0–1.5 g/kg (Latium)[1] > 0.10% (Andalusia)[2] 0.80–1.22% (Basilicata)[3] 0.06–0.12% (Tuscany)[4]
Exchangeable potassium	Soils with coarse texture 30–100 ppm; average texture 175–300 ppm; soils with fine texture 300–500 ppm (Andalusia)[5] 50–155 ppm (Siena)[6] 218–480 ppm (Basilicata)[3] 128–344 ppm (Florence, Grosseto, Leghorn)[4]
Assimilable phosphorus	50–100 ppm (Sardinia)[7] 12–23 ppm (Basilicata)[3] 8–25 ppm (Andalusia)[2] 8–45.8 ppm (Florence, Grosseto, Leghorn)[4]
Active limestone	2.9–7.4% (Basilicata)[3] 20–50% (Latium)[1] 0–1.4% (Florence, Grosseto, Leghorn)[4] Soils with coarse texture 500–800 ppm, soils with average texture 600–2400 ppm, soils with fine texture 3000–4000 ppm (Andalusia)[2]
Exchangeable magnesium	116–263 ppm (Basilicata) Soils with coarse texture: 25–60 ppm, soils with average texture 80–180 ppm, soils with fine texture 120–300 ppm (Andalusia)[2]
Exchangeable iron	20–150 ppm (Andalusia)[5]

[1] *Tittarelli et al., 2002;* [2] *Troncoso, 1998;* [3] *Lacertosa et al., 1998;* [4] *Toma, 1999;* [5] *Navarro and Parra, 2001;* [6] *Costantini et al., 2006b;* [7] *Bandino et al., 1994.*

Organic Matter

Organic matter plays a role in providing inorganic nutritive elements (e.g., N, K, P, Ca, Mg) and in improving the soil physical properties. Organic matter mineralization results in a gradual release of nitrogen; therefore, nitrogen is less subject to loss due to leaching and is available over a longer span of time. Other indirect functions of humic substances are the chelating properties of many functional groups (carboxylic, carbonyl and amide) that help protect microelements from insolubilization. Organic matter also increases soil cationic exchange capability, thanks to the sites with negative charges; in this way a higher quantity of mineral elements can stay longer in the root zone, and therefore be absorbed.

Organic matter also takes part in the formation of soil structure. It favours the presence of aggregates that reduce plasticity and tenacity and increase permeability in clayey soils and water retention in sandy soils. In the latter, even a low content of organic matter is often able to increase structural stability. Finally, in soils rich in silt, aggregation among particles reduces the risk of formation of surface crusts and of sub-superficial compact horizons.

pH

It has been demonstrated that the olive tree finds its *optimum* in soils with pH values between 6.8 and 7.5, but also that this plant can keep on growing in a broader range of pH (5.5–8.5). In fact, the olive tree is not a forced calciophilous species, since it is also able to grow in neutral and sub-acid soils. In Tuscany, the olive tree is present both in alkaline soils rich in lime (Costantini et al., 2006b) and in neutral and sub-acid soils on sandstone. In Calabria, the olive tree often grows on soils with an acid reaction; in Sardinia, it is present both on Gallura (acid soils) and on the Planargia (calcareous soils). In fact, the olive tree has a great adaptability to lime, dependent on other factors, particularly climatic and pedoclimatic ones like soil water content.

However, with extreme pH values (pH < 5.1 and > 8.4) there is the risk of problems concerning toxicity or deficiency of elements. In acid soils, toxic ions, such as aluminium and manganese, may be present in the circulating solution; in strongly alkaline soils, macro- and microelements, which are important for plant development, may be immobilized.

Salinity

All soils hold a certain quantity of soluble salts, but if their concentration is high plants may have problems in absorbing water and, at the same time, may be damaged by the toxicity of specific ions (Cresti et al., 1994). Soil salinity is expressed through electrical conductivity (EC). The olive tree is resistant to salinity more than other fruit-bearing trees. Usually, a soil is considered saline if the EC is higher than 1 or 2 dS/m, but it is thought that the olive tree is able to produce at higher values as well, although oil production decreases notably (Vega et al., 2001). Therefore, it is better to avoid new plantations in soils characterized by salinity.

Excess salinity may derive from the parent material of the soil, as well as from irrigation water. In order to avoid the latter, it is absolutely necessary to analyse the water or to adapt irrigation quantities to chemical and physical properties of the soil.

Although soils with salinity problems are not recommended, they may be managed through an appropriate choice of varieties. Literature shows strong differences among varieties regarding resistance to salinity in the olive tree. Rather resistant cultivars are *Frantoio, Coratina, Carolea, Maurino* and *Moraiolo*, while *Leccino* trees must be considered scarcely resistant (Gucci and Tattini, 1997; Dettori et al., 1999).

Boron, Chlorine and Magnesium Toxicity

Boron and chlorine are necessary to the correct development of the olive tree, but in low quantities; excessive quantities of these two elements may result in toxicity. The toxicity limit to chlorides for the olive tree, associated with a 10% cut in production, varies between 10 and 15 mmoles/L; the limit for boron is equal to 2 ppm. Like salinity, toxicity may also result from irrigation water or, as far as boron is concerned, from excessive supplies to the plant. Toxicity due to excess magnesium may be found in soils developed from green rocks (ophiolitic rocks); in these soils we can also find a surplus of other elements, particularly metals such as chrome and zinc.

Mycorrhizae

In the rhizosphere (the area of contact between roots and soil), there are microbial populations that set up several types of symbiotic interactions with the plant roots. These symbioses have an influence on several aspects of plant physiology, such as mineral nutrition, plant development and protection. They also indirectly influence the stabilization of soil structural aggregates through humic substance accumulation and limit weed invasion. Research has demonstrated that use of inoculant with arbuscular mycorrhiza associations (AM) brought about improvement in nutrient absorption, increase in resistance to pathogenic fungus and tolerance of biotic and abiotic stress (Barea et al., 2000).

27.3.3. Ion Absorption

Ion absorption by the plant does not depend on roots alone. In fact, the olive tree is able to receive nutrients also through epigeal organs, such as limbs, branches, and, above all, leaves.

There are two mechanisms that allow nutritive elements to reach the root surface: convection and diffusion. The former (convection) is the consequence of water percolating through the soil and the flow determined by plant transpiration, which acts as a suction pump. This mechanism may be insufficient to meet plant requirements since concentrations in the circulating solution are lower than the required element quantities. Therefore, the second type of flow is diffusion, due to a concentration gradient that is established between the depleted solutions of the portions of soil in contact with roots and the surrounding solutions, which tend to recover concentration.

For some ions (e.g., calcium and magnesium), flow is dominated by convection; therefore, these primary elements must be present in the soil solution in rather high concentrations and must not interact with the exchange complex of the soil. For other ions (phosphorus, potassium, boron, iron, zinc, manganese and chlorine), flow is mainly controlled by diffusion. Nitrogen supply is provided by both processes, given the different forms it takes during the transformation of organic substances in the soil.

Nutritive ion absorption is regulated by synergies and antagonisms that are set up between the nutrients themselves and takes place at different speeds. For example, it is quicker for the nitric ion than for phosphorus and quicker for potassium than for calcium. Antagonism associating potassium to manganese, for example, ensures that a surplus of one of these two elements reduces the absorption of the other and vice versa.

Environmental conditions, by modifying the transpiration demands of the plant, also have an impact on absorption: low relative air moisture, such as that occurring in very arid and windy soils, far from the sea, reduces absorption of potassium but not of calcium; therefore, the Ca/K ratio in tissues increases. Calcium and potassium absorption is reduced also when light intensity is low. Finally, a pedoclimate that is made very dry by drought or by excessive summer harrowing aids the absorption of one element rather than another. In such a situation, roots find the best conditions in depths where the temperature is lower and absorption of calcium is greater than that of potassium.

Here follows some data concerning macro- and micronutrient absorption.

Nitrogen

Nitrogenous substances come in contact with roots due to diffusion and are absorbed as nitrate ion (NO_3^-) and as ammonium ion (NH_4^+). In the olive tree, nitrate ion absorption is low and belongs to the active type, causing the plant to waste energy at the expense of photosynthetic products. Whenever energy availability is not a limiting factor, the plant tends to absorb the nitrate in quantities that are higher than its needs. Absorption of the element depends both on its relative quantity in the soil solution (concentration) and on the presence of calcium and potassium ions (positive action) as well as ammoniac nitrogen (inhibiting action). The ammoniac form is, on the contrary, absorbed through a passive diffusion process and with a lower energy demand. The key factor regulating the ammonium ion is its concentration in the soil solution. The absorption of nitrogen in its ammoniac form (NH_4^+) is easier in soils with neutral or alkaline pH.

Potassium

The absorption of potassium, as K^+ ion, mainly takes place through diffusion and therefore is tightly linked with the soil water regime. Water shortages, although modest, may seriously jeopardize its absorption. Potassium, more than the other nutrients, has a low mobility through the soil, but is easily translocated within the plant. After stone lignification, drupes need increasing quantities of potassium, which is readily translocated from leaves (Ortega Nieto, 1969).

Phosphorus

Phosphorus is absorbed in the form of a phosphated ion (HPO_3^{2-}; HPO_4^{2-}). Its availability and absorption depend basically on soil pH, as well as on the presence of other ions. This is very important, considering its poor mobility through the soil. High nitrogen availability, as nitrate ion (NO_3^-), reduces the absorption of phosphorus. The presence of mycorrhizae apparently has a significantly positive effect on phosphorus absorption.

Calcium

For calcium absorption, which takes place as Ca^{2+} ion, a stable water regime in the soil is essential. Water shortages have a negative impact on the absorption of this nutrient.

Magnesium

Magnesium is absorbed as Mg^{2+} ion. Magnesium in high quantities competes with calcium both for absorption and for translocation; vice versa, in the case of shortages, magnesium favours calcium absorption.

Boron

Boron is absorbed as undissociated boric acid (H_3BO_3). It favours the passage of calcium from the soil to the plant. For boron, as is the case for calcium, water regime is important; water shortage reduces absorption (Demetriades, 1968). Boron accumulates in old leaves and is not translocated again; therefore, its supply to the soil must always be ensured.

Iron

Iron is mainly absorbed by the root as inorganic ion Fe^{2+} (ferrous), whereas Fe^{3+} (ferric) plays a minor role. In sub-alkaline soils, roots have difficulty in finding the ferrous ion since it tends to oxidize and to precipitate in the form of hydroxides. Iron is one of the micronutrients most required by the plant, but its tendency to form different types of insoluble compounds within the soil may reduce its real availability.

Other Elements

The circulating solution directly absorbs other elements as ions, including silicon, sodium, sulphur, manganese, chlorine, copper, zinc and molybdenum.

27.3.4. Water Absorption

Arboreal trees have efficient mechanisms for water absorption and endogenous movement. Since their nutrition is primarily gaseous (photosynthesis, photorespiration), they are equipped with a system for gas exchange that determines loss of water by transpiration through leaves, fruits and roots; water is constantly reintegrated thanks to root absorption. Along with environmental requirements (temperature, air hydrometric state, ventilation, light and soil available water supply), metabolic demand is the most important factor that has an impact on the transpiration mechanism. The plant absorbs nutrients and water in order to meet specific requirements of vegetation development involving crown change and fruit growth.

Water uptake occurs by "active" and "passive" mechanisms. The former occurs whenever there is good water availability in the soil and when the plant has low transpiration intensity. In these conditions, absorption is ensured by the osmotic gradient that is created between soil and root cells. This mechanism ensures the distribution of the water absorbed to the main conductive system (vessels). During the "active" absorption phase, which coincides with the renewal of vegetation activity (budding), a real root pressure is set and reaches high values (10–12 atm). This also contributes to the transportation of water going up through the plant conductive system. As soon as transpiration intensity increases, the "active" absorption is followed by the "passive" one that prevails in the plant water balance: the water rises up along the stem, drawn by the tension caused by leaf water loss (transpiration).

27.3.5. Water Erosion

Many olive trees are located in areas with a high risk of water erosion. Under the same conditions in terms of vegetation cover, cultivation type and soil conservation management practices, erosion increases with slope, maximum length of the plot, and increase in soil erodibility. Soil erodibility mainly depends on soil type and structural stability, but also on its organic matter content, permeability and texture.

Steep soil slopes, lack of organic matter, agricultural husbandry that usually does not care for the surface during rainy periods, all cause serious damage in many olive tree cultivation areas. In fact, soil water erosion determines the loss of the most fertile portion of the soil, since it removes the finest and richest nutrient particles of the soil. Erosion control is not an easy practice to be implemented: with slopes of less than 14%, this phenomenon can be controlled through correct management. With steeper slopes, soil preservation entails a special plantation setup, such as by traversing or contouring fields, or by creating terraces, connected terraces, embankments or lunettes.

27.4. Olive Grove Design and Management

During the planting of a new olive grove it is necessary to assess the pedological suitability of the chosen sites, i.e., lands where there are good chances of finding soils that produce in a satisfactory manner most years, with a successful qualitative result, that do not need expensive agrotechnical crop-husbandry and that have no geomorphological risks or damage in terms of soil conservation. Below we report some agronomic aspects of planting and management techniques for the olive grove, where identifying soils is very important indeed.

27.4.1. Bush Clearing and Stone Removal

The soil must be freed from plant remains from previous growth, including the root apparatus of arboreal plants or shrubs and hedges existing in the plot. For replanting, it is necessary to eradicate old olive trees, paying attention to remove roots and uncover areas invaded by parasites that often establish themselves on what is left of the old trees. Soils with unfavourable properties, such as low permeability, hydromorphy, or high acidity, may suffer the attack of root parasites more frequently.

Often too much stone is removed when clearing a site. The existence of stones is a very limiting factor, particularly for manual olive picking from the ground. On the other hand, since the use of grass cover is spreading in automated olive cultivation, the presence of a small quantity of stones is more an advantage than a limit, aiding soil preservation against surface erosion, water conservation in the soil, drainage improvement, thermal regulation, and supply of mineral elements. Nevertheless, wherever stones have a diameter larger than 25 cm, fine earth is limited, or soils have a limited thickness (lithic soils), their presence may be detrimental (Costantini, 1985).

27.4.2. Levelling

In hilly areas, the building of new and specialized plantations is usually preceded by levelling operations and by slope regularization. The main goal is to model the soil surface so as to remove morphological asperities and allow vehicles to circulate without obstacles. However, these operations to obtain wide areas with a homogeneous slope must be executed with special care, especially during the rainy season, so that they do not cause the movement of huge quantities of soils along the slopes, thus speeding up massive erosions (Fig. 27.12). This phenomenon may lead to the outcropping of non-pedogenized substrata, with very limiting characteristics for olive cultivation (Fig. 27.13).

It is possible to solve this problem by setting aside the fertile layer of the soil, in order to redistribute it over the surface afterwards, once levelling is done. However, during levelling it is advisable to adapt the fields to soil morphology, to limit plot length along the maximum slope and to avoid the total removal of the sub-surface horizon of the soil (B horizon) (Fig. 27.14).

Fig. 27.12. Water erosion by gully erosion on an excessively levelled slope.

Fig. 27.13. Outcropping of a non-pedogenized layer in a soil. This soil, originally wooded with olive trees, was left without vegetation for about five years. After several sowing operations with legumes and green manure, a specialized vineyard was built.

Fig. 27.14. Olive groves planted by adapting fields to morphology, and avoiding excessive land movements.

27.4.3. Surface and Deep Water Drainage

Ditches and pipe drainage are used to limit the risk of landslides, landslips and surface erosion as well as to ensure the quick runoff of surplus water. In addition to drainage systems with impluvia and contour ditches at the boundaries of the plantation, in hilly soils it is necessary to pay the greatest attention to lithological discontinuities existing between the soil and underground sediments, which cause a change of water permeability. Layers with a different texture or structure, even a few decimetres thick, may determine undersurface water movement and thus resurgences along the slope and landslides.

Resurgence of water is accentuated by excessive levelling. In fact, poorly pedogenized soils of the substratum hold much less water than the soil removed during levelling; therefore, hypodermic circulation of water tends to saturate the remaining soil much more quickly and to cause landslides (Fig. 27.15).

For soils with water permeability problems along the profile, classified in the category "moderately well drained" or worse (see specifications for soil properties and qualities in the Appendix), the creation of an appropriate drainage network is absolutely necessary in order to allow normal farming techniques and plant growth.

27.4.4. Manuring

In soils where intensive agriculture takes place there is often a high average annual deficit in organic matter. Pre-planting manuring of an intensively overworked soil is therefore a unique chance to reset the appropriate level of organic matter available. Organic fertilizer is spread in spring and on bare soils; afterwards the material is earthed in by ploughing or by more superficial tilling, followed by scarifying. For loose soils, i.e., soils characterized by conditions favouring a high mineralization of the organic substance, it is better to divide the supplies over the following years.

Fig. 27.15. In close-up and in the background, landslips from resurgences in a plantation of specialized tree crops built with excessive earth moving.

Mature fertilizer (ovine or bovine) still represents the best stock of nutrients for growth and the best slow-release fertilizer. A worthy substitute is the use of other organic products (for instance, fowl-manure, compost, meal) or green manure of fodder grasses and legumes (burial in April–May). However, fresh plant products show a poor formation of humus and mineralization takes place too quickly. This is why the use of green manure before planting often yields a limited supply of stable organic substances.

For soils in which internal drainage is moderately good or worse, deep burial of organic substances, particularly if fresh, may trigger reductive phenomena that are highly detrimental to the olive tree (Costantini et al., 2006a). Special attention must be paid to the execution of pre-planting tilling: deep ploughing by inverting the tilled horizons may cause a sudden increase of organic matter in depth. This may increase the intensity of the reduction phenomena, at the expense of iron and manganese. The burial of the superficial layer also causes a reduction of irregular and elongated pores in the soil, consequently compromising soil internal drainage. For soils with drainage difficulties it is therefore better to use well-decomposed organic matter, limiting burial to the first decimetres and favouring the mixing action through tilling without overturning the soil. As a substitute for ploughing, deep scarifying may be advisable, because it increases macro-porosity without changing the organization of natural horizons in the soil. The volume of soil tilled is almost equal to that obtained by deep ploughing if scarifying is crisscrossed over the whole surface. Another solution may be "double layer" tilling, done by deep scarifying (80–100 cm) with a simple vertical cut and then by ploughing in fertilizers and possible organic correctors (40–50 cm deep).

27.4.5. Planting Density and Grafting Distances

The olive grove plantation may be square, rectangular or triangular with varying distances between the trees depending on variety, cultivation system, and pedological and climatic characteristics. It is also necessary that the olive tree crowns, once maximum development is reached, do not touch each other.

A high number of olive trees per hectare greatly reduces the initial fruitless period and makes is possible to reach "full" production quickly. If cultivar energy is such that it results in a development greater than the space available to each plant, the initial advantages are lost, because as soon as the trees cover the available space, they cease growing and do not yield well. A judicious organization of the plantation must ensure a correct balance between soil fertility and tree growth. In fact, too much distance between plants represents a waste of soil, so that olive tree output sharply drops. The best density for rational olive cultivation, using suitable cultivation practices and in a climate and soil conditions without extremes, varies between 300 and 400 plants per hectare, with planting distances varying between a 6×8 m and 6×4 m rectangular planting space, in order to facilitate the movement of vehicles.

27.4.6. Row Orientation

Once planting density and planting distance are chosen, before planning the layout of the plantation, it is necessary to decide row orientation. North-South orientation ensures the best solar radiation for plants, but it is not always possible to achieve. With hilly soils in particular, it is necessary to follow a sloping line that reduces run-off phenomena to a minimum and facilitates mechanization.

27.4.7. Cultivar Choice

The selection of variety may depend on different project choices, because each variety may imply the implementation of a different cultivation model and a different mechanization level and therefore lead to different economic results. This choice has an important impact on production quantity and quality and must be linked with other preliminary choices for the olive grove, as well as with cultivation techniques. Autochthonous varieties ensure a greater adaptation to the specific pedological and climatic conditions of each soil. Their use is also necessary to increase the commercial value of typical oils and to comply with regulations concerning the production of Protected Geographical Indication (PGI) oils and the ones with Protected Denomination of Origin (DOP). Otherwise, it is possible to plant some varieties that have shown excellent adaptability to different environments even outside their own areas of origin. However, it is necessary to take into account the chemical and organoleptic differences of the resulting oils.

The selected varieties must be functional to the harvest system chosen: for an automated harvest with shakers, cultivars with small fruits are penalized as well as those with ripening by degrees. For manual harvesting or harvesting with tools, cultivars with moderate vegetation development are to be preferred, with a gradual ripening. It is better to use cultivars according to their final product destination: that of oil or table olives.

Given the biological characteristics of the olive tree, it is necessary to bear in mind that it is impossible for many cultivars to self-fertilize. It is also necessary to plan the presence of at least 10–15% of pollinator trees, with a distance of 30 m between the producing variety and the pollinator (Alfei and Pannelli, 2002). In any case, it may be suitable to associate two or three interfertile varieties with concordant flowering. During the planting phase, it may be advantageous to arrange the varieties in separate rows in order to prepare a differentiated harvest depending on ripening and on the genotype for mono-varietal oil production. However, it is necessary to consider that the nature of the soil may be more important than the variety in determining oil quality (Costantini et al., 2006b).

27.4.8. Choice of Farming Material

At present, debate on the advisability of using grafted or self-rooted plants seems to be over. Experience showed that these two systems of propagation are equivalent, although self-rooted plants offer the advantage of a better genetic and health control of the propagated plant material. Grafted olive trees must be at least 2 years old from grafting time and those obtained from cuttings must be 18–24 months old. The plantlet must be taller than 130 cm and must be covered with branches up to 60–70 cm from the base. The certification CAC (Conformitas Agraria Communitatis) pertains to both phytosanitary and cultivar identity requirements in addition to the phenological requirements for plantlets placed on the market; it is possible to identify this through a label. This certification promises the absence of some dangerous pathogens such as the following: *Verticillium dahlia*; *Pseudomonas syringae* pv. Savastanoi, a bacterium that causes the olive-knot disease; phytophagous insects such as the black scale bug (*Saissetia oleae*); and nematodes belonging to the *Meloidogyne* type. Additional certifications include the categories "virus controlled materials" and "virus-free material"; they ensure lack of virus by laboratory testing on the planting material.

27.4.9. Bedding

The layout of the plantation must be planned according to the spacing chosen. In areas with a milder climate (like in central and southern Italy) it is possible to plant an olive grove in autumn, so that the youngest plants may take advantage from autumn rains and be less vulnerable during the winter period. On the contrary, in northern Italy and at higher altitudes it is advisable to plant in spring in order not to incur damage caused by intense cold, but it is necessary to intervene with urgent irrigation during the summer.

The dimensions of the holes must accommodate the root ball wrapped in burlap, so that roots are 5 cm below the level of the land. Before planting, fertilizers must not touch the roots. When the plant is set, it is necessary to place a tree stake about 2.5 m high. This support must follow plant growth for at least 4 years from planting.

27.4.10. Choice of the Plant Growing Forms

The olive tree is grown in very different environments and some climatic variables (temperatures, rainfall, and radiation) have a strong impact on plant development and output, thus strongly influencing growing shapes and pruning techniques. Other variables such as size at planting, age and harvesting method must be added. The most common growing shapes are: vase, bush, cone-shaped and globe.

In new olive groves, the growing form must be chosen before plants are bought in the nursery, considering above all the harvesting method to be used. Generally, the more precise and regular the growing form, the greater the need for frequent and expensive interventions by qualified personnel (for pruning).

Vase

The plant has a trunk with variable height (30–120 cm); three or more primary limbs branch off from there. From these limbs, the secondary and tertiary branches will sprout leading to production branches. This shape is suitable to all environments and is useful for automated harvest.

Bush and Vase Bush

These forms are obtained by letting the plant grow freely, without a main trunk but with several limbs starting from the ground. They are suitable for small and medium plantations with a planting distance of at least 6×6 m. These forms are suitable for harvesting with harvesting tools, but not for automated harvesting.

Single-cone

Vegetation is distributed along one single central axis from which secondary limbs sprout, shorter and shorter from the bottom to the top and planked high (1–1.3 m from the ground) in order to allow the shaker connection. To obtain the single-cone form, it is necessary to employ olive trees that have already been trained in the nursery into one single very high axis and attached to an appropriate stake. The single-cone shape is suitable for automation and large plantations; for this type of form it is possible to use spacing of between 6×6 m and 6×3 m.

Globe

The crown, which is on a trunk whose height may be variable, is allowed to grow freely. It is typical of the hottest areas and of those with a high light intensity: vegetation takes a more or less spherical form that protects the primary and secondary limbs against possible damage caused by high temperatures.

27.5. Crop Management

27.5.1. Fertilization

Fertilization has to be planned carefully for each farming unit, taking into account a series of factors: e.g., nutritional state and age of plantation, soil properties, climate, fertilization time, agronomic practices. Then the best fertilization plan must be selected given that the synergistic and antagonistic relationship between elements must be evaluated as well as the total quantities of the single elements. Moreover, it will be necessary to assess the proportion of fertilization being used on an annual basis by the plant for branch, leaf and fruit development, which are removed through pruning and harvesting.

Nutritive elements may be added to the soil in many ways: on the overall surface or in a localized manner around each plant, on the surface or in depth. Choices depend on the agricultural techniques that are generally used (e.g., harrowing, tilling, green manure, grassing over), the irrigation method, seasonal rainfall, the practicality of fertilizing operations and, last but not least, an accurate economic evaluation.

Fertilizers must be spread on the soil near the areas mainly explored by absorbing roots and, consequently, under the crown projection, avoiding the areas closest to the trunk. It is not strictly necessary to respect this criterion for high density plantations since, in this case, the trees still make use of the fertilizer, but it is essential whenever rows are spaced far from each other and soils are not very fertile. For some fertilizers, manure efficiency varies widely. For instance, in soils with a clayey texture and a low pH that are poor in phosphorus, just distributing perphosphate to the most superficial layer of the soil helps its absorption.

Even fertigation and leaf fertilization may be used to supply nutritive elements to the plant. Through fertigation, mineral elements dissolved in a water solution are distributed through the irrigation system and according to specific phases of the growing season. This technique, in fact, permits treating the plants with small quantities of different elements during plant development, thus meeting its real needs and limiting excessive absorption and percolation of the more mobile elements.

Leaf fertilization, exploiting the capacity of the leaf to quickly absorb both macro- and microelements, allows a direct distribution of fertilizers to the plant through its top part. Plant nutritional needs are met, simultaneously reducing the quantities of elements to be added to the soil. Since the nutrients do not need to be translocated through the root apparatus, they may be provided in relatively small quantities and during times crucial to the plant (florescence, setting, stone hardening, summer aridity phases).

Growth Fertilization

These operations may be planned for the first 3 to 4 years after planting and basically aim at speeding up the formation of the root apparatus of the young olive tree, therefore supporting a good growth phase. Fertilizer dosage must be gauged to plant age, to its increasing dimension over the years and, of course, to the gradual development of roots. Development is greater in sandy, scarcely fertile soils and is limited in clayey, very fertile soils. Attempting to provide excessive quantities of nutrients to young olive trees should be avoided, because young roots seldom reach them. An appropriate and correctly balanced fertilization reduces the fruitless juvenile phase of the plant. For the olive tree, the demand for nitrogen prevails; this is true particularly for plantlets. The demand for phosphorus and potassium is very low, at least during the first 4 years after planting (Palese et al., 1996).

Production Fertilization

Production fertilization coincides with the time the first phase of plant growth development ends and the plant really starts bearing fruits. It is rather difficult to define the quantities of fertilizer to be given to the plant when the plantation is fully productive. First of all, this is due to the fact that we must describe or approximate an olive scenario that is very heterogeneous by definition; secondly, we must take into account different climatic conditions and know the agronomic response of each fertilizer; finally, the annual supply of fertilizers should never be "standardized", but follow the production trend of the plantation (e.g., rich and poor crops, annual average output) as well as the production specificity (olives for oil, table olives).

Nitrogen supply is essential for growth development and for plant production, but it varies depending on rain distribution and intensity; total fertilization may be divided into two periods of the year: two thirds just before the vegetative flush (February–March) and the remaining part before flowering (May–June). Whenever florescence is poor, the rule is to avoid applying the second part in order not to excessively favour tree growth development, leading to the formation of sterile branches at the foot (buds) or on the limbs (young shoots) of the tree. The suggestion of dividing the application of nitrogenous fertilizer into two periods is particularly suitable where emergency irrigation is provided. In other situations, it is advisable to apply the nitrogenous fertilizer during a period when a plentiful rainfall is expected (end of winter to early spring).

It has been shown that the lack of some microelements, such as boron, cancel out the effects of nitrogen-potassium fertilization, even though the latter may be rich. In this case, it is possible to apply sodic borate toward the end of winter (Delgado et al., 1994) by burying the fertilizer beneath the crown, 10–15 cm deep, or through the leaves.

Given that phosphorus and potassium are elements fixed in the soil, it is clear that for their assessment one must remember that the more nutrients given and incorporated (20–30 cm deep tilling) into a soil rich in organic substances, the more important their fertilizing effects.

Therefore, even if phosphorus-potassium fertilization usually depends on the supply given before soil tillage, it is advisable to introduce phosphorus and potassium in the form of mixed fertilizers every 2 or 3 years.

Magnesium is generally given to the plant through its leaves as magnesium sulphate. Finally, it is possible to rectify lack of iron, which is typical of the very calcareous soils of Andalusia, by using chelates (Fernandez Escobar et al., 1993).

Organic fertilizers can be applied alone or as a supplement to synthetic fertilizers; in both cases it is necessary to respect the maximum quantities allowed for nitrogen. Undoubtedly, bovine or ovine manure is the best product, but if manure is not available, as is often the case, other organic products may be applied (for instance poultry manure, plumage of birds, leather wastes) taking care to combine them with supplies of materials poor in nitrogen (straw or stems of maize in doses of up to 25–33 t/ha). The most effective option is green manure through herbaceous covers, given their property of providing a huge quantity of easily decomposable organic material to the soil. This procedure should be implemented during the autumn-winter period, in order to avoid possible competition for water between trees and herbaceous species. Of course, the choice of type depends on the objective of the operation. Vetch green manure, for example, can give up to 150 units of nitrogen per hectare. Crucifers have a remarkable capacity of absorbing the most soluble mineral phosphates; huge quantities of phosphorus are made available by ploughing in colza or mustard. Several studies state that at least 50% of nutritive elements provided by green manure are readily released and are therefore immediately available for root absorption, while the remaining percentage will be available a year later (Guet, 1997).

Whenever there is a water shortage, an alternative to manuring is partial green manure, in alternating rows, or the use of compost resulting from organic material of different natures (e.g., urban solid residues, organic wastes of industrial processing).

27.5.2. Pruning

Pruning, along with other agronomic operations, helps keep the right physiological balance between the vegetative part and the reproductive part of the plant, striving to achieve a good and regular production over the years. Pruning has different goals and intensities according to the phase of the plant's vital cycle.

Regulatory pruning is done following the fruitless period of the olive tree by adapting it to the chosen growing form; it must commence in the very early years if specific plant characteristics (cone-shaped) are desired, while it must be performed to a lesser degree with free growing forms (free vase, bush vase). Production pruning aims at maintaining the productive capacity over time and favouring air and light penetration inside the crown. This decreases the impact of parasitic diseases and improves fruit size and, therefore, qualitative features. Finally, renewal pruning drastically cuts back foliage in order to recover the efficiency that is depleted with plant ageing.

Decisions concerning production pruning regard the choice of period and intensity, decisions that must take into account several factors. Cutting intensity must be related to tree age (too intensive pruning during the first years jeopardizes output or brings on forms that are not suitable to agronomic needs) and to tree production capacities: in the case of poor energy and of a high ratio between old shoots (older than 2 years) and new shoots (1 or 2 years old), it is necessary to prune considerably, while in anticipation of a poor crop, it is advisable to decrease cutting intensity.

At present, the pruning period system has been greatly reorganized, since this procedure represents 15–20% of the overall cost of farming. In fact, it is necessary to simplify pruning operations in order to limit costs, as well as problems concerning the availability of a skilled workforce. One of the methods to cut costs consists in prolonging the pruning period: this choice may be appropriate, without problems, for well-watered olive groves or whenever pedological and climatic conditions (fertile soils, mild climate) allow some production or growth renewal even without the encouragement of pruning (Gucci, 2005). At the end of 2 or 3 years, pruning will be more severe than when using the annual method, in order to ensure a sufficient growth of vegetation and the renewal of fruit-bearing branches for the whole period. Still, depending on growth and production behaviour and on biological characteristics, minimal pruning methods have also been suggested; instead of obtaining a definite form, they tend to manage the growing form as freely as possible, in order to cut pruning costs without lowering output. In this case, interventions with a few major cuts are provided in order to recover the right height of the plan, to rectify the structure of the limbs and to reduce vegetation intensity, above all inside the crown.

27.5.3. Irrigation

Irrigation is very important to increase production in olive cultivation, to decrease crop alternation, to ensure more constant yields and a greater and more stable product quality. In new plantations with limited spacing and pruning, water needs are higher than in the traditional olive groves and the water supply provided by rainfall is often insufficient. Irrigation is particularly advised for table olive varieties, for which fruit size represents a key marketing feature. The effects of irrigation have been compared in order to assess their impact on fruit production and oil yield (Lavee et al., 1990). Generally, irrigation increases the number of inflorescences, decreases the ovary abortion percentage, reduces fruit falling, increases and fruit size and the pulp-stone ratio (Costagli et al., 2003; Goldhamer et al., 1993; Inglese et al., 1996). Apparently, acidity and the fatty acid content are not affected by irrigation, while there are discordant opinions on the content of polyphenols, which seems to decrease when water is provided (Alegre, 2001).

In order to calculate the water volume with which to irrigate, it is necessary to quantify the water needs of the olive grove. Water uptake varies both during the day and during the season and is influenced by factors such as temperature, relative humidity, solar radiation, and water availability in the soil. On a summer day, the evapotranspiration (ET) of an olive grove may be equal to 5 mm/d, while in winter it decreases to 1 mm/d (Fereres, 2005). A reference point for calculating the quantity of water to be supplied is equal to the highest demands of ET for the olive grove, but the present trend is to provide lower water quantities; in other words, the farmer uses a strategy of reduced deficit irrigation (RDI). Applicable strategies are the following:

- supporting irrigation, with water supplies slightly lower than the olive grove ET;
- controlled irrigation, with a constant application rate over the whole season;
- controlled irrigation using a maximum deficit in summer, but almost saturation in spring and autumn.

In the case of reduced deficit irrigation, the olive grove uses the soil's available water supply, at least partially, in order to meet its needs. It is therefore better to avoid excessive or prolonged deficit to avoid compromising fruit development and oil storage, as well as jeopardizing the water reserves in the soil.

In areas where rainfall is lower, olive growers have obtained successful results by using constant and very low water quantities, because production significantly increases at once with the first increase in irrigation water (20 to 30 kg/ha per mm of water) (Fereres, 2005). In areas characterized by prolonged drought (soils with xeric and dry-xeric pedoclimate), irrigation may even double oil production, while in more humid climates, where summer aridity lasts for about 2 months and is interrupted by short rains (soils with ustic and udic pedoclimate), the production increase is more moderate.

In choosing and planning the water system, it is necessary to take into consideration the soil volume that is available to the tree as well as soil physical characteristics. With limited planting distances, irrigation is necessary to give the tree a suitable water supply in the limited soil volume it explores.

In sandy surface soils or in those rich in rock fragments, water storing capacity is modest and irrigation is able to greatly increase production. Since the olive tree is able to produce even in dry areas, it is better to assess the productive benefits resulting from the installation of an irrigation system and to compare them with amortization and management costs of the irrigation system. Generally, the estimated production increase is at least equal to 20% (Gucci, 2003).

27.5.4. Protection

The ecosystem of the olive grove hosts about 250 species of parasites among phytophagous and entomophagous species, fungi, bacteria and viruses, but only some of them are really detrimental to the olive tree and therefore are the target of specific control. The wide geographical spread of the olive tree and the numerous morphological and microclimatic differences existing, together with differences in cultivar, determine a huge variability in the extent and in the impact of the damage caused by the parasites. Therefore, protection must be diversified both as to the targets to be fought and as to the number of interventions to be carried out. The farmer must make a choice between the available fighting techniques, depending on the type of olive cultivation area (protected or not), the type of production (organic oil), and the economic threshold of the damage. For table olive production it is also necessary to apply the principle of tolerable damage threshold and intervention; the latter must be carried out with the first symptoms of infestation or of infection (in particular for phytophagous species and pathogens that directly affect the drupe).

The organic or integrated pest control systems used in olive cultivation are agronomic, biotechnical and microbiological. Agronomic pest control strives to reset the agro-ecological balance, where parasites and pathogens may coexist to the extent that they imply very little economic damage; resistant cultivars are planted, fruit is harvested earlier, environmental diversity is increased (e.g., herbaceous cover of the olive grove, hedges, windbreaks), pruning is carried out in a targetted manner. The biological and biotechnical control methods applied in fighting phytophagous parasites are based on the use of natural antagonists (parasites or predators) and on mass capture using different types of traps.

27.5.5. Harvest

The harvest is a moment of paramount importance for olive cultivation: its correct performance influences product quantity and quality as well as production costs. A simple classification of harvesting methods divides them into the manual and the mechanical.

Harvesting in the Tree or Hand-picking

Hand-picking is the oldest system, but it is still the most widespread wherever an alternative option for harvesting is not available and, above all, whenever the product must have a qualitative value. Simple tools (manual harvesting tools) are available on the market: if they are suitably employed, they improve operator efficiency during the fruit detachment phase. Generally, nets are placed under the tree to catch the drupes and to make it easier to transfer them into boxes.

Gathering after Natural Fruit Abscission

Olives that spontaneously fall are gathered manually or by machine. Abscission may take place because fruits are ripe or because of accidental causes such as wind or intensive parasite infestation. This procedure is sometimes facilitated by previously placing nets under the plants. The olives are gathered in stages and over longer or shorter time intervals, depending on the falling sequence. Excessive drupe ripening, however, jeopardizes the chemical features of the resulting oils (increase in acidity, in peroxides, in K_{232} and K_{270}, as well as a decrease in chlorophyll and polyphenol contents) and their organoleptic features.

Mechanical Systems

Olives can be picked from the plant by machine (shakers, harvesting poles, vibrating mechanisms). Mechanical picking mainly detaches the olives with a high unit weight, those with the greatest colour change and those with the highest content of oil. The fruits that stay on the plant sometimes represent a production loss, estimated at 10–15%. The quality of the oil obtained with mechanical picking does not suffer significant variations when olives are transferred for extraction that same day. This is important, because these olives generally show damage and therefore may undergo accelerated oxidation phenomena. For the cultivar whose fruits are meant for the table, automated picking by shaking is not a feasible option.

Harvesting Tools

Pneumatic combs cause fruit detachment both by direct action and by the vibrations transmitted to limbs and branches. These devices allow the harvesting of almost all the hanging products and increase work productivity by 2–4 times that of manual picking.

When choosing the harvesting method, it is necessary to take into account that high quality standards involve a supply of whole and undamaged olives to the oil mill. Surface lacerations not only cause the loss of fatty substances, but above all expose fruits to the microbic action that is responsible for an increase in acidity and for the deterioration of organoleptic features. Compatibly with the farm reality and the production area, the most suitable harvesting system would be the one that minimizes damages to the drupes.

27.5.6. Soil Tillage

Tillage represents the main soil management procedure and its goal is at least temporarily to increase water infiltration and storage in depth, particularly for less structured soils, as well as to eliminate weeds and turn fertilizers into the soil. A common tilling calendar usually plans cultivation to about 20–30 cm deep for favouring rain infiltration to be executed before autumn. In spring and summer cultivation it is more superficial (15–20 cm) for eliminating weeds. It is generally advisable not to execute more than three cultivations a year and not to go deeper than 15–20 cm (Mazzoncini et al., 2002). Intense and frequent cultivations, in fact, may cause an excessive destruction of soil aggregates and thus accentuate packing of surface horizons and cause interruptions in the continuity of the pore system. These changes hamper humification processes and, particularly in clayey and silty soils, degrade soil structure and increase the risk of surface crust formation. For soils with a lot of fine sand and silt, tilling that is always repeated to the same depth may determine the formation of the so-called "plough pan" and consequent reduction in soil permeability (Fig. 27.16). These asphyctic conditions, in turn, hamper gas exchanges and result in inhibiting conditions for micro-organic activity.

Fig. 27.16. Evident plough pan about 40 cm deep in an olive grove at Montepulciano (Siena).

Fig. 27.17. Profile with evident surface crust caused by too-frequent cultivation in an olive grove at Strove (Siena).

Mechanical cultivation with tilling machines, hoeing machines and harrowers breaks up soils that are poor in organic matter and clay and may cause the collapse of the structure of the tilled horizon, as well as the formation of a superficial crust (Fig. 27.17).

27.5.7. Grass Cover

Soil grassing implies the presence of a permanent or temporary grass cover in the olive grove. This procedure gives greater stability to the agro-ecological system: it increases organic matter, limits loss of nitrates by leaching and regulates soil nitrogen availability. In terms of soil conservation, a grass cover reduces erosion, develops soil structure and improves drainage; it also improves physical and hydrological conditions, with advantages for trafficability. In calcareous soils, for example, grassing represents a good agronomic procedure, because it is able to limit problems of ferric chlorosis. Grasses produce root exudates that are able to render iron soluble, therefore making it more readily available to roots. In areas with sufficient rainfall, a stable meadow that is managed with a hay-cutter without grass removal may increase the supply of organic matter by 1% in a soil stratum of about 10 cm over a decade (Zucconi, 1996).

The limit of this technique is the competition for water and nutrition that occurs between the olive tree and the weed mixture present in the meadow. However, there are artificial meadows with autochory characteristics and an autumn-summer cycle that make this procedure feasible even when rainfall is low, because they dry up in summer. In any case, legumes are better than other groups, given their capacity to fix nitrogen and draw water from the deepest soil horizons.

In Mediterranean climates, this competition has been verified in young plantations and particularly during the first years of grafting. During that time, the root system of the young trees develops mainly at the surface and is able to explore the strata of the soil already occupied by the roots of herbaceous plants. In any case, it is advisable to apply the grassing technique to soil management from the fifth year after grafting; this choice prevents water and nutritional competition between the plant and the herbaceous cover, which would keep olive seedlings from growing regularly.

For soils with a good capacity to store water available to plants (more than 150 mm), the rate of rainfall may help choose the grassing method: with an annual average rainfall of 700–800 mm, there are no obstacles to permanent grassing (natural or artificial); with a rainfall lower than these values, down to a minimum of about 500 mm per year, or with a rainfall of about 150 mm during the period between May and August, temporary grassing is feasible, in alternate rows or beds, integrated with control techniques for weeds during the period of greatest water competition.

27.6. Soil Suitability for Cultivation of the Olive Tree

There are only a few experiments that have linked soil and land parameters to quantitative and qualitative results in olive production. Among these, the following are indicated: Lulli and collaborators in Calabria and in Sardinia (Lulli, 2004), Tittarelli and others in Latium (Tittarelli, 2002), and Cimato and collaborators in the province of Siena (Costantini et al., 2006b). The evaluations reported here make reference in particular to this last experience, but are basically analogous to the results of the other two research experiments mentioned.

Soil functional factors for olive cultivation belong to two categories: site and profile. In the former case, elevation and aspect must be taken into consideration, following the values reported in Table 27.2.

Table 27.2. Matching table to olive cultivation (soil site parameters).

Parameters	Suitability class			
Site character	Very suitable (S1)	Suitable (S2)	Moderately suitable (S3)	Not suitable (N)
Elevation (m a.s.l.)	200–400	400–500	500–800	> 800
Aspect	south; south-west	east-west	north-west; north-east	north

Experimental results show that the olive tree root system requires a good balance between available water and oxygen capacity; therefore, soils with a good or moderate drainage are more suitable. Soils with poor drainage are very unsuitable, if the drainage difficulties are not reduced by good external drainage (runoff). Several experiences demonstrate that the water retention of the soil (AWC) is the parameter that most influences olive tree output (Chaves et al., 1976). According to literature and to experimental results, AWC intervals that are considered satisfactory (coinciding with category S2) are fixed on values of at least 70 mm, while for category S1 the limit is 110 mm.

The literature shows that olive tree tolerance to salinity is comparable to that of the vine (Sbaraglia and Lucci, 1994); given the wide root system of the olive tree, the analysis of this property must be 1 m in thickness.

Following these indications, soil functional factors for cultivation of the olive tree may be divided into two categories: the former determines the conditions unsuitable to growth (presence of a water table within 1 m or soil types categorized as "Vertisols"); the latter measures different conditions of suitability (AWC, internal and external drainage and salinity, expressed as electrical conductivity) (Table 27.3).

Table 27.3. Matching table of soil suitability for olive tree cultivation.

Parameter	Suitability			
Quality and characteristics	Very suitable (S1)	Suitable (S2)	Not very suitable (S3)	Not suitable (N)
Water table depth (cm)	> 100	> 100	> 100	< 100
Classification	No Vertisols	No Vertisols	No Vertisols	Yes Vertisols
Electrical conductivity dS/m (1 m)	< 1	1–2	3–4	> 4
AWC (mm/100 cm)	> 110	110–70	69–30	< 30
Internal drainage	Good, or moderate, or somewhat poorly drained if runoff > average, or somewhat poorly drained if runoff = average and skeleton ≥ 35%	Sometimes excessive, or somewhat poorly drained if runoff ≥ average or skeleton ≥ 35%, or somewhat poorly drained if runoff = average and skeleton ≥ 35%	Excessive, or imperfect, or poorly drained if runoff > average and skeleton ≥ 35%	Poorly drained and very poorly drained

27.7. Land Evaluation

Generally a land evaluation, drawn up by applying tables like the above, does not provide information on the reliability of the assessment itself. The reader's judgment is based above all on the quality of the methodological approach as well as the quantity of data processed. In fact, it is important to make clear the uncertainty of the information provided and make it easy to read, mainly to allow users to be aware of the limits of a correct use of the cartography and to identify areas where further studies are needed.

The uncertainty about soil geography can be expressed as a percentage of inclusions inside polygons, i.e., soils that have different typology and/or suitability (pedological "purity" and suitability "purity"). Otherwise, it may be made implicit in the name of the soil map unit, i.e., "complex", "association", "undifferentiated group", "unassociated group" (Van Wambeke and Forbes, 1986). Besides "purity", two other indices can be used to define the quality of the information: "confidence" of the suitability evaluation and "reliability" of the "soil survey paradigm" (Hudson, 1992), which is the relationship linking the kind of landscape to the type of soil. The confidence index expresses the reliability of the estimation made about soil behaviour for a certain purpose in a specific place or for a specific polygon. The reliability index tells us if the generalization of the relationship found between a type of soil and a specific landscape is consistent for all the landscapes with similar characteristics (Costantini and Sulli, 2000).

In a study carried out in the Province of Siena (central Italy), algorithms commonly used in ecology, Shannon's index and the evenness indices, as well as a non-parametric accuracy analysis (Kappa statistic), were performed to assess land evaluation reliability.

27.7.1. Indices of Diversity

The Administration of the Province of Siena financed an olive tree zoning project. A detailed explanation of the methodology used in the soil survey and land evaluation is reported in Costantini and Sulli (2000). It provided for the creation of a soil typological unit and subunit database, aimed at allocating each soil observation on the basis of its classification, landscape characteristics and management problems. Soil typological subunits were the basic units for the suitability evaluation. Soil geography was described at the 1:100,000 scale following the Land System methodology (Tuscan Agriculture and Forestry Dept., 1992). The main constituents of the methodology were land systems, land units and land elements. Land units were geographic objects (polygons) recognized and delineated by photo interpretation (aerial photos at the 1:33,000 scale), having a hydrological pattern, a range of lithology (1:100,000 geological map overlaid) and geomorphology, and a characteristic arrangement of land elements. Land elements were the smallest components that could be recognized by photo interpretation of a characteristic lithology, landform and land use pattern, and could be described and codified, but not delineated on a 1:100,000 map. In addition to their identification, their approximate proportion inside each delineated polygon was also recorded. Land systems grouped land units together, on a physiographic and geological basis.

The entire Province of Siena was described with the Land System approach, and each soil profile stored was referred to a land element typology. During the field survey, the soil percentage distribution in each land element was recorded in the field file, and then in the database. Similar land elements formed land element typologies and each land element typology had a set of soil typological subunits.

To sum up, every map polygon belonged to one land unit and was made up of one or more land elements, and every land element was made up of one or more soil typological subunits. In each soil typological subunit, a set of observations (profiles, auger holes, mini pits) similar for soil classification, landscape, management requirements and crop suitability were allocated. As we knew the area of each polygon and the proportion of the land elements inside it, as well as the proportion of soil typological subunits forming each land element, we were able to estimate the area covered by each land element and by each soil typology.

As regards the land evaluation methodology, each observation was singularly evaluated and attributed to a land suitability class. Four suitability classes (0 = not suitable, 1 = moderately suitable, 2 = suitable, 3 = very suitable) were assessed on the basis of the results obtained in experimental plots, where the reference olive tree variety "*Frantoio*" was cultivated. The suitability class of a soil typological subunit was the modal value of all its observations. In turn, the suitability class of a land element was the modal value of the suitability classes found in it. All information was stored in a soil information system.

To manage soil uncertainty, data were first of all submitted to descriptive statistics (observations by hectare, observations by soil typological unit and subunit, observations by polygon, observations by land element). Then the Shannon index was applied to assess pedodiversity of land elements, polygons, soil typological unit and subunits. The Shannon index is well known and widely used in ecology, but it has also been proposed in pedology (Ibañez et al., 1995):

$$H' = \sum_{i=1}^{n} p_i * \ln(p_i) \tag{1}$$

where H' is the (negative) entropy and p_i is the proportion of soil individual i in the area studied, p_i can be estimated by ni/N, with ni = the number of individuals i and N the total number of samples. In our case, p_i was the percentage of surface area occupied by the i object, that is, the class of soil suitability (pi = class/polygon (km²/km²).

The Evenness index is derived from the Shannon index and can range from 0 (maximum equality) to 1 (maximum diversity):

$$E = H'/H \max = H'/\ln S \tag{2}$$

where S is the numerical richness.

Figure 27.18 shows the results of the first phase assessment of the Siena soils for olive cultivation. The legend provides some indications on the frequency in each polygon of soils with different aptitude (categories S1, S2, S3, N).

The uniformity index was employed in order to assess the differences in assessments for each polygon. As Fig. 27.19 shows, the index values distribution shows that there is one small

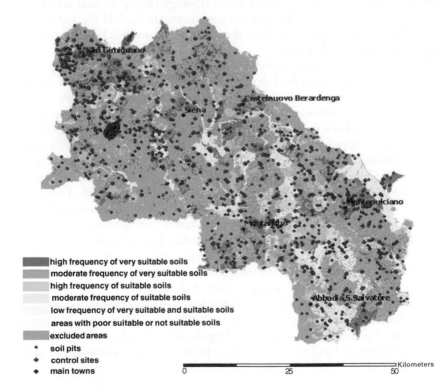

Fig. 27.18. Land suitability map for intensive olive tree cultivation in the Siena province. Soil pits and control sites.

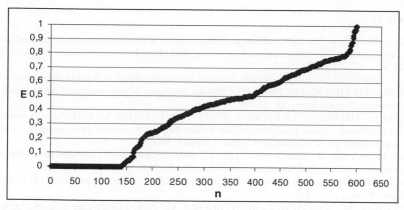

Fig. 27.19. Purity of the land suitability. $E_{mean} = 0.365$, $n_{total} = 601$.

group of homogeneous polygon only, i.e., with soils having all the same aptitudes; on the contrary, most of the polygons of the chart include soils with different aptitudes. The average index value is, however, rather low (E = 0.365). This means that only relatively few polygons have soils with three or four different categories; whenever there are soils with different aptitudes, there is, in any case, a soil that prevails from a territorial viewpoint.

27.7.2. Indices of Accuracy

The reliability of the Land System model to spatialize land evaluations was submitted to a validation test using an independent data set consisting of 170 observations, placed on croplands (about 2000 km²) on an approximate grid during the years 1998–2002 (control sites in Fig. 27.18). Each observation collected the information needed to assess workability, by means of field instruments and hand augering.

The Kappa statistic was chosen to measure the agreement between predicted and observed classes, since it allows correction for agreement that occurs by chance (Jenness and Wynne, 2005). According to this method, Overall Accuracy is simply the number of correctly classified sample points divided by the total number of sample points. Overall Specificity is the general ability of the model to avoid misclassifying sample points as X if they are not X. Kappa statistic is the chance-corrected measure of model accuracy, based on the actual agreement between predicted and observed values and the chance agreement between the row and column totals for each classification. The P-value reflects the probability that this model performs better than random chance at predicting the distribution on the landscape of classes of crop workability.

The statistical test of accuracy of the Land System model to spatialize the indicator gave an Overall Accuracy of 0.59 and a Kappa value of 0.36, with P-value < 0.00001; Overall Specificity was 0.86. These results indicated a rather low chance-corrected measure of the Land System spatial model accuracy, but good ability to avoid misclassifying sample points.

References

ALEGRE, S. 2001. Efecto de diferentes estrategias de riego deficitario controlado durante la epoca estival sobre la produccion de olivo (Olea europaea L.) cv Arbequina. Tesis doctoral, Universitat de Lleida, Leida, 224 p.

ALFEI, B., PANNELLI, G. 2002. *Guida alla razionale coltivazione dell'olivo*. ASSAM, Regione Marche.

AMIOT, M.J., FLERIET, A., MACHEIX, J.J. 1989. Accumulation of oleuropein derivatives during olive maturation. Phytochemistry 28, 67–69.

BANDINO, G., DELRIO, G., DETTORI, S., PASCHINO, F., PROTA, U. 1994. *Impianto dell'oliveto, tecniche colturali e aspetti qualitativi dell'olio d'oliva*. Regione Autonoma della Sardegna, Consorzio Interprovinciale per la frutticoltura, Cagliari, Oristano, Nuoro.

BAREA, J.M., ACZON, R., ACZON AGIULAR, C. 2000. *Micorrizazione e tecniche microbiologiche per migliorare l'attività radicale in vivaio*. Frutticultura 2, 71–75.

BARGIONI, G. 1992. *Manuale per l'olivicoltore*. Centro I.R.I.P.A., Federazione Provinciale Coltivatori Diretti, Verona, pp. 1–84.

BERTUCCIOLI, M., ANICHINI, F., MONTEDORO, G. 1978. Maturazione delle olive e qualità dell'olio: costituenti volatili dello "spazio di testa". Tecnologie Alimentari 1, 97–104.

BONGI, G., PALLIOTTI, A. 1994. *Olive*. In: Schaffer, B., Andersen, P.C. (Eds.), Handbook of Environmental Physiology of Fruit Crops. CRC Press, Boca Raton, Florida, pp. 165–187.

CAMERA, L., ANGEROSA, F. 1976. *Gli alcoli superiori (alifatici e tripenici) degli oli d'oliva: loro evoluzione nel corso della maturazione delle olive*. Ann., Istituto, Sper., Elaiotecnica, Pescara.

CHAVES, M., TRONCOSO, A., ROMERO, R., PRIETO, J., LINAN, J. 1976. *Estudio de los caracteres optimos de los suolos de olivar en diferentes zonas de andalucia occidental*. Nutricion del olivo, Sevilla.

CIMATO, A. 1988. *Variazioni di parametri durante la maturazione delle olive: Influenza delle tecniche colturali*. P.S. "Nuovi orientamenti dei consumi e delle produzioni alimentari". Consiglio Nazionale delle Ricerche Roma (Ed.), June, pp. 6–65.

CIMATO, A., BALDINI, A., MORETTI, R. 1997. *L'Olio di Oliva, Cultivar, Ambiente e Tecniche Agronomiche*. Regione Toscana, CNR, ARSIA, pp. 1–93.

CIMATO, A., BALDINI, A., MORETTI, R. 2001. *Cultivar, Ambiente e Tecniche agronomiche*, 2nd ed. Regione Toscana, CNR, ARSIA, January, pp. 1–168.

CIMATO, A., FIORINO, P. 1985. L'alternanza di produzione dell'olivo: 2. Influenza dei frutti sulla differenziazione a fiore e sulla nutrizione minerale. Rivista Ortoflorofrutticoltura Italiana 69, 413–424.

CIMATO, A., FIORINO, P., POLI, P. 1986. Induzione e differenziazione a fiore nell'olivo. 1. Defogliazione ed ombreggiamento strumenti per studiare il processo. L'Informatore Agrario 31, 47–50.

CIMATO, A., LAPUCCI, C., SANI, G., CORTEZ EBNER, S. 2002. L'utilizzazione del boro nella fertilizzazione fogliare dell'oliveto. Frutticoltura 10, 57–61.

CIMATO, A., MARRANCI, M., TATTINI, M. 1989. *The use of foliar fertilization to modify sinks competition and to increase yield in olive (Olea europaea L. cv Frantoio)*. International Symposium on Olive Growing, Cordoba, 26–29 September. Acta Horticulturae, 286, 175–178.

CIMATO, A., SANI, G., MARZI, L., MARRANCI, M. 1994. *Efficienza e qualità della produzione in olivo: riflessi della concimazione fogliare con urea*. Olivae 54, 48–55.

CORTI, G., CONIGLIO, R. 2000. *Studio pedologico dei suoli coltivati a vite e olivo nel Comune di Montespertoli*. It. Comm., Firenze.

COSTAGLI, G., GUCCI, R., RAPOPORT, M.F. 2003. Growth and developement of fruits of olive "Frantoio" under irrigated and rainfed conditions. Journal of Horticultural Science and Biotechnology 78(1), 119–124.

COSTANTINI, E.A.C. 1985. La lavorazione del suolo sui terreni pietrosi e in pendenza. Genio Rurale XLVlll, 6, 23–33.

COSTANTINI, E.A.C., PELLEGRINI, S., VIGNOZZI, N., BARBETTI, R. 2006a. Micromorphological Characterization and Monitoring of Internal Drainage in Soils of Vineyards and Olive Groves in Central Italy. Geoderma 131, 3–4, 388–403.

COSTANTINI, E.A.C., BARBETTI, R., BUCELLI, P., CIMATO, A., FRANCHINI, E., L'ABATE, G., PELLEGRINI, S., STORCHI, P., VIGNOZZI, N. 2006b. *Zonazione viticola ed olivicola della provincia di Siena*. Grafiche Boccacci Editore, Colle val d'Elsa (SI), 224 p.

COSTANTINI, E.A.C., BARBETTI, R., RIGHINI, G. 2002. *Managing the Uncertainty in Soil Mapping and Land Evaluation in Areas of High Pedodiversity*. Methods and Strategies Applied in the Province of

Siena *(Central Italy)*. Acts of the Int. Symp. on Soils with Medium Type of Climate, CIHEAM IAMB, Bari, pp. 45–56.

COSTANTINI, E.A.C., SULLI, L. 2000. Land evaluation in areas with high environmental sensitivity and qualitative value of the crops: the viticultural and olive-growing zoning of the Siena province. Bollettino della Società Italiana della Scienza del Suolo 49(1–2), 219–234.

COUTTS, M.P., PHILIPSON, J.J. 1980. *Mineral Nutrition of Fruit Trees*. Butterworths, London.

CRESCIMANNO, F.G, SOTTILE, I., AVERNA, V., BAZAN, E. 1975. Ricerche sulla nutrizione minerale dell'olivo. Variazioni del contenuto in N,P,K,Ca e Mg in piante autoradicate ed innestate. Ortoflorofrutticoltura 59, 48–68.

CRESTI, M., CIAMPOLINI, F., TATTINI, M., CIMATO, A. 1994. Effect of salinity on productivity and oil quality of olive (Olea europaea L.) plants. Advances in Horticultural Science 8, 211–214.

DELGADO, A., BENLLOCH, M., FERNÀNDEZ-ESCOBAR, R. 1994. Mobilization of boron in olive trees during flowering and fruit development. Hort Science 29 (6), 616–618.

DEMETRIADES, S.D., HOLEVAS, C.D. 1968. *Toxicité de bore chez l'olivier*. 2nd Col., Eur., Med., Contr., Fert., Sevilla.

DETTORI, S., FILIGHEDDU, M.R., IBBA, M., VIRDIS, F. 1999. Irrigazione con acque saline dell'olivo in vasca lisimetrica. Frutticoltura 61, 73–77.

DI MATTEO, M., SPAGNA MUSSO, S., GRASSO, G., BUFALO, G. 1992. Caratterizzazione agronomica e merceologica in relazione al grado di maturazione delle produzioni di alcune cultivar olearie della provincia di Avellino. Rivista della Società Italiana di Scienza dell'Alimentazione 21, 35–36.

DOORENBOS, J., PRUITT, W.O. 1977. *Las necessidades de agua de los cultivos*. Estudio, FAO, Riego y drenaje, no. 24, Roma.

FERNANDEZ ESCOBAR, R., BARRANCO, D., BENLLOCH, M. 1993. Overcoming iron chlorosis on olive and peach trees using a low-pressure trunk-injection method. HortScience 28, 192–194.

FERERES, E. 2005. Strategie d'irrigazione in olivicoltura: uno studio spagnolo. Phytomagazine 41.

FERRINI, F. 1996. *Danni da venti e da freddo in olivo*. L'olivicoltura mediterranea verso il 2000. Atti VII°, International Course on olive growing, Scandicci, 6–11 May, pp. 239–245.

FRIAS, L., GARCIA, A., FEFFEIRA, J. 1975. *Composition en acidos grasos del aceite di oliva en frutos con distinto grado madure*. Il Sem., Oleic., Int., Cordoba, 6–17 October.

GOLDHAMER, D.A., DUNAI, J., FERGUSON, L. 1993. Water use requiremens of Manzanillo olives and responses to sustained deficit irrigation. Acta Horticulturae 474, 369–372.

GONZALES, F., CATALINA, L., SARMENTO, R. 1976. *Aspectos bioquimicos de la floración del olivo, var. "Manzanillo", en relación con factores nutricionales*. IV Coll., Int., CNRS, Ghent, pp. 409–426.

GUCCI, R. 1998. *Effetti dei fattori ambientali sulla produttività dell'olivo*. Atti VII, International Course on Olive Growing, Scandicci, 6–11 May, pp. 207–211.

GUCCI, R. 2003. *L'irrigazione in olivicoltura*. ARSIA, Regione Toscana.

GUCCI, R. 2005. Linee guida per la semplificazione della potatura in olivicoltura. Phytomagazine 41.

GUCCI, R., TATTINI, M. 1997. Salinity tolerance in olive. Horticultural Reviews 21, 177–214.

GUERRIERO, R., SCARAMUZZI, F., CRESCIMANNO, F.G., SOTTILE, I. 1972. *Ricerche comparative fra olivi innestati ed autoradicati*, Tecnica Agricola, 4.

GUERRIERO, R., VITAGLIANO, C. 1973. Influenza dell'ombreggiamento sulla fruttificazione dell'olivo. L'Agricoltura italiana 73, 85–115.

HARTMANN, H.T., URIU, K., LILLELAND, O. 1966. *Olive nutrition. Fruit nutrition*. Childers, Brunswich N., pp. 252–261.

HUDSON, B.D. 1992. *The soil survey as paradigm-based science*. Soil Science Society of America Journal 56, 836–841.

IBAÑEZ, J.J., DE-ALBA, S., BERMUDEZ, F.F., GARCIA-ALVAREZ, A. 1995. Pedodiversity: concepts and measures. Catena 24, 215–232.

INGLESE, P., BARONE, E., GULLO, G. 1996. The effect of complementary irrigation on fruit growth, ripening pattern and oli characteristics of olive (Olea europea L.) cv. Carolea. Journal of Horticultural Science 71(2), 257–263.

IUSS-ISRIC-FAO-ISSDS. 1999. *World Reference Base for Soil Resources*. Versione italiana edited by Costantini, E.A.C., Dazzi, C. ISSDS, Firenze, 98 p.

JENNESS, J., WYNNE, J.J. 2005. *Cohen's Kappa and classification table metrics 2.0: an ArcView 3x extension for accuracy assessment of spatially explicit models*. U.S. Geological Survey Open-File Report OF 2005-1363. U.S. Geological Survey, Southwest Biological Science Center, Flagstaff, Arizona.

LA CERTOSA, G., MONTEMURRO, N., PALAZZO, D., PIPINO, V. 1998. Stato nutrizionale dell'olivo e fertilità dei terreni. L'Informatore Agrario 15, 109–144.

LAVEE, S. 1977. The grow potential of the olive fruit mesocarp in vitro (Olea europaea). Acta Horticulturae 78, 115–122.

LAVEE, S., NASHEF, M., WONDER, M., HARSHEMESH, H. 1990. The effect of complementary irrigation added to old olive trees (*Olea europaea* L.) cv Souri on fruit characteristics, yeld and oil production. Advances in Horticultural Science 4, 135–138.

LAVEE, S., WODNER, M. 1991. Factors affecting the nature of oil accumulation in fruit of olive (*Olea europaea* L.) cultivars. Journal of Horticultural Science 66(5), 583–591.

LULLI, L. 2004. Il suolo e la qualità dei prodotti. Bollettino della Società Italiana della Scienza del suolo 53, 21–24.

MAZZONCINI, M., BONARI, E. 2002. Adattare le lavorazioni al tipo di agricoltura. L'Informatore Agrario 24, 35–39.

MODI, G., SIRIANI, G., NIZZI GRIFI, F. 1992. Diminuzione di clorofilla e di betacarotene in olive della cultivar Frantoio durante la maturazione. Bollettino Chimici Igienisti 43, 141–146.

MONTEDORO, G. 1981. *Raccolta Meccanica delle olive.* C.N.R., Progetto Finalizzato, Roma.

MORETTINI, A. 1972. *Olivicoltura.* R.E.D.A, Rome.

MOSQUERA, M.I., FERNANDEZ-GARRIDO, J. 1989. Chlorophyll and carotenoid presence in olive fruit (Olea europaea). Journal of Agricultural and Food Chemistry, 37(1), 1–7.

NAVARRO, C., PARRA, M.A. 2001. Plantación. In: Barranco, D., Fernandez-Escobar, R., Rallo, L. (Eds.), El cultivo del olivo. Junta de Andalusia, pp. 173–213.

NUZZO, V., DICHIO, B., XYLOYANNIS, C. 1995. *Effetto cumulato negli anni della carenza idrica sulla crescita della parte aerea e radicale della cv. Coratina.* Atti del convegno Olivicoltura Mediterranea, Rende (CS), 26–28 January, pp. 512–518.

OLIAS, J.M., GUTIERREZ, F., DOBARGANES, M.C., GUTIERREZ, R. 1980. Componentes volatiles en el aroma del aceite de oliva. IV. Su evolucion e influencia en el aroma durante el proceso de maduracion de los frutos en las variedades Picual y Hojiblanca. Grasas y Aceite 31, 391–401.

ORTEGA NIETO, J.M. 1969. *La poda dell'olivo.* Min. Agric., Madrid.

PALESE, A.M., CELANO, G., XILOYANNIS, C. 1996. L'importanza del terreno nell'oliveto. Olivo & Olio XX, 22–25.

PATUMI, M., BRAMBILLA, I., LUCHETTI, A., 1990. *Indagini sulla biosintesi lipidica in drupe di olivo. Possibili correlazioni con aspetti quantitativi e qualitativi.* Atti Conv., Problematiche qualitative dell'olio di oliva, Sassari, 6 November, pp. 307–316.

PATUMI, M., FONTANAZZA, G., BALDONI, L., BRAMBILLA, I. 1998. *Determination of some precursors of lipid biosynthesis in olive fruits during ripening.* International Symposium on Olive Growing, Cordoba, 26–29 September, Acta Horiculturae 286, 199–202.

ROMANI, A., 1996. *Aspetti del metabolismo dei polifenoli in frutti di olivo.* International Course On Olive Growing, Scandicci, 6–11 May.

SANCHEZ, RAYA, A.J., DONAIRE, J.P., MARTIN, E., RECALDE, L. 1975. An. Edafologia Agrob. 34, 98–104.

SANCHEZ, J., HARWOOD, J.L. 1992. Fatty acid synthesis in soluble fracions from olive (Olea europaea L.) fruits. Journal of Plant Physiology 140(4), 402–408.

SBARAGLIA, M., LUCCI, E. 1994. *Guida all'interpretazione delle analisi del terreno ed alla fertilizzazione.* Studio Pedon, Rome, 123 p.

SERVILLI, M., BALDIOLI, M., PERRETTI, G., COSSIGNANI, L., SANTINELLI, F., MONTEDORO, G.F. 1995. *Evoluzione di alcuni componenti minori dell'oliva dal frutto all'olio.* Atti Convegno Internazionale "L'olivicoltura Mediterranea", Rende, (Cs), 26–28 January, pp. 695–699.

SOLINAS, M., DI GIOACCHINO, L., CUCURACHI, A. 1975. I polifenoli delle olive e dell'olio. 1. Variazioni dei polifenoli con il procedere della maturazione. Annali Istituto Sperimentale Elaiotecnica, Pescara 5, 105–129.

TISCORNIA, E., BESTINI, G.C., PAGANO, M.A. 1978. La composizione della frazione sterolica di oli ottenuti da olive campionate in differenti momenti del loro ciclo di maturazione. Rivista Italiana Sostenibile Grasse 55, 277–282.

TITTARELLI, F., NERI, U., POLETTI, P., LA CERTOSA, G., RAUS, R. 2002. Monitoraggio dello stato nutrizionale dell'olivo. L'Informatore Agrario 44, 37–51.

TOMA, M. 1999. Dieci anni di sperimentazione olivicola in Toscana. Manuale Arsia, Regione Toscana, ARSIA, pp. 1–167.

TOMBESI, A. 1994. Olive fruit growth and metabolism. Acta Horticulturae 356, 225–229.

TOMBESI, A., STANDARDI, A. 1977. Influenza dell'Alar su alcune caratteristiche delle drupe di tre cultivar di olivo. Rivista Ortoflorofrutticoltura Italiana 58, 4, 265–273.

TOSCANA. DIP. AGRICOLTURA E FORESTE 1992. *I sistemi territoriali della Comunità montana Alto Mugello Mugello Val di Sieve: studio per una caratterizzazione fisica dell'ambiente mugellano (Progetto sistemi territoriali).* Giunta Regionale Toscana, Firenze, 198 p.

TRONCOSO, A. 1998. Nutrition mineral del olivo. In: International Course on Olive Growing. COI, CNR, ARSIA, Regione Toscana, Scandicci, 6–11 May, pp. 185–196.

TURA, D., FAILLA, O., PEDÒ, S., GIGLIOTTI, C., SERRAIOCCO, A., BASSI, D. 2005. Andamento meteorologico e qualità dell'olio extravergine di oliva. L'Informatore Agrario 8, 53–56.

VAN WAMBEKE A. AND FORBES T. 1986. *Guidelines for using Soil Taxonomy in the names of soil mapping units.* SMSS Technical Monograph No. 10, USDA, Washington DC.

VASQUEZ RONCERO, A., MAESTRO DURAN, R., GRACIANI COSTANTE, E. 1971. Cambions en los polifenoles de la aceituna durante la maturacion. Grasas y Aceite 5, 366–370.

VEGA, V., FERNANDEZ, M., PASTOR, M., ROJO, J., HIDALGO, J., ZAPATA, A. 2001. *Riego con aguas salinas de olivar joven en crescimiento en condiciones de campo. Efectos sobre la production.* XIX Congreso National de Riegos. Zaragoza, 12–14 June 2001, pp. 95–96.

VINCIERI, F. 1997. *Polifenoli nelle olive: caratteristiche analitiche, nutrizionali ed attività per la tutela della salute.* Convegno Internazionale "Arauco '97", La Rioja, (Argentina), 6–11 May.

WILLIAMS, M., SANCHEZ, J., HANN, A.C., HARWOOD, J.L. 1993. Lipid biosynthesis in olive cultures. Journal of Experimental Botany 44, 1717–1723.XILOYANNIS, C., DICHIO, B., SOFO, A., PALESE, A.M. 2004. Capacità di adattamento dell'olivo agli ambienti siccitosi. L'Informatore Agrario 40, 43–45.

ZUCCONI, F., NERI, D., SABBATICI, P. 1996. Fertilità del suolo e concimazioni minerali. In: Corso internazionale "Gestione dell'acqua e del territorio per un'olivicoltura sostenibile".

28. Stone Fruits: Apricot, Cherry, Peach, Prune and Plum

Elvio Bellini[1], Edgardo Giordani[1] and Daniele Morelli[1]

28.1. Apricot (*Prunus armeniaca* L.)

28.1.1. Taxonomy and Range

Apricot (*Prunus armeniaca* L.) originated in China or in East Asia. It was introduced to the territories of the Roman Empire via Greece and Armenia and is cited by Pliny the Elder (23–79 AD) in his *Naturalis Historia* (AA. VV., 1991; Nasi et al., 1994).

In the past 40 years, production of apricots has constantly increased (Fig. 28.1). In 2006, apricot production was 3,251,254 t (FAOSTAT, 2007); the main countries producing apricots are reported in Fig. 28.2.

28.1.2. Rootstocks

The issue of apricot rootstocks requires attention, given the large number of rootstocks that make apricot cultivation possible under various conditions. It must be said that affinity of several rootstocks and varieties is not known. It is therefore difficult for nurseries to supply appropriate rootstock-cultivar combinations for new varieties without more research. For these reasons, rootstock innovation is slow in apricot (AA.VV., 1991, 1996; Nasi et al., 1994; Pirazzini, 2002).

---

[1] Università di Firenze—Facoltà di Agraria, Dipartimento di Ortoflorofrutticoltura, Firenze, Italy.

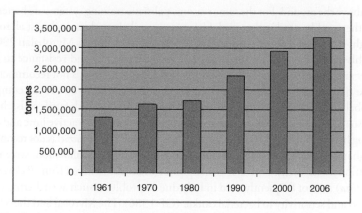

Fig. 28.1. Trend of world production of apricots since 1961 (FAOSTAT, 2007).

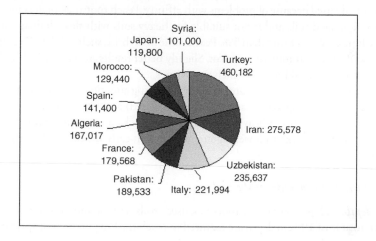

Fig. 28.2. The 10 leading countries producing apricots (tonnes) in 2006 (FAOSTAT, 2007).

- *Seedlings and seedling selections.* Seedlings are the most commonly used rootstocks in Italy. Trees have high vigour and satisfactory longevity but are slow to fruit compared to myrobalan plum or peach. Apricot seedlings are recommended in dry and calcareous soils (< 8% active lime) and soils with moderate salinity (0.8 dS/m). These rootstocks prefer fertile and permeable soils and do not prosper with waterlogging, where bacterial canker (*Agrobacterium tumefaciens*), root rot (*Armillaria mellea* and *Rosellinia necatrix*) and crown rot (*Phytophthora cactorum*) can be a problem. Additionally, apricot seedlings are subject to nematodes. Nurseries prefer Manicot 1236, a selected seedling with greater uniformity of growth.
- *Plum.* Myrobalan (*P. cerasifera*) seedlings are probably still the most commonly used despite several problems including heterogeneous population and disaffinity with swelling and breakage at graft union. Given these problems, in windy locations grafting is done

high above the ground. For the same reasons, palmette training systems are better avoided since they act like a sail in the wind. However, myrobalan tolerates almost all soils and is drought- and lime-tolerant. Among the myrobalan selections is Myrobalan 29C, which is highly recommended for its good production efficiency and resistance to bacterial canker and some nematodes. On the other hand, Myrobalan B is not recommended due to the disaffinity with several cultivars, excessive vigour (+20–40%) compared to Myrobalan 29C, medium productivity and small fruit size. With fertile soils, the MrS 2/5 is also recommended though it is only moderately resistant to active lime and drought. Among hybrid rootstocks, the Marianna (*P. cerisifera x P. munsoniana*) is not recommended though it is resistant to root anoxia, lime and salinity. It has a shallow root system and has low affinity with many cultivars. Other plum rootstocks (derived from *P. domestica* and *P. insititia*) are not frequently used in Italy due to problems such as suckering (Damson P 1869) and sensitivity to bacterial canker (GF 1380). The Myrocal and the Myrobalan P1254 have been proposed as alternatives to Myrobalan B in France.

- **Peach.** Peach rootstock is commonly used in the United States while in Italy it is only occasionally used because of problems with affinity. Peach rootstock prefers moist, deep and well-aerated soils and is not suitable for heavy soils with slow drainage and with active lime content greater than 5%. Peach-almond hybrids such as GF 677 are not used because of their disaffinity with apricot. Slightly better results are obtained with Montclar and Rubirà (with red leaves), French peach lines.

- **Hybrid.** Multipurpose hybrids are currently attracting attention, such as Ishtara (suitable for heavy soils as long as they are moist and irrigated) and Citation (which appears to have affinity problems in heavy soils. These rootstocks do not sucker much and have good graft affinity, impart lower vigour, and present early fruiting and very good fruit quality.

28.1.3. Functional Properties of the Soil

- **Soil depth.** Independent of the rootstock used, soils with no limitation to root growth down to at least 100 cm depth are required to reach the full potential of apricot growth and productivity.
- **Texture and cracking.** Apricot prefers medium to slightly coarse textures. Adaptability depends on the grafting combination. Apricot seedling rootstocks are more suited to loose soils including rocky ones due to they deep root system, while in fine soils quick drainage of excess water must be guaranteed due to their sensibility to root anoxia. Plum rootstocks have more superficial roots and are therefore more intolerant of drought and not suitable for coarse soils. Fine soils cause growth limitations due to cracking, which is caused by wetting and drying cycles and consequent expansion and contraction. Only with low cracking are root growth and functionality not impaired.
- **Sodium and salinity.** The best soils for apricot cultivation have electrical conductivity values and exchangeable sodium percentage (ESP) of 0.4 dS/m and 8% respectively. With increasing values, growth is increasingly limited and becomes unacceptable when values are above 0.8 dS/m and 10% respectively.
- **Risk of flooding.** Apricot-growing areas should be chosen where there is limited or no risk of flooding. Apricots seedlings suffer particularly from root anoxia, so much that more

than 7 d of waterlogging seriously compromises growth and even survival of the trees and makes the trees vulnerable to fungal diseases. Plum rootstocks (myrobalan seedlings, Myrobalan 29 C and MrS 2/5) better tolerate asphyctic soil conditions probably because of lower hydrogen cyanide concentrations in the roots with anoxia.

- *Availability of oxygen.* The above information about waterlogging applies to availability of oxygen, since oxygen availability is inversely related to the water content of the soil. Excess water and the consequent insufficient soil aeration are the major limiting factors for apricot cultivation. Apricot trees, especially when grafted on apricot seedlings, require soils where water is promptly drained or soils where waterlogging does not occur during the vegetative season.
- *Soil pH.* Apricot prefers soil of pH near neutral. Optimal pH values are between 6.5 and 7.5, but strong limitations to growth and yield do not occur until extreme values are reached (pH < 5.4 or pH > 8.5).
- *Active lime content.* Apricot has average resistance to active lime. When grafted on apricot seedling it grows well in soils with less than 8% active lime content, though severe limitations occur only with more than 12%. When grafted on plum it is more sensitive to lime, tolerating maximum values of 10% (AA.VV., 1991, 1996; Landi, 1999; Nasi et al., 1994).
- *Evaluation table.* See Table 28.1 for a summary of soil limitation factors for apricot growth on different rootstocks, compiled by I.TER (1998) project "Catalogo dei suoli della pianura emiliano romagnola" (Soil catalog of the Emilia-Romagna plain) funded by the Servizio Sviluppo Sistema Agro-alimentare (Service for Food and Agriculture System Development) of Emilia-Romagna, based on L.R.52/92 (regional policy 52/92) and revised by Paola Pirazzini (CISA M. Neri) and Carla Scotti (I.TER) in December 2005.

Table 28.1. Evaluation of limiting properties of the soil for apricot growth on different rootstocks.

Soil properties	Groups of rootstocks	Limitation		
		None or weak	Moderate	Strong
Available rooting depth (cm)	Apricot rootstock, Myrobalan (seed), Myrobalan 29C, Mr.S. 2/5	> 100	50–100	< 50
Texture	Apricot rootstock	Medium, moderately coarse	Moderately fine, fine, coarse	–
	Myrobalan (seed), Myrobalan 29C, Mr.S. 2/5	Medium, moderately coarse	Moderately fine	Coarse, fine
Cracks	Apricot rootstock	Low	Medium	Strong
	Myrobalan (seed), Myrobalan 29C, Mr.S. 2/5	Low	–	Medium, strong
Electrical conductivity (salinity) (EC 1:5 dS/m)	Apricot rootstock, Myrobalan (seed), Myrobalan 29C, Mr.S. 2/5	< 0.4	0.4–0.8	> 0.8
Exchangeable sodium percentage (ESP)	Apricot rootstock, Myrobalan (seed), Myrobalan 29C, Mr.S. 2/5	< 8	8–10	> 10
Flooding risk	Apricot rootstock, Myrobalan (seed), Myrobalan 29C, Mr.S. 2/5	None or rare	Occasional	Frequent
Flooding risk, duration	Apricot rootstock	Extremely brief, very brief	Brief	Long, very long
	Myrobalan (seed), Myrobalan 29C, Mr.S. 2/5	Extremely brief, very brief, brief	Long	Very long
Available oxygen supply	Apricot rootstock	Good	Moderate	Imperfect, poor, very poor
	Myrobalan (seed), Myrobalan 29C, Mr.S. 2/5	Good, moderate	–	Imperfect, poor, very poor
Chemical response (pH in water)	Apricot rootstock, Myrobalan (seed), Myrobalan 29C, Mr.S. 2/5	6.5–7.5	5.4–6.5; 7.5–8.5	< 5.4; > 8.5
Content of exchangeable calcium (%)	Apricot rootstock	< 8	8–12	> 12
	Myrobalan (seed), Myrobalan 29C, Mr.S. 2/5	< 7	7–10	> 10

28.2. Cherry (*Prunus avium* and *Prunus cerasus* L.)

28.2.1. Taxonomy and Range

Both sweet cherry (*Prunus avium* L.) and sour cherry (*Prunus cerasus* L.) originated in an area between the shores of the Black Sea and the Caspian Sea, from which they have spread to several countries (AA.VV., 1991). In the past 40 years, cherry alternated with a growing trend (Fig. 28.3) and in 2006 the cherry production was 1,872,774 t (FAOSTAT, 2007). The main cherry-producing countries are shown in Fig. 28.4.

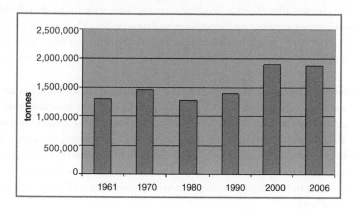

Fig. 28.3. Trend of world production of cherries since 1961 (FAOSTAT, 2007).

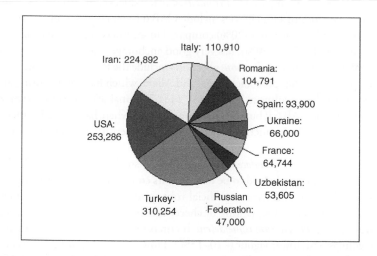

Fig. 28.4. The 10 leading countries producing cherries (tonnes) in 2006 (FAOSTAT, 2007).

28.2.2. Rootstocks

Among the traditional cherry rootstocks most commonly used today there are three species: *Prunus avium* (sweet cherry), *Prunus mahaleb* (mahaleb cherry), and *Prunus cerasus* (sour cherry). Other hybrid rootstocks exist, the most important of which is Colt (De Salvador and Lugli, 2002; Lugli and Sansavini, 1999; Lugli et al., 2005; Nasi et al., 1994)

- *Prunus avium.* Sweet cherry seedlings are the most vigorous rootstocks and have upright habit, good affinity with all cultivars and reduced suckering. They have robust root systems both extended and superficial, providing good anchorage. They are not advisable for replanting cherries after cherries. Seedling rootstocks are resistant to *Armillaria mellea* and *Phytophthora* but sensitive to *Verticillium* and nematodes. They are very hardy and are used when mahaleb cherries are allowed. The seedling selection Mazzard F 12/1 has been abandoned because of its high vigour and unsatisfactory productivity and fruit quality.

- *Prunus mahaleb.* This rootstock has great affinity with sour cherries while it often shows disaffinity with sweet cherries. It is very sensitive to *Agrobacterium tumefaciens* and many root diseases such as *Armillaria* and *Verticillium*. It has a taproot system with optimal anchorage and low suckering. Vigour is less than in cherry seedlings but varies with seed origin. Yield efficiency and productivity are greater than with cherry seedlings. Clonal selections include SL 64, preferable to the mahaleb seedlings. This rootstock has lower vigour (−20%) and more homogeneous plants while maintaining all the positive features of mahaleb seedlings.

- *Prunus cerasus* and *Prunus canescens.* *Prunus cerasus* is less vigorous and tends to have a bushy multi-stem growth habit, but it is the least preferred rootstock for sweet cherry owing to frequent disaffinity and tendency to sucker. It requires moist soils and has good resistance to parasites. Compared to *Prunus avium*, it induces earlier fruiting and smaller size. Clonal selections of sour cherries include CAB 6 P, which is suitable for many soil types and induces lower vigour (−20%) compared to seedlings. Tabel® Edabriz* induces lower vigour than F 12/1 (−50%), has very good anchorage but is sensitive to drought and *Phytophthora*. Interesting *P. cerasus* selections include the P-HL series, which induces high productivity but has poor anchorage, and Adara, which has good grafting affinity with many sweet cherry cultivars. The Camil® GM 79 clonal selection of *P. canescens* has similar vigour to the CAB 6 P, but with less compact trees where harvesting is easier. It has high fertility and high production efficiency, but it is sensitive to diseases including *Agrobacterium tumefaciens*.

- *Hybrid.* Colt (*P. avium* x *P. pseudocerasus*) is a clonal rootstock obtained in East Malling. It was the first interspecific hybrid to be used and its control of sweet cherry vigour was only partial. Though Colt has a superficial root system it gives good anchorage and tolerates *Phytophthora* and replanting of cherry after cherry. It is not very resistant to water stress and *Agrobacterium tumefaciens*. It enters into production earlier than sweet cherry seedlings with lower vigour (−10–15%). However, in deep fertile soils, at times the vigour is actually greater. Worth mentioning are the MaxMa DELBARD® 14 Brokforest* and MaxMa DELBARD® 97 Brokgrove*, both hybrids between *P. mahaleb* and *P. avium*. The former is characterized by good graft affinity, low suckering, and good adaptability to pedoclimatic conditions. It is resistant to bacterial canker and ferric chlorosis and induces the lowest vigour and earliest fruiting. The latter has the same features but is slightly more vigorous and is very hardy, has good affinity, early entrance into bearing and low suckering. Other rootstocks are not propagated in Italy though they are currently under evaluation in experimental fields. Among these are Weiroot® 158 (*P. cerasus x P. avium*), which is interesting for high-density orchards, and GISELA®5 (*P. cerasus x P. canescens*), which is of medium vigour, conferring early bearing, high productivity and production efficiency in addition to large fruit size.

28.2.3. Functional Properties of the Soil

- *Soil depth.* Research carried out by the Servizio Sviluppo Sistema Agro-alimentare (Service for Food and Agriculture System Development) of Emilia-Romagna suggests that an optimal depth for cherry root growth when grafted on cherry seedling or Colt is greater than 1 m. It is necessary, especially in flat fields, to conduct a soil profile analysis to detect the presence of impermeable soil layers that might hinder root growth and drainage of excess water. Other research carried out in Italy indicates the following rootstocks are promising for their more superficial root system: CAB 6 P, Weiroot® 158, GISELA® 4, GISELA® 5, GISELA® 7, and Edabriz. These rootstocks could be used in soils with a relatively high impermeable layer (50–100 cm).
- *Texture and cracking.* In Emilia-Romagna, the best results are obtained when cherries are grafted on cherry seedling or on the hybrid Colt in soils with medium, moderately fine or moderately coarse texture. Severe growth limitation occurs in heavy or excessively loose soils. Experimental trials carried out in different areas of Italy (De Salvador and Lugli, 2002) as part of the project "Liste di orientamento varietale dei fruttiferi" (Fruit variety lists), funded by the Italian Ministry of Agriculture (MiPAAF), found that the best rootstocks for heavy soils were CAB 6P, MaxMa DELBARD® 14 Brokforest*, GISELA®7 clone 148/8, and GISELA®12 195/2, and for loose or rocky soils mahaleb cherry seedlings and mahaleb selctions SL 64 and Pi-Ku 1* clone 4,20.
- *Sodium and salinity.* Independent of rootstock, and similarly to pear, optimal conditions for cherry cultivation are found at salinity values lower than 0.4 dS/m and ESP lower than 8%. Values above 0.8 dS/m and 10% ESP make the soil unsuitable for cherry.
- *Risk of flooding.* Research carried out in Italy shows that cherry is very sensitive to even short-term waterlogging. Trees on cherry seedling rootstock are more sensitive than trees on Colt. Soils with risk of waterlogging for more than 48 h are to be excluded.
- *Availability of oxygen.* Cherry is particularly sensitive to root anoxia especially when on sweet cherry seedling, mahaleb seedling, and SL 64 rootstocks. Compared to these rootstocks, Colt, GISELA® 6 clone 148/1, and GISELA® 12 clone 195/2 confer greater resistance to anoxia (De Salvador and Lugli, 2002). Only soils in which water is promptly drained and without excessive humidity during the growing season allow good growth and production in these species.
- *Soil pH.* Independent of the rootstock used, cherry prefers pH values between 6.5 and 8.5. Higher or lower values bring about increasing difficulties for cherry cultivation.
- *Active lime content.* When grafted on sweet cherry seedlings or Colt, cherry does not encounter growth limitations with active lime content below 7%. Rootstocks such as

mahaleb seedling, SL 64, the hybrid Avima ®-Argot and Weiroot® 158 appear more tolerant to lime (AA.VV., 1991; De Salvador and Lugli, 2002; Landi, 1999; Nasi et al., 1994).

- *Evaluation table.* See Table 28.2 for a summary of soil limitation factors for cherry growth on different rootstocks, compiled by I.TER (1998) project "Catalogo dei suoli della pianura emiliano romagnola" (Soil catalog of the Emilia-Romagna plain) funded by the Servizio Sviluppo Sistema Agro-alimentare (Service for Food and Agriculture System Development) of Emilia-Romagna, based on L.R.52/92 (regional policy 52/92).

Table 28.2. Evaluation of limiting properties of the soil for cherry growth on different rootstocks.

Soil properties	Groups of rootstocks	Limitation		
		None or weak	Moderate	Strong
Available rooting depth (cm)	Cherry rootstock (seed), Hybrid Colt	> 100	50–100	< 50
Texture	Sweet cherry rootstock (seed), Hybrid Colt	Medium, moderately fine, moderately coarse	–	Coarse, fine
Cracks	Sweet cherry rootstock (seed), Hybrid Colt	Low	Medium	Strong
Electrical conductivity (salinity) (EC 1:5 dS/m)	Sweet cherry rootstock (seed), Hybrid Colt	< 0.4	0.4–0.8	> 0.8
Exchangeable sodium percentage (ESP)	Sweet cherry rootstock (seed), Hybrid Colt	< 8	8–10	> 10
Flooding risk	Sweet cherry rootstock (seed), Hybrid Colt	None or rare	Occasional	Frequent
Flooding risk, duration	Hybrid Colt	Extremely brief, very brief	Brief	Long, very long
	Sweet cherry rootstock (seed)	Extremely brief	Very brief	Brief, medium, long, very long
Available oxygen supply	Sweet cherry rootstock (seed)	Good	–	Moderate, imperfect, poor, very poor
	Hybrid Colt	Good	Moderate	Imperfect, poor, very poor
Chemical response (pH in water)	Sweet cherry rootstock (seed), Hybrid Colt	6.5–8.5	5.4–6.4	< 5.4; > 8.5
Content of exchangeable calcium (%)	Sweet cherry rootstock (seed), Hybrid Colt	< 7	7–12	> 12

28.3. Peach [*Prunus persica* (L.) Batsch.]

28.3.1. Taxonomy and Range

The peach [*Prunus persica* (L.) Batsch.] is thought to have originated in China, and was brought to Europe over 2000 years ago (AA. VV., 1991; Baldini and Scaramuzzi, 1981; Nasi et al., 1994).

In the past 40 years, the production of peaches has increased notably (Fig. 28.5; in 2006 production was 17,188,840 t (FAOSTAT, 2007). Main peach-producing countries are listed in Fig. 28.6.

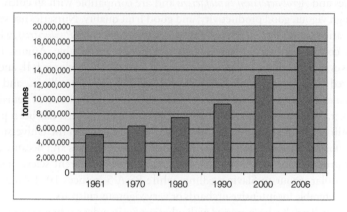

Fig. 28.5. Trend of world production of peaches since 1961 (FOSTAT, 2007).

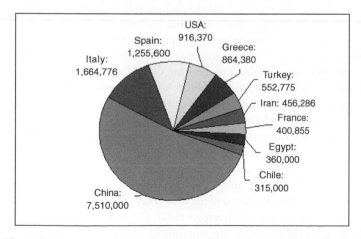

Fig. 28.6. The 10 leading countries producing peaches (tonnes) in 2006 (FAOSTAT, 2007).

28.3.2. Rootstocks

The cultivation of peaches in diverse environments has spurred the development and introduction of new rootstocks. The working group "Valutazione dei portinnesti dei fruittiferi" (fruit tree rootstock evaluation), promoted by the Italian Ministry of Agriculture (MiPAAF), has been active in recent years (AA. VV., 1996; Baldini and Scaramuzzi, 1981; Giovannini et al., 2005; Loreti and Massai, 2003; Nasi et al., 1994).

- *Peach seedlings.* *Prunus persica* seedlings are the most commonly used rootstock for peach, though they are being replaced by GF 677. Peach seedlings are subject to nematodes and *Agrobacterium tumefaciens* and are compatible with all cultivars. They induce high vigour, good productivity, and good fruit quality. Nurseries use seeds from selected varieties such as peach seedlings from the Balkan area (*P. sylvestris*), which were the most commonly used in Italy as long as seeds were available. Among the many selections evaluated, only a few are widely used: Missour (and its clones I, II, and III, etc.) is preferred in Italy, owing to lack of availability of Slavic seedlings, and in other Mediterranean countries, also because this rootstock is slightly tolerant of nematodes and root canker. However, it cannot be used in soils in which peach has been previously grown or in heavy and wet soils. The rootstock series PS, used at the University of Pisa, is also important. It is interesting for the uniformity of the trees and because it confers constant productivity while the vigour changes for the different selections: PS B 2 induces similar vigour to Slavic seedling, while PS A 5 induces 10–15% less vigour, which is preferable for intensive orchards or very vigorous cultivars.

- *Hybrid peach.* Peach has been crossed with other species in order to obtain newer rootstocks adaptable to pedological conditions unsuitable for peach seedlings. The most important hybrid is GF 677 (*P. persica* x *P. amygdalus*), which is the most commonly used. It induces greater vigour than peach seedlings by 10–15%. This rootstock has partly solved the problem of replanting peach after previous peach cultivation and the problem of chlorosis in soils with active lime greater than 8%, which occurs frequently in southern Italy and in some other Mediterranean areas. GF677 appears very compatible with all cultivars of peaches and nectarines, though it does not tolerate clay soils or soils with frequent waterlogging. It is susceptible to *Armillaria mellea*, *Agrobacterium tumefaciens*, *Phytophthora cactorum* and *Stereum purpureum*.

  The Hansen® hybrids are rather susceptible to *Agrobacterium* but are highly resistant to lime and are immune to gall nematodes. Among the new hybrids, Sirio® (I.S. 5/22) is particularly interesting. It and other clones were selected by the Dipartmento di Coltivazione e Difesa delle Piante Legnose (Department of Cultivation and Pest and Disease Control of Woody Plants) of the University of Pisa. Sirio® (I.S. 5/22) induces a 40–50% reduction in vigour with respect to GF 677 and it is quite resistant to lime, up to 13%, though it has a negative effect on fruit size. Trials are currently underway to test its possible use in peach after peach replantings. In heavy soils where GF 677 cannot be planted after peach, a hybrid that is worth experimenting with is Barrier® 1 (*P. persica* x *P. davidiana*). This hybrid is also resistant to nematodes because of its large and deep roots and is suited for many soil types including ones that tend to induce chlorosis and have a lack of oxygen.

- ***Plum.*** Plum rootstocks are not commonly used for peaches, though they can be used in certain situations where neither peach seedlings nor GF 677 can be used. At present only MrS 2/5 appears useful when replanting peach after peach. These rootstocks represent a valid alternative to peach seedlings and GF 677 in fertile or very fertile soils where it permits the reduction of spacing and the use of training systems such as spindle and delayed vase. However, nursery problems limit its use.

Ishtara-Ferciana is a multipurpose interspecific hybrid characterized by good affinity and reduced suckering. It imparts medium vigour, limiting tree growth in fertile and moist soils, though it suffers with an active lime content greater than 5%. Other new prune rootstocks include Penta and Tetra, which were obtained at the Istituto Sperimentale per la Frutticoltura in Rome. Penta was obtained with free pollination with Imperial Epineuse while Tetra was produced by free pollination of Regina Claudia Verde. These rootstocks have been tested with several cultivars of peaches and nectarines over several years and at five Italian locations with different pedoclimatic conditions. Results were always excellent both in heavy clay soils and in light and medium soils. The main differences between the two rootstocks is the greater tree size that Penta induces (10–15% more than Tetra) and early ripening in Tetra compared to Penta and peach seedling (by a few days). Further, though both rootstocks have very high fruiting efficiency, Penta has higher fruiting efficiency than Tetra.

Since 1994, the Italian Ministry of Agriculture (MiPAAF) has been testing possible rootstocks for the cultivar Suncrest within the project "Valutazione dei portinnesti" (Rootstock evaluation). Field trials have been established in several areas of the north (Cuneo and Forlì), centre (Pisa, Ancona and Roma), and south (Caserta, Matera, and Catanzaro) and in Sardinia (Cagliari). The research yielded the following results: GF 677 and Barrier® 1 uniformly induced high vigour while peach seedlings or prune clones such as Ishtara and Julior (Ferdor) imparted minor differences in vigour compared to the general average. The greater vigour imparted by GF 677 and Barrier® 1 was even more accentuated under low fertility, high active lime content, lack of irrigation, field slope, and heavy soil (such as that in Forlì where the soil had the following characteristics: clay-loam with active lime of 6.5%, neutral pH, low organic matter [< 2%] and total nitrogen [< 1.2‰], high cation exchange capacity [> 30 meq/100 g]), as well as in Ancona, Pisa, Roma, Caserta, and Cagliari. In contrast, under high fertility the differences between peach seedlings and prune clones Ishtara and Julior are less evident.

The increased vigour of plants on GF 677 and Barrier® 1 resulted in more biomass being removed by green pruning. These rootstocks gave the highest cumulative yield, followed distantly by PS A 5, PS A 6, and Ishtara, but with some exceptions in some locations where peach seedlings and Ishtara gave yields equal to or greater than Barrier®1, as in Cuneo, Matera and Caserta. On the whole, yield differences were not as great as vigour differences. This was due to the fact that the low vigour of some rootstocks improved light penetration and photosynthetic efficiency. In fact, the lowest photosynthetic efficiency occurred in GF 677. Fruit size was more dependent on the environment than on the rootstock. In locations with reduced fertility and especially in cases of non-irrigated fields, fruit size was reduced.

28.3.3. Functional Properties of the Soil

- *Soil depth.* The soil volume is explored by the roots of the rootstock. The most commonly used rootstock in peach is GF 677. This rootstock quickly occupies the available volume of soil quickly reaching a soil depth of 170 cm. Because of this, peach on GF 677 is very drought resistant. Water extraction capacity is high, with maximum extraction at 60 cm depth. Up to 20 years ago, the majority of peach orchards in Emilia-Romagna were not irrigated.

- *Texture and cracking.* Peach is highly adaptable to different soils though it prefers loose soils (60–70% sand, 20–30% clay). It prefers fertile soils (2–3% organic matter) with low or absent gravel. Peach is particularly susceptible to problems when replanting peach after peach (symptoms vary, from stunted growth to delayed fruiting). This problem is particularly evident in loose soils, corresponding to the presence of nematodes (specifically *Pratylenchus vulnus*), lack of micronutrients, and toxins released by the previous root system. It is preferable to avoid replanting peach after previous peach cultivation. Low soil cracking is preferable since heavy cracking compromises peach cultivation.

- *Sodium and salinity.* Independent of rootstock, peach is best cultivated with salinity levels of less than 2 mS/cm and exchangeable sodium (ESP) of less than 5%. With greater values, peach growth is impaired and peach cannot be grown with salinity levels over 3 mS/cm and ESP of 10%. Irrigation is an important tool to control these parameters and is indispensable for modern peach cultivation. Paradoxically, irrigation contributes to desertification especially in areas with high water deficit and scarce rainfall. In the Metaponto area, irrigation combined with intensive cultivation techniques has caused structural deterioration of the soil mainly due to loss of organic matter, salinization, and increased alkalinity. Loss of soil fertility has long been a problem in arid and semi-arid areas such as those of the Mediterranean. Comparing characteristics of irrigation water used in the orchards in the Metaponto and Emilia-Romagna areas, the risk of fertility loss in Metaponto is evident. In fact, irrigation with high salinity water damages soil structure (the sodium cation has a deflocculating action on clay particles), altering size and morphology of soil pores. This hinders water absorption by plants due to an increase in osmotic potential of the soil water. Additionally concentration of some ions (boron, sodium, magnesium, and sulfate and carbonate ions) may reach toxic levels.

- *Drainage.* Proper drainage allows the elimination of excess water from the soil. This guarantees sufficient depth of soil free from waterlogging, allowing deep root penetration. This is especially important in gravel-free soils. Drainage is ensured by placing corrugated plastic pipes at above 100–120cm depth, at variable distances depending on soil nature: shorter intervals are necessary in compact soils (5–10 m) than in loose soils (15–20 m). In loam-clay soils, drain length should not exceed 100–120 m.

- *Waterlogging.* Waterlogging occurs when water fills the soil pores, surpassing field capacity. Clay or loam soils that are not sufficiently drained or have poor water management may suffer superficial waterlogging. Peach does not tolerate root anoxia or impermeable, compact soils and is not suitable in those areas at risk of medium- to long-term flooding. When grafted on GF 677, peach is slightly more tolerant of waterlogging than when grafted on peach seedlings. Flooding may favour some root diseases while it may reduce others that prefer lower humidity levels in the soil; in fact, waterlogging may resolve problems with tracheofusariosis and several nematodes.

- *Availability of oxygen.* As with other fruit trees, peach roots require moderate soil gas exchange to thrive. Soil oxygen content less than 2% results in interruption of plant growth; 10% of oxygen is the adeguated value.
- *Soil pH.* Peaches prefer soils with pH close to neutrality. Optimal values are believed to be between pH 6.5 and 7.5 and up to 8 if grafted on GF 677 and Cadaman. At more extreme pH values, soil conditions become increasingly unsuitable for peach cultivation. At high pH, especially with high levels of calcium carbonate (more than 5%), alkalinity induces foliar chlorosis. This is aggravated by low nitrogen and low iron in the soil, a problem that occurs at pH greater than 7.5. Acidic soil, on the other hand, induces a phenomenon called short life, which occurs in some areas of the United States. This has been associated with the presence of the nematode *Criconomella xenoplax*, which fortunately is not yet encountered in Italy.
- *Active lime content.* Active lime is one of the factors responsible for leaf chlorosis. Peach does not tolerate more than 12% of active lime if grafted on Mirabolan MrS 2/5 or 13% if grafted of GF 677. With peach seedlings and seedling selections, peach susceptibility increases greatly with strong symptoms at 4%.
- *Replanting peach after peach.* When a new peach orchard is planted on the site of a previous peach orchard, trees may have stunted and non-uniform growth, especially in the beginning, and delayed fruiting. This is a phenomenon of increasing importance given the limited area suitable for peach cultivation. Among the many causes are pathogens such as nematodes but also fungi and bacteria, the toxins from roots remaining from the previous culture or their decomposition, or low micronutrient levels. Studies have reviewed the correlation between these symptoms and the presence of nematode *Pratylenchus vulnus*. Some symptoms may be resolved with soil fumigation or appropriate fertilization. Among the best agronomic remedies are an appropriate rotation of crops and use of resistant or tolerant rootstocks (AA. VV., 1991; Baldini and Scaramuzzi, 1981; Landi, 1999; Nasi et al., 1994).
- *Evaluation table.* See Table 28.3 for a summary of soil limitation factors for peach growth on different rootstocks, compiled by I.TER (1998) project CRPV "Rilancio della peschicoltura" (Re-launching peach cultivation) funded by the Servizio Sviluppo Sistema Agro-alimentare (Service for Food and Agriculture System Development) of Emilia-Romagna, based on L.R.28/98 (regional policy 28/98); the chart has been validated by experts in peach cultivation.

Table 28.3. Evaluation of limiting properties of the soil for peach growth on different rootstocks

Soil properties	Groups of rootstocks	Evaluation		
		None or weak	Moderate	Strong
Available rooting depth (cm)	Hybrid GF 677, Hybrid Cadaman, Myrobalan MrS 2/5, Hybrid Ishtara	> 100	50–100	< 50
Texture	Hybrid GF 677, Hybrid Cadaman	Medium, moderately fine, moderately coarse	–	Fine, coarse
	Myrobalan MrS 2/5, Hybrid Ishtara	Medium, moderately coarse	Moderately fine	Fine, coarse
Cracks	Hybrid GF 677, Hybrid Cadaman	Low	Medium	Strong
	Myrobalan MrS 2/5, Hybrid Ishtara	Low	–	Medium, strong
Electrical conductivity (salinity) (EC 1:5 dS/m)	Hybrid GF 677, Hybrid Cadaman, Myrobalan MrS 2/5, Hybrid Ishtara	< 0.2	0.2–0.4	> 0.4
Exchangeable sodium percentage (ESP)	Hybrid GF 677, Hybrid Cadaman, Myrobalan MrS 2/5, Hybrid Ishtara	< 5	5–10	> 10
Flooding risk	Hybrid GF 677, Hybrid Cadaman, Myrobalan MrS 2/5, Hybrid Ishtara	None or rare	Occasional	Frequent
Flooding risk, duration	Hybrid GF 677, Hybrid Cadaman, Hybrid Ishtara	Extremely brief, very brief	Brief	Long, very long
	Myrobalan MrS 2/5	Extremely brief, very brief, brief	Long	Very long
Available oxygen supply	Hybrid GF 677, Hybrid Cadaman, Hybrid Ishtara	Good	Moderate	Imperfect, poor, very poor
	Myrobalan MrS 2/5	Good, moderate	–	Imperfect, poor, very poor
Chemical response (pH in water)	Myrobalan MrS 2/5 Hybrid Ishtara	6.5–7.5	5.5–6.5; 7.5–8.5	<5.5; >8.5
	Hybrid GF 677, Hybrid Cadaman	6.5–8	5.5–6.5; 8–8.5	<5.5; >8.5
Content of exchangeable calcium (%)	Hybrid Ishtara	< 5	5–9	> 9
	Hybrid GF 677, Hybrid Cadaman	< 8	8–13	> 13
	Myrobalan MrS 2/5	< 7	7–10	> 10

28.4. Plum and Prune [(*Prunus salicina* Lindl.) and (*Prunus domestica* L.)]

28.4.1. Taxonomy and Range

There are about 10 species of plum and prune of agronomic interest, as either rootstocks or cultivars. These species are quite diverse botanically, in terms of growth habit and environmental adaptation. Four of these species are of greater cultural interest: *Prunus domestica* L. (prune, or European plum), *Prunus triflora* Roxbgh = *Prunus salicina* Lindl. (Japanese plum), *Prunus insititia* Jusl. (bullace or damson), and *Prunus cerasifera* Ehrh. (myrobalan plum). Watkins and other researchers conclude that the genus *Prunus* is of Central Asian origin. Currently, four geographic areas of origin are recognized: central-western Asia, Europe, China, and North America. Each one characterized by the presence of one or more plum-type species of the *Prunus* genus. *Prunus salicina* is recognized as the most representative species among those that belong to the Japanese plums. This species originated long ago in China, where nowadays it is very widespread. Its cultivation, however, appears to have originated less than 2000 years ago. Knowledge of this crop in Europe is evident in the writings of Pliny the Elder in the first century. Pliny gives information about several varieties of this plum with different shapes and colours in his *Naturalis Historia* (AA.VV., 1991, 1996; Baldini and Scaramuzzi, 1981, Nasi et al., 1994; Morettini, 1963).

After a long period of stability, plum and prune production has increased since 1990 (Fig. 28.7) and in 2006 yield was 9,431,322 t (FAOSTAT, 2007). Main plum-producing countries are shown in Fig. 28.8.

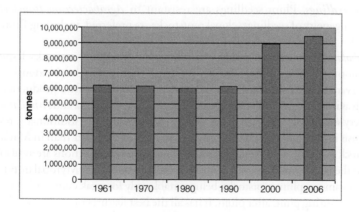

Fig. 28.7. Trend of world production of plums since 1961 (FAOSTAT, 2007).

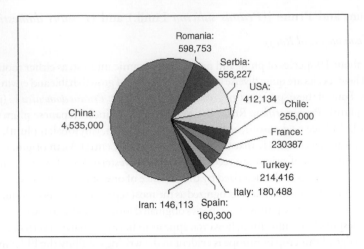

Fig. 28.8. The 10 leading countries producing plums (tonnes) in 2006 (FAOSTAT, 2007).

28.4.2. Rootstocks

Several rootstocks are available for plums, each with some advantages and disadvantages, making rootstock choice particularly complex (AA.VV., 1991, 1996; Baldini and Scaramuzzi, 1981, Nasi et al., 1994; Morettini, 1963).

- **Plum seedlings.** Plum seedlings are resistant to *Agrobacterium tumefaciens*. Due to heterogeneity of seedlings, clonal rootstocks are preferred, including Brompton and GF 43.
- **Myrobalan.** Myrobalan (*P. cerasifera*) seedlings are important rootstocks despite the lack of uniformity among plants. They have a strong root system that is extensive and deep, giving very good anchorage. Myrobalan is slow to enter into fruit production but yield is high as is affinity with all scion cultivars. It is subject to Sharka and *Armillaria rot*, while it is very sensitive to *Agrobacterium tumefaciens* (crown gall). Additionally, resistance to chlorosis is insufficient while resistance to root anoxia is lower than in Marianna. This rootstock is suitable for vegetative propagation and several selections are available, among which the most commonly used are Myrobalan B and 29C. Myrobalan B is the most vigorous among plum rootstocks, being suitable only for weak cultivars in poor soils and when replanting plum after plum. It has all the best features of the Myrobalans as far as adaptability, affinity, productivity and disease resistance. Myrobalan 29C has a less extensive root system with less anchorage imparting lower vigour than Myrobalan B (−30%). It has good affinity and induces early fruiting and greater production efficiency. It is recommended for vigorous cultivars in fertile soils. Among the Myrobalan hybrids, one that can be used also for apricot and peach is the MrS 2/5 (*P. cerisifera x P. spinosa*).
- **Marianna.** Marianna is a group of hybrids between *P. cerisifera* x *P. munsoniana*, which includes Marianna GF 8/1. These rootstocks induce vigour intermediate between Myrobalan B and Myrobalan 29C, though their productivity and fruit quality is closer to the 29C. The root system is shallow but gives sufficient anchorage. These rootstocks are recommended for their resistance to cold winters and nematodes.

- **Damsons.** This term is used to describe seedling rootstocks usually belonging to *P. insititia*, though often also to *P. domestica*. Several clonal damson plums have been selected, among which the most interesting is St. Julien A, which is vigorous but not tolerant of low temperatures and Damson 1869. A clonal selection within the St. Julien group yielded Pixy®. This rootstock has a shallow root system of medium expansion with insufficient anchorage and is not drought tolerant. It has the advantage of reducing vigour 30–50% less than Myrobalan B. Its affinity is good with all tested varieties; yield and yield efficiency are high, as is fruit quality when irrigated. It is recommended for intensive orchards with vigorous cultivars and fertile soils, while it is not advisable for dry soils or without irrigation. Pixy® showed good resistance to root anoxia but is not recommended when replanting plum after plum.

These rootstocks have not yet been used in commercial orchards.

28.4.3. Functional Properties of the Soil

- *Soil depth.* Under normal circumstances most of the roots develop within the first 60–80 cm depth, but soils of more than 1 m depth ensure ideal conditions for root development of trees grafted on the rootstocks shown in Table 28.5, all of which confer extensive root systems (with the partial exception of Myrobalan 29C).
- *Texture and cracking.* Generally speaking, plums grafted on Myrobalan prefer medium, moderately fine, or moderately coarse soils. Moderate limitations of growth occur in soils of fine or coarse texture. Trees grafted on Ishtara have reduced growth also in soils with moderately fine texture and are not suited to fine or coarse soils. Plums experience severe growth limitations with soils rich in expandable clays. Root growth and functionality are very sensitive to cracking.
- *Sodium and salinity.* Plums tolerate salinity values up to 0.4 dS/m, which maximum exchangeable sodium (ESP) is about 8–10%. It is always necessary to check irrigation water quality since prolonged use of brackish water can cause serious problems (Table 28.4), especially when rainfall is insufficient to leach the salts that accumulate in the soil.

Table 28.4. Limits of resistance to salinity (kg/ha in 1.2 m depth) of plum.

Sulphate (sodium sulphate)	10,200
Carbonate (sodium carbonate)	1,530
Chlorides (sodium chloride)	1,360
Total salinity	13,300

- *Risk of flooding.* Waterlogging severely impairs growth of plums. Only in areas with low or no risk of flooding can growth limitation be avoided. With regard to duration of flooding, myrobalans are more resistant than, for example, peach and apricot to root anoxia. Their greater resistance appears to be related to a lower production of hydrogen cyanide in the roots under anaerobic conditions. On the other hand, plums on Ishtara, peach/almond (GF 677) and peach are more sensitive to root anoxia. Soil saturation for more than 48 h must therefore be avoided, especially during the vegetative season, since in addition to reducing vegetative growth such conditions may prompt the occurrence of diseases such as crown rot (*Phytophthora cactorum*).

- *Availability of oxygen.* Oxygen availability is inversely related to the water content of the soil. Therefore, the above information about waterlogging applies to availability of oxygen. Trees grafted on the interspecific hybrid Ishtara are extremely sensitive to lack of oxygen and will face growth limitations unless planted in soils with high water conductivity and no signs of anoxic discoloration in the root zone.
- *Soil pH.* Plums prefer soils of approximately neutral pH (6.5–7.5). Soils with pH lower than 5.5 or greater than 8.5 must be avoided.

Table 28.5. Evaluation of limiting properties of the soil for plum growth on different rootstocks.

Soil properties	Groups of rootstocks	Evaluation		
		None or weak	Moderate	Strong
Available rooting depth (cm)	Myrobalan (seed), Myrobalan 29C, Mr.S. 2/5, Hybrid Ishtara	> 100	50–100	< 50
Texture	Myrobalan (seed), Myrobalan 29C, Mr.S. 2/5	Medium, moderately fine, moderately coarse	Coarse, fine	–
	Hybrid Ishtara	Medium, moderately coarse	Moderately fine	Coarse fine
Cracks	Myrobalan (seed), Myrobalan 29C, Mr.S. 2/5, Hybrid Ishtara	Low	–	Medium, strong
Electrical conductivity (salinity) (EC 1:5 dS/m)	Myrobalan (seed), Myrobalan 29C, Mr.S. 2/5, Hybrid Ishtara	< 0.2	0.2–0.4	> 0.4
Exchangeable sodium percentage (ESP)	Myrobalan (seed), Myrobalan 29C, Mr.S. 2/5, Hybrid Ishtara	< 5	5–10	> 10
Flooding risk	Myrobalan (seed), Myrobalan 29C, Mr.S. 2/5, Hybrid Ishtara	None or rare	Occasional	Frequent
Flooding risk, duration	Myrobalan (seed), Myrobalan 29C, Mr.S. 2/5	Extremely brief, very brief, brief	Long	Very long
	Hybrid Ishtara	Extremely brief, very brief	Brief	Long, very long
Available oxygen supply	Myrobalan (seed), Myrobalan 29C, Mr.S. 2/5	Good, moderate	–	Imperfect, poor, very poor
	Hybrid Ishtara	Good	Moderate	Imperfect, poor, very poor
Chemical response (pH in water)	Myrobalan (seed), Myrobalan 29C, Hybrid Ishtara	6.5–7.5	5.5–6.5; 7.5–8.5	< 5.5; > 8.5
Content of exchangeable calcium (%)	Myrobalan (seed), Myrobalan 29C	< 7	7–10	> 10
	Hybrid Ishtara	< 5	5–9	> 9

- *Active lime content.* Plums have average resistance to active lime. Myrobalan does not experience limitation up to 7% active lime content. Ishtara is more sensitive, requiring soils with less than 5% (AA.VV., 1991, 1996; Baldini and Scaramuzzi, 1981; Landi, 1999; Nasi et al., 1994; Morettini, 1963).
- *Evaluation table.* See Table 28.5 for a summary of soil limitation factors for plum and prune growth on different rootstocks, compiled by I.TER (1998) project "Catalogo dei suoli della pianura emiliano romagnola" (Soil catalog of the Emilia-Romagna plain) funded by the Servizio Sviluppo Sistema Agro-alimentare (Service for Food and Agriculture System Development) of Emilia-Romagna, based on L.R.52/92 (regional policy 52/92) and revised by Paola Pirazzini (CISA M. Neri) and Carla Scotti (I.TER) in December 2005.

References

AA. VV. 1991. *Frutticoltura speciale*. Reda, Roma.

AA. VV. 1996. Linee tecniche di produzione integrata. Albicocco. Suppl. Terra e Vita 20.

AA. VV. 1996. Linee tecniche di produzione integrata. Susino. Suppl. Terra e Vita 26.

AA. VV. 1996. Pesco: Linee tecniche di produzione integrata. Suppl. Terra e Vita 46.

BALDINI, E., SCARAMUZZI, F. 1981. *Frutticoltura anni 80 - Il pesco -*. REDA, Roma.

BALDINI, E., SCARAMUZZI, F. 1981. *Frutticoltura anni 80 - Il susino -*. REDA, Roma.

DE SALVADOR, F.R., LUGLI, S. 2002. I portinnesti del ciliegio. Suppl L'Informatore Agrario 51, 9–16.

FAOSTAT. 2007. Internet: http://www.fao.org

GIOVANNINI, D., LIVERANI, A., MERLI, M., BRANDI, F. 2005. Comportamento agronomico di portinnesti del pesco: risultati di uno studio condotto in Romagna. Frutticoltura 7–8, 43–46.

LANDI, R. 1999. *Agronomia e ambiente*. Edagricole—Edizioni Agricole, Bologna.

LORETI, F., MASSAI, R. 2003. Orientamenti per la scelta dei portinnesti del pesco. Frutticoltura 7–8, 26–30.

LUGLI, S., CORREALE, R., GAIANI, A., GRANDI, M., MUZZI, E., QUARTIERI, M., SANSAVINI, S. 2005. Nuovi portinnesti di ciliegio validi per impianti intensivi. Frutticoltura 3, 41–47.

LUGLI, S., SANSAVINI, S. 1999. Nuovi portinnesti clonali del ciliegio adatti per impianti intensivi a media densità: risultati positivi di una prova decennale condotta nella zona tipica di Vignola. Frutticoltura 6, 57–64.

MORETTINI, A. 1963. *Frutticoltura generale e speciale*. REDA, Roma.

NASI, F., LAZZAROTO, R., GHISI, R. 1994. *Coltivazioni arboree*. Liviana Editrice, Padova.

PIRAZZINI, P., 2002. I portinnesti dell'albicocco. Suppl. L'Informatore Agrario 51, 5–8.

REGIONE EMILIA ROMAGNA—I.TER. 1998. *Catalogo dei suoli della pianura emiliano romagnola*. Internet: http://www. gias.regione.emilia-romagna.it/suoli/guidaindice.asp

29. Pome Fruits: Apple and Pear

Elvio Bellini,[1] Edgardo Giordani[1] and Daniele Morelli[1]

29.1 Apple (*Malus pumila* Mill.)

29.1. 1. Taxonomy and Range

The apple (*Malus pumila* Mill.) is native to the south Caucasian region though it was introduced to and spread within Europe early enough that it is mentioned in the writings of ancient civilizations. The origin of *Malus pumila* was probably a hybridization between *Malus sylvestris* and other *Malus* species (AA.VV., 1991).

In the past 40 years, the production of apples has increased notably (Fig. 29.1); in 2006, production was 63,804,534 t (FAOSTAT, 2007); the main countries producing apples are reported in Fig. 29.2.

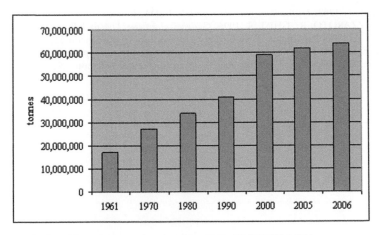

Fig. 29.1. Trend of world production of apples since 1961 (FAOSTAT, 2007).

---

[1] Università di Firenze—Facoltà di Agraria, Dipartimento di Ortoflorofrutticoltura, Firenze, Italy.

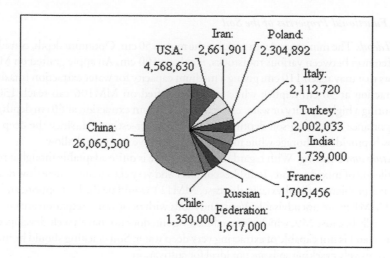

Fig. 29.2. The 10 leading countries producing apples (tonnes) in 2006 (FAOSTAT, 2007).

29.1.2. Rootstocks

Ninety percent of the modern apple production in Italy uses a single rootstock type: M9 (*Malus p. paradisiaca*). Other more vigorous rootstocks are useful only for weak or spur-type scion varieties, or where pedoclimatic conditions are challenging.

M9 was selected from a population of "Yellow Paradise of Metz" obtained at East Malling in 1914. Heat therapy was then used to eliminate viruses and the least vigorous clones were re-selected. M9 induces low vigour, requires staking or other support and is thus well adapted for high-density orchards. It produces few root suckers, is generally compatible with cultivated scion varieties, comes into production early, and has high productivity and efficiency of production. Fruit quality is high in terms of flavour, size and colour. M9 is damaged by temperatures below −20°C and is susceptible to *Agrobacterium tumefaciens* and to the wooly aphid though it is resistant to *Phytophthora*. Since it is a dwarfing rootstock, M9 has a relatively superficial root system with weak anchorage. Among virus-free clones used in Italy, there is the slightly more vigorous clone M9 EMLA, propagated in Emilia-Romagna. Other clones worth mentioning include Pajam 1, Pajam 2, NAKB 337, KL 29 and Burgmer 751.

M26 is best suited for use in new plantings of spur-type Red Delicious. It copes well with dry soils though it cannot tolerate waterlogged soils and is very susceptible to bacterial fire blight. M26 induces medium-low vigour, although even that is too much for modern orchard management from the ground. Among the still most commonly used rootstocks, there is MM106, well adapted for non-waterlogged soils of medium fertility (it suffers from root rot caused by *Phytophthora cactorum*) and noted for stimulating early fruit production. MM106 induces moderate vigour and is thus not compatible with modern needs. MM111 is suited to all soil types, but it induces high vigour. M7 is no longer used.

Recently introduced rootstocks, undergoing experimentation and widespread in Europe, include M9 T337 and M9 T339, obtained from clonal selections of "Yellow Paradise of Metz". They induce slightly less vigour than M9 EMLA and produce few root suckers. Fruiting and fruit quality are good (AA.VV., 1991, 1996; Nasi et al., 1994).

29.1.3. Functional Properties of the Soil

- *Soil depth.* The minimum depth useful to the roots is 50 cm. Optimum depth, considering differences between various rootstocks, is at least 100 cm. An apple grafted on M9 has roots that may reach 110 cm, giving a medium capacity for water extraction (maximum extraction at 50 cm depth); while an apple grafted on MM106 can reach 130 cm, ensuring a high capacity or water extraction (maximum extraction at 60 cm depth). The importance of rooting soil depth should not be over-estimated, since the deep roots mostly provide anchorage while most of the absorptive roots are shallow.

- *Texture and cracking.* With regard to soil, the apple is quite adaptable though it prefers fertile soil of mixed texture. It dislikes compact and very clayey soils where slow drainage creates problems of asphyxia; in this case, MM111 would be the best option, and M26 and MM106 are not advisable. Very sandy soils with poor water retention are not ideal, especially because M9, which gives reduced vigour, does not have a well-developed root system and is not capable of extracting very deep water. Soil cracking should be limited, as extensively cracking soils are not ideal for cultivation.

- *Sodium and salinity.* Soil salinity of around 0.4 dS/m or more causes tree suffering and fruit production decrease. Apple trees can tolerate exchangeable sodium percentage (ESP) up to 10%. Irrigation is indispensable in modern apple cultivation, and the quality of irrigation water is critical for good production.

- *Drainage.* Internal drainage, or the capacity to eliminate excess water, of an agricultural soil suitable for growing fruit trees depends on the permeability of the soil and its horizons, the depth of the water table, and other factors. For cultivation of apples, the best soils are those with adequate but not excessive drainage: these are soils of mixed texture, in which water is released readily but not too rapidly so that it is available for absorption by roots. Subterranean drainage permits the elimination of excess water in the soil, allowing deep root growth; this is much more important where the soil is compact and lacks gravel.

- *Waterlogging.* With respect to waterlogging, which arises when the level of water in the soil remains above the field capacity until it reaches saturation of the pores, the apple, when grafted on M9, is very susceptible. M27, M26, and MM106 do not tolerate clay and anoxic soils, the latter preferring soils not prone to waterlogging because it is susceptible to root rot caused by *Phytophthora cactorum*.

- *Risk of flooding.* Proneness to flooding represents the statistical probability of an area being covered by floods, within a defined time interval, due to overflow of a river or body of water. For the above-mentioned reasons, cultivation of apple in mountainous or flat terrain must be in areas with low or no risk of flooding.

- *Availability of oxygen.* As with cultivation of other fruit trees, the apple needs good availability of oxygen in the soil in order for respiration and root growth to take place. When volume of oxygen falls below 2%, plant growth ceases.

- *Soil pH.* With respect to pH, optimal values fall between 6.5 and 8.5. Soil values below pH 5.4 or above pH 8.5 are considered too extreme for apple cultivation.

- *Active lime content.* Active lime content expresses approximately the percentage by weight of finely subdivided and easily soluble carbonates. More precisely, it corresponds to the percentage of Ca^{++} ions that react with ammonium oxalate (determined with the

Drouineau-Gallet calcimeter method): active lime content is considered to be low when equal to or less than 0.5%, moderate at 0.5–10%, and high above 10%. Above the threshold of 10%, fixation of P often occurs and there is reduced availability of microelements (especially Fe, causing chlorosis). The apple is moderately resistant to active lime when grafted on M9, tolerating up to 10%, while notable resistance is obtained with MM106, which tolerates active lime up to 12–15% (AA.VV., 1991, 1996; Landi, 1999; Nasi et al., 1994)

- *Evaluation table.* A table of the soil limitations of apple cultivation is reported in Table 29.1, developed by I.TER (1998) as part of the project "Catalogo dei suoli della pianura emiliano romagnola" (Soil catalog of the Emilia-Romagna plain) funded by Servizio Sviluppo Sistema Agroalimentare (Service for Food and Agriculture System Development) of Emilia-Romagna, based on L.R.52/92 (regional policy 52/92) and revised by Ugo Palara (CISA M. Neri) and Carla Scotti (I.TER) in December 2005.

Table 29.1. Evaluation scheme of limiting properties of the soil for apple growth on different rootstocks.

Soil properties	Groups of rootstocks	Limitation		
		None or weak	Moderate	Strong
Available rooting depth (cm)	M9 and clones	> 100	50–100	< 50
Texture	M9 and clones	Moderately coarse, medium, moderately fine	Fine, coarse	–
Cracks	M9 and clones	Low	Medium	Strong
Electrical conductivity (salinity) (EC 1:5 dS/m)	M9 and clones	< 0.4	0.4–0.8	> 0.8
Exchangeable sodium percentage (ESP)	M9 and clones	< 8	8–10	> 10
Flooding risk	M9 and clones	None, rare	Occasional	Frequent
Flooding risk, duration	M9 and clones	Extremely brief, very brief	Brief	Long, very long
Available oxygen supply	M9 and clones	Good	Moderate	Imperfect, poor, very poor
Chemical response (pH in water)	M9 and clones	6.5–8.5	5.4–6.4	< 5.4; > 8.5
Content of exchangeable calcium (%)	M9 and clones	< 8	8-10	> 10

29.2. Pear (*Pyrus communis* L.)

29.2.1. Taxonomy and Range

Most of the cultivated pear varieties originated from *Pyrus communis* L. This species originated in Europe: its seeds, fruit remnants and woody remains have been found in ancient pile dwellings in Central Europe and much information exists in classical Greek and then Roman texts. In Europe the real development of pear cultivation dated back to the 1700s, following the impressive work of the "seeders" mostly in Belgium, France and England (AA.VV, 1991;

Bellini, 1993). In the past 40 years, pear production has increased notably (Fig. 29.3). In 2006, it was 19,539,533 t (FAOSTAT, 2007); the most relevant countries involved in pear production are reported in Fig. 29.4.

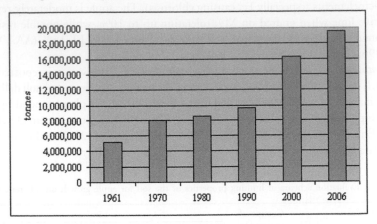

Fig. 29.3. Trend of world production of pears since 1961 (FAOSTAT, 2007).

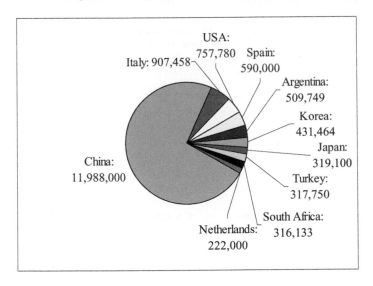

Fig. 29.4. The 10 leading countries producing pears (tonnes) in 2006 (FAOSTAT, 2007).

29.2.2. Rootstocks

Choice of rootstock is of paramount importance when planning a new orchard, allowing the possibility of helping a cultivar to be productive given the pedoclimatic conditions and chosen spacing and training systems. Pear rootstocks currently used in Italy can be grouped into seedlings (selections of *Pyrus communis*) and quinces (selections of *Cydonia oblonga*) (Bellini 1993; Bellini and Nin 1999; Nasi et al., 1994; Sansavini and Musacchi, 2000).

- *Pear seedlings and seedling selections.* The current trend for ever-higher orchard densities has increased the need for more dwarfing rootstocks. Consequently, the use of seedlings is decreasing since they tend to induce high vigour, slow entrance into production, smaller fruit size and lighter coloration. Seedlings are susceptible to the wooly aphid, fire blight, and *Agrobacterium tumefaciens*. Further, trees are not uniform, above all with respect to vigour and growth habit. Another problem with seedling rootstocks is the difficult propagation with classical nursery techniques, in fact micropropagation is often used. On the other hand, seedlings offer many advantages, including predictable affinity with most cultivars and high and constant productivity and hardiness, that is, tolerance of cold and torrid climates, and diverse soil types (seedling rootstock tolerates up to 10–12% of active lime). Among the clonally propagated seedling rootstocks are the American series OHF, derived by crossing *Old Home* x *Farmingdale,* known for their resistance to serious diseases (namely bacterial fire blight), the French Brossiér series, the Retuziére series, the BP and B series of South African origin, and lastly the Italian Fox (11 and 16) series, which was obtained from free pollination of two very hardy local cultivars of low vigour and easy propagation (Pera Volpina and Mora di Faenza). Lastly, new German selections appear interesting, most importantly Pyrodwarf rootstock, which has been introduced also in Italy.

- *Quince selections.* Both in Italy and in Europe, growers have been using quince rootstocks for several decades, so much so that almost all new pear orchards are grafted on moderately vigorous rootstocks such as BA29, the most vigorous and hardy, or intermediately vigorous, such as EM A and Sydo, which is increasingly used because of its tolerance to active lime relative to EM A, or weakly vigorous rootstocks such as MC and Adams, which are considered the only truly dwarfing quince rootstocks suitable for medium- to high-density orchards. Hitherto not frequently used are dwarfing seedlings and some insufficiently tested hybrids.

  Compared to the seedling rootstocks, the quince rootstocks induce lower vigour and have more superficial root system, which does not always guarantee adequate anchorage and can confer susceptibility to water and temperature stresses. Quince rootstocks have other problems, including serious ones, including rootstock disaffinity with some of the most common cultivars (Williams, Abate Fétel, and Kaiser), which can be accentuated in the presence of viruses and phytoplasm infections. Another problem of the quince rootstock is its poor performance in alkaline soils when active lime is greater than 4–5%, which causes leaf chlorosis. Quince rootstocks used in unsuitable soils or with incompatible scion cultivars can cause new orchards to fail within a few years.

- *Self-rooted pear.* An alternative to pear rootstock use is micropropagated self-rooted pear. Self-rooted William, Abate Fétel, Kaiser, and Max Red Bartlett are commercially available though these materials have not proven to be problem-free. The best cultivars of self-rooted pear have initial problems with anchorage, high vigour, and late entrance into fruit bearing and are able to reach their usual productivity only under particular environmental conditions. Management of self-rooted pear orchards, especially in fertile soils, can be difficult. However, self-rooted pear does have some advantages, such as no graft affinity problems, notable hardiness, good tolerance of active lime, and eventual high production.

29.2.3. Functional Properties of the Soil

- *Soil depth.* Usually pears prefer deep soils free of waterlogging to a depth of more than 1 m. The volume of soil occupied by roots depends on many factors, the most important of which is rootstock. Seedling rootstocks having deep taproots prefer deeper soils, while more shallow-rooted quince rootstocks adapt better to shallow soils (Table 29.2) (Bellini, 1993). The latter extract water mostly from the top layers of the soil and therefore are less tolerant of water stress.
- *Texture and cracking.* Pear is less demanding than other fruit tree crops. Research carried out in Emilia-Romagna suggests that all pear rootstocks have limited growth in soils with very sandy, sandy-loam, or coarse loamy-sand textures. Quinces appear more sensitive than seedlings and self-rooted pears to soil cracking in fine-textured soils.
- *Sodium and salinity.* Independent of the rootstock, pears do best with electrical conductivity of less than 0.4 dS/m and percentage of exchangeable sodium (ESP) less than 8%. Higher values reduce pear performance, becoming unacceptable with values greater than 0.8 dS/m and 10% ESP.
- *Risk of flooding.* Pears grow best in areas with low or no flooding risk (waterlogging events no greater than once every 10 years for less than 48 h). Areas with higher risk of flooding (duration of waterlogging greater than 7 d) are unsuitable for pear cultivation.
- *Availability of oxygen.* Pear roots, like those of other fruit trees, require moderate soil gas exchange to thrive. Soil oxygen content less than 2% results in interruption of plant growth; 10% of oxygen is the optimum value. Presence of the water table at less than 1 m of depth reduces oxygen availability to the roots and impairs growth. Oscillating depth of the water table is more detrimental than a permanent level just below the root zone: this can cause anoxic conditions, seriously damaging the roots and potentially killing the trees.
- *Soil pH.* Pears prefer soils with pH close to neutrality. With quince rootstocks BA 29, Sydo, and MC, optimal values are considered to be between 6.5 and 7.5. Self-rooted cultivars and seedling rootstocks, including seedling selections, do not show growth limitations up to pH 8. Beyond these limits, pear growth is increasingly impaired.
- *Active lime content.* Different pear rootstocks have varying tolerance to active lime in the soil, which is one of the causes of foliar chlorosis. Pear seedlings and quince CTS 212 are the most resistant, tolerating levels of active lime up to 12%. Among the other quinces, BA29 has intermediate tolerance (it is not suitable above 8% active lime) and MA, MC, and Sydo are more sensitive (values greater than 6% induce severe growth limitations).

Table 29.2. Depth and lateral expansion of roots of 8-year-old pears grafted on pear and on quince rootstocks.

Root expansion	Dr. J. Guyot		Conference
	Pear	Quince A	Quince A
Amount of roots in the layer:			
0–30 cm	32.4%	69.2%	71.3%
30–100 cm	47.3%	22.4%	18.3%
>100 cm	20.3%	8.4%	10.4%
Maximum root length (cm)	314	240	247
Mean lateral expansion of roots (cm)	335	315	290

Table 29.3. Evaluation scheme of limiting properties of the soil for pear growth on different rootstocks.

Soil properties	Groups of rootstocks	Evaluation		
		None or weak	Moderate	Strong
Available rooting depth (cm)	Pear, selections (OHF 40, OHF 69) and self-rooted, Quince BA 29, Sydo, Quince MC	> 100	50–100	< 50
Texture	Pear, selections (OHF 40, OHF 69) and self-rooted	Fine, medium, moderately fine, moderately coarse medium	–	Coarse
	Quince BA 29, Sydo, Quince MC	Moderately fine, fine, moderately coarse	fine	Coarse
Cracks	Pear, selections (OHF 40, OHF 69) and self-rooted	Low, medium, strong	–	–
	Quince BA 29, Sydo, Quince MC	low	Medium, strong	–
Electrical conductivity (salinity) (EC 1:5 dS/m)	Pear, selections (OHF 40, OHF 69) and self-rooted, Quince BA 29, Sydo, Quince MC	< 0.4	0.4–0.8	> 0.8
Exchangeable sodium percentage (ESP)	Pear, selections (OHF 40, OHF 69) and self-rooted, Quince BA 29, Sydo, Quince MC	< 8	8–10	> 10
Flooding risk	Pear, selections (OHF 40, OHF 69) and self-rooted, Quince BA 29, Sydo, Quince MC	None or rare	Occasional	Frequent
Flooding risk, duration	Pear, selections (OHF 40, OHF 69) and self-rooted, Quince BA 29, Sydo, Quince MC	Extremely brief, very brief	Brief	Long, very long
Available oxygen supply	Pear, selections (OHF 40, OHF 69) and self-rooted, Quince BA 29, Sydo, Quince MC	Good, moderate	–	Imperfect, poor, very poor
Chemical response (pH in water)	Pear, selections (OHF 40, OHF 69) and self-rooted	6.5–8	5.4–6.5; 8–8.8	< 5.4; > 8.8
	Quince BA 29, Sydo, Quince MC	6.5–7.5	5.4–6.5; 7.5–8.8	< 5.4; > 8.8
Content of exchangeable calcium (%)	Pear, selections (OHF 40, OHF 69) and self-rooted	< 10	10–12	> 12
	Quince BA 29, Sydo	< 5	5–8	> 8
	Quince MC	< 4	4–6	> 6

- *Iron deficiency and induced chlorosis.* The issue of foliar chlorosis due to iron deficiency deserves attention because it is common in pear-growing areas that often have considerable levels of lime. The main cause of chlorosis is the lack of assimilable iron uptake with consequent decrease in or total impairment of photosynthesis. Symptoms include foliar yellowing and decreased quality and quantity of yield. Excessive content of lime increases pH, causing iron to be in the non-assimilable oxidated form. Iron chlorosis can be avoided or limited by using resistant rootstocks (seedlings, OHF 40, OFH 69) and by cultural practices such as permanent green mulch and the use of soil amendments that increase soil organic matter by allowing iron chelation. Foliar sprays of iron chelate and other organic acid or synthetic compounds are also used (Bellini 1993; Landi, 1999; Nasi et al., 1994).

- *Evaluation table.* See Table 29.3 for a summary of soil limitation factors for pear growth on different rootstocks, compiled by I.TER (1998) as part of the "Coordinamento Settore Suolo" (Coordination of Soil Sector) project, funded by the Servizio Sviluppo Sistema Agro-alimentare (Service for Food and Agriculture System Development), based on L.R.28/98 (regional policy 28/98); the table has been validated by experts in pear cultivation (www.suolo.it).

References

AA. VV. 1991. *Frutticoltura speciale*. Reda, Roma.

AA.VV. 1996. Linee tecniche di produzione integrata. Melo. Suppl. Terra e Vita 34.

BELLINI, E. 1993. La coltivazione del pero. Edizioni L'Informatore Agrario, Verona.

BELLINI, E., NIN, S. 1999. La coltivazione del pero in Italia. Scelta del portinnesto e delle cultivar. L'Informatore Agrario 41.

FAOSTAT. 2007. Internet: http://www.fao.org

LANDI, R. 1999. *Agronomia e ambiente*. Edagricole, Bologna.

NASI, F., LAZZAROTO, R., GHISI, R. 1994. *Coltivazioni arboree*. Liviana Editrice, Padova.

REGIONE EMILIA ROMAGNA—I.TER. 1998. *Catalogo dei suoli della pianura emiliano romagnola*. Internet: http://www.gias.regione.emilia-romagna.it/suoli/guidaindice.asp

SANSAVINI, S., MUSACCHI, S. 2000. Nuovi impianti di pero: densità, portinnesti e forme di allevamento. Frutticoltura 9.

30. Kiwi Fruit [*Actinidia deliciosa* (Chev.) Liang & Ferguson]

Elvio Bellini,[1] Edgardo Giordani[1] and Daniele Morelli[1]

30.1. Taxonomy and Range

The kiwi [*Actinidia deliciosa* (Chev.) Liang & Ferguson and other species] is native to China, specifically to the valley of the Yang-Tze, where it is found in the wild. The most ancient descriptions date back to the Min dynasty (1200 BC) and kiwi was introduced in Europe towards the mid-1800s (Zuccherelli, 1994). In the past 20 years, kiwi production has increased (Fig. 30.1), reaching almost 1,200,000 t in 2006 (FAOSTAT, 2007); the main producing countries in 2006 are reported in Fig. 30.2.

30.2. Rootstocks

Kiwi can be propagated with self-rooted cuttings or micropropagated, or grafted. Self-rooted material is preferred in cold areas. If rooted cuttings are used, it is advisable to plant at least 2-year-old plants. In areas not subject to cold damage, it is possible to use rootstocks. The most

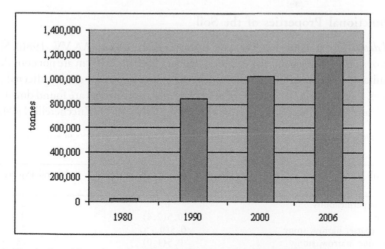

Fig. 30.1. Trend of world production of kiwi fruits since 1980 (FAOSTAT, 2007).

[1] Università di Firenze—Facoltà di Agraria, Dipartimento di Ortoflorofrutticoltura, Firenze, Italy.

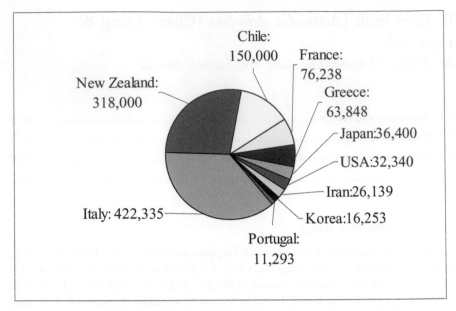

Fig. 30.2. The 10 leading countries producing kiwi fruits (tonnes) in 2006 (FAOSTAT, 2007).

commonly used rootstocks today are seedlings of the Bruno and Hayward varieties, and D1. Bruno seedlings are very common in New Zealand and in Italy and are resistant to chlorosis. Hayward seedlings are less common because they are more liable to show chlorosis. D1 is a clonal selection from Vitro Plant of Cesena. It has good agronomic qualities, allowing good vegetative homogeneity, good yield, and high tolerance to active lime (up to 8%) (Zuccherelli, 1994; Zuccherelli and Zuccherelli, 1987).

30.3. Functional Properties of the Soil

- *Soil depth.* Optimal soil depth for kiwi is greater than 100 cm (AA.VV., 1986). Shallow soils or soils with mechanical obstacles at depths of less than 50 cm are not considered to be suitable for kiwi cultivation. Root extension and depth are greatly affected by soil nature. Generally, in the absence of obstacles, most of the roots are found down to 1 m depth and in a radius of 2–3 m from the trunk (Table 30.1) (Zuccherelli, 1994).

Table 30.1. Depth and lateral expansion of the root system of adult female plants of kiwi fruit in different soil types.

Types of soil	Depth (m), maximum depth reached in parentheses	Lateral expansion (m)
Coarse, slow permeability	0.7	2.2
Sandy, deep	0.5(2.4)	2.0
Moderately coarse, narrow, stony	0.5(0.75)	2.5
Moderately fine, narrow, stony	0.5(1.0)	3
Moderately coarse, deep, scarcely stony	0.75(1.0)	1.5–3.0*
Moderately fine, deep	1.5(2.0)	1.8–2.8*

*Non-homogeneous distribution (first figure across the row, second figure along the row)

- *Texture and cracking.* Ideal soils are those with medium to moderately fine to moderately coarse texture, permeable with high water-holding capacity and rich in organic matter. Kiwi is adaptable to light and gravelly soils as long as they are abundantly irrigated and fertilized and contain organic matter. Kiwi adapts to fine soils as well, as long as they are well drained. The presence of gravel is considered advantageous up to 10–15% by volume enabling good water infiltration and drainage. Kiwi is rather sensitive to soil cracking, thus requiring soils with low levels of expandable clays and reduced tendency to cracking.
- *Sodium and salinity.* Optimal values for kiwi are below 0.4 dS/m for electric conductivity and below 8% for exchangeable sodium (ESP). At higher values, growth is increasingly impaired and soils with more than 0.8 dS/m and 10% ESP must be discarded. Given the high water demand of kiwis, it is very important to control the quality of irrigation water (Table 30.2). Water with high amounts of salts is not suitable, interfering with osmotic mechanisms involved in water absorption, and increasing symptoms of water stress. Additionally, Ca^{2+}, Fe^{2+}, $P_2O_4^{3-}$, and S^{2-} may precipitate, plugging irrigation pipes, sprinklers and drippers. Independent of the rootstock used, kiwi is very sensitive to waterlogging; therefore, it is of critical importance to evaluate waterlogging risk before planting. Only where waterlogging risk is absent or low and the duration of the water submersion is less than 4 h can kiwis be planted.

Table 30.2. Classification of water for irrigation in relation to the total electrical conductivity (salinity).

Water quality	Electric conductivity (dS/m)	Salt content (mg/l)	Salinity
Very good	< 0.25	< 160	Low
Good	0.25–0.75	160–480	Medium
Poor	0.75–2.25	480–1.470	High
Unsuitable	> 2.25	> 1.470	Very high

- *Availability of oxygen.* Kiwi roots are very demanding in terms of oxygen availability, indispensable for root respiration. Soils with high hydraulic conductivity in which water is promptly removed and no waterlogging occurs during the vegetative season are considered optimal. Air content of the soil, in volume, must be at least 10%.
- *Soil pH.* Kiwi prefers mildly acidic soils (6 < pH < 6.5). Growth and productivity can be moderately limited with values between 5.4 and 6 and between 6.5 and 8. When the pH values exceed 7.5–7.6, plants begin showing symptoms of chlorosis. These symptoms become more severe with increasing alkalinity. Soils with pH greater than 8 are not suitable.
- *Active lime content.* Kiwi tolerates active lime content of 6% when self-rooted and up to 8% when grafted on D1. Soils are considered optimal when active lime is below 4% (AA.VV., 1986; Landi, 1999; Zuccherelli, 1994; Zuccherelli and Zuccherelli, 1987).
- *Evaluation table.* See Table 30.3 for a summary of soil limitation factors for kiwi cultivation, compiled by I.TER (1998) project "Catalogo dei suoli della pianura emiliano romagnola" (Soil catalog of the Emilia-Romagna plain) funded by the Servizio Sviluppo Sistema Agro-alimentare (Service for Food and Agriculture System Development) of Emilia-Romagna, based on L.R.52/92 (regional policy 52/92).

Table 30.3. Evaluation of limiting soil properties for kiwi fruit growth on different rootstocks.

Soil properties	Groups of rootstocks	Limitation		
		None or weak	Moderate	Strong
Available rooting depth (cm)	Self-rooted, D1	> 100	50–100	< 50
Texture	Self-rooted, D1	Medium, moderately fine, moderately coarse	Coarse, fine	–
Cracks	Self-rooted, D1	Low	Medium	Strong
Electrical conductivity (salinity) (EC 1:5 dS/m)	Self-rooted, D1	< 0.4	0.4–0.8	> 0.8
Exchangeable sodium percentage (ESP)	Self-rooted, D1	< 8	8–10	> 10
Flooding risk	Self-rooted, D1	None or rare	Occasional	Frequent
Flooding risk, duration	Self-rooted, D1	Extremely brief	Very brief	Brief, medium, long, very long
Available oxygen supply	Self-rooted, D1	Good	Moderate	Imperfect, low, very low
Chemical response (pH in water)	Self-rooted, D1	6.0–6.5	5.4–5.9; 6.6–8.0	< 5.4; > 8.0
Content of exchangeable calcium (%)	Self-rooted, D1	< 4 < 6	4–6 6–8	> 6 > 8

References

AA. VV. 1986. *La coltura dell'actinidia*. Atti del convegno, Verona 29 aprile.

FAOSTAT. 2007. Internet: http://www.fao.org

LANDI, R. 1999. *Agronomia e ambiente*. Edagricole, Bologna.

REGIONE EMILIA ROMAGNA-I.TER. 1998. *Catalogo dei suoli della pianura emiliano romagnola*. Internet: http://www. gias.regione.emilia-romagna.it/suoli/guidaindice.asp

ZUCCHERELLI, G. 1994. *L'actinidia e i nuovi kiwi*. Frutticoltura moderna, Manuali tecnici Edagricole.

ZUCCHERELLI, G., ZUCCHERELLI, G. 1987. *L'actinidia pianta da frutto e da giardino*. Frutticoltura moderna, Manuali tecnici Edagricole.

31. Soil Knowledge Applied to the Cultivation of Arboreal Crops

Carla Scotti[1]

31.1. Introduction[2]

When a new arboreal plantation is planned, the soil is one of the factors to be taken into consideration. It is possible to gather some basic information from soil documentation, such as texture, active lime content, pH and problems of poor water drainage, which should be collected prior to any decision-making. In the divulgation methods project,[3] a work method was fine-tuned to create application maps deriving from Soil Maps on a scale of 1:250,000. The method used is based on the collaboration of experts in the fields to which the maps will be applied and on the definition of evaluation parameters correlating the edaphic requirements of the plants with the potential growth classes. In Emilia Romagna, this method was applied to the elaboration and updating of evaluation charts regarding the growth of the principal types of pear and peach rootstocks, used to produce the application charts deriving from the Plainlands Soil Maps on a scale of 1:50,000.[4]

Shown in the previous chapters (see Chapters 28, 29, 30) are the matching tables for the evaluation of pear, peach, apple, cherry, kiwi, plum, and apricot, deriving from the cooperation between the pedologists of I.TER, the research technicians of the CISA M. Neri of Imola and of the CRPV of Cesena, and experts in arboreal cultivations belonging to the technical assistance service for the agricultural businesses in the Region Emilia Romagna. In particular, the evaluation charts of pear and peach have been followed up by processes of advanced

---

[1] I.TER-Cooperative, Bologna, Italy.

[2] This text was written with the contribution of Andrea Giapponesi (Region Emilia Romagna), Giampaolo Sarno (Region Emilia Romagna), Ugo Palara (CISA M. Neri), Roberto Colombo (CISA M. Neri), Paola Pirazzini (CISA M. Neri), and Sandro Bolognesi (M. Marani Agronomy Research Center).

[3] Project "Methodological definition and expeditious updating of the 1:250,000 region soil maps and divulgation methods for data" (Resolution no. CARp990882, 15/9/99, by the Regional Council Emilia Romagna) with which the I.TER collaborated through a research agreement with the Geologic, Seismic and Soil Service.

[4] Soil map of the Emilia Romagna Plainlands, Geologic, Seismic and Soil Service, 1998, www.regione.emilia-romagna.it/cartpedo.

validation and sharing that has led to the production of application charts that are also used by both technicians and experts in the cultivation of pear and peach.

31.2. Matching Tables

The evaluation takes into consideration those characteristics of the soil capable of directly influencing plant growth and does not take into consideration climatic characteristics and those (usually external to the soil) that can influence management operations and risks of soil and environmental degeneration; furthermore, it does not consider the interaction of the characteristics examined. The evaluation constitutes a descriptive synthesis of the edaphic requirements of the plants and represents a "transparent" and "sharable" methodological instrument for the production of application maps deriving from Soil Maps.

Therefore, matching tables presented come directly from the updating and re-elaboration of those shown in the Catalogues of soil types for the Region Emilia Romagna[1] and are the results of an interdisciplinary research group whose members were:

- Andrea Giapponesi (Agriculture-foodstuffs System Development Service-RER): regional representative responsible for the project "Coordination of soil sector";[2]
- Giampaolo Sarno (Agriculture-foodstuffs System Development Service-RER): regional representative co-responsible for the project "Coordination of soil sector";
- Carla Scotti (I.TER): scientific technician responsible for the project "Coordination of soil sector" and pedologist of reference for the project "Revival of Peach Cultivation in Emilia Romagna";[3]
- Province coordinating technicians for technical assistance: Maurizio Fiorini and Guido Ghermandi for Bologna, Fausto Grimaldi for Ferrara, Gabriele Marani for Ravenna and Massimo Fornaciari for Modena;
- Agronomic experts in pear cultivation: Sandro Bolognesi, Stefano Vergnani, Stefano Bretta, and Sergio Crovini;
- Experts from the Research Center for Plant Production (CRPV): Giampaolo Mascalzoni;
- Experts from the CISA M Neri: Ugo Palara, Roberto Colombo and Paola Pirazzini.

The group based updating and re-elaboration of the charts on reference to experience carried out in the region in the fruit-producing sector and on suggestions found in literature. The threshold values were "fine-tuned" by visiting some arboreal plantations and observing both soil characteristics and the vegetative-productive state of the plants. In particular, the evaluation charts are the results of a process of active "sharing" and "validation" also through the creation of application maps; these maps derive from the application of the evaluation method to characteristics of soils described in the Region catalogue of the plainlands and its soil map on

---

[1] "Region catalogue of the principal agricultural soil types of hillsides and mountains", Agriculture-foodstuffs System Development Service-RER, Geologic, Seismic and Soil Service-RER, I.TER, 1996 and "Region catalogue of soils of the Emilia Romagna Plainlands", Agriculture-foodstuffs System Development Service, Geologic, Seismic and Soil Service, I.TER, 1998; www.ermesagricoltura.it.

[2] Project I.TER "Coordination of soil sector" financed by the Agriculture-foodstuffs System Development Service, RER abridged version of the Regional Law 28/98, 2000, 2001, 2002, 2003 and 2004.

[3] Project CRPV "Revival of Peach Cultivation in Emilia Romagna", Agriculture-foodstuffs System Development Service, RER abridged version 2004-2005 of Regional Law 28/29.

a scale of 1:50,000. Validation was carried out by means of the revision of the applied cartography by the group, through successive approximations based on a progressive sharpening of the interpretive outline, which established the relationships between pedological parameters and crop response. In fact, with the use of the application map some "dubious zones" were identified, i.e., zones in which experts knew of plantations that potentially gave crop response not in alignment with the information on the map. Through joint inspection the group discovered the critical aspects as they emerged in the representative orchards, opportunely circumstantiating the interpretive hypotheses to adjust the outcome. These moments of encounter in the field were found to be fundamental to reciprocal understanding between pedologists and agronomists, overcoming the obstacles determined by different terminology and disciplinary approaches. Thanks to these controls and other encounters for comparison, these evaluation charts were revised (method for future approximations), leading to the production of the chart and the final map that was "validated" by the experts (Fig. 31.1).

The evaluation charts include three classes of limitation intensity and are referred to soils managed according to sustainable agronomic criteria. The definition of the three classes is as follows:

- *Absent or slight limitations*: the soils do not present limitations or are suitable to hosting the rootstocks, favouring full achievement of the quality-quantity production potential; soils may be cultivated with ordinary techniques and do not require specific operations to improve their natural potential, besides those depending on eventual specific requirements of the individual rootstocks.

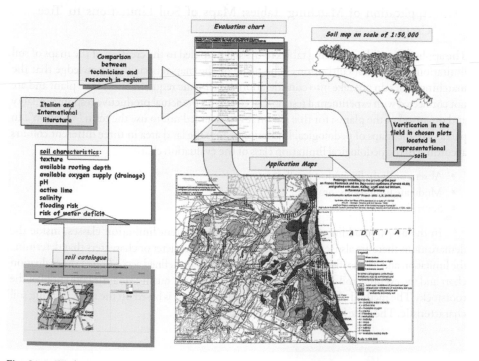

Fig. 31.1. Work methodology used.

- *Moderate limitations*: the soils present some limiting factors that require agronomic corrective operations to recover the full quality-quantity production potential that the rootstock can achieve.
- *Severe limitations*: the soils present factors that severely limit cultivation; eventual agronomic corrective operations may be too expensive or insufficient to recover the full quality-quantity potential that the rootstock can achieve.

To evaluate the degree of intensity of the limitations of the single pedological characteristics, the following depths were examined:

- 100 cm for texture of the most limiting horizon;
- 60 cm for chemical response (pH);
- 60 cm for active lime;
- 120 cm for salinity.

Threshold values of the soil characteristics attributed to the three classes of pedological limitations refer to the classes derived from the manual for the description of soils in the Region Emilia Romagna so that the outline can be applied to the characteristics of soils in the region (defined according to the aforesaid manual) and so that the application maps can be created. This aspect obviously required sharing of pedological terminology and adjustment of its parameters. In the appendix of this volume are described the pedological characteristics taken into consideration in the evaluation maps.

31.3. Application of Matching Tables: Maps of Soil Limitations to Tree Crops

The application of the matching tables to the soil maps led to the creation of the maps of soil limitations to arboreal cultivations. The name of the map reflects the knowledge that the matching tables do not take into consideration the climatic requirements of the plant and are not connected with experimental testing that correlates the actual productive response (quality and quantity) of the plants; for this reason it was decided not to use the term "vocation". In particular, the maps of pedological limitations show the land area in three different colours according to the pedological limitation class of the evaluation chart:

- Absent or slight limitations: yellow;
- Moderate limitations: orange;
- Severe limitations: red.

In the maps, in addition to the colours of the different limitation classes, inside the demarcations there are also letters representing the soil character or characters that determine the limitation and its classification. The limitation classes are defined according to the evaluation chart and the eventual soil characters that may influence the quality-quantity potential of the rootstocks. The method of attribution of the limitation class is based on the most limiting soil characteristic. The code letters used are in Table 31.1.

Table 31.1. Code used for the limiting factors.

Code	Soil limitations
c	Active lime
d	Available oxygen
f	Cracks
i	Flooding risk
o	Sodicity
r	Chemical response
s	Salinity
t	Texture
u	Available rooting depth

Tables outlining the parameters used for the soils present in the Soil Map on a scale of 1:50,000 (Fig. 31.2) were created to make visible and more easily understandable the method used for the evaluation chart and therefore the assignment of each soil type to its limitation class. In particular, the chemical response (pH) was not taken into consideration for the assignment of the classes; the interdisciplinary work group decided that the typical pH values of the soils planted with pear and peach in the plains of Emilia Romagna, averaging between 7.8 and 8.2, are easy to correct with ordinary operations of fertilizing with iron chelate.

Soil	Available rooting depth (cm)	cracks	Available oxygen supply	salinity (EC 1:5 dS/m)	(ESP)	Texture_lim100	Flooding risk	Content of exchangeable calcium (%) Medium	Content of exchangeable calcium (%) Dev.st andard	pear, selections (OHF 40, OHF 69) and self-rooted	Quince BA29
AGO1	50-100 cm	low	imperfect	>2 dS/m)	<8	coarse; moderately coarse	none	0.9	0.7	3 d-s-t	3 d-s-t
BTR1	50-100 cm	medium	moderate	<0,15 dS/m	<8	moderately fine, fine	none	9.7	0.6	2 u	3 c
CTL1	>150 cm	low	good	<0,15 dS/m	<8	medium moderately fine	none	1.5	0.7	none	none

Fig. 31.2. Soil characteristics present in the Soil Map on a scale of 1:50,000.

In the case of delineations of the Soil Map characterized by the simultaneous presence of soils with different plant growth potential, it was decided to use the following colour and codes: "background colour" in which the primary colour is related to the quality-quantity potential class attributed to the principal soil and the secondary colour is related to the potential class attributed to the secondary soil; "composite codes", for example, "d/c" signifying available oxygen supply that limits the principal soil type and active lime that limits the secondary soil type (Figs. 31.3 and 31.4).

31.4. Orientation Inferred from the Tables and Resulting Soil Maps

The application maps constitute an orientation in the choice of the rootstock most suitable to the soil present in the lot; it is, however, always essential to verify the characteristics of the soils in the field by soil coring. In fact, in the field soil samples removed with a drill may be studied up to a depth of 120 cm through simple evaluations that principally regard:

- Texture, i.e., the content of sand, silt and clay: through manipulation of a mixture of soil with water, a trained technician can determine the texture with a 5% margin of error (therefore, totally marginal to a correct agronomical evaluation).
- The presence of eventual problems of poor drainage: by verifying the presence or absence of grey colours coupled with reddish colours in the soil sample, it is possible to identify the presence or absence of this type of problem, because in conditions of anaerobiosis due to excess water, iron assumes grayish colorations.
- Total calcium content: it is possible to determine the total calcium content based on the degree of reaction (effervescence from absent to violent) when pouring a few drops of a 10% hydrochloric acid solution on a soil sample.
- pH: by pouring a few drops of universal indicator on the soil sample placed in a porcelain tub, it is possible to determine the pH according to the chromatic variation of the indicator.

The evaluation charts can be improved by involving other experts and by visiting a larger number of plantations and, most of all, by maintaining the exchange of information between researchers so that new experimentation is carried out in soils that are known and representational.

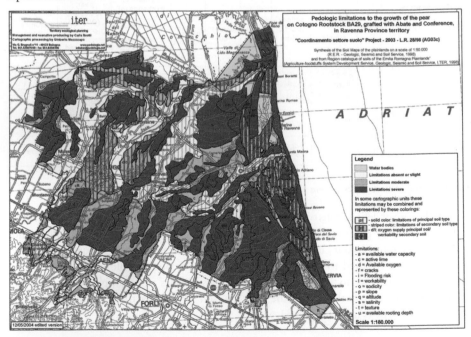

Fig. 31.3. Map of pedological limitations to growth of pear on quince rootstock BA29, Sydo.

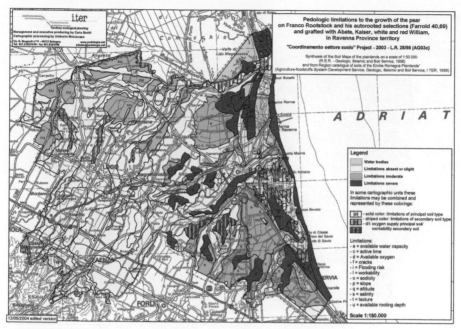

Fig. 31.4. Map of pedological limitations to growth of pear on free rootstock.

31.5. Supporting Maps

In Emilia Romagna, the work group decided that the application charts should be accompanied by information regarding practical advice for agronomic management according to the principal pedological environments. For this reason the "Land Map" was created, illustrating the principal pedological environments of the territories of the province (see, as an example, Fig. 31.5). The "Land Map" represents a synthesis of the Soil Maps of the plainlands on a scale of 1:50,000 in which were grouped environments and soils that have a similar behaviour regarding soil capability for the growth of the various rootstocks taken into consideration. It also contains a synthetic caption that describes a first analysis of the considerations on the agronomic management of the soils as a function of the rootstocks. In fact, the purpose of the map is to highlight the pedological differences, and the consequent agronomic operations, existing in different environments that fall into the same limitation class. For example, in well-drained soils of a medium texture, in which there are no pedological limitations for the vegetation response of the plants, it may be necessary to use cultivation techniques that can restrain vegetative vigour, because the plant tends to grow too much to the detriment of fruit production; on the other hand, in moderately fine soils, even though still considered free of limitations, there is a more balanced growth response between vegetation production and fruit production. The creation of the "Land Map" follows the same process of successive approximations and the caption represents the initial work stage.

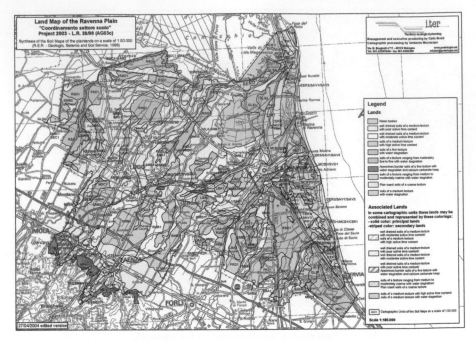

Fig. 31.5. Land Map of the Ravenna plain.

31.6. Discussion of Results

In synthesis, the work process used was the following:

- The soil technicians were aware that a definition of useful evaluation charts is successful only if there is a joint effort of those who produce Soil Maps and those who use them: this leads to the participation, besides that of pedologists, also of cooperating experts, such as technicians specializing in the specific production sector and researchers.
- In order to "share" and "validate" the evaluation chart, it is essential to supply "successive approximations" of prototypes for application "Land Maps" that are subjected to verification by technicians of the potential users each time.

It is important to specify that the interdisciplinary work group made it possible:

- to share experience and knowledge, integrating information about soils and the response of the production plantations;
- to share the evaluation charts that correlate the soil characteristics with the production response (in terms of quality and quantity);
- to test and then to validate the evaluation chart and the corresponding final application map;
- to make of the use of soil maps easier to understand and therefore to promote their use; in fact, since the principal users of the charts and the maps derived were involved in the process and recognize them as their own, they may not only use them themselves, but also share them with their colleagues in the sector.

The results of the work group have allowed the identification, by territorial areas, of the growth and production response potential of pear and peach as a function of the principal rootstocks and varieties used in the region. This allows us:

- to identify the soil characteristics that most influence the growth of peach (for example, active lime);
- to create new plantations in optimal pedological conditions and with the use of suitable rootstocks;
- to single out preferential areas for the cultivation of pear and peach and therefore contribute to the improvement of the territory;
- to contribute to the strategic planning on a large-scale production level (for example, planning of the choice of rootstock or variety as a function of the pedological environments).

Last of all, it should be said that the principal objective of these application maps is that of making information readily available for use, to allow the optimization of entrepreneurial decisions in the agricultural businesses. It is therefore a useful tool that suggests technical advice without imposing limits or obligations.

Examples of application maps produced may be seen at INFOSUOLO (www.suolo.it) in the section "Carte applicative dei Suoli a supporto dell'agricoltura". Other reference sites are: www.ermesagricoltura.it and www.regione.emilia-romagna.it/cartpedo.

References

A.A. V.V. 1996. Catalogo regionale dei principali tipi di suolo agricoli della collina e montagna, Regione Emilia Romagna (Servizio Sviluppo Sistema Agro-alimentare, Servizio Geologico, Sismico e dei Suoli), I.TER.

A.A. V.V. 2005. Catalogo regionale dei suoli della pianura emiliano romagnola, (www.ermeseagricoltura.it sezione suoli), Regione Emilia Romagna (Servizio Sviluppo Sistema Agro-alimentare, Servizio Geologico, Sismico e dei Suoli), I.TER.

A.A. V.V. 2005. Carta dei suoli della pianura emiliano romagnola in scala 1:50.000, www.regione.emilia-romagna/cartpedo, Regione Emilia Romagna (Servizio Geologico, Sismico e dei Suoli).

Scotti C. 2001. Conoscere il suolo, Il divulgatore 8–9 Agosto-Settembre 2001.

Scotti C. 2005. I portinnesti delle piante da frutto, Notiziario Centro Ricerche Produzioni Vegetali di Cesena 71 Novembre 2005.

PART III

LAND SUITABILITY AND LAND ZONING

Land suitability for forestry and grazing

32. Land Evaluation for some Important Italian Conifers

Andrea Giordano[1]

[1] Università di Torino—Facoltà di Agraria, Dipartimento di Economia e Ingegneria Agraria Forestale e Ambientale, Torino, Italy.

32.1. Introduction

Indications concerning the requirements of silver fir, Norway spruce, European larch and Scotch pine when used for afforestation or plantations may be inferred observing the plant formations around or near the considered site. However, generalizations on the presence of single species must be avoided because a given species may have a different behaviour, even with the same climate, according to the edaphic conditions or, on the other hand, on the same soil the plants may present dissimilar development due to bioclimatic differences.

In the scientific literature there are a number of syntheses on the climatic and edaphic requirements of forestry species but the inherent degree of generalization does not always fit the local situation. Such generalization is bound to the concept of orthotype that, especially at the boundary of a climax area, does not seem to be appropriate. A well-known example is provided by the presence of silver fir and beech in humid sites located far from the typical climax area of these two species.

Being aware of the difficulties, if not of the impossibility, of identifying the "orthotype" condition, the U.S. Forestry Service evaluates the land suitability for the different forestry species by attributing them to the land types (or land units). To each cartographic unit is assigned a given number of forestry species, selected among the existing ones or among those belonging to the potential vegetation: These species are ranked in three classes: desirable, acceptable and less desirable. The judgement is formulated on the base of the site index, i.e., the height of the dominant or codominant trees at the reference age. Examples of use of land units in the land evaluation for forestry are found in Smalley (1982, 1991); Booth and Saunders (1985) and Wenger (1998). The site level is foreseen for larger scales (Barnes et al., 1982; Steele et al., 1983). The described approach is close to that of the "land capability": each portion of the terrestrial surface is evaluated to verify the compatibility with groups of sustainable uses.

FAO (1984) presents four main groups of forestry sustainable uses:

- for growth
- for management
- for conservation
- for recreation

The purpose of this research is the definition of the land and soil characteristics and qualities that determine the success of forestry species. This approach corresponds to the purpose of the land suitability for specific use and in particular to the suitability for forestry species growth. Among the several examples of land suitability to forestry species here are mentioned Baker and Broodfoot (1979); Sindou and Madesclaire (1986); Becker and Levy (1988) and Giordano et al. (1988).

The environmental factors important for tree growth have been critically revised by Dotta and Motta (2000), IPLA (200)1, ERSAL (2001) and Giordano (2002).

In the light of the concepts and methodologies mentioned, an attempt has been made to define, even in a general way, the requirements of some single forestry species. They are discussed from the point of view of the autoecology, keeping in mind that in many cases the synecology is as important as the autoecology, if not more.

32.2 Silver Fir (*Abies* spp.)

32.2.1. Systematic Framework and Main Botanical Characteristics

The genus *Abies* is present throughout the world with more than 50 species often interfertile. Only two species are found in the boreal forest: *Abies sibirica* Lebed. in Russia and *Abies balsamea* Mill. in North America. In other regions, *Abies* is typically mountainous with a temperate climate, alternate periods of physiological activity and rest and absence of dry conditions. In Europe and especially in the Mediterranean basin, there are 11 species: only one (*Abies alba* Mill.) for the centre and southern centre, 5 for the south (*nordmanniana*, *Boris-regis, cephalonica, nebrodensis* and *pinsapo*), 3 in in the Near East (*cilicica, bornmulleriana*, and *equi-troiana*) and 3 in North Africa (*numidica, marocana* and *tazaotana*).

The domain of silver fir in Europe and in the Mediterranean basin is entirely mountainous (Pyrenees, Central Massif, Jura, Black Forest, Apine Chain, Vosges, Dinaric Alps, Carpathians, Corsica and Apennines).

In the post-glacial era, the silver fir spread northward. The Apennines range then played a very significant role for all of Europe. Afterwards, a withdrawal period occurred (2000–600 years ago) to the advantage of Norway spruce and beech. In the Alps, the silver fir is present in Trentino, Asiago high plateau, Cadore and Alps Carniche. In the rest of the Alps, it is discontinuous, being mixed with Norway spruce in the centre and with European larch in the western part. In the Apennines, the silver fir stands are dispersed and sometimes have been protected and expanded by monasteries (e.g., Vallombrosa in Tuscany) and/or by governments. The most important sub-natural stands are those of Abetone (Foce di Campolino) and Foreste Casentinesi. The Calabrian stands are particularly important for their surface and good behaviour.

Silver fir is a tree of prime size and can reach 40–50 m in height. The stem is straight, slightly tapering. With age the top becomes flat and is shaped into a "stork nest". The bark is whitish or ashy with silver shading on the side facing the sun, where lichen colonies develop (Pavari, 1952).

The branches are verticillate and tend to be horizontal. The needles are persistent (lasting even 10 years) and dark green with two silver lines beneath. There is a difference between the light needles that are curled and rigid and the dark needles that are more flexible. The yellowish male flowers are located in the low part of the foliage and the green female flowers are concentrated in the upper part. The pollination is anemophilous; in certain years it is particularly evident and is known as "sulphur rain". The cones stand erect, they are cylindrical, up to 18 cm long, green turning to brown before disarticulation. The root system is strong and made up of several roots going deep into the soil, allowing the plant to profit from a considerable volume of earth and at the same time to resist the wind.

32.2.2. Varieties

The silver fir was long believed to be a stable species, i.e., lacking in geographical races that could be differentiated. Trials involving silver fir of southern and northern origin (Pavari, 1951) have proved that the growth period of the southern fir is longer than that of the northern one. This and other differences induced Giacobbe (1973) to propose the Apennine variety for southern Italy. Further researches suggested that from Molise southward, rather than a change in characteristics, there is an increase in the population variability that demonstrates genetic richness (Ciampi et al., 1973, quoted in Bernetti, 1995).

Trials on 38 different Middle-Europe silver fir materials have shown that the elements of southern origin are more resistant because they are genetically richer (Larsen, 1986 in Bernetti, 1995).

Undoubted facts related to silver fir, besides the genetic richness of the southern fir, are:

- existence of a geographic particularity of the Pyrenees populations,
- existence of a continuous variation of the populations of central Italy northward and southward, due to the selection induced by climate.

32.2.3. Regeneration

Silver fir regenerates only by generative way. Cone production begins between 30 and 60 years of age. The cones reach maturity during the summer and disarticulate in the period from September 15 to October 15. For nurseries, the seeds must be collected when they are still inserted into the cones. These are exposed to air and then the seeds are extracted by hand or by hulling machine with gentle strokes. The seeds cannot be preserved for long and must be used within the spring following the harvest. Nevertheless, the seeds may be preserved under vacuum at −10 or −15°C. The germinability is inferior to 50%. In the nursery each square metre receives 8–10 hg of seeds, from which may arise approximately 800 seedlings that need to be shaded. At transplanting there are around 100 seedlings. Seedlings are planted 2+2 with naked roots.

32.2.4. Pedoclimatic Requirements

Silver fir, like the Norway spruce, resists cold but, unlike the latter, needs more heat. As a consequence, it is found in sites similar to those of the beech but with less atmospheric humidity: as a matter of fact in the Alps it penetrates in some internal valleys too continental for the beech. On the Apennines and in southern Italy, it seems to be more thermophilous than the beech, which occupies the crests; silver fir grows at a lower altitude, mixing with chestnut and Turkey oak. Silver fir has a high water requirement that is satisfied partly by rainfall and partly by the root system, which is able to exploit a large volume of soil. During dry years, the growth is consistently reduced (Corona, 1983 and Spiecker, 1986, quoted in Bernetti, 1995).

Silver fir is indifferent to the lithology as well as the pH, which becomes a limiting factor only in case of very acid soils, such as those of some heathlands. The edaphic parameters especially important for silver fir are the following (Duchaufour, 1988):

- depth
- moisture but not water stagnation
- nitrogen especially in ammonium form
- aluminium absence
- type of humus (mull-moder or moder)

It must be noted that the humus of mull type is not satisfactory because it enhances nitrophilous weeds and pathological agents

A good silver fir regeneration is ensured by mineral soils when horizons rich in un-decomposed organic matter are unfavourable for the following reasons (Mancini, 1959 and Burschel, 1964):

- young plants may easily dry out,
- water absorption from the inner layers is moderate,
- seed adherence to the organic matter is imperfect.

Regeneration is hampered by the presence of *Rubus idaeus, Rubus hirsutum, Epilobium angustifolius* and *Senecio fuchsii*. It is good when, beside a favourable climate, silver fir is mixed with beech and when there are other species able to produce a litter easily decomposable (maple, hazelnut, sorb and *Cytisus laburnum*).

In the Piedmont Region, stands of silver fir are subdivided into five types characterized by different edaphic conditions (IPLA, 1996):

- eutrophic on Eutrodepts and sometimes on Mollisols
- mesotrophic on Dystric Eutrodepts
- oligotrophic on Dystrudepts
- high mountains with megaforbs on Inceptisols and Haplorthods in a few places
- endalpic with Norway spruce on Inceptisols of different suborders according to the lithology

32.2.5. Cultural Techniques

Without entering into the details of the cultural techniques (see Bernetti, 1995), only the two most important are here mentioned.

Felling and regeneration distributed over the whole area (selection system).

This system is especially applied for management of woods where silver fir is mixed with other species. The system is designed to obtain an uneven wood structure with high yield. Silver fir meets this objective very well because it has the tendency to regenerate by small dispersed patches. It is tolerant of shade and can easily live with pioneer species (larch, Scotch pine) as well as with more demanding species (maple, ash, sorb) that ameliorate the humus composition.

Old crop cleared by single felling (clear-cutting system).

The result is a wood of even age, usually obtained with artificial regeneration. In Italy, the clear-cutting system has been applied only to forests belonging to ancient monasteries (Vallombrosa, Camaldoli) or to ancient State lands (Abetone, Foreste Casentinesi). The result is a high and uniform production of wood. New management concepts have since arisen: on the one hand natural regeneration and on the other landscape protection. In case of preservation of the clear-cutting system, the suggested spacing is 2×2 m or slightly more (Paganucci, 1989 in Bernetti, 1995).

32.2.6. Adversities

Among the harmful insects must be mentioned those that eat the bark and aphids that installed on the needles are responsible for the sweet dew that weakens the plants, depriving them of sugar and liquid. Among the fungi have to be mentioned those responsible for the dumping off disease (*Armillaria mellea* and *Heterobasidion annosum*).

The German term *Tannensterben* is used to indicate a set of pathological symptoms of unclear origin. The main symptoms are early needle drop, colour change of needles, liquid flow along the stem, rotten heart wood, loss of fine roots and predisposition to fungal attack. Hypotheses about the origin of this new disease range from atmospheric pollution to climatic global changes.

Consistent damage may be done by ungulates that eat the tender buds. Considering all the environmental and ecological aspects, an evaluation of the wildlife carrying capacity seems absolutely necessary.

32.2.7. Land Suitability for Silver Fir

Since silver fir is unlikely to be present in artificial plantations, at least in Italian environments, land suitability is evaluated from a naturalistic point of view without heavy human interventions.

In Table 32.1, the main environmental factors are evaluated in terms of suitability for silver fir. Table 32.2 is devoted to the pedological and pedo-climatic parameters. The data of the tables are mainly qualitative and must be intended as having a large range of discretion. Such data must periodically be revised following up-to-date scientific researches.

Table 32.1. Silver fir climatic and geomorphological requirements.

Climate	Luminosity	Sciaphilous species
	Atmospheric humidity	High but not exceeding 85%
	Precipitation	1000–1800 mm
	Wind	Resistant
	Summer drought	Not resistant
	Summer heat	Fairly demanding
	Spring frost	Fairly resistant
Phytoclimatic belt	*Fagetum*, sub-zone warm and cold. *Picetum*, sub-zone cold.	
	Castanetum, cold belonging to type 1 (precipitation > 700 mm)	
Geomorphology	Parent material	Any
	Stone and boulder deposits	Not suitable
	Altitude	1000–1600 m

Table 32.2. Silver fir edaphic characteristics.

Soil moisture regime	Udic
Soil temperature regime	Mesic and frigid
Useful depth	Moderate
Rock fragments	Common
Texture	All textures except clay
Macroporosity	High
Drainage	Good
pH	Acid to slightly alkaline
Active calcium carbonate	Absent to slightly present
Chemical fertility	Fairly high
Sensitivity to compaction	High
Humus	Mull-moder

32.2.8. Soil Functional Characteristics.

Soil functionality under silver fir canopy must satisfy two conditions (Table 32.2):
- it must guarantee soil moisture conditions fitting the udic regime,
- it must maintain a reasonably good soil fertility to allow balanced development of the seedlings and of the herbaceous vegetation.

The parameters presented in Tables 32.1 and 32.2 are necessary for the assessment of the following qualitative functions:

atmospheric humidity,
sensitivity to frost,
sensitivity to drought,
sensitivity to wind,
possibility of growth on the crest,
possibility of growth in the depressions,
possibility of growth on the stony colluvium.

Other parameters can be evaluated in a quantitative way:

- useful depth cm (total soil depth—% volume particles > 2 mm)
  high   > 75
  moderate   25–75
  low   < 25
- % particles diameter > 2 mm
  high   > 50
  moderate   20–50
  low   < 20
- macroporosity % (empty spaces visible)
  high   > 5
  moderate   1–5
  low   < 1
- internal drainage cm/h
  rapid   > 12.5
  moderately rapid 6.0–12.5
  moderate   2.0–6.0
  moderately slow 0.5–2.0
  slow   0.1–0.5
  very slow   < 0.1
- chemical fertility (% bases saturation on the total cation exchange capacity, as proposed by Varley in Landon, 1984)
  high   > 60
  moderate   20–60
  low   < 20
- sensitivity to compaction (bulk density g/cm³, as proposed by Taylor, 1966)
  high   < 1.3
  moderate   1.3–1.6
  low   > 1.6

32.3. Norway Spruce (*Picea abies* L. Karst)

32.3.1. Systematic Framework and Main Botanical Characteristics

Norway spruce (*Picea abies* L. Karst., sometimes called *Picea excelsa*; Pignatti, 1982) is the most important forestry species in the world. Its importance is mainly due to three factors:

- the size of its area of distribution,
- the uses of its wood,
- its spread in artificial plantations (Bernetti, 2005).

The genus *Picea* comprises 36 species (Schmidt-Vogt, 1977). In the boreal belt, there are only a few species: *P. abies* in Eurasia and *P. glauca* and *P. mariana* in North America. Towards the South, the number of species increases but at the same time their areas become smaller. The maximum concentration with 20 species occurs along the Chinese border of Himalaya. The most distant species are *Picea abies* and *Picea Omorika*, which cannot generate any hybrids. The American *Piceae* are widely cultivated in Europe for different purposes: *P. sitchensis* for ligniculture in countries characterized by Atlantic climate, *P. glauca*, *P. mariana* and *P. pungens* as ornamental species in gardens and parks.

The huge area of Norway spruce can be subdivided into two sub-areas: one is the Baltic Siberian that goes from the Pacific Ocean to Fennoscandia, the other is the central European area with several mountain ranges (Sudetes, Carpathians, Transylvanian Alps, Rodope mountains and Dinaric Alps).

The Italian area of Norway spruce is quite extensive on the Alps, where the *Picetum* phyto-climatic belt is found in pure or mixed stands. Norway spruce may also be present in the lower phyto-climatic belts of the *Fagetum* and *Castanetum* and in the higher belt of the *Larici-Cembretum*. Two interesting facts pertain to Norway spruce in Italy: the first is its scarcity in the western Alps, especially in the Cozie and Maritimes, because of the influence of the Atlantic climate, which is more suitable to silver fir; the second is the presence of remaining areas of Norway spruce in the northern Apennines, where the most important site is Alpe delle Tre Potenze across the provinces of Pistoia, Lucca and Modena.

Norway spruce may reach a height of 60 m, the stem is straight, the root system is superficial, and the shape of branches is usually conical. The branch shape may be of three different types (Bernetti, 2003): going up like a brush, flat or going down. The leaves are persistent, needle-shaped, acuminate; the verticils, 15–25 mm long, are inserted around the twigs. Norway spruce has male flowers as pink-yellow amenta and female flowers as purple amenta from which hanging strobili originate; they are 7–15 cm long without desquamation at maturity and they are yellow-brown. The reddish-brown bark can be detached in more or less thick facets. The undifferentiated resinous wood is particularly valued for its regularity and behaviour when used for construction and carpentry (Giordano and Passet-Gros, 1962).

32.3.2. Varieties

Norway spruce has a notable polymorphism that can be seen in the shape of cone scales, the secondary branches, the cone colours, the twig tomentosity, and the length and density of the needles. The main Norway spruce varieties are based on the shape and scale of the cones. The recognized varieties are *acuminata, Europea, montana, rotundata, fennica* and *obovata*.

In the Alps, all the varieties coexist but *acuminata* and *montana* are scarcely represented. The distribution of the characteristics quite closely follows the altimetry: at high altitudes the cone scales are ovate, the foliage is close to avoid damage from the snow, the secondary branches are complanate and the cones are reddish. At low altitudes the foliage is open and the secondary branches are hanging. The ecotypes of the high altitude when cultivated at low altitude undergo a fast vegetative phase with the risk of late frost. On the other hand, low-altitude ecotypes when cultivated at high altitude have the vegetation phase deleted but more prolonged during the summer (Bernetti, 1995). The quality of the wood does not seem to be related to a precise variety. As an example, the characteristic "resonance wood" is not inherited but it is strongly associated with the sites where the resonance characteristic is adsorbed in the internal variability of the populations.

32.3.3. Regeneration

The strong genetic and morphological variability allows the selection of many cultivars (suitable for wood, ornamental trees and Christmas trees) propagated by graft or even by scion with special techniques (Bernetti, 2005).

The years rich in seed have different frequency according to the altitude: every 2 years in the mountain belt, every 10 years in the upper belt corresponding to the limit of trees. The seed are normally preserved for 2 or 3 years. Seeds may be better conserved at 1°C. The sprouting rate is high: 75–80%. Seeding in the nursery is done during the spring (100 g/m²), to prevent damping-off it is suggested that seeding be done on a sand bed. The 2+2 transplants give the best results. The seedling can be transplanted with the soil pan around the roots or with the naked root. The use of pots is recommended at high altitudes where, due to the local climate, the seedling must be transplanted during the summer.

32.3.4. Pedoclimatic Requirements

Norway spruce is a shade-tolerant, mesophytic, microthermal plant, sensitive to aridity and to wind. It is highly resistant to cold. At high altitude the limiting factor is the shortness of the vegetative period rather than the low temperatures. The relationships between Norway spruce and water are delicate because this conifer demands a large amount of water but its root system, being superficial, is helped in absorbing water by the dense foliage during rainfalls. The aridity limit corresponds to less than 300 mm of rain during the summer.

Norway spruce grows on several different types of soil provided they are not clayey, compact, impermeable and scarcely developed as gravely or stony. The pH may range from 4 to 7, and active calcium carbonate must be absent. Owing to its superficial root system, Norway spruce adjusts itself to shallow soils provided the summer is not arid. Norway spruce likes acid soils where the competition of other plant species is reduced; it can withstand peaty and swampy soils.

As for regeneration, mull humus is not favourable because of the strong herbaceous competition and the presence of harmful fungi. Mor and moder humus are favourable if they are not too thick and if they are not covered by litter (Piussi, 1965). It has been noted that mor implies scarcity of N, P, and Mg.

Tree stumps and big clods that upset when windthrow occurs are preferred regeneration sites for the following reasons (Piussi, 1986):

- good water balance,
- elimination of the competitive vegetation,
- reduction of the snow period,
- high C/N ratio favourable to mycorrhizal activity.

On podzolic soils, the regeneration of Norway spruce is difficult because the young seedlings must pass through the row humus to reach the E horizon, which in turn is deprived of nutrients. For these reasons, Norway spruce prefers the mineral soils where it prevails over the larch (Mancini, 1959). Old stands of Norway spruce may take advantage of the water retained by the cemented spodic horizon (Giordano, 2002). As a matter of fact, many of the successful stands of the German Black Forest are growing on soils with spodic horizon (*orthstein*).

On the Italian mountains the types of Norway spruce are fundamentally the sub-alpine and the mountainous. The former are mostly on acid or very acid soils, the moisture regime is always udic, and the regeneration is slow. The maximum height is 30 m. The botanical species of this environment are: *Vaccinium myrtillus*, *Luzula* sp., *Avenella flexuosa*, *Melampyrum sylvaticum* and *Oxalis acetosella*. The mountainous stands may be subdivided into acidophilous, xeric and eutrophic. In this last case, the height can reach 40 m but the spruce has to compete with silver fir and beech. The most representative species are *Cardamine, Sanicula europea* and *Galium odoratum*.

Everywhere on the Italian alpine range and on silicate rocks with continental climate, stands of Norway spruce tend to accumulate row humus at the surface with two consequences: (1) sharp separation between organic and mineral fraction and then soil drought during the summer and (2) the removal of bases, iron and aluminium from the top layers and the deposition with the help of the organic acids contained in the row humus of these substances in the deep layers, which become impermeable, creating favourable conditions for blueberry and rhododendron.

32.3.5. Cultural Techniques

Norway spruce is considered, especially in middle Europe, a fast-growing species well suited to woody arboriculture.

In case of Norway spruce arboriculture in the mountainous belt, some rules are suggested (Burschel and Huss, 1986):

- plantation with 1,800–2,500 seedlings per hectare,
- thinning when the stand has reached a dominant height of 12 m (usually at 20–25 years),
- selection of 200–400 trees pruned at 8 m,
- further thinning, so at the age of 70 years only the 200–400 selected trees remain for other 20–30 years.

In case of sylvicultural practices in semi-natural stands it is necessary to keep separated the subalpine and the mountainous formations.

Sub-alpine Stands Devoted to Soil Protection

For soil protection we rely on two characteristics of Norway spruce: its regulating power with respect to rainfall and its tendency to grow as a bush on an earth mound. Its major disadvantage is the reduced root depth. The sylvicultural systems foresee cuts of 0.1–0.4 ha with successive artificial regeneration. Trees must be thinned not singly but in groups.

Sub-alpine Stands Devoted to Production

The main sylvicultural systems are:

- Cutting in strips of a width 1.5 times the tree height and oriented according to the maximum slope so the trailing of the woody material doesn't cross the living stand.
- Clear cutting by parcels of 1.0–3.0 ha with artificial regeneration. This system, traditional in Val di Fiemme (Italy), must be evaluated also from the point of view of landscape protection.
- Clear cutting by circular areas of diameter 1.0–1.5 times the tree height.
- Successive cutting by groups of diameter 4–2 times the tree height in which a small number of trees are left as a source of seed for natural regeneration. This system of cutting provides the best guarantee of success.

Mountainous Stands

The irregular shelter wood system is the most appropriate because it foresees reciprocal integration among the species (silver fir and beech). A recurrent problem concerns the Norway spruce stands that because of mismanagement are on the way to becoming invaded by blueberry or rhododendron. To prevent this situation, one or more of the following operations must be carried out:

- thinning so that increasing light and heat decreases organic matter accumulation,
- constitution of mixed stands to improve regeneration of Norway spruce,
- removal of humus and shrubs by small parcels or in strips,
- encouragement of shrubs and plants with foliage rich in bases and with a neutral reaction (e.g., willow, elder, sorb, alder)

In case of intensive development of nitrophilous weeds in the cut areas, it must be kept in mind that Norway spruce seldom competes successfully against other species.

32.3.6. Adversities

Damping-off produced by *Heterobasidium annosum* is particularly harmful to Norway spruce.
Humid environments with Atlantic climate are suitable for this pathogenic agent. Another harmful fungus is *Herpotichia juniperi*, which attacks young seedlings in sites where snow cover is prolonged. Among the noteworthy noxious insects are some *Scolitidi* and *Curculionidi*, which dig galleries under the bark. *Cephalcsia arvensi*, which is responsible for leaf drop, *Epinotia tedella*, and *Dasineura abietiperda*.

32.3.7. Land Suitability for Norway Spruce Culture

Several studies undertaken during the 20th century have produced evaluation systems for Norway spruce. Some of the best known are the following:

- yield curves based on site index (i.e., the height of the dominant and co-dominant trees at a reference age) of the English Forestry Service (Hamilton and Christie, 1971),
- productivity of the uneven stands of Cadore (Italy), evaluated on the base of dominant height and the current annual increment (Susmel, 1966),
- average yield prediction (mc/year/ha) of large groups of sites in Scotland, considering altimetric belts and five soil types (Toleman, 1975),
- productivity of different site typologies on the calcareous high plateau of Loraine (Becker and Levy, 1988).

From the above research, it appears that Norway spruce is a continental species, adapted to different substrata but requiring soil moisture, especially during the summer.

Table 32.3 presents a synthesis of the main environmental aspects.

Table 32.3. Norway spruce climatic and geomorphological requirements.

Climate	Luminosity	Moderately sciaphilous
	Atmospheric humidity	Not much tolerant
	Precipitation	1,000–1,800 mm
	Wind	Scarcely resistant
	Summer drought	Scarcely resistant. It requires at least 300 mm rain during the summer.
	Summer heat	It requires 3.5 months with an air temperature > 14.5°C
	Spring frost	Very resistant
Phytoclimatic belt	*Picetum* subzone warm and cold and possible descent in the *Fagetum* (Pavari, 1916)	
Geomorphology	Altitude	1,200–1,600 m
	Parent material	Any
	Stone and boulder deposits	Fairly tolerant

Besides the above-mentioned requirements it must be added that Norway spruce requires a chilling period during the winter (120 d of frost per year).

32.3.8. Soil Functional Characteristics

According to a number of studies, (Wilde et al., 1965; Hagglund e Lundmark, 1977; Levy, 1978; Jokela et al., 1988), it has been observed that the height and diameter growth of Norway spruce can be correlated with the following parameters: drainage, water balance, aeration degree and amount of nutrients (Table 32.4).

Norway spruce has proves more exigent for physical parameters than for chemical ones. This was observed by Duchaufour as early as in 1953. As a matter of fact, the success of this species is linked to soil-water and soil-air relationships. The main limiting factors for the regeneration are the compacted humus layers and the herbaceous turf (Table 32.4).

Table 32.4. Norway spruce edaphic functional characteristics.

Soil moisture regime	Udic and ustic
Soil temperature regime	Frigid and mesic
Useful depth	Medium
Rock fragments	Frequent
Texture	Tendentially sandy
Macroporosity	Medium
Drainage	Good to fairly slow
pH	Acid or sub-acid
Active calcium carbonate	Absent
Chemical fertility	Moderate
Sensitivity to soil compaction	Moderate
Humus	Mor and moder

32.4. European Larch (*Larix decidua* Mill.)

32.4.1. Systematic Framework and Main Botanical Characteristics

The genus *Larix* has only 10 species, of which the most important for Europe is the European larch *(Larix decidua Mill.)*. The difference among species is not very pronounced owing to their interfertility. In Russia, *Larix sybirica Lebed.* is dominant and in the Far East *Larix leptolepis Gord.* (Japanese larch), which may hybridize with European larch, creating *Larix eurolapis Henry.* In North America, there is *Larix laricina (Tamarack)* and *Larix occidentalis*

The European larch, having a long life span, may reach 40 m height and 1 m diameter. On steep slopes larch trunks frequently present a sabre shape because of the action of snow. The bark is thin and grey when young and becomes rough and thick at maturity. The foliage is rather sparse and organized by a ramification of big branches ending in thin twigs that in old trees often are hanging.

The needles are 2–5 cm long, soft green and spirally arranged on the tender twigs. On woody twigs, the needles form clusters of more than 30 short brachyblasts.. During the autumn, the needles turn from green to golden yellow and they fall during the winter. The larch is monoecious: the male flowers are small while the female ones are big and purple amenta. The cones are clear grey and have a leather-like consistency. The grain is light and winged and may be transported by wind at long distance. Volumes and increments are summarized in Table 32.5.

Table 32.5. European larch volumes and increments in Italian forests (Sala, 1969).

Altitude (m)	Climate	Soil	Rotation (years)	Volume (m³/ha)	Increment (m³/year/ha)
1,000–1,500	Slightly cold	Good	100	280	5.00
1,500–2,000	Cold	Fairly good	120	175	2.5
2,000–2,300	Very cold	Fairly bad	140	130	1.00

The wood is well differentiated, with thin whitish sap-wood and duramen brown or purple red, highly resistant to wear as well as to mechanical actions. Taking into consideration these aspects, larch wood is used for heavy constructions, bridges, trusses, foundations, maritime works, fixtures and lumber.

A very high productivity (average increment of 8.2 m^3/ha/year at the age of 20 years) has been recorded in a larch afforestation 1000 m high, with 1000 mm rain well distributed along the year, on a deep soil developed on calcareous schists (Giordano et al., 1973).

32.4.2. Varieties

Among the different regions of origin, the best larch for shape and increment are found in north-eastern Europe (Sudetes Mountains, Poland and Wienerwald), while those of inferior quality belong to the Piedmontese and French Alps. In plantations it is suggested that ecotypes of the same altitude be used, or those that vary at the most by 200 m (Mayer and Ott, 1991 quoted in Bernetti, 1995). The reason is to prevent an early vegetative period with the inherent risk of late frost.

32.4.3. Regeneration

Larch has vegetative reproduction. At the age of 30–40 years the trees may produce seeds that germinate the next year, being maintained in the cones until March. The extraction of the seeds from the cones is less easy than in other conifers because artificial heat must be given gradually to prevent the production of a resinous paste of seeds and cones. Seed conservation is easy but germination is low.

Artificial plantations use almost nursery seedlings (2 kg seed/100 m^2, i.e., 50,000–70,000 seedlings). The seedlings are transplanted with a density of 40 for each square metre. The seedlings do not require shade. The transplant may be done with normal seedlings or with transplantings 2+1 or 1+2. If containers are used it is advisable that the seedlings do not stay too long in the containers, to prevent root deformation.

32.4.4. Pedoclimatic Requirements

European larch has two natural ranges: the Alps and the Carpathians. In Italy it is present in the *Picetum* and in the *Alpinetum*. It is a pioneer species, microthermal, heliophilous, characterized by strong evapotranspiration. For this reason it prefers fresh soils but not humid climates. It can be found on different types of soils (Duchaufour, 1952; Ozenda, 1985) with variable productivity. As far as regeneration is concerned, the larch, being a non-exacting species, prefers mineral soils: coarse-textured colluvial deposits, moraines, torrent beds, and areas that have suffered for fires or avalanches.

The compacted herbaceous felt is negative for the larch regeneration due to the hydrological disequilibrium resulting in a soil permeability decrease and in accentuated summer drought (Giordano, 2002). This fact is particularly severe in the sub-alpine belt of the Western Alps, which is drier than the Eastern Alps. As a consequence, larches have difficulties in competing with *Pinus uncinata* in the Western Alps (Motta and Dotta, 1995).

Larici-Cembretum

This is the only botanical area in which the larch, often predominant, remains for a long time. The soils range from superficial to medium deep, and the texture from sandy to loamy. The organic matter s usually well integrated into the soil. The great soil groups are generally Eutrudepts and Haplusteps on carbonatic rocks and Dystrudepts and Dystrustepts on siliceous rocks

Larch Stands on Prairie Soils

Prairie soils originate from the abandonment of grazing lands. In these areas the larch grows well because it is a pioneer species. The altitudinal belt suitable for larch ranges from 1,600 to 2,300 m on southern exposure and from 1,000 to 2,000 m on northern exposure. In these areas the larch often coexists with invasion species, especially with rhododendron and blueberry. Soils are generally medium–deep and sandy or loamy and have a fair amount of organic matter even if the organic horizons are not clearly pronounced. Often the soils are quite developed because animal dung has accelerated the pedogenetic processes.

Larch Stands in Subalpine Xeric Environment

Compared with stands on prairie soil, larch in a subalpine xeric environment is found at a lower altitude and coexists with xeric vegetation, the most typical expression of which is the prairie of *Calamagrostis villosa* with juniper and in certain sites with *Pinus mugo* and *Pinus uncinata*. Soils are medium-deep or deep, texture ranges from sandy to clay-loam, and rock fragments are more or less important in accordance with the lithology and the geomorphology. The litter is always consistent, as is the organic matter. In very dry environments, *Brachypodium cespitosum* is substituted for *Calamagrostis*.

Larch Stands in Wet Sites

On wet sites, together with larch, there are *Alnus viridis*, *Acer pseudoplatanus* and *Fraxinus excelsior*. Soils are deep, often skeletic, and very rich in organic matter that sometimes gives rise to a form of humus called Humimor by Green et al. (1993).

Scotch Pine Stands with Larch

Both Scotch pine and larch are pioneer species that may be found together on southern exposures in the mountain belt. Soils are shallow and scarcely developed and have a reduced amount of organic matter, which forms a type of humus classified as Leptomoder by Green et al. (1993). In some sites of the Western Alps, birch replaces Scotch pine (IPLA, 1996).

Larch Stands at Low Altitude

There are larch formations down to an altitude of 600 m in coppices of beech, chestnut and hornbeam. These formations, always temporary, are the consequences of heavy cuts that have degraded the soil, making way for less exigent pioneer species. Soils are always shallow with variable texture and with many coarse rock fragments. Organic matter is scarce.

32.4.5. Cultural Techniques

The cultivation techniques are different according to the desired uses of the larch stands.

Productive Larch Stands

Productive stands are those of nearly equal age (20–30 years), of the subalpine belt on the grazed prairie. The most convenient system is clear cutting if the area is less than 1.5 ha. In larger areas, cuts are done by strips or by groups that can be enlarged (G. Giordano, 1955). The "Swiss Femelschlag" system is similar but foresees more groups of trees to be cut (Bernetti, 1995). The cut and the consequent increase in light help regeneration, especially if the soil is worked. In certain cases wild boars breaking up the ground have a beneficial influence on larch regeneration.

A larch plantation usually has 500–600 seedlings/ha, which is reduced to 250–300 after the required thinning. In places where there is a danger of trees falling, trees must not be thinned regularly but anchorage groups formed by dominant and vigorous trees must be left in place (Bernetti, 1995).

Larch Mixed Stands

A larch stand invaded by Norway spruce has a high productivity. In order to maintain this high productivity a cut of the intermediate Norway spruce following an irregular shelter wood system is advisable, particularly if there is an undergrowth of blueberry (G. Giordano, 1955). In the mountainous larch stands invaded by broadleaf trees (beech, maple, ash and chestnut), it seems convenient to intervene with clear cuttings and artificial regeneration, leaving the broadleaf trees as a coppice layer (Bernetti, 1995).

In artificial afforestation characterized by consociation with Norway spruce, Scotch pine and sometimes with silver fir, larch may be considered as temporary species for creating better conditions for the other species or as a permanent species for setting up an uneven stand.

Amelioration of Protective Stands

At high altitudes, the *Cembro-Laricetum* is the best protection against avalanches and rock falls. At lower altitudes a fair protection can also be afforded by the larch-Norway spruce formation: the former provides the soil with anchorage while the latter develops rapidly (Bernetti, 1995). Soil in small areas or portions of terraces must be worked to prevent creeping and to create discontinuity in the felted humiferous top layer.

Enrichment of Coppices

The introduction of larches in the coppices (80–100 larches/ha) doubles the total productivity of the stand (Leibundgut, 1967).

Arboriculture Designed for Wood Production

The role of larch in arboriculture is mainly economical, anticipating the final product with some intercalary wood material.

32.4.6. Adversities

Cryptogams

Dasyscypha willkommii is a destructive agent of larch canker. It finds its optimum conditions in humid soils or in sites with an oceanic climate. Japanese larch and its hybrids are immune from this pathology. *Phytophthora omnivora* attacks the young seedlings. *Polyporus sulfureus* produces red-brown decay of the wood, while *Polyporus borealis* produces yellow-brown decay. *Fomes annosus* makes the wood spongy and friable. *Armillaria mellea* is responsible for damping-off. *Melampsora laricis-tremulae* is the agent of larch-tree rust.

Insects

There are many pathogenic agents, from non-specific agents such as *Grillotalpa vulgaris,* active on larch roots, to the very specific such as *Ips laricis,* active only on larch wood. The favourite targets of the more than 30 insects affecting larch are bark, cambium, buds, roots, needles and wood.

32.4.7. Land Suitability for Larch Culture

The larch site index and the mean annual increment have been investigated (A. Giordano et al., 1972) in the municipality of Chiomonte (Valle di Susa, province of Torino, Italy). These data combined with the soil survey of the same area have given rise to a map of land suitability for larch, silver fir and Scotch pine. Further research on the factors influencing larch growth was carried out in Valle di Susa in the late 1980s. Table 32.6 presents a synthesis of the main environmental factors determining land suitability for larch culture.

Table 32.6. Larch climatic and geomorphological requirements.

Climate	Luminosity	Larch is eliophilous
	Atmospheric humidity	Scarcely tolerant due to its high evapotranspiration needs
	Precipitation	At least 700 mm
	Wind	Resistant
	Summer drought	Resistant
	Summer heat	Required for evapotranspiration
	Spring frost	Resistant
Phytoclimatic belt	*Picetum* and *Alpinetum*	
Geomorphology	Altitude	1400–1900 m
	Parent material	Any
	Stone and boulder deposits	Tolerant

32.4.8. Soil Functional Characteristics

Larch is highly tolerant of soil characteristics but two aspects are peculiar. The first is its need for a consistent amount of water deep in the soil. Being highly exigent as far as evapotranspiration is concerned, larch does not thrive in places where there is an elevated atmospheric humidity, preferring dry sites but with the possibility of water in the deep layers that can be reached by the deep roots. The second peculiar aspect is the negative effect of a felted herbaceous top layer or of humiferous layers too thick on the larch regeneration; both of them produce a hydrological disequilibrium.

Table 32.7 illustrates the main functional characteristics favourable for larch growth.

Table 32.7. European larch edaphic functional characteristics.

Soil moisture regime	Udic and ustic
Soil temperature regime	Cryic, frigid and mesic
Useful depth	Moderate
Rock fragments	Frequent
Texture	All except clay
Macroporosity	High
Drainage	Moderate
pH	Acid to neutral
Active calcium carbonate	Absent or very low
Chemical fertility	Reduced
Sensitivity to compaction	High
Humus	Moder and mull-moder. Regeneration is hampered by too-thick humus layers

32.5. Scotch Pine (*Pinus sylvestris* L.)

32.5.1. Systematic Framework and Botanical Characteristics

In the subgenus *Pinus*, the *sylvestris* section encompasses the following species (Debazac, 1977): *P. sylvestris* L., *P. mugo Turra, P. uncinata Ramond, P. pumilio Haenke, P. nigra Arn, P. leucodermis Ant.*, several species from Asia and two from North America.

Pinus sylvestris is a Euro-Asian conifer that during the post-glacial period had in Europe a consistent diffusion determining the "Scotch pine epoch". The subsequent milder climate allowed the spread of competitive species and the consequent reduction of the Scotch pine area. Today Scotch pine is established in a number of sites that are unfavourable to the broadleaf species. The trunk is usually straight, with a reddish-yellow bark. The foliage is sparse, regular in the juvenile phase, then globose, asymmetrical and untidy. The needles are joined by two in small clusters that are spiral, rigid and green-grey. The male flowers are *amenta* yellow red, the female ones are globose, solitary and green. The greyish *strobili* have a short stalk and ripen in 2 years. The seeds are small and brown with rather large, thin wings. The root system is deep and strong. At maturity the annual increment is 2–5 m³/ha but may double in good heathlands (G. Giordano and Passet-Gros, 1962).

The resinous wood having pinkish-white *alburnum* and reddish-brown *duramen* is largely used for poles, trusses, boards, carpentry, packing, and paper mill. The value of the material depends on variety and origin; wood produced in Scandinavia, Finland and Russia has fine and regular texture and very narrow annual rings and for this reason is highly valued for fastenings and fixtures. In a continental climate, Scotch pine may reach a height of 40 m and a diameter of 80 cm. In warmer climates, the height is reduced and the form is stouter.

32.5.2. Varieties

Scotch pine is a very variable species according to the different populations. The European Flora of 1964 quotes 27 botanical varieties fundamentally established on geographic base (Bernetti, 1995).

32.5.3. Regeneration

The years good for seed are frequent. Germination rate is high and seeds need not be treated to wake them up from a dormant phase. Seeding in nursery foresees 0.7–0.9 kg of seed/100 m². The seeding depth is 0.5–1.0 cm. Shade is not necessary. The material for transplanting is almost 1+2 or 2+2 with naked roots (Bernetti, 1995). The cenosis most suited to Scotch pine regeneration is characterized by the presence of slightly competitive species such as birch, aspen, juniper and larch (Bernetti, 1995).

32.5.4. Pedoclimatic Requirements

Scotch pine is light-demanding, xerophilous, and indifferent to parent material. Taking into account its great site variability, it is convenient to analyse the main environments in which Scotch pine grows (Ozenda, 1985). We limit the analysis to Italy.

Xeromorphic Pine Forests

These are continental endalpic valleys, scarcely rainy, sunny, at an altitude between 900 and 1,600 m, called "Scotch pine valleys" (Fenaroli, 1967). They are zonal stands, i.e., they are in equilibrium with the climate subject to important thermic differences, limited precipitation (600–900 mm), scarce and short snowfall, and rather constant winds. Inside the xeromorphic stands there are two different typologies: basiphilous and acidophilous.

Basiphilous Pine Forests

Basiphilous pine forests are found at an altitude between 1,000 and 1,600 m on carbonatic rocks that may appear as outcrops, talus colluvial deposits, and dejection cones. Soils are shallow, scarcely developed and rich in rock fragments, almost sandy, poor in organic matter, well drained, seasonally dry, basic or neutral, and abundant in calcium carbonate. According to the USDA (1998) the soils are Entisols, Eutrudepts, Haplustepts, and eventually Haploxerepts and Dystroxerepts and Dystrustepts. The characteristic botanical species are *Ononis rotundifolia* for the herbaceous layer and *Arctostaphylos uva-ursi* and *Prunus mahaleb* for the shrub layer (IPLA, 1996)

Acidophilous Pine Forests

Acidophilous pine forests can be found, excluding the carbonate rocks, on several and different lithotypes. The soils mostly belong to the order of the Orthents, which are skeletic, sandy, poor in organic matter, acid and excessively drained. Their botanical composition often includes larch, juniper and *Amelanchier ovalis*. *Avenella flexuosa* is characteristic of the herbaceous layer.

Mesoxeromorphic Pine Forests

These constitute a polymorphic whole at an altitude of 600–1,000 m in the sub-mountainous belt characterized by a climate with Atlantic tendency. They also present two phases: basiphilous and acidophilous.

Mesophilous Pine Forests

They are located in the esalpic environment at an altitude of 400–1,000 m, sometimes mixed with beech, mountain maple and other broadleaf trees. In such an environment, Scotch pine is not a climax species, so its regeneration is made possible especially through fires.

Pine Forests Influenced by Edaphic Conditions

Pine forests of heathland on fluvio-glacial terraces. The climate is fundamentally sub-Atlantic, with rainfall ranging from 1,000 to 1,400 mm. A quarter of the precipitations fall during the summer as storms. Soils are generally deep, loamy or silt-loamy with deeply weathered stones, iron and manganese concretions and unsatisfactory drainage. The most widespread soils are Fragiudalfs (Fragic Luvisols, FAO 1998). Scotch pine is a pioneer species and so transitory in this environment where the original botanical formation is the *Querco-carpinetum*, which has been destroyed during the centuries. Scotch pine, being limited in its regeneration by a dense cover of *Molinia coerulea* and fern, has almost mature or overripe individuals of bad shape and with thick bark due to frequent fires.

   Pine forests on moraines. Soils are quite deep, with loamy or sandy-loam texture, abundance of stones, scarcity of organic matter and rapid drainage. Soils belong to the orders of Inceptisols or Orthents (Cambisols or Regosols following FAO, 1998).

   Endalpic pine forests torrent beds. They are located on level areas where there are old river beds at an altitude of 800–1,200 m. The cenosis is stable and does not present signs of dynamism towards a more well-developed cenosis. The soils (Orthents and Fluvents) are superficial, stony, dry, and rapidly or excessively drained, with a humus of mor type. Among the arboreal species may be observed birch and black poplar, among the shrubs *Salix purpureus* and *Salix eleagnos*, and among the herbaceous plants *Calamagrostis pseudophragmites*.

32.5.5. Cultural Techniques

In natural or semi-natural situations, Scotch pine is managed with clear cutting on strips or small areas. If there is no concurrence of broadleaf trees it may be managed with irregular shelter wood system (Bernetti, 2005). Afforestation on silicate rocks foresees 40×40×40 cm pits prepared some months before planting. On calcareous soils, it is advisable to form terraces in order to reduce the aridity. Direct seeding has been abandoned in favour of transplanting at 2+2 or 2+1 with naked roots. The transplanting season may be the fall or the spring, with advantages or disadvantages often compensated for by the weather during the year (Bernetti, 1995).

32.5.6. Adversities

The pathologies are determined by fungi and insects common to many conifers: the ascomycetes *Lophodermium seditiosum* is the agent of needle cast, the basidiomycetes *Melampsora pinitorqua* is responsible for pine twisting rust and damping-off is produced by *Heterobasidion annosum* and by *Armillaria* spp. A simple and effective control against damping-off is elimination of diseased stumps. A suggested prevention practice is cutting during the summer because at a temperature higher than 30°C the fungus stops growing and producing basidiospores (Moriondo, 1989). In case of afforestation with Scotch pine, it is better to rule out environments with *Vincetoxicum* sp. and *Gentiana asclepiadea* because these species are necessary for the beginning of the cycle of *Cronartium flaccidum*, agent of two-needle pine stem rust (Moriondo, 1989). Among the harmful insects, the cochineal *Matsucoccus pini* deserves to be mentioned.

32.5.7. Land Suitability for Scotch Pine

Land suitability for Scotch pine has been particularly investigated through the "site index" method by Ilvessalo (1923, quoted in Wilde, 1958), by the Forestry Service of England (Hamilton and Christie, 1971), and by Alex (1964, quoted in Purcelean, 1970). For a comprehensive view of the environmental factors influencing land suitability for Scotch pine, see Table 32.8.

Table 32.8. Scotch pine climatic and geomorphological requirements.

Climate	Luminosity	Well lit
	Atmospheric humidity	Indifferent
	Precipitation	500–1000 mm
	Wind	Resistant
	Summer drought	Resistant
	Summer heat	Fairly exigent
	Spring frost	Resistant
Phytoclimatic belt	*Picetum* and *Fagetum*	
Geomorphology	Altitude	800–1800 m
	Parent material	Indifferent
	Stone and boulder deposits	Tolerant

32.5.8. Functional Soil Characteristics

The indifference of Scotch pine to the parent material is nearly absolute. It can stand aridity, economizing on a small amount of water and waiting for photosynthetic activity when water level in soil is increased. Bare soils or rocky or stony soils on southern exposure are suited to Scotch pine, taking into consideration also the fact that in these conditions the choice is limited to a few forestry species. Table 32.9 illustrates the main soil functional characteristics for Scotch pine afforestation.

Table 32.9. Scotch pine edaphic characteristics.

Soil moisture regime	Any
Soil temperature regime	Mesic and frigid
Useful depth	Reduced
Rock fragments	Frequent
Texture	Any
Macroporosity	High to reduced
Drainage	Rapid to slow
pH	Acid to alkaline
Active calcium carbonate	Absent to high
Chemical fertility	Medium to reduced
Sensitivity to compaction	Reduced
Humus	Moder

References

BACKER, J.B., BROADFOOT, W.M. 1979. A practical field method of site valuation for commercially important southern hardwood. Forest Service USDA, Gen. Tech. Rep., SO-26, South. For. Exp. Stn., New Orleans (USA).

BARNES, B.V., PREGITZER, K.S., SPIES, T.A., SPOONER, T.H. 1982. Ecological Forest Sites Classification. Journal of Forestry 80(8), 493–500.

BECKER, M., LEVY, G. 1988. A propos du dépérissement des forêts: climat, sylviculture, et vitalité de la sapinière vosgienne. Revue Forest. Franç. XL(5), 345–358.

BERNETTI, G. 1995. Selvicoltura speciale. UTET, 414 p.

BERNETTI, G. 2005. Atlante di selvicoltura Edagricole, 496 p.

BOOTH, T.H., SAUNDERS, J.C. 1985. Applying the FAO guidelines on land evaluation for forestry. Forest Ecology and Management 12, 129–142.

BURSCHEL, P., HUSS, J. 1987. Gundriss des Walbaus. P. Parey, 362 p.

BURSCHEL, P., HUSS, J., KALBHENN, R. 1964. Die Natürliche Verjüngung der Buche. Schriftenreihe der Forstlichen Fakultat der Univ. Göttingen, Vol. 34 I.D.Sauerländer's Verlag, Frankfurt.

CIAMPI, C., DI TOMMASO, P.L. 1973. Osservazioni morfo-anatomiche sul comportamento in vivaio dei semenzali di abete bianco (Abies alba Mill) di differenti origini gografiche. Annali Acc. Ital. Scienze Forest. 12, 61–90.

CORONA, E. 1983. Ricerche dendrocronologiche preliminari sull'abete bianco di Vallombrosa. Annali Acc. Ital. Scienze Forest. 32, 149–164.

DEBAZAC, E.F. 1977. Manuel des conifères. Ecole Nationale des Eaux et Forêts, 172 p.

DOTTA, A., MOTTA, R. 2000. Indirizzi selvicolturli dei boschi di conifere montane. Direz. Economia Montana e Foreste, Regione Piemonte, 191 p.

DUCHAUFOUR, PH. 1952. Etude sur l'écologie et la sylviculture du Melèze. Ann. Eaux et Forêts XIII(1), 1–203.

DUCHAUFOUR, PH. 1988. Les pessièes d'altitude: physiologie de la nutrition et problèmes de régéneration. Chambery 20–22 September 1988. Université de Savoie.

ERSAL. 2001. Carta di orientamento pedologico per l'arboricoltura da legno della pianura lombarda. Regione Lombardia, Azienda Regionale Foreste, 13 p.

FAO. 1984. Land evaluation for forestry. Forestry Paper 48, Rome.

FAO. 1998. Word Reference Base for Soil Resources. Word Soil Resources Report 84, Rome.

FENAROLI, L. 1967. Gli alberi d'Italia. Martello, Milano, 320 p.

GIACOBBE, A. 1973. A proposito della var. apennina Giac.dell'Abies alba. Italia Forest. e Montana 28(1), 30–32.

GIORDANO, A. 2002. Pedologia Forestale e Conservazione del Suolo, UTET, 600 p.

GIORDANO, A., LILLELUND, H., MONDINO, G.P. 1972. Carta della feracità forestale per il larice, abete bianco e pino silvestre in Comune di Chiomonte, Torino. Ann. Ist Sper. per la Selvicoltura.

GIORDANO, A., GIOVANETTI, G., DE VECCHI, P.G. 1988. Definition of land types and site index of black locust coppices (Piedmont, Italy). EC seminar on wood technology, Munich, 14–15 April, 1987.

GIORDANO, A., MONDINO, G.P., PALENZONA, M., ROTA, L., SALANDIN, R. 1973. Ecologia ed utilizzazioni prevedibili nella Valle di Susa. Ann. Ist. Sper. per la Selvicoltura 5, 83–196.

GIORDANO, G. 1955. Distribuzione e caratteristiche del Larice sulle Alpi italiane. Atti Congr. Naz. Selvicoltura, Firenze (Acc. It. Sci. For.).

GIORDANO, G., PASSET-GROS, M. 1962. Dizionario enciclopedico agricolo-forestale e delle industrie del legno. Ceschina Edit, 1216 p.

GREEN, R.N., TROWBRIDGE, R.L., KLINKA, K. 1993. Towards a taxonomic classification of forms of humus. Soc. of Amer. Foresters, Bethesda, Maryland (USA), 51 p.

HAGGLUND, B., LUNDMARK, J.E. 1977. Site index estimation by means of site properties. Scots pine and Norway spruce in Sweden. Studia For. Suecica 138, Stockolm.

HAMILTON, G.J., CHRISTIE, J.M. 1971. Forest management tables. Forest Commission Booklet, 34. London, 200 p.

IPLA. 1996. I tipi forestali del Piemonte. Assessorato Econ. Montana e Foreste, Regione Piemonte, 372 p.

IPLA. 2001. Guida alla realizzazione ed alla gestione degli impianti di arboricoltura da legno. Assess. Economia Montana e Foreste, Regione Piemonte, 111 p.

JOKELA, E.J., WHITE, E.H., BERGLUND, J.V. 1988. Predicting Norway spruce growth from soiland topographic properties in New York. Soil Sci. Soc. Am.

LANDON, J.R. 1984. *Booker Tropical Soil Manual*. BAI, Longman Group, England, 450 p. LARSEN, J.B. 1986. Die geographische Variation der Weisstanne (Abies alba Mill.) Wachstmentwicklung unt Frostresistenz. Forstwiss, Centralblatten 105(5), 396–407.

LE TACON. 1976. La présence de calcaire dans le sol, influence sur le comportement de l'épicé et du pin noir. Thèse de Doctorat, Nancy.

LEIBUNDGUT, H. 1967. Untersuchungen über Ergbnisse des Lärchenbestandes im schweizerischenMittelland Schweizer. Zeitsc. F. Forstwesen 118(4), 183–208.

LEVY, G. 1978. Nutrition et production de l'Epicéa commune adulte sur sols hydromorphes en Lorraine: liaisons avec les caractéristiques stationelles. Ann. Sci. Forest. 35(1), 33–53.

MAGINI, E. 1973. Esiste sull'Appennino una varietà dell'abete bianco?. Italia Forest. e Montana 28(5), 173–175.

MANCINI, F. 1959. I terreni della foresta di Panveggio (TN). Ann. Acc. It. Sci. Forest. VIII, 373–454.

MAYER, H., OTT, E. 1991. *Gebirgswadbau Schutzwaldpflege*. Fischerm, Berlin.

MORIONDO, F. 1989. Introduzione alla Patologia Forestale. UTET, 251 p.

MOTTA, R., DOTTA, A. 1995. Les mélézeins des Alpes Occidentales: un paysage à défendre. Rev. For. Franç. XLVII, 329–342.

OZENDA, P. 1985. *La végétation de la châine alpine*. Masson, Paris.

PAGANUCCI, L. 1989. Sperimentazione sulla densità di impianto delle abetine di Vallombrosa. Italia Forest. e Montana 44(1), 30–44.

PAVARI, A. 1916. Studio preliminare sulla coltura delle specie forestali esotiche in Italia. Annali Regio Istituto Superiore Forestale Nazionale I, 159–379.

PAVARI, A. 1951. Esperienze e indagini sulle provenienze e razze dell'abete bianco (Abies alba Mill.). Staz. Sperim. di Selvicoltura, Pubbl. 8, 96 p.

PAVARI, A. 1952. *Abete bianco*. Voce dell'Enciclopedia Agricola REDA, Roma.

PIGNATTI, S. 1982. *Flora d'Italia*. Edagricole, Bologna.

PIUSSI, P. 1986. La rinnovazione delle peccate subalpine. Le Scienze 37(215), 58–67.

PURCELEAN, S. 1970. Importanza dello studio della vegetazione e della stazione forestale per la valutazione della produzione potenziale delle foreste. Acc. It. Sc. For. XIX 305–328.

SALA, G. 1969. *Larice*. Voce dell'Enciclopedia Agraria REDA, Roma.

SCHMIDT-VOGT, H., 1977. *Die Fichte*. Vol. I, P. Parey, 647 p.

SINDOU, C., MADESCLAIRE, A. 1986. Le melèze d'Europe sur les plateaux calcaires de Lorraine. Rev. For. Franç. XXXVIII(5), 439–449.

SMALLEY, G.W. 1982. Classification and evaluation of forest sites on the Mid Cumberland Plateau. Exp. Stn., New Orleans (USA).

SMALLEY, G.W. 1991. Classification and evaluation of forest sites on the Nantchez trace State Forest. USDA South. For. Exp. Stn, New Orleans (USA).

SPIECKER, H. 1986. Das Wachstum der Tannen und Fichten im Plenterwald, Versuchflachen des Schwarzaldes in der Zeit von 1950bis 1984. Allgemeine Forst und Jagd Zeitung 157(8), 152–164.

STEELE, R., COOPER, S.V., ONDOV, D.M., ROBERTS, D.W., PFISTER, R.D. 1983. Forest habitat type of Eastern Idaho, Western Wyoming. Tech. Rep. INT 144, Intermountain Forest and Range Exp. Stn, Ogden (US).

SUSMEL, L. 1966. Metodi di stima della fertilità stazionale nei climi temperati. Monti e Boschi 4, 3–19.

TOLEMAN, R.D.L. 1975. *National Soil Classification*. Forestry Commission Research and Development Division, UK.

USDA. 1998. Keys to soil taxonomy. Natural Resources Conservation Service, Washington, DC.

WILDE, S.A. 1958. *Forest Soils*. Ronald Press, New York, 538 p.

WILDE, S.A., IYER, J.G., TANZER, C., TRAUTMANN, W.L., WATTERSON, K.G. 1965. Growth of Wisconsin coniferous plantations in relation to soils. Research Bulletin 262, Univ. of Wisconsin.

33. Land Suitability for Grazing

Paolo Baldaccini[1] and *Andrea Vacca*[2]

33.1. Introduction

Pastures are defined as tracts of land used for grazing, where natural vegetation is the main forage resource, and which are never or almost never mown. In Italy, pastures occur in a large variety of geographic, ecologic and socio-economic situations. The considerable differences existing among pastures in the Alps, in the Apennines, in Southern Italy, and on the islands paint a picture of their variety in climate, flora, soils, livestock rearing, and productivity. Land evaluation for grazing is therefore strongly affected by this variety. Consequently, the information provided here is to be considered as general.

33.2. General Principles

Most of the general principles discussed in this paper are drawn from the FAO guidelines (FAO, 1991), which follow the methodology adopted by the FAO framework for land evaluation (FAO, 1976). These general principles, which concern extensive grazing, can also be applied to Italian pastures but their relative importance depends on the local situation.

The following should be borne in mind in the evaluation of grazing suitability:

- land use requirements differ according to grazing intensity;
- soil survey and land evaluation are generally less intensive than for other land uses;
- land evaluation for grazing must include both primary production of forage and secondary production of livestock;
- planned stocking rate, i.e., the number of livestock per unit area per unit time, is varied in order to match the amount of forage available over space and time;
- livestock have continuous and relatively constant nutrient and water requirements throughout the seasons.

Therefore, four sets of land use requirements need to be provided for all grazing land use types. These requirements are related to:

---

[1] Formerly Università di Sassari, Italy.

[2] Università di Cagliari-Dipartimento di Scienze della Terra, Cagliari, Italy.

- the growth and botanical composition of forage at the primary production level;
- livestock requirements at the secondary production level;
- the land characteristics affecting livestock rearing;
- requirements for conserving the land by maintaining a sustainable grazing system.

33.3. Functional Factors

33.3.1. Primary Production Level

Factors Affecting Growth and Botanical Composition of Forage

The factors affecting the growth and botanical composition of forage are climate (radiation regime, temperature regime, moisture regime, frost hazard), soil (drainage, oxygen availability to roots, nutrient availability, rooting conditions, salinity, sodicity, toxicity), environmental hazard (fire hazard, flood hazard), and genetic potential of natural vegetation. Climate conditions can be assessed by means of total radiation, number of hours of sunshine per day, slope aspect and steepness, elevation above sea level, average annual and seasonal temperature, rainfall and evapotranspiration, and their extreme values for the coldest and warmest months during the growing season. Special attention should be paid to possible early or late frost. Soil conditions may be assessed by means of survey descriptions (rock outcrops, coarse surface fragments, soil depth, presence of limiting horizons, horizon types), physical properties (texture, structure, consistency, infiltration rate, permeability, water retention capacity), and chemical properties (pH, exchangeable cations, base saturation, cation exchange capacity, organic carbon content, total nitrogen content, available phosphorus, available potassium and other nutrients, microelements, electrical conductivity, soluble salts content, aluminium excess, calcium carbonate content, gypsum and other toxic substances).

Factors Affecting Management

The factors affecting management are ease of controlling undesirable plant species, soil workability and the potential for mechanization. These factors can be assessed by means of slope steepness and length, rock outcrops, coarse surface fragments, drainage density and surface morphology.

Factors Affecting Land Conservation

The factors affecting land conservation are erosion hazard, soil tolerance of trampling, and vegetation tolerance of degradation. Erosion hazard can be assessed by means of rainfall intensity, slope steepness and length, soil erodibility, percentage of vegetation cover, and presence of mass movements. Soil tolerance of trampling can be assessed by means of soil texture, soil drainage class and soil structural stability. Vegetation tolerance of degradation can be assessed by means of the characteristics of the vegetation type and conditions.

33.3.2. Secondary Production Level

Factors Affecting Growth

The factors affecting growth are availability of forage and drinking water, biological hazards, climatic limitations, accessibility for animals, and hay and silage production potential. The availability of drinking water can be assessed by means of moisture surpluses or deficits, rainfall variability, presence and regime of springs, streams and rivers, groundwater hydrology and depth, presence of aquifers and quality of drinking water. The biological hazards can be assessed by means of the observed incidence of harmful plants and insects and of diseases and disease vectors. The climatic limitations can be correlated to total and solar radiation, number of hours of sunshine per day, extreme temperatures, frost incidence, relative humidity, and wind speed. Accessibility for animals can be estimated by measurements of the elevation above sea level, slope angle, the percentage of coarse surface fragments and rock outcrops, vegetation type and density, flood severity and frequency and land morphology. The conditions for hay and silage production depend on climatic characteristics such as rainfall variability, moisture surplus, relative humidity and wind speed, and on the characteristics of the vegetation such as carbohydrate assimilation type and type of flora.

Factors Affecting Management

The factors affecting management are the ease of fencing or hedging and the distance to markets. The ease of fencing or hedging mainly depends on soil depth, rock outcrops, and coarse surface fragments.

33.4. Examples of Application in Italy

In Italy, the reported and known applications of soil suitability evaluation for grazing are based on the general principles outlined by FAO (1976, 1991). In particular, these evaluations have been performed in Sardinia (Aru et al., 1989; Madrau, 1992, 1993; Baldaccini et al., 1995, 2002; Baldaccini and Madrau, 1996; Madrau et al., 1998), in Lombardy (Comolli, 1996), and in Molise (Litterio et al., 1993). In all these applications, the results achieved are not empirical but can be considered successful attempts, aimed at rationalizing grazing management, at interpreting the different geographical situations examined relying on the authors' experiences and information reported in the literature. Comolli (1996) proposed a reference table for determining soil suitability classes for cattle grazing in an Alpine mountain pasture (Val Gerola, Lombardy) (Table 33.1). The characteristics affecting the technical aspects of management (slope, rock outcrops, and coarse surface fragments), forage production (soil and climate characteristics) and soil degradation (erosion and trampling) were examined. The productive potential of pastures in Molise was evaluated using a land information system, overlapping and processing the different layers of data concerning the main environmental factors (climate, soil, and vegetation) influencing that potential (Litterio et al., 1993). The land qualities considered influence both the quality and the quantity of production (seasonal aridity hazard and rooting depth), management techniques (potential for mechanization), and soil conservation (sheet erosion and slumping hazard).

For evaluating the seasonal aridity hazard (Table 33.2), the authors considered the available water capacity of the soil and the duration of the water deficit period, defined by FAO (1983) as that period of the year during which precipitation is less than half the real evapotranspiration. The potential for mechanization was evaluated, adapting the FAO proposal (FAO, 1983) to the Molise region, taking into consideration slope and the presence of rock outcrops and coarse surface fragments (Table 33.3). Table 33.4 shows the final interpretive scheme.

Table 33.1. Criteria and limits proposed for determining soil suitability classes for cattle grazing (Comolli, 1996).

Characteristic	S1 highly suitable	S2 moderately suitable	S3 marginally suitable	N1 actually not suitable	N2 permanently not suitable
Slope gradient (%)	0–20	20–40	40–60	60–80	> 80
Rockiness surface stoniness (%)	0–10	10–30	30–60	60–80	> 80
Soil depth (cm)	> 50	25–50	25–50	10–25	< 10
Rock fragments (%)	0–15	15–35	35–70	70–85	> 85
Texture	L	SL, SIL, CL, SICL, SCL	LS, SIC, SC	S, C, SI	
Chemical fertility*	high	medium	low	very low	
Hydromorphy	slight or none	moderate	severe	very severe	
Water deficit**	none	moderate	severe	very severe	
Erosion hazard	none	moderate	severe	very severe	extreme

The "chemical fertility" concept considers the organic matter content, soil reaction, and available P and K content. The critical limits of the chemical fertility classes are determined according to the following criteria (Comolli, 2005, personal communication):

pH (H_2O) (in the topsoil)
S1: pH > 5.6
S2: 5.1 < pH ≤ 5.6
S3: 4.4 < pH ≤ 5.1
N1: pH ≤ 4.4

Organic matter content (in the topsoil)
S1: OM > 5%
S2: 3.5% < OM ≤ 5%
S3: 2% < OM ≤ 3.5%
N1: OM ≤ 2%

Available P content (measured in the topsoil by the Olsen method, expressed in mg/kg)
S1: P > 15
S2: 10 < P ≤ 15
S3: 5 < P ≤ 10
N1: P < 5

Available K content (measured in the topsoil, expressed as a percentage with respect to CEC)
S1: K > 2.5
S2: 2 < K ≤ 2.5
S3: 1.5 < K ≤ 2
N1: K < 1.5

**Water deficit is determined by comparing real evapotranspiration (RET) with potential evapotranspiration (PET) (sensu Thornthwaite) (Comolli, 2005, personal communication). The critical limits of the water deficit classes are the following:*
S1: RET > 0.9 PET
S2: 0.8 PET < RET ≤ 0.9 PET
S3: 0.7 PET < RET ≤ 0.8 PET
N1: 0.5 PET < RET ≤ 0.7 PET
N2: RET ≤ 0.5 PET

The framework for the evaluation of soil suitability for grazing and pasture improvement in Sardinia (Madrau et al., 1998) is based on the reference tables published by Aru et al. (1989), with the introduction of some new elements. The reference tables drawn up by Aru et al. (1989) differed according to the different land units in Sardinia, taking into account the wide variety of pasture lands. The framework proposed by Madrau et al. (1998) contains an

Table 33.2. Reference table for assessing aridity hazard (Litterio et al., 1993).

Available water capacity of soil (mm)	Duration of water deficit period (d)	Aridity hazard (class)
< 100	< 70	moderate
	70–110	severe
	> 110	extreme
100–200	< 70	low
	70–110	moderate
	> 110	extreme
> 200	< 70	low
	70–110	moderate
	> 110	severe

Table 33.3. Reference table for evaluating the potential for mechanization (Litterio et al., 1993).

Land characteristics	Potential for mechanization (class)			
	High	Moderate	Low	Very low
Slope gradient (%)	0–20	20–35	35–50	> 50
Rockiness (%)	0–2	2–10	10–25	> 25
Surface stoniness (%)	0–3	3–15	15–50	> 50

Table 33.4. Final interpretive scheme for evaluating the productive potential of pastures (Litterio et al., 1993).

Land qualities	Productive potential class			
	A High productive potential	B Moderate productive potential	C Marginally productive potential	D Very marginal productive potential
Rooting depth	> 50 cm	30–50 cm	30–10 cm	< 10 cm
Aridity hazard	slight to moderate	severe	extreme	extreme
Potential for mechanization	high	moderate	low	very low
Erosion hazard	slight to moderate (< 10 to 50 t/ha/year)	slight to moderate (< 10 to 50 t/ha/year)	severe (50–200 t/ha/year)	extreme (> 200 t/ha/year)
Disturbance class (% of area affected by disturbance)	low to moderate (0–2 %)	low to moderate (2–10 %)	high (10–30 %)	very high to extreme (30–60 %, > 60%)

evaluation table for each of the eight land units used for grazing. These land units were distinguished according to soil parent material and include: Palaeozoic metamorphites; Palaeozoic intrusive complex; marly limestone and sandy marls; Palaeozoic and Mesozoic limestone and dolomite; tuffs and ignimbrites alternating with rhyolites and trachytes; basalts and andesitic basalts; Pleistocene alluvial deposits, terraces and glacis; Holocene alluvial deposits. Table 33.5 is an example of one of these evaluation tables. The differences between the proposed evaluation tables concern the soil depth, texture, base saturation and available water

Table 33.5. Criteria and limits proposed for Sardinia for determining soil suitability classes for pasture improvement for the "Palaeozoic intrusive complex" (Madrau et al., 1998).

Characteristic	S1	S2	S3	N1	N2
Elevation (m a.s.l.)	< 600	600–800	600–800	800–1,000	> 1,000
Slope gradient (%)	0–2	2–6	6–15	15–55	> 55
Aspect < 1,000 m a.s.l.	S	E-W	N		
> 1,000 m a.s.l.	S	E-W			N
Vegetation cover (%) with predominance of shrubs*	< 2	2–10	10–25	25–50	> 50
Vegetation cover (%) with predominance of trees	< 2	2–10	10–20		> 20
Rockiness (%)	0	< 2	2–5	5–10	> 10
Surface stoniness (%)	< 0.1	0.1–3	3–15	15–50	> 50
Drainage (duration of wet periods, percentage of area with drainage system)	absence of ponded or free water	ponded or free water for short periods; < 20% of area with drainage system	ponded or free water for long periods; 20–50% of area with drainage system	ponded or free water for long periods; > 50% of area with drainage system	aquic moisture regime everywhere or almost everywhere
Number of consecutive days after summer solstice during which soil MCS** is dry (d)	< 45	45–75	75–90	> 90	> 90
Frost (duration, frequency)	none	rare	rare, in consecutive years	common, in consecutive years	frequent, in consecutive years
Soil depth (cm)	> 60	60–40	40–20	20–10	< 10
Texture	L, CL	fine SCL, SL	coarse SCL	LS	S
Structural stability	high	medium	low	faint structure	structureless
Base saturation (%)	> 50	50–30	< 30		
Available water (%)	> 15	15 10	< 10		
Morphogenetic processes***	a-l	d	e-f-b	g	c-i-h-m

* For mixed vegetation cover, with tree cover ≥ 20% the suitability class is N1.
** MCS = moisture control section (sensu Soil Survey Staff, 1997).
*** a = steady state, b = unsteady state, c = dynamic state, d = sheet erosion, e = sheet and rill erosion, f = rill and gully erosion, g = mass erosion, h = bank erosion, i = bed erosion, l = deposition, m = gravity erosion.

classes. The evaluation table for the alluvial deposits also takes into account extent of the area and flood hazard. Table 33.6 shows, for each suitability class within the land units considered, the typical soils classified at great group level according to the Keys to Soil Taxonomy (Soil Survey Staff, 2003). The selected soils come from the soil map "Carta dei suoli della Sardegna" (scale 1: 250,000) (Aru et al., 1990).

Table 33.6. Classification at great group level (Soil Survey Staff, 2003) of the typical soils for each suitability class for grazing and pasture improvement within the land units considered in Sardinia.

Land unit	S1	S2	S3	N1	N2
Palaeozoic metamorphites and Palaeozoic intrusive complex	Typic Xerorthents Typic Haploxerepts	Dystric Xerorthents Typic Dystroxerepts Lithic Xerorthents Lithic Haploxerepts Lithic Dystroxerepts		Lithic Xerorthents Lithic Haploxerepts Lithic Dystroxerepts	Lithic Xerorthents
Marly limestone and sandy marls	Typic Haploxerepts Typic Xerorthents Typic Rhodoxeralfs Calcic Haploxerepts Typic Calcixerepts	Lithic Haploxerepts Vertic Haploxerepts Vertic Calcixerepts	Lithic Xerorthents Typic Haploxererts Entic Haploxererts	Lithic Xerorthents Lithic Haploxerepts	Lithic Xerorthents
Palaeozoic and Mesozoic limestone and dolomite	Typic Haploxerepts Typic Rhodoxeralfs Typic Xerorthents	Lithic Xerorthents Lithic Haploxerepts Lithic Rhodoxeralfs		Lithic Xerorthents	
Tuffs and ignimbrites alternating with rhyolites and trachytes	Typic Haploxerepts Calcic Haploxerepts Typic Xerorthents	Andic Haploxerepts Lithic Haploxerepts Vertic Haploxerepts	Lithic Xerorthents Lithic Haploxerepts	Lithic Xerorthents Lithic Xerorthents Lithic Haploxerepts	Lithic Xerorthents
Basalts and andesitic basalts	Typic Haploxerepts Typic Xerorthents	Lithic Haploxerepts	Lithic Xerorthents	Lithic Xerorthents Lithic Haploxerepts	Lithic Xerorthents
Pleistocene alluvial deposits, terraces and glacis	Typic Haploxeralfs Calcic Haploxeralfs	Typic Palexeralfs Calcic Palexeralfs Petrocalcic Palexeralfs	Ultic Palexeralfs Aquic Haploxeralfs Aquic Palexeralfs Petrocalcic Palexeralfs	Ultic Palexeralfs Aquic Palexeralfs	–
Holocene alluvial deposits	Typic Xerofluvents Mollic Xerofluvents	Vertic Xerofluvents	Aquic Xerofluvents Typic Haploxererts Chromic Haploxererts	Typic Fluvaquents Vertic Fluvaquents	–

References

ARU, A., BALDACCINI, P., LOJ, G. 1989. I suoli: caratteristiche che determinano la loro marginalità e la loro valutazione per il pascolo. In: Idda, L. (Ed.), Sistemi Agricoli Marginali. Lo scenario Margine-Planargia. CNR-Prog. Final. IPRA, Sassari, pp. 29–52.

ARU, A., BALDACCINI, P., DELOGU, G., DESSENA, M.A., MADRAU, S., MELIS, R.T., VACCA, A., VACCA, S. 1990. *Carta dei suoli della Sardegna, in scala 1:250.000*. Dipartimento Scienze della Terra, Università di Cagliari, Assessorato Regionale alla Programmazione Bilancio ed Assetto del Territorio, SELCA (Firenze).

BALDACCINI, P., MADRAU, S. 1996. La valutazione del territorio come base per la difesa delle aree collinari e montane. Un esempio nella Sardegna centro-settentrionale. Atti del Convegno annuale S.I.S.S. "Contributi della Scienza del Suolo allo studio e alla difesa dei territori montani e collinari", Milano, 17–19 June 1996. Bollettino della Societa Italiana della Scienza del Suolo 8, 261–290.

BALDACCINI, P., MADRAU, S., DEROMA, M.A. 1995. I suoli del bacino del rio d'Astimini-Fiume Santo (Sardegna nord-occidentale). Valutazione della loro attitudine al miglioramento pascoli. Atti del Convegno annuale S.I.S.S. "Il Ruolo della Pedologia nella Pianificazione e Gestione del Territorio", Cagliari, 6–10 June 1995, pp. 289–298.

BALDACCINI, P., MADRAU, S., PREVITALI, F., DESSÌ, G., DEROMA, M.A. 2002. *I suoli del bacino del rio d'Astimini-Fiume Santo (Sardegna nord-occidentale)*. Dipartimento di Ingegneria del Territorio, Nucleo Ricerca Desertificazione, Università degli Studi di Sassari, 159 p.

COMOLLI, R. 1996. Lo studio dei suoli nell'ambito della pianificazione d'utilizzo dei pascoli: il caso della Val Gerola. Atti del Convegno annuale S.I.S.S. "Contributi della Scienza del Suolo allo studio e alla difesa dei territori montani e collinari", Milano, 17–19 June 1996. Bollettino della Societa Italiana della Scienza del Suolo 8, 235–256.

FAO. 1976. *A framework for land evaluation*. FAO Soils Bulletin no. 32, Rome, 72 p.

FAO. 1983. *Guidelines: land evaluation for rainfed agriculture*. FAO Soils Bulletin no. 52, Rome, 239 p.

FAO. 1991. *Guidelines: land evaluation for extensive grazing*. FAO Soils Bulletin no. 58, Rome, 158 p.

LITTERIO, M., DI GENNARO, A., FRESCHI, P., LEONE, A., MOTTI, R., SANTOPOLO, A., TERRIBILE, F. 1993. Un sistema informativo territoriale per la gestione delle aree a pascolo del territorio della regione Molise. Genio Rurale 6, 29–40.

MADRAU, S. 1992. Valutazione della attitudine al pascolo dei suoli della Sardegna. I territori comunali di Gavoi e Lodine (Nu). Atti dell'Istituto di Geopedologia e Geologia Applicata dell'Università degli Studi di Sassari 6(1985–1992), 125–165.

MADRAU, S. 1993. Valutazione della attitudine alla utilizzazione agronomica e al miglioramento pascoli dei suoli della Sardegna. Il territorio comunale di Sindia (Nu). Studi Sassaresi, Sez. III-Ann. Fac. Agr. Univ. Sassari (1) 35(1), 41–61.

MADRAU, S., LOJ, G., BALDACCINI, P. 1998. Modello per la valutazione della attitudine al miglioramento dei pascoli dei suoli della Sardegna. Nucleo Ricerca Desertificazione—Università di Sassari, Ente Regionale di Sviluppo ed Assistenza Tecnica in Agricoltura, Cagliari, 51 p.

SOIL SURVEY STAFF. 1997. *Keys to Soil Taxonomy*, 7th ed. USDA—Soil Conservation Service, Pocahontas Press, Inc. Blacksburg, Virginia.

SOIL SURVEY STAFF. 2003. *Keys to Soil Taxonomy*, 9th ed. USDA–NRCS, 332 p.

APPENDIX
SOIL PROPERTIES AND QUALITIES

Appendix: Soil Properties and Qualities
Edoardo A.C. Costantini

Introduction

In this manual, the following data about the properties and quality of soil were used. Unless specified otherwise, the sources used were Constantini (2006), Ministero delle Politiche Agricole e Forestali (2000), Sbaraglia and Lucci (1994), Servizio Geologico, Sismico e dei Suoli-Regione Emilia Romagna (2002), and USDA (1993).

1. Active Lime Content

This only approximately expresses the percentage weight of finely subdivided and easily soluble carbonates. More exactly, it corresponds to the percentage of Ca^{++} ions that react with ammonium oxalate (determined with the Drouineau-Gallet calcimeter method). Soils rich in salts, particularly gypsum, do not provide reliable values.

Evaluation	Values (%)
Low	< 5
Medium and high	5–10
Very high	> 10

2. Available Rooting Depth

This indicates the depth to layers that are impenetrable by roots. An impenetrable horizon is considered one that presents a rooting capacity of 30% or lower. Rooting capacity is estimated by measuring compactness, dimensional distribution of pores, aeration, water retention capacity, chemical conditions.

Evaluation	Values (cm)
Very shallow	< 25
Shallow	25–50
Moderately deep	51–100
Deep	101–150
Very deep	> 150

3. Available Water Capacity (AWC)

The available water capacity (AWC) specifies the amount of water retained in soil and available to the reference plant (sunflower).

Evaluation	Values (mm)
Very low	< 50
Low	50–100
Moderate	101–150
High	151–200
Very high	> 200

4. Base Saturation

Evaluation	Values (%)
Very low	< 35
Low	35–50
Medium	51–60
High	61–75
Very high	> 75

5. Cationic Exchange Capacity (CEC)

Evaluation	Values (meq/100 g)
Very low	< 5
Low	5–10
Moderately low	11–15
Moderately high	16–24
High	25–50
Very high	> 50

6. Chemical Response (pH in Water)

Evaluation	Values
Extremely acid	< 4.5
Strongly acid	4.5–5.0
Moderately acid	5.1–6.0
Slightly acid	6.1–6.5
Neutral	6.6–7.3
Slightly alkaline	7.4–7.8
Moderately alkaline	7.9–8.4
Strongly alkaline	8.5–9.0
Extremely alkaline	> 9.0

7. COLE, Coefficient of Linear Extensibility (Shrink-Swell Coefficient)

This is the linear dimension ratio between a wet field sample and the same sample when air-dried. It describes the tendency of the soil to swell and shrink following the absorption or loss of water by the silicate clays.

Evaluation	Values (%)
Low	< 3.0
Moderate	3.0–5.9
High	6.0–9.0
Very high	> 9.0

8. Content of Assimilable Microelements (Fe, Mn, Cu, Zn)

Evaluation	Fe (mg/kg)	Mn (mg/kg)	Cu (mg/kg)	Zn (mg/kg)
Very low	< 2	< 1	< 0.2	< 0.5
Low	2–5	1–3	0.2–1	0.5–1
Medium	6–20	4–20	2–10	2–10
High	21–50	21–50	11–20	11–20
Very high	> 50	> 50	> 20	> 20

9. Content of Exchangeable Carbon

Evaluation	Values (meq/100 g)
Very low	< 1.5
Low	1.5–3.0
Medium	3.1–5.0
High	5.1–10.0
Very high	> 10.0

10. Content of Exchangeable Magnesium

Evaluation	Values (meq/100 g)
Very Low	< 0.4
Low	0.4–0.8
Medium	0.9–1.2
High	1.3–1.6
Very High	> 1.6

11. Content of Exchangeable Potassium

Evaluation	Values (meq/100 g)
Very low	< 0.13
Low	0.13–0.26
Medium	0.27–0.39
High	0.40–0.52
Very high	> 0.52

12. Content of Organic Carbon

Evaluation	Values (%)
Very scarce	< 0.45
Scarce	0.45–0.90
Medium	0.91–1.36
High	1.37–1.81
Very high	> 1.81

13. Content of Sodium and Exchangeable Sodium Percentage (ESP)

Evaluation	Absolute value (meq/100 g)	ESP (%)
Normal	< 1.0	< 5.0
Slightly high	1.1–2.0	5.1–10.0
High	2.1–3.0	10.1–15.0
Very high	> 3.0	> 15.0

14. Cracking

This describes the soil tendency to fissure or crack following cycles of desiccation-shrinking and wetting-swelling. Soils that crack have large quantities of high activity clays.

Evaluation	Description
Strong	Soils subject to vertic movement (due to repeated, successive cycles of expansion and contraction of clays) with such intensity and frequency that it seriously damages root systems, buildings and roads.
Medium	Soils subject to vertic movement (due to repeated, successive cycles of expansion and contraction of clays) with such intensity and frequency that it moderately damages root systems, buildings and roads.
Low	Soils not subject to vertic movement or subject to movement with so little intensity and frequency that it does not significantly interfere with root systems, buildings and roads.

15. Electrical Conductivity (Salinity)

Evaluation	Values (dS/m–1:2.5 at 25°C)
Marginal	0–0.5
Moderate	0.6–1.0
Strong	1.1–2.0
Very strong	2.1–4.0
Excessive	> 4.0

16. Evaluation of the Mg/K Ratio

Evaluation	Value (meq/meq)
Very low	< 0.5
Low	0.5–1.0
Slightly low	1.1–2.0
Optimal	2.1–6.0
Slightly high	6.1–10.0
High	> 10.0

Manual of Methods for Soil and Land Evaluation

17. Flooding Risk

Frequency and duration

Flooding is the temporary covering of the soil surface by water flowing from any type of source. Shallow stagnant or flowing water present for more or less time after a rainfall is not included in this definition. Still water or water that forms a permanent covering is excluded from this definition. The following classes express the probability of an event over a time span of 100 years (recurrence rate based on historical documentation or based on oral testimony of inhabitants).

Evaluation	Description
None	Event not predictable or with frequency and intensity insignificant to agro-forestry use
Rare	1–5 times in 100 years
Occasional	6–50 times in 100 years
Frequent	> 50 times in 100 years
Very frequent	classes (2) and (3) for certain purposes may be grouped
Unknown	Data not available

Evaluation	Duration
Extremely brief	< 4 h
Very brief	4–48 h
Brief	2–7 d
Long	7 d to 1 month
Very long	> 1 month

18. Internal Drainage, Oxygen Availability, Hydraulic Behaviour

For the relationship between internal drainage and oxygen availability, this is a description of the potential hydraulic behaviour of soils regarding some of the principal agronomic practices.

Evaluation		Description of hydraulic behaviour
Internal drainage	Oxygen availability	
Excessively drained	Good	These soils have a high hydraulic conductivity (between 10 and 100 µm/s) or very high (> 100 µm/s) and a low available water capacity (low or very low AWC < 100mm). Unless irrigated, crops may not be grown. Soils have no redox mottles.
Somewhat excessively drained	Good	These soils have a high hydraulic conductivity (between 10 and 100 µm/s) and a higher available water capacity (low or moderate AWC, > 50 mm, but <150 mm). Unless irrigated only a restricted variety of crops may be grown with a low yield. Soils have no redox mottles.
Well drained	Good	These soils hold an optimum quantity of water (high or very high AWC, >150 mm), water drains rapidly from soil, and/or there are no periods during growing season characterized by excessive wetness limiting plant development. Soils usually have no redox mottles in the rooting zone.
Moderately well drained	Moderate	Water is at or near the soil surface for periods long enough to compromise operations of planting and harvesting mesophytic crops, unless artificially drained. Moderately well-drained soils commonly have a layer of low hydraulic conductivity (between 0.1 and 0.01 µm/s) within 1 m depth, flow of infiltrated water or a combination of the two. They have common redox features (over 4–5 %) at least in the lower part of the rooting zone.

Somewhat poorly drained	Imperfect	Water is at or near the soil surface for periods long enough to severely compromise operations of planting, growth and harvesting of crops unless artificially drained. Somewhat poorly drained soils commonly have a layer of low hydraulic conductivity, a high water table, flow of infiltrated water, or a combination of these conditions. They generally have common to abundant redox features in the rooting zone; may show temporary stagnation mottles due to tillage.
Poorly drained	Poor	Water is at or near the soil surface for a considerable part of the year, making field cultivation of crops impossible under normal conditions. Conditions of poor drainage are due to a saturated layer, a horizon with low hydraulic conductivity, flow of infiltrated water or a combination of these conditions. They generally have common to abundant redox features within the first 50 cm.
Very poorly drained	Very poor	Water is at or near the soil surface for a considerable part of the year. Cultivation of most crops (except rice) is impossible unless artificially drained. Usually have abundant chroma mottles ≤ 2 on the soil surface and inside peds.

19. Oxygen Availability

This refers to the oxygen in the soil available for biological activity. It is evaluated by the presence of free water, capillary imbibition, or traces of hydromorphy.

Evaluation	Description
Good	Water is readily drained, and/or there are no periods during growing season characterized by excessive wetness limiting plant development.
Moderate	Water drains slowly during some periods and soil is wet only for a brief period during growing season, but not long enough to interfere negatively with the growth of mesophytic crops.
Imperfect	Water drains slowly and soil is wet for significant periods during growing season; excess water inhibits development of mesophytic crops.
Poor	Water drains so slowly that soil is periodically saturated during growing season; excess water makes it impossible to grow the majority of mesophytic crops.
Very poor	Water drains so slowly that it remains on the surface during most of the growing season.

20. Permeability and Saturated Hydraulic Conductivity

Permeability is the field estimate of saturated hydraulic conductivity that must instead be measured in the laboratory.

Permeability	Code	Saturated hydraulic conductivity class Ksat (μm/s)	Horizon properties
Fast	1	Very high > 100	Fragmentational particle-size class, ashy or pumiceous. Many medium or coarse vertical pores spread throughout the layer.
	2	High(100–11)	Medial pumiceous particle-size class, medial skeletons, pumiceous ash, ashy skeletons, hydro-pumiceous, that is very friable, friable, soft or loose. Moderately humid or wet soil, moderate or highly developed granular structure, strongly developed polyhedrals of every size, or prismatic smaller than the most coarse, without pressure faces or slickensides. Common vertical pores, medium or coarse, spread throughout the layer.

Medium	3	Moderately high (10–1.1)	Very coarse polyhedral or prismatic structure and no pressure faces or slickensides. Clay 35% or more, soft, slightly hard, friable or very friable, without pressure faces of slickensides. Clay is sub-active after having subtracted from the CEC the quantity = 2 (CO 1.7).
Medium	4	Moderately low (1.0–0.2)	Few pressure faces and/or slickensides. Firm and very or extremely rigid, or weakly cemented. Moderately cemented.
Slow	5	Low (0.1–0.01)	Many or common pressure faces and/or slickensides.
	6	Very low (< 0.01)	Hard or very strongly cemented.

21. Risk of Water Deficit

This refers to the risk of periods with insufficient water supply in the soil.

Class	Phenomenon
Very severe	Thin soils, on slopes facing southward and/or soils with very low available water capacity (AWC ≤ 50 mm).
Strong	Shallow soils, on slopes facing southward and/or soils with low AWC (50 < AWC ≤ 100 mm). Tolerated by the more xerophytic species.
Moderate	Moderately thick soils, sloping or with moderate AWC (100 < AWC ≤ 150 mm). Tolerated by many all plant species, but subject to dry periods.
Slight	Thick soils with high AWC (150 < AWC ≤ 200 mm). Possible presence of deep or very deep ground water or perched water table.
Absent	Very thick soils with very high AWC (AWC ≤ 200 mm), or with shallow or very shallow ground water or perched water table.

22. Rock Fragments (Skeleton)

Lithoidal fragments over 2 mm in diameter present in profile.

Evaluation	Values (%)
Absent	0
Scarce	0.1–5.0
Common	5.1–15.0
Frequent	15.1–35.0
Abundant	35.1–70.0
Very abundant	> 70.0

23. Rockiness

Represents the percentage of stone outcropping on surface (material with diameter > 500 mm) that is not removable by ordinary cultivation methods.

Evaluation	Values (%)
Absent	0
Scarcely bouldery	0.1–2.0
Bouldery	2.1–10.0
Very bouldery	10.1–25.0
Extremely bouldery	25.1–90.0
Boulder outcropping	> 90

24. Slope Gradient

Evaluation	Values (%)
Level surface	< 0.2
Almost level surface	0.2–2
Slightly inclined surface	3–5
Moderately inclined surface	6–13
Highly inclined surface	14–20
Strongly inclined surface	21–35
Very strongly inclined surface	36–60
Steeply inclined surface	61–90
Very steeply inclined surface	> 90

25. Soil Texture

Texture refers to the distribution by size class of a soil's elementary particles. It is measured by field estimate and/or by laboratory determinations. The relative proportions of the principal particle sizes in fine earth (diameter < 2 mm) are described according to the following terms.

General terms	Fundamental classes (USDA, 1993)
Coarse	Sands (CS)
	Loamy sands (LS)
Moderately coarse	Corse sandy loam (CSL)
	Sandy loam (SL)
	Fine sandy loam (FSL)
Medium	Very fine sandy loam (VFSL)
	Loam (L)
	Silt loam (SIL)
	Silt (SI)
Moderately fine	Clay loam (CL)
	Sandy clay loam (SCL)
	Silty clay loam (SICL)
Fine	Sandy clay (SC)
	Silty clay (SIC)
	Clay (C)

26. Structural Stability or Mean Weight Diameter (MWD)

There is no universal interpretive matching table accepted for the agronomic evaluation of the structural stability of soil aggregates measured by water sieving; in fact, to date, there is no official method of measuring structural stability. We quote a matching table created by research carried out during the project: *Evaluation of territories in the Siena province for viticulture and olive-growing capability* (Costantini et al., 2006).

Evaluation	Values (mm)
Low	< 0.7
Medium	0.7–1.4
High	> 1.4

27. Surface runoff

Surface runoff (also called over-land flow or surface drainage) is the flow of water from an area that occurs over the surface of the soil.

Evaluation
Marginal (T)
Very low (MB)
Low (B)
Medium (M)
High (A)
Very high (MA)

To determine the surface runoff class, the slope of the site and the saturated hydraulic conductivity of the most limiting horizon within 1 m depth must be defined. Concavity is the characteristic of an area from which water may not leave by runoff.

Slope (%)	Saturated hydraulic conductivity					
	very high	high	moder. high	moder. low	low	very low
Concavity	T	T	T	T	T	T
<1	T	T	T	B	M	A
1–5	T	MB	B	M	A	MA
6–10	MB	B	M	A	MA	MA
11–20	MB	B	M	A	MA	MA
≥ 20	B	M	A	MA	MA	MA

28. Surface Stoniness

Lithoidal fragments over 75 mm in diameter present on the soil surface.

Evaluation	Values (%)
Absent	0
Very scarce	0.1–0.3
Scarce	0.4–1.0
Common	1.1–3.0
Frequent	3.1–15.0
Abundant	15.1–50.0
Very abundant	50.1–90.0
Stone outcropping	> 90.0

29. Total Calcium Carbonate Content

Evaluation	Values (%)
Non-calcareous	< 0.5
Scarcely calcareous	0.5–1.0
Low calcareous	1.1–5.0
Moderately calcareous	5.1–10.0
Very calcareous	10.1–20.0
Strongly calcareous	20.1–40.0
Extremely calcareous	> 40.0

30. Total Nitrogen Content

Source of data is Giardini (1986).

Evaluation	Values (g/kg)
Poor	< 1.0
Medium content	1.0–1.5
Good content	1.6–2.2
Rich	2.3–5.0
In excess	> 5.0

References

COSTANTINI, E.A.C., BARBETTI, R., BUCELLI, P., CIMATO, A., FRANCHINI, E., L'ABATE, G., PELLEGRINI, S., STORCHI, P., VIGNOZZI, N. 2006. *Zonazione viticola ed olivicola della provincia di Siena (2006)*. Grafiche Boccacci editore, Colle val d'Elsa (SI), 224 p.

COSTANTINI, E.A.C. (ed.). 2006. Metodi di valutazione dei suoli e delle terre. Cantagalli, Siena, 922 p.

GIARDINI, L. 1986. *Agronomia generale*. Patron editore, Bologna, 597 p.

MINISTERO DELLE POLITICHE AGRICOLE E FORESTALI. 2000. *Metodi d'analisi chimica del suolo*. Franco Angeli, Milano.

SBARAGLIA, M., LUCCI, E. 1994. *Guida all'interpretazione delle analisi del terreno ed alla fertilizzazione*. Studio Pedon, Pomezia (Roma), 123 p.

SERVIZIO GEOLOGICO, SISMICO E DEI SUOLI—REGIONE EMILIA-ROMAGNA. 2002. *Guida alla descrizione delle Unità Tipologiche di Suolo*. Bologna, 66 p.

USDA. 1993. *Soil Survey Manual*. Handbook 18, USDA, Washington D.C., 438 p.

Index

Color Plate Section

Chapter 2

Fig. 2.2. Example of soils with different land capability classification classes. Class I includes soils of the lower river terraces that are level, deep, without limitations. Because of limitations related to the nature of the soil, river terraces at a higher level are in classes II and III. Class IV includes soils on moderately sloping land that have a high erosion risk. Those of class V have no erosion risk, but have a high risk of frequent flooding, being near watercourses. In class VI there are thin soils on a slope, left for pasture. The soils with the strongest slope and highest risk of erosion are in class VII and require conservative silviculture. The areas found in class VIII are not suited for forestry or agriculture, but are set aside for non-productive purposes (e.g., natural ecosystem, wildlife habitat, water collection, recreation).

Chapter 3

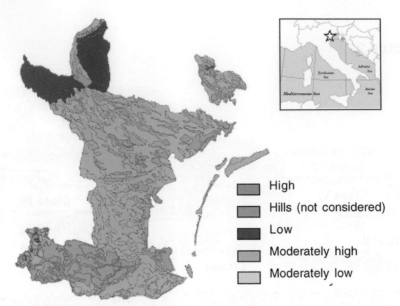

High

Hills (not considered)

Low

Moderately high

Moderately low

Fig. 3.2. Venice lagoon watershed: soil protective capacity (ARPAV-Regione del Veneto, 2004).

Printed and bound by CPI Group (UK) Ltd, Croydon, CR0 4YY

18/10/2024

01776208-0018

Chapter 5

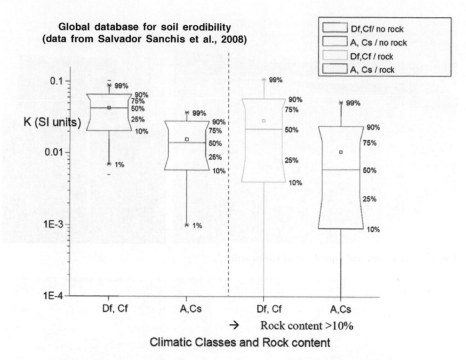

Fig. 5.2. Box and Wiskers plot of soil erodibility values distributions measured for various groups of climatic class following Koppen-Geiger climate classification and rock content %: left < 10%, right > 10%.

Chapter 14

Fig. 14.1. Landscape and typical soil of Cecita series.

Fig. 14.2. Landscape and typical soil of Sila series.

Fig. 14.3. Landscape and typical soil of Croce della Palma series.

Fig. 14.4. Differences in potato growth between (a) the Sila and (b) the Croce della Palma series.

Fig. 19.3. Soil monoliths from the River Mirna valley (Bragato et al., 2005). Unproductive soils are characterized by a recurrent periods of water stagnation. In some horizons, they otherwise display a well-developed structure where pore size distribution has been modified, decreasing the degree of oxygenation of the soil system. Truffle-producing areas of the Mirna valley are located in recurrently overflooded soils characterized by chaotic sedimentation of solid particles.

Chapter 21

Fig. 21.1. Young orchard of cactus pear on sandy terrains in Tunisia.

Chapter 24

Fig. 24.1. View of walnut plantation in Castilenti Hill area. The presence of strong hydric erosion features (the "calanchi") represent a strong constraint on correct planning of walnut plantation.

Fig. 24.2. Karst depression with walnut stand in the S. Demetrio de' Vestini area.

Fig. 24.3. Common walnut and wild cherry mixed plantation 10 years old. S. Demetrio de Vestini, Abruzzo Region.

Fig. 24.4. Excellent wild cherry single plant 10 years old. S. Demetrio de Vestini, Abruzzo Region.

Fig. 24.5. Common walnut and Neapolitan alder mixed plantation 10 years old. S. Demetrio de Vestini, Abruzzo Region.

Chapter 26

Fig. 26.5. South-east slope at San Gimignano: traditional agro-ecotopes, with mixed cultivation.

Fig. 26.6. North-west slope at San Gimignano, the new agro-ecotope, with specialized vineyards saturating the landscape unit.

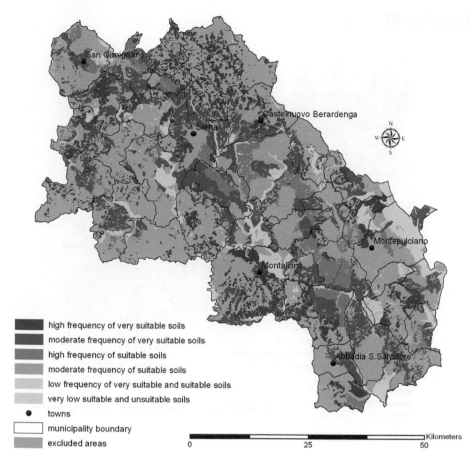

Fig. 26.13. Example of a suitability map for wine production (Sangiovese vine) in the Province of Siena (Italy). Legend accounts for the probability of finding more or less suitable soils.

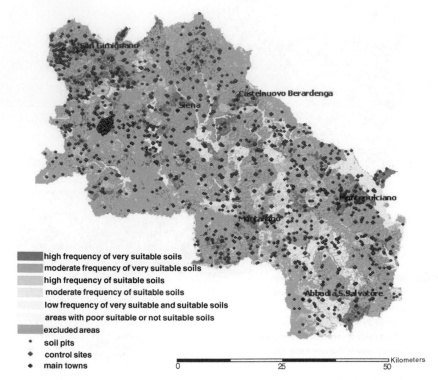

Chapter 27

Fig. 27.18. Land suitability map for intensive olive tree cultivation in the Siena province. Soil pits and control sites.

Chapter 31

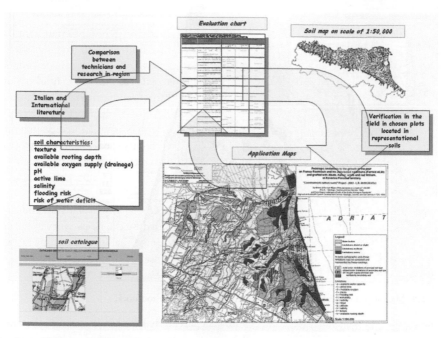

Fig. 31.1. Work methodology used.

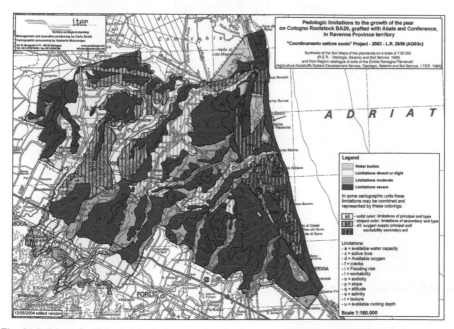

Fig. 31.3. Map of pedological limitations to growth of pear on quince rootstock BA29, Sydo.

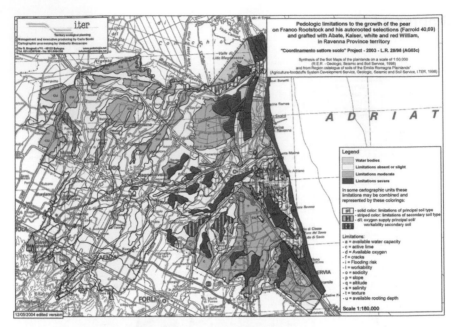

Fig. 31.4. Map of pedological limitations to growth of pear on free rootstock.

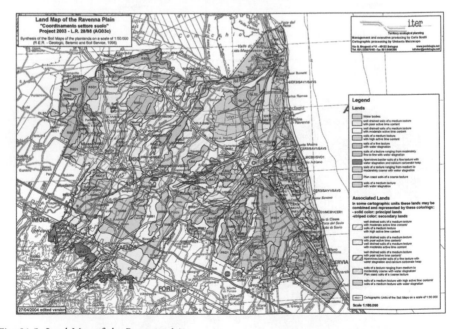

Fig. 31.5. Land Map of the Ravenna plain.